COMPLEXITY
METAPHORS, MODELS, AND REALITY

COMPLEXITY
METAPHORS, MODELS, AND REALITY

Editors

George Cowan
Santa Fe Institute
Sants Fe, New Mexico

David Pines
University of Illinois at Urbana-Champaign
Urbana, Illinois

David Meltzer
Southeastern Louisiana University
Hammond, Louisiana

The Advanced Book Program

PERSEUS BOOKS
Cambridge, Massachusetts

Library of Congress Catalog Card Number: 99-66279

ISBN 0-7382-0232-0

Perseus Books is a member of the Perseus Books Group

Cover design by Suzanne Heiser

1 2 3 4 5 6 7 8 9 10–   –02 01 00 99

Find us on the World Wide Web at
http://www.perseusbooks.com

About the Santa Fe Institute

The *Santa Fe Institute* (SFI) is a multidisciplinary graduate research and teaching institution formed to nurture research on complex systems and their simpler elements. A private, independent institution, SFI was founded in 1984. Its primary concern is to focus the tools of traditional scientific disciplines and emerging new computer resources on the problems and opportunities that are involved in the multidisciplinary study of complex systems—those fundamental processes that shape almost every aspect of human life. Understanding complex systems is critical to realizing the full potential of science, and may be expected to yield enormous intellectual and practical benefits.

All titles from the *Santa Fe Institute Studies in the Sciences of Complexity* series will carry this imprint which is based on a Mimbres pottery design (circa A.D. 950–1150), drawn by Betsy Jones. The design was selected because the radiating feathers are evocative of the outreach of the Santa Fe Institute Program to many disciplines and institutions.

Santa Fe Institute
Studies in the Sciences of Complexity

Lectures Volumes

Vol.	Editor	Title
I	D. L. Stein	Lectures in the Sciences of Complexity, 1989
II	E. Jen	1989 Lectures in Complex Systems, 1990
III	L. Nadel & D. L. Stein	1990 Lectures in Complex Systems, 1991
IV	L. Nadel & D. L. Stein	1991 Lectures in Complex Systems, 1992
V	L. Nadel & D. L. Stein	1992 Lectures in Complex Systems, 1993

Lecture Notes Volumes

Vol.	Author	Title
I	J. Hertz, A. Krogh, & R. Palmer	Introduction to the Theory of Neural Computation, 1990
II	G. Weisbuch	Complex Systems Dynamics, 1990
III	W. D. Stein & F. J. Varela	Thinking About Biology, 1993

Reference Volumes

Vol.	Author	Title
I	A. Wuensche & M. Lesser	The Global Dynamics of Cellular Automata: Attraction Fields of One-Dimensional Cellular Automata, 1992

Proceedings Volumes

Vol.	Editor	Title
I	D. Pines	Emerging Syntheses in Science, 1987
II	A. S. Perelson	Theoretical Immunology, Part One, 1988
III	A. S. Perelson	Theoretical Immunology, Part Two, 1988
IV	G. D. Doolen et al.	Lattice Gas Methods for Partial Differential Equations, 1989
V	P. W. Anderson, K. Arrow, D. Pines	The Economy as an Evolving Complex System, 1988
VI	C. G. Langton	Artificial Life: Proceedings of an Interdisciplinary Workshop on the Synthesis and Simulation of Living Systems, 1988
VII	G. I. Bell & T. G. Marr	Computers and DNA, 1989
VIII	W. H. Zurek	Complexity, Entropy, and the Physics of Information, 1990
IX	A. S. Perelson & S. A. Kauffman	Molecular Evolution on Rugged Landscapes: Proteins, RNA and the Immune System, 1990
X	C. G. Langton et al.	Artificial Life II, 1991
XI	J. A. Hawkins & M. Gell-Mann	The Evolution of Human Languages, 1992
XII	M. Casdagli & S. Eubank	Nonlinear Modeling and Forecasting, 1992
XIII	J. E. Mittenthal & A. B. Baskin	Principles of Organization in Organisms, 1992
XIV	D. Friedman & J. Rust	The Double Auction Market: Institutions, Theories, and Evidence, 1993
XV	A. S. Weigend & N. A. Gershenfeld	Time Series Prediction: Forecasting the Future and Understanding the Past
XVI	G. Gumerman & M. Gell-Mann	Understanding Complexity in the Prehistoric Southwest
XVII	C. G. Langton	Artificial Life III
XVIII	G. Kramer	Auditory Display
XIX	G. Cowan, D. Pines, and D. Meltzer	Complexity: Metaphors, Models, and Reality

Contributors to This Volume

Philip W. Anderson, Princeton University
Kenneth Arrow, Stanford University
W. Brian Arthur, Stanford University and Santa Fe Institute
Per Bak, Brookhaven National Laboratory
James H. Brown, University of New Mexico
Leo W. Buss, Yale University
Luigi L. Cavalli-Sforza, University of California
George A. Cowan, Santa Fe Institute
James P. Crutchfield, University of California
Marcus Feldman, Stanford University
Walter Fontana, Santa Fe Institute
Hans Frauenfelder, University of Illinois
Murray Gell-Mann, Santa Fe Institute
Brian Goodwin, The Open University
John H. Holland, University of Michigan
Peter T. Hraber, Santa Fe Institute
Alfred Hübler, University of Illinois
Erica Jen, Los Alamos National Laboratory
Stuart A. Kauffman, Santa Fe Institute
Alan Lapedes, Los Alamos National Laboratory
Ben Martin, Stanford University
John Maynard Smith, University of Sussex at Brighton
David Meltzer, Southeastern Louisana University in Hammond
Melanie Mitchell, Santa Fe Institute
Alan S. Perelson, Los Alamos National Laboratory
David Pines, University of Illinois at Urbana–Champaign
Tom S. Ray, ATR Human Information Processing Research Laboratories
Peter Schuster, Universität Wien
Charles Stevens, The Salk Institute
Lev A. Zhivotovsky, Russian Academy of Sciences

Contents

Complexity: Metaphors, Models, and Reality
Eds. G. Cowan, D. Pines, and D. Meltzer, SFI Studies in the
Sciences of Complexity, Proc. Vol. XIX, Addison-Wesley, 1994   **xi**

Editor's Foreword

Perseus Books's *Advanced Book Classics* series has been designed to make available, at modest cost and in an attractive format, graduate-level lecture notes, texts and monographs that are classics, in that the depth and insight they have provided in the past may be expected to continue to play a significant role in the education of the present and future generations of graduate students, postdoctoral research associates, and their more senior scientific colleagues. It is hoped that although books in the series may, in some cases, have been written twenty-five or more years ago, the unique perspective and pedagogical clarity provided by the authors will make them as attractive and useful to today's reader as they were to the generations of readers who received them enthusiastically at the time of their original publication.

In the five years since its publication, with very few reviews, and based mainly on word-of-mouth recommendations, *Complexity* has become something of an underground classic. Considered by many the best introduction to the ideas underlying attempts to construct theories and models of complex adaptive systems, it contains not only the lectures presented at the first major conference devoted to the search for integrative themes in complex adaptive systems, but a quite complete account of the frequently lively, and often argumentative, discussions that followed each lecture. These provide the reader with an "insider's perspective" on the thinking of many of the major contributors to this emerging field. Originally published as part of the Santa Fe Institute Series on Complexity, it is our hope that reprinting the volume as one of the *Advanced Book Classics* will make it once again available to the increasingly broad audience of scientists and laymen alike interested in complexity.

David Pines
Tesuque, NM
August 1999

George Cowan† and David Pines‡
†Santa Fe Institute, 1660 Old Pecos Trail, Suite A, Santa Fe, NM 87501
‡Department of Physics, UIUC, 1110 W Green Street, Urbana, IL 61801

Foreword

The founding workshops of the Santa Fe Institute[1] played an important role in charting its initial intellectual directions, and in identifying potential godparents, participants and supporters alike, who might nourish our infant enterprise. Some seven years later, both the founding members, and the many godparents who had helped steer the Institute through its early childhood, decided the time was right for an examination of the intellectual heart of our enterprise, the common features of complex adaptive systems. Thus, the suggestion by Stuart Kauffman in the fall of 1991 that SFI hold a workshop on integrative themes of the sciences of complexity fell on fertile ground, and SFI President Ed Knapp appointed a small committee chaired by one of us to examine how this might best be done.

The planning for the workshop was in many ways characteristic of the SFI scientific style, a somewhat chaotic, bottoms-up affair, in which the Steering Committee of the SFI Science Board, augmented by a number of visiting scholars, held a series of meetings to identify potential integrative themes and key discussion leaders. Like the systems it studies, the SFI scientific community is a complex adaptive system, and the group went through a process best described as self-organization in arriving at a surprising degree of consensus on both topics and participants. We agreed to co-chair a small committee which took day-to-day responsibility for organizing the workshop along the lines set forth by the augmented Science Board Steering Committee.

Complexity: Metaphors, Models, and Reality
Eds. G. Cowan, D. Pines, and D. Meltzer, SFI Studies in the
Sciences of Complexity, Proc. Vol. XIX, Addison-Wesley, 1994   **xvii**

A key element in the design of the program was the provision for periods of extended informal discussion following each presentation. We also decided to set aside blocks of time during the meeting in which participants could both assess what we had learned and plan how best to organize the remaining days of the meeting. At times during the meeting the participants split into small subgroups, but for the most part we functioned as a committee of the whole.

We decided early on that for these proceedings it was important not only to secure written manuscripts from the discussion leaders, but also to prepare and edit a transcript of the informal discussions. In developing this material we were fortunate to secure the assistance of David Meltzer, who has had extensive transcription and editorial experience, and who took a keen interest in the topics under discussion. It gives us pleasure to thank him here for his valuable contributions to this volume. We hope the reader will share our view that it is in these discussions, perhaps even more than in the prepared manuscripts, that one can see the SFI community in action, as it attempts to define and refine candidate integrative themes and lists of questions to be addressed, returning again and again to improve upon themes and questions, much in style of Jasper Johns. To convey more effectively the flavor of the discussions, the transcripts have undergone only modest editing; thus much of the redundancy has been left in. The alert reader will notice that more than one participant, in attempting to focus group attention on a particular theme or approach, was inclined to follow the lead of the Bellman in *The Hunting of the Shark*, "What I tell you three times must be true."

Adaptation also played a role in our choice of a title for this volume. The initial title for the workshop was "Integrative Themes: Common Features of Complex Adaptive Systems." However, in the course of the workshop we realized that what we were doing was to propose and refine candidate metaphors for the description of complex adaptive systems, and then seeking to identify which metaphors survived when subject to the reality check of experiment, observation, or computer simulations based on models which incorporate essential features of complex adaptive systems. In our Afterwords we return to this perspective, as we search for order in the chaos that occurs when an unusually lively, articulate group of scientists "works at the blackboard" on problems that are often ill-defined and far from solution. It is our hope that on the whole the workshop participants managed to avoid the scientific approach attributed by the legendary Soviet theoretical physicist, Lev D., Landau, to cosmologists, "always uncertain, but never in doubt."

In many ways, the most significant scientific contribution made by the Santa Fe Institute in its first seven years is the creation of a "complexity community," a network of scholars drawn from a remarkable variety of disciplines who share a common interest in adaptive complex systems. The range of scientific interests and the quality of the participants illustrate the convening power of SFI, while the discussions make evident the effort made by participants from disparate fields to communicate with one another, and to work together to build something new. We hope the publication of these proceedings will not only give our readers a sense of the resulting group dynamics, but will encourage many to participate in future SFI activities.

We were singularly fortunate in the setting for the workshop which through the generous support of Beth and Charles Miller was held at Sol y Sombre, a remarkable environment, which proved equally conducive to reflection and interaction. On behalf of all participants we thank Beth and Charles for their support and encouragement.

REFERENCE

1. Pines, D., ed. *Emerging Syntheses in Science.* Santa Fe Institute Studies in the Sciences of Complexity, Proc. Vol. I. Reading, MA: Addison-Wesley, 1987.

George A. Cowan
Santa Fe Institute, 1660 Old Pecos Trail, Suite A, Santa Fe, NM 87501

Conference Opening Remarks

It is my great privilege to welcome this distinguished group to what is surely an historic event in our common enterprise. It comes at a critical and promising time, a time when the need has become especially urgent for more comprehensive approaches to a global agenda of shared problems and threats of catastrophe, a time when new tools have appeared for examining the complex systems that are shaping our future.

We all know that the behavior of complex systems cannot be understood by simply adding up the behavior of their separate parts. However, despite our inability to predict the precise behavior of truly complex systems, we share a hope that we can become better able to use the new tools to predict and perhaps even learn to affect the probabilities of various outcomes.

In pursuing a variety of activities at the Santa Fe Institute, we have assumed that they all have a common conceptual framework which we have chosen to call "complexity." As used here, the word refers to systems with many different parts which, by a rather mysterious process of self-organization, become more ordered and more informed than systems which operate in approximate thermodynamic equilibrium with their surroundings. The Institute was organized eight years ago to nurture research on such systems and has been intensively active for the past five

Complexity: Metaphors, Models, and Reality
Eds. G. Cowan, D. Pines, and D. Meltzer, SFI Studies in the
Sciences of Complexity, Proc. Vol. XIX, Addison-Wesley, 1994

years. We propose now to tell each other what we have learned to date about the kind of "ordered complexity" in which we are most interested.

The modern awakening of interest in complexity as a science began in Vienna in 1928 with von Bertalanffy's largely descriptive graduate thesis on living organisms as systems. A few years earlier Alfred North Whitehead had described his similar vision of a "philosophy of organism" in *Science and the Modern World.* In the following two decades essential contributions were made with the publications of McCulloch and Pitts on neural networks, von Neumann on cellular automata and complexity, and Wiener on cybernetics.

A growing unease with elegant but overly constrained physical science became apparent about in the '50s. It was expressed most succinctly by George Miller in 1956 when he wrote that scientific journals are catalogs of parts for a machine they never build. However, despite the important contributions of Prigogine and Haken among others, interest in complexity seemed to lag in following decades. It required the explosive development and sharply falling costs of computer hardware and software and the development of new mathematical techniques and concepts to make the notion of research on truly complex systems much more than wishful thinking. I believe that our discussions will demonstrate that in the '90s the elements of a science of complexity are beginning to come more clearly into view.

During the past five years at the Santa Fe Institute many of the people here have explored a variety of seemingly disparate topics dealing with complex processes. As we discuss what has been learned about such processes and what they may have in common, we should try, if possible, to establish common dimensionalities and to define the units of energy, matter, information, interaction, and time scales which are characteristic of each system. We shall ask whether these systems share a sufficient number of fundamental properties to support the premise which encouraged the establishment of the Institute, that research on them defines a large, coherent, and important area of scientific study.

We begin with the basic idea that complex systems contain many relatively independent parts which are highly interconnected and interactive and that a large number of such parts are required to reproduce the functions of truly complex, self-organizing, replicating, learning, and adaptive systems. In our discussions we shall examine to what extent a consensus exists on these and other questions.

The program committee has endeavored to assure that this program includes all of the interests pursued at SFI plus a few of the many other related subjects that should be pursued here as resources become available. We shall read about the properties of cellular automata which are not self-organizing or adaptive but which can exhibit very complex behavior shaped by relatively simple local rules. The program deals with the appearance of folded proteins and the beginning of highly interconnected, self-organizing, and adaptive systems. It ranges from the formation of cells and organs, particularly including the brain, to organisms, particularly man, and to the enormously interactive systems studied in social science. The human dimension really begins with nature's invention of the human cortex, a prerequisite for the invention of symbols, language, culture, electronic communication, and the evolving behavior of collective social units which have increased in size until they now embrace a truly global community.

We shall read various definitions of complexity which will reveal the ambiguities that exist in our efforts to include all of the essential properties that are found in complex systems. We tend to use definitions that focus either on static properties such as program size or on dynamic properties such as time and expandable memories. Presumably this ambiguity can be resolved by imposing a rate of exchange between run time and program size. We should explore the connection between definitions of complexity which include time and the fact that components at different levels in the natural hierarchy of complexity can be assigned to the appropriate level by characteristic interaction times. The time scale at each higher level of complexity is frequently orders of magnitude greater than at the level below. Systems at the very beginning of time operated at perhaps 10^{-50} seconds, increased to 10^{-20} seconds at the nuclear level, jumped again to 10^{-12} seconds at the simple chemical level, to 10^{-3} at the neural level, and so on to considerably longer times at the level of individual human interactions and longer still for the development of cultures and institutional behavior. Social and cultural structures that were developed in a more leisurely age are now unable to adjust readily to the tremendous shortening of time scale in human interactions at a distance. These have been produced largely in our generation by the science and technology which gave rise to the information explosion and the electronic media that are now replacing the printed media.

It is important to note a particular characteristic of the systems which we will explore. They never stop at fixed points. They are forever dynamic and can be considered dead and of little interest when they come to thermodynamic equilibrium. It is really the dynamic properties of complexity, the motion pictures, not the snapshots, which characterize the systems in which we are interested and it would seem appropriate if our definition of complexity explicitly recognizes this emphasis.

We are masters of the art of abstracting from the physical world and creating conceptual systems which capture some of the essential features of real world processes. These abstractions are at the heart of physical science. But today we are engaged in a somewhat more complicated exercise. We wish to define both the extent to which phenomenological descriptions of apparently different real world systems actually resemble one another in fundamental ways and the extent to which our metaphors and abstract concepts of such systems, if we have them, may also resemble one another. At the same time we do not wish to overlook the differences between complex systems. Nature has many ways to construct systems that metabolize; receive, process and store information; and respond to stimuli from surrounding environments. But, as these various systems organize themselves and learn and remember and evolve and adapt and persist and eventually disintegrate and disappear, what common patterns and fundamental principles, if any, shape their remarkable behavior and trajectories?

We cannot assert that we have correctly and fully described any process that occurs in the real world. We can only describe our perceptions of an external world, that part to which we are able to attend and which we somewhat arbitrarily call reality because it is all we know. But perhaps we can hope to account for a larger part of what is happening around us, to achieve a deeper understanding of those aspects of our world's future behavior which can be anticipated, and to admit the existence of important parts which may be unknowable.

Fortunately, the chairman is not expected to provide answers to his own questions but to seek consensus or to comment on its absence. Given this privilege, I confess to having an endless supply of questions. I feel a profound sense of wonderment when I contemplate the remarkable behavior of complex systems. One of the sources of my wonderment is the variety of forms in which information is generated, conveyed, processed, stored, and retrieved. I marvel at the way information remains quiescent for great lengths of time and suddenly is reacquired. I wonder why we pay so much attention at SFI to information functions and much less to metabolic functions which may enter into and be deeply intertwined with the generation, acquisition, retention, and retrieval of information. Why is it that, at each level in the hierarchy of complexity, new properties emerge which are not presently predictable from the properties of the systems at the hierarchical level just below and which may prove, in practice, to be too demanding of resources to predict? What is the nature of learning and, if optimality is never truly achievable, what guides adaptation?

The ascending levels of the hierarchy of complexity demonstrate emergent properties at each level which appear to be nonpredictable from the properties of the component parts. Thus the commonly held expectation that we should be able to derive macroeconomics from microeconomics is probably unrealistic. It follows that we must almost surely fail to predict the behavior of a global economy from the behavior of nations. Does the notion of emergent properties tell us in general that, contrary to a commonly held assumption, societal behavior cannot be adequately described by any practically achievable integration across the behavior of individuals? The tension between the behavior of individuals and the behavior of what Rosenau calls "collectivities" has been greatly amplified by the information revolution and increased exposure of individuals to the beliefs and practices of other people and societies. This tension is accompanied by a widening rift between what informed individuals say and profess to feel and what more slowly adapting societies actually do. The nature of this tension and the possibility of dealing with it more effectively may be the most important topic of our time. Clearly we need a more complete catalogue of the processes of evolution, learning, and adaptation and the ways in which they differ from one hierarchical level to another.

I now have the honor to introduce Phil Anderson who has thought deeply for many years about the nature of complexity.

Fundamental Concepts

Philip W. Anderson
Princeton University, Department of Physics, Jadwin Hall, Princeton, NJ 08544

The Eightfold Way to the Theory of Complexity: A Prologue

Abstract: I will focus on some of the sources of theories of complexity, especially those which are rooted in the physical sciences. There will be some attempt to be comprehensive and perhaps more to be critical, and of course I will focus on those lines which I see as leading towards some of the main themes we will discuss. I may sketch some history in some cases. A preliminary list:

1. The "mathematical" → Turing → von Neumann → modern complexity theory. I have a physicists' prejudice against the Lemma Proof structure and the nonquantitative nature of these ideas, and a sense as a complexologist that any problem worth studying is necessarily incomputable in one of their senses. But these categories have often brought us up with a jerk to the realization that naive optimism on such problems as protein folding and the SK spin glass is probably out.
2. Information theory. This theory of limits is still often ignored to our peril. We cannot hold more bits than our number of synapses in a neural network, and we better not forget it. Hamming space is a glorious concept which we use every day. But information theory per se is not

Complexity: Metaphors, Models, and Reality
Eds. G. Cowan, D. Pines, and D. Meltzer, SFI Studies in the
Sciences of Complexity, Proc. Vol. XIX, Addison-Wesley, 1994

very enlightening. The more complex measures of complexity which Charlie Bennett and his ilk are studying are of much more interest, but perhaps measuring complexity is not as important as doing something about it. It has yet to be seen whether these measures have as much usefulness as the more primitive measures of information theory.

3. Ergodic theory $\rightarrow$ dynamic maps $\rightarrow$ chaos, attractors, etc. Rooted in the ergodic theory and in the study of conservative trajectories by Kolmogorov, etc., and of area-preserving maps, this field has bifurcated into two fields, one of extraordinary interest to the public: the field of dissipative dynamical maps, called "chaos" but as far as I can tell not having the essential feature of convergence to a limited subset or "attractor" in the space of solutions which give classical "chaos" theory its attraction.

   The ideas of relevant vs. irrelevant variables and of the consequences of Liouville's theorem for dynamical systems enlighten every concept from the Big Bang and the arrow of time through the origin of life to chaos proper and conservative dynamical systems. The concept is a simple and very general one: through the "fanning out" of entropy into irrelevant variables, which accompanies all dissipative processes, the trajectory in "relevant" variables occupies a smaller and smaller region in phase space, tending eventually towards an "attractor"; which may be a point, a limit cycle, or a complex manifold such as a "strange attractor," or even more complex behavior. It is these "even more complex behaviors"—as yet not at all completely explored—which is the point at which this line touches the kind of problem we attempt here.

4. Random manifolds and broken ergodicity, percolation, localization, spin glass, neural nets, etc. Of course, this has been my major entry point into the field of complex systems and I may tend to give it some extra weight for that reason. It has two origins: simple natural random physical systems, where it represents the first step in statistical mechanics beyond simple ergoidc theory towards dealing with a system with a complex system of attractors—in the optimum case, a system extending out in number to the limits set by information theory. Thus it is at the very least a counterexample which disproves the conjecture that attiactors can only be of the relatively simple sort provided by few dimensional dynamical systems. But with the advent of neural nets— basically, the union of this line of thinking with other ideas about neural networks of Cooper and Rumelhart and others by Hopfield—one came to realize that one could tailor one's spin glass to any desired problem, so that one could begin to think in terms of problem space rather than solution space. At this point one begins to impinge on the idea of the complex adaptive system (CAS) and I should leave the field "to be continued."

5. A line that is as yet somewhat ill-formed and that relates a little bit to the above is the scheme of "self-organized criticality (SOC)." The self-organized critical system does not necessarily contain any random elements—indeed, its best known examples, turbulence and avalanche models, are explicitly nonrandom—but the outcome is almost necessarily a random fractal system. Fractals enter the problem of complexity in two places: as possible attractors of complex dynamical systems, which seems never to give a random fractal, and in this kinds of system. You will note that I classify turbulence as SOC systems—in my opinion any definition of SOC which does *not* encompass turbulence has thrown out the phenomenon of SOC; turbulence is in fact the canical example of SOC. I feel that, rather like the spin glass, SOC is something which may be used or may appear as a feature of a CAS; it is, of course, in the way the simple CAS in that it is a consequence of adapting to a major difference in scales—but it is not the essential feature.

6. AI. Under this heading one has two lines, one, in my opinion, a dead end, one very much alive. The simple attempt to "imitate life" by means of a von Neumann-style computer and an "expert system" is of course the approach taken by great numbers of people but is of little interest here. Quite why it is so repugnant is hard to define but I think primarily it is the idea that this is a "copout": it has no possibility of growing or of helping one in the search for how real systems work.

   Also under AI, but possibly logically much more closely related to neural network and to dynamical system theory, is the direct computer science approach: the idea that one hopes to use the computer to build a system *ab initio* which has in some sense artificial intelligence. Doyne Farmer and his like do this with a complex dynamical system background; John Holland and others with an approach which has elements of all of our points of view.

7. A final approach could be described as "wetware": the attempt to understand how the brain, human or animal, actually does work, without attempting to provide any particular set of underlying principles. This is again to some extent John Holland's approach. It is certainly Jack Cowan's and Murray Gell-Mann's; and I think all of us have this point of view in the back of our minds. At this point it is appropriate for the introduction to end and for the curtain to go up, and I retire.

INTRODUCTION

In my abstract there are seven Ways mentioned; I added an eighth in honor of our Science Board Chairman, to show that I know what he does even if he doesn't reciprocate. It turned out, according to Erica Jen, also to have a Confucian meaning—the eight virtues: Fidelity, Courtesy, ..., and Shame. The following, which she passed to me, should be a cautionary homily for our whole enterprise:

> A scholar was requested by an Evil Emperor to provide a piece of calligraphy. He presented the Emperor with an elegant scroll on which were written the first seven of the Virtues; in other words "omitting the eighth," implying shamelessness on the part of the recipient and also conferring on him the label of "tortoise" that in Chinese has the connotation of impotence.

My prologue is meant to survey the sources of theories of complexity, especially those that have provided some background for the ideas we are about to discuss. I mean specifically to stop short of what I hope the remaining general speakers will describe: ways of understanding systems that have the full degree of complexity which we call "adaptive," in which we hope to find the "common themes" of our title. So I shall tantalize you by breaking off, in each case, just when the story gets interesting.

In addition to trying to be rather comprehensive I was rather critical: on my transparencies I marked in red some Ways that, I felt, were not very fruitful, and in blue those that were in my opinion especially hopeful. This, you will see, caused a lot of discussion inside and outside of the meeting, probably beneficially, so I shall not apologize for it, just remind you that I am little less fallible than the next guy. This represents my own specific background and bias. So: to the eight Ways.

The first two are part of our gift from the mathematicians.

(1) There is a mathematical/computer science subject called "The Theory of Complexity." Why don't we just close up shop and let the mathematicians tell us what can be known? The roots of this are in the mathematical tradition and come from what I call their "lemma-theorem-proof" structure. The background is in Turing's general computing machine, von Neumann's ideas about automata, and goes on to the theory of computer languages and grammars; on the way stopping by at a concept of equivalence classes of problems, of machines, etc. From this came such things as the "NP completeness" of a problem, "equivalence to a general computer," "undecidability," etc. NP is "not polynomial," i.e., with the time necessary to solve the problem increasing faster than a polynomial N^i with the size N of the problem.

I marked much of this in red. "Complexity Classes" are a concept that is not very useful: "NP complete" problems include the "golf green problem," with only one solution in 2^N that is more favorable than all the others, which are equivalent: a problem on which one makes no headway at all; and linear programming, which is soluble quickly in almost all cases. Somehow, the concept fails to catch the fact that our problems come from the real world which has, itself, a structure and a statistics that is not random and not arbitrary. There is a statistics of problems in which we only see the overwhelmingly probable ones. A school including Traub is improving things somewhat but, I feel, not with the right point of view, yet. I am also not impressed with the computer mathematicians' ideas of "equivalence," which is often slow and difficult to actually implement. Another difficulty is that the mathematicians are only concerned with exact unique solutions, not with good enough approximations which is all we usually need.

I marked in blue the theories of grammars and of computer linguistics; clearly we need to understand these, as you will hear from Walter Fontana.

(2) A second gift from the mathematicians is information theory and its generalization to measures of complexity as well as of information.

From information theory we gladly accept the limits it puts around us, and the ideas of Hamming space. If we try to get a genome or a memory to read back more bits than these limits, we are in trouble. This is a real plus. I marked in red, very controversially, the idea of measures of complexity. My question is, do we need to know? Isn't everything we will study so complex as to be off scale? Or, conversely, if it isn't off scale, are we interested in it? In so far as such measures can set limits like information theory limits, they are useful; has it been shown that they can be used this way?

(3) A source field that spans physics, computer science, and mathematics is the remarkable field that has grown up from the idea of dynamical maps: orbits, attractors, deterministic dynamical systems.

In the public mind, this field is identified as "chaos," which, properly, denotes the field of dynamical systems with few degrees of freedom, with low-dimensional attractors or only a few dimensions with positive Lyapunov exponents. This area I marked in red because to my mind it is basically not complex in a true sense: it has not settled down to a simple fixed point, indeed, but the number of bits necessary for specification of where you are is highly limited.

Two other developments which involve dissipationless systems also seem somewhat irrelevant: first, the idea of dissipationless computing via energy-conserving dynamical systems is an amusing but, I think, pointless game—it seems obvious that such devices, much as they may have intrigued Feynman, are fragile, even when they do not depend on unphysical idealizations. By their very nature they allow no error-correcting coding, no coarse-graining, none of the things Murray will tell us are the very soul of a CAS.

The subject of "quantum chaos" again allows none of the robustness that we get by using the attractor properties of a dissipative system, and is another red herring; I marked it so.

Liouville's theorem comes into its own when coupled with the attractor mechanism in a truly complex system: since phase space volume is conserved, the tossing of entropy into irrelevant variables (via dissipation, cooling, radiation) compresses the trajectories in the relevant variables onto attractors that are more ordered and therefore more useful. This is the underlying physical mechanism that allows the whole ugly mess to get started: the Big Bang, the origin of life, etc. This is the infectious mechanism of the arrow of time, when we recognize that it is the expanding universe that eats all of the entropy thrown away in these processes. This is very much a blue line.

Another blue line is the discovery that even simple, passive physical systems can have N-dimensional attractors: the key contribution of spin-glass physics—see Way #5.

(4) A very computer science, hacker-oriented Way is via artificial entities, typically cellular automata. Here is where complex automaton processes like the Game of Life appear; and it is in this field that Norm Packard first intuited the idea of complexity as highest at the "edge of chaos" between turbulence and quiescence or regularity. This is not a field that satisfies my personal biases: I have never in person met a computer that was not my enemy, and I have a strong professional bias towards hardware or wetware. I simply am not in tune with the sense that computer output is real, in which I am rather *out* of tune with many here at SFI; I mark this as a neutral Way, one for others but deeply mysterious to me.

(5) For the next Way, mine is also a biased view, in this case positive, as one of its originators: large, random physical systems (or pseudorandom or complexly determined). Spin glass is the basic model. These systems give us a statistical mechanics of complexity, and the concept of non- or broken ergodicity. One finds very complex attractors ($\sim N$ dimensions) very large in number ($\sim N$ also), which pushes against the information theory limits; these are robust, also, for reasons not as well understood.

I marked also in blue the mathematics of these systems, "replica theory," which, I feel, has not yet reached its limits and may be useful elsewhere in our enterprise— as, for example, in Way #6.

This Way has already been related to complexity theory, to neural net theory, and to evolutionary biology. Neural net theory, with its collective models, originated with Marr, Rumelhart, Cooper, and others but, I think, gained great insight from the tie-in to statistical mechanics via Hopfield. It was quickly enough broken away by Sompolinski and especially by Gardner. The brilliant inversion of the problem by Gardner leaves one's hands free of the limitation to symmetric functions: "find a network of whatever connectivity that has the given solution as attractor," not "find an attractor that solves the given network."

The tie-in to evolution itself was forcefully further evolved by Stu Kauffman; perhaps here the most useful idea is the concept of "frustration" which is the clue defining a truly "rugged" landscape. Stu insists on climbing hills rather than falling into attractors, but otherwise his N-K models evolved in parallel to spin glasses. He has carried our early ideas about cooperative coevolution (the nominal earliest goal being to understand proteins interacting with RNA) up to the point where it is necessary for me to drop this Way at this point as leading us too close to CAS proper.

(6) Self-Organized Criticality (SOC). This idea is roughly that a system driven by some conserved or quasi-conserved quantity uniformly at a large scale, but able

to dissipate it only to microscopic fluctuations, may have fluctuations at all intermediate scales and, hence, exhibit scale-independent behavior. (The alternative is a cycling of stability and catastrophe, which, of course, happens in some cases but is surely not "adaptive" in any clear sense.) The "SOC" behavior leads to random fractals as descriptions of the state, and to scaling laws for the distributions of "avalanches." Where regular fractal structures are beautiful but artificial structures, random fractals immediately strike us as a valid description of much of the natural world around us. The canonical case of SOC is turbulence; in this case SOC suffers from the Moliére syndrome that it merely broadens the perspective on the phenomenon without adding any new methodology. SOC has an intriguing feature: it seems to generate true randomness from regular initial conditions. Is SOC a feature of some or all CAS? Or is it a generator of the CAS? It seems peripheral but important, and certainly CAS systems often lead to scaling behavior.

Also, SOC has the feature of tuning robustly to the critical "point": adapting to the edge of chaos?

(**7**) We must, in the end, bring out the two most direct Ways to the truth about CAS: to try to build them and to try to study them, or AI and Wetware.

In AI, the standard method of Expert Systems I marked firmly in red, because, from our point of view, this is surely a cop-out: you are using the adaptive ability of real systems simply as a lookup table to tell you what to do in a new situation: in other words, there is an absence of "compression" in Murray's sense.

After that, there are several more promising lines that constitute in some sense real "learning machines." That which, here, is identified with Doyne Farmer is the assumption that the interesting, relevant variables are sufficiently low in number that their behavior may be predicted from some few-dimensional dissipative dynamical systems, i.e., as a finite-dimensional attractor. I mark this questionable, not negative, because it does involve great compression with a real-world rationale for it; but I feel I would be disappointed if the true system we are aiming for should have so little true complexity. We should be "computing with attractors" in Huberman's sense, but not with few-dimensional ones. Our system should, instinctively, have many more degrees of freedom, for complex adaptive behavior.

A more real attempt is the kind of thing that our own John Holland is attempting, along, apparently, with a number of other of the more sophisticated AI workers such as Minsky. The characteristic here which I, as an outsider, identify is that, at this level, we have computers programming computers at two or several levels; rules or programs that can themselves adapt or learn. Here the protagonists are ready and eager to tell you all about them, and I must drop out.

(**8**) Finally, there is the Wetware or Naturalist's approach. We have examples of CAS at our disposal, in the shape of real organisms which we suppose to be the outcome of some CAS and, themselves, to be one or more CAS. The primary CAS is the brain itself, which we can study with an eye to learning how it works and imitating it; or, as natural scientists, just because it is the least understood object around. This is my own natural M.O.; and I resonate with it. We set up no artificial worlds, nor a priori principles, but just try to learn how Nature works. J. Cowan and Edelman are well-known exponents of the natural history of the brain; a recent comprehensive, fascinating, but surely not complete attempt was made by David Dennett in his book on Consciousness. We will certainly, in the end, have to deal with the C-word, because that seems to be crucial to real complexity. His view of C is that it is, in some sense, a "virtual" von Neumann architecture operating as a system manager for a set of parallel computers; there are many fascinating concepts attached, for example, that the manager itself is to an extent programmed by external data.

At that point I am encroaching heavily on the material of our key lecturers, and I must close. Let the party begin!

DISCUSSION

JEN: Would you agree that your seven folds are qualitatively different from your eighth fold? The first seven appear to represent methodologies or world views for approaching the question of complexity in generic form, without involving an understanding of the microscopic details of the particular physical phenomena being studied; whereas the eighth represents a very large set of detailed approaches, with the brain representative of the type of problem that people are studying. Most of traditional science is focused on the approach that you described as the eighth way; in other words, trying to figure out how something works, with complexity and emergent properties then arising in the process of understanding how something works. Is that a reasonable statement, and also can you make some comments on how to balance the first seven versus the eighth, if you agree that there is a dichotomy?

ANDERSON: I don't know that I totally agree with you. Many people just started out with the brain in a naive way and said "Let's do the best we can with it." If you start with the wetware approach, you still have to have the sophistication built up in these other approaches. On the other hand, it represents a separate approach of sorts.

I think you can invent a CAS that works like a brain but may not work at all like our brain. It may be quite different. There's no reason why brains have to work like our brain. Maybe there *is* a reason, but we certainly don't know it yet. So it is a different approach, because it's looking at this particular example which may not be the only example.

JEN: But the brain is only one example. You could look at the economy and say, "I want to understand how the economy works. I don't want to necessarily develop a general black-box method for studying *other* complex systems. I just want to understand economics"...

KAUFFMAN: You mentioned Doyne [Farmer] and the Prediction Company which made me think about grammar models that started with theorem-lemma-theorem. The interesting thing here is: We know from theory that you can have a computation having the property that the only way to understand what it will do is just to watch it grind forward; there's no foreshortened description of it. So "true" complexity may mean that the world unfolds as if it were a computation and the only way you can find out what it will do is to watch it. I suspect that that's true.

ANDERSON: You do? Then you think the whole thing is a cop out.

KAUFFMAN: Yes. If it all unfolds in such a way that there's no foreshortened description of it, then all we can do with complex adaptive agents is to deal with an undecidable world, and that must have a profound effect on the kinds of predictions that we can do and how locally accurate they can be, and so on.

ANDERSON: Well, one of the things that the spin glass story tells you is that although you may be totally unable to specify the actual solution of the problem, you can still find out a lot about it from general principles. You can find the energy of the apparent ground state of a spin glass—there are several apparent ground states—to as many significant figures as you like, yet you don't know anything about the configuration. The self-organized criticality system tells you a lot about its statistics even though you don't know the specific state. So there's a lot you can do even if, in some sense, we have to let it grind ahead, but in other senses you can bound and describe and limit it in very interesting ways. *In these systems, quite literally, there is no foreshortened description of the solution, but there is one of the statistics of the solution.* You may even know exactly how well it can do without knowing how it will get there.

LLOYD: On your transparencies you used three colors of pen. Let us say black was for what you wanted to say, and blue for what was good, and red for what was bad—you should have used a fourth color, and I think green is the correct one, because it's for things that are a mixture of good and bad. It seems that many of your complaints—what you had in red—are of two different kinds which you failed to distinguish. There are problems that are intrinsically unsolvable by a particular approach, and that's bad; they ought to be in red. But then there are problems that an approach may solve, which have not yet been solved. It's bad that it hasn't solved them yet, but good that it might solve them; so those ought to be green.

For example, I can understand why you have reservations about a proof-and-lemma structure for dealing with science, because that hasn't traditionally, necessarily, given all the things you want. But in the theory of computation, surely the fact that the formal mathematical theory of computation has not *yet* given us an average case or more useful notion of what problems can be solvable or tractable or intractable, that's something which you can approach in the future. There's nothing wrong with the fact that it hasn't done it yet.

ANDERSON: Yes, I should have put the name "Traub" in green, and the words under that. Because he is trying to do that, and a number of other people are. I don't know if you are.

LLOYD: And similarly: "Why do we have measures of complexity?" Certainly, if you have a measure of complexity that's just a number that says, "This object is so much complex," that doesn't tell you anything about it. On the other hand, assume you have a particular state that you want to attain—say, a slightly better state for the economy than we have now—and you want to know how complicated that problem is to solve, and you're able to measure complexity. One of Murray Gell-Mann's proposals for how complicated a problem is, is "what's the minimum amount of money you'd need to solve it?"

ANDERSON: Well, that's proportional to the computer time.

LLOYD: Perhaps the unit of complexity ought to be "money." And if you are able, in some sense, to formalize the difficulty of solving this problem, of getting the economy better, and you find you can measure its complexity in terms of dollars, or maybe yen, then that sort of measure could be extremely useful.

ANDERSON: Yes, I agree. I was actually being purposely provocative. I hoped to be, anyway.

Murray Gell-Mann
Santa Fe Institute, 1660 Old Pecos Trail, Suite A, Santa Fe, NM 87501 and Los Alamos National Laboratory, Los Alamos, NM 87545

Complex Adaptive Systems

INTRODUCTION

The various groups at the Santa Fe Institute studying complex adaptive systems (CAS) have somewhat different points of view and have adopted different vocabularies. Some of us speak of "artificial life" or "artificial social life" or "artificial worlds," while others, of whom I am one, prefer to consider natural CAS and computer-based systems together. The latter include methods for adaptive computation as well as models and simulations of natural CAS.

Even the term CAS has different meanings for different researchers. As one distinguished professor remarked, "a scientist would rather use someone else's toothbrush than another scientist's terminology." For example, my nomenclature differs from that of John Holland, from whom I have learned so much. He calls something a CAS only if it is a collectivity of interacting adaptive agents, each of which I would refer to as a CAS. Likewise, John uses the term "internal model" to mean what I call a schema.

There are additional possible sources of misunderstanding as well, stemming from the relation between computer-based and natural systems. At one of our Science Board Symposia, a speaker asked, "Are we using computation as an aid in understanding biology (e.g., evolution, thinking, etc.) or are we using biology as

Complexity: Metaphors, Models, and Reality
Eds. G. Cowan, D. Pines, and D. Meltzer, SFI Studies in the
Sciences of Complexity, Proc. Vol. XIX, Addison-Wesley, 1994   **17**

a metaphor for work on computation?" That is an important question. At some institutions where computation and neural systems are studied, there is real confusion on this issue. For example, success in designing a computing system based on "neural nets" is sometimes taken as evidence that such nets furnish a serious model of the human brain, with the units or nodes corresponding to individual neurons.

I favor a comprehensive point of view according to which the operation of CAS encompasses such diverse processes as the prebiotic chemical reactions that produced life on Earth, biological evolution itself, the functioning of individual organisms and ecological communities, the operation of biological subsystems such as mammalian immune systems or human brains, aspects of human cultural evolution, and adaptive functioning of computer hardware and software. Such a point of view leads to attempts to understand the general principles that underlie all such systems as well as the crucial differences among them. The principles would be expected to apply to the CAS that must exist on other planets scattered through the universe. Most of those systems will of course remain inaccessible to us, but we may receive signals some day from a few of them.

As to successful adaptive computational methods and devices, we have examples such as neural net systems, based on a perceived similarity, even though it may be rather remote, to the functioning of the human brain, and genetic algorithms, based on a resemblance to evolutionary processes. Surely these sets of methods belong, together with many others, mostly as yet undiscovered, to a huge class of computational CAS, with common features that will be well worth identifying and understanding. Some of the new computational methods may exhibit similarities to the operation of natural CAS that we know, such as the immune system, but others may be quite unlike any natural process familiar to us.

A CAS gathers information about its surroundings and about itself and its own behavior, at a certain level of coarse graining. The time series that represents this information can sometimes be approximated by a steady one, although in general it is changing with time, frequently in ways that depend on the system's behavior, and the surroundings are often coevolving. The following[1,2] are general characteristics of a CAS:

1. Its experience can be thought of as a set of data, usually input $\rightarrow$ output data, with the inputs often including system behavior and the outputs often including effects on the system.

2. The system identifies perceived regularities of certain kinds in the experience, even though sometimes regularities of those kinds are overlooked or random features misidentified as regularities. The remaining information is treated as random, and much of it often is.

3. Experience is not merely recorded in a lookup table; instead, the perceived regularities are compressed into a schema. Mutation processes of various sorts give rise to rival schemata. Each schema provides, in its own way, some combination of description, prediction, and (where behavior is concerned) prescriptions for action. Those may be provided even in cases that have not been encountered

before, and then not only by interpolation and extrapolation, but often by much more sophisticated extensions of experience.

4. The results obtained by a schema in the real world then feed back to affect its standing with respect to the other schemata with which it is in competition.

Now the feedback process need not be a clear-cut one in which success is well defined and leads to survival of the schema while failure, equally well defined, results in its disappearance. Fitness may be an emergent or even an ill-defined feature of the process; the effect on the competition among schemata may be only a tendency; and a demoted schema may be kept for use in a subordinate capacity or retained in memory while not utilized (it might, after all, produce useful variants). The important thing is the nature of the selection pressures exerted in the feedback loop, whether or not they are expressible in terms of a fitness function. (Similarly, physical forces may or may not be derivable from a well-defined potential.)

An excellent example of a CAS is the human scientific enterprise, in which the schemata are theories, giving predictions for cases that have not been observed before. There is a tendency for theories that give successful predictions (and exhibit coherence with the body of successful theory) to assume a dominant position, although that is by no means a simple, mechanical procedure. Older, less successful theories may be retained as approximations for use in restricted sets of circumstances. Even wrong theories are not necessarily wholly forgotten, since they may inspire some useful theoretical work in the future.

In its application to the real world, a schema is in a sense reexpanded, reequipped with some of the arbitrariness of experience, some of the random material of the kind that was stripped away from the data when regularities were identified and compressed. For instance, a theory must be combined with boundary conditions in order to give a prediction. The additional data adjoined to the schema may simply be part of the continuing stream of incoming data, which contain, in general, the random along with the regular.

In most CAS the level of the schemata and the level at which results are obtained in the real world are entirely distinct. In the realm of biological organisms, that is the distinction between genotype and phenotype, where the phenotype depends not only on the genotype but on all the accidents of development that intervene between the DNA and the adult organism. However, in some cases, such as Tom Ray's world of digital organisms, the genotype and phenotype are not physically different, but distinguished only by function. His sequences of machine instructions play both roles. As Tom Ray remarks, certain theories of the origin of life on Earth assert that RNA once behaved that way, both as bearer of information and as agent of chemical activity, before the appearance of organisms exhibiting separate genotype and phenotype.

Some new computer simulations of evolution try to include distinct genotypic and phenotypic levels. One that is under development at UCLA even simulates sexual reproduction, with haploid and diploid generations, and tries to test William Hamilton's idea that the principal utility of the male lies in helping to outrace enemies, especially parasites, by providing the offspring with genetic diversity that would be lacking in parthenogenesis.

Complex adaptation is to be contrasted with simple or direct adaptation, as in a thermostat, which just keeps mumbling to itself, "It's too cold, it's too cold, it's too hot, it's just right, it's too cold," and so forth. In the 1940s, the chemist Cyril (later Sir Cyril) Hinshelwood put forward a direct adaptation theory of the development of bacterial resistance to drugs. Genetic variation and selection were rejected in favor of a straight negative feedback process in chemical reactions in the cell. The drug interfered at first with the chemistry of the cell, but then the deleterious effects were mitigated as a result of reaction dynamics, and the mitigation was transmitted mechanically by the bacteria to their progeny in the course of cell division. There was no compression of regularities, no competition of schemata.

Hinshelwood's theory lost out, of course, but it has not been totally forgotten, and it now serves my purpose as an example of direct adaptation rather than the operation of a CAS. Direct control mechanisms are common in nature and in human industry, and they formed the subject matter of cybernetics half a century ago.

The cybernetic era was followed by the era of the expert system, employing a fixed "internal model" designed using the advice of experts in a field, for instance a decision tree for medical diagnosis. The expert system did not learn from the results of its work, however. It remained fixed until it was redesigned. (Only if the human redesigners are included can the expert system be regarded as a CAS, of the kind that involves "directed evolution" or "artificial selection," with humans in the loop.) The new era of CAS in robotics and other such fields is the age of constructed systems that actually learn, by formulating schemata subject to variation and to selection according to results in the real world.

It is useful to distinguish various levels of adaptation. In particular, we can take the example of human societies, where a schema is a set of customs, traditions, myths, laws, institutions, and so forth, what Hazel Henderson calls "cultural DNA." (The biologist Richard Dawkins has invented the word "meme" for a unit of that DNA analogous to a gene.)

The schemata include prescriptions for collective behavior. A culture operating on the basis of a given schema reacts to altered circumstances such as climatic change, invasion, and so forth, in ways prescribed by that schema. If the climate turns warmer and drier, the response of a group of villages may be to move to higher elevations. In the event of attack by outsiders, the inhabitants of all the villages may retire to a fortified site, stocked with food and water, and sustain a siege. What happens at this level is something like direct adaptation.

On the next level, the society may change its schema when the prevailing one does not seem to have given satisfactory results. Instead of migration to the highlands, the villagers may try new crops or new methods of irrigation or both. Instead of retreating to a fort, they may respond to invasion with a counterattack aimed at the enemy's heartland.

Finally, there is the level of Darwinian survival of the fittest (as in population biology). In some cases, not only does a schema fail, but the whole society is wiped out. (The individual members need not all die, but the society ceases to exist as a functioning unit.) At this level the successful schemata are the ones that permit the societies using them to survive.

Not only are these three levels of adaptation distinct, but the time scales associated with them may be very different. Nevertheless, discussions of adaptation in the social science literature sometimes fail to discriminate among the levels, with unfortunate results for clarity.

The disappearance of societies is somewhat analogous to the death of organisms or to the forgetting of ideas. Such phenomena are, of course, universal and not unrelated to the second law of thermodynamics. Still, over a given period of time, the importance of mortality can vary from one domain to another.

In cases where death is very important at the phenotypic level, a crucial measure of success for a schema is phenotypic survival, and reproduction assumes great significance. Moreover, population can then supply a rough quantitative measure of fitness. In biology, one often follows the population of a cluster of genotypes such as a species or subspecies, and the clustering phenomenon is itself of very great interest. One can also follow subpopulations characterized by particular alleles of certain genes.

By contrast, there are situations where death is comparatively unimportant, whether at the genotypic or the phenotypic level. One schema can dominate another without the losing one disappearing; reproduction is not of overwhelming interest; and population is not of critical importance as a measure of fitness. Consider individual human thinking, for example. If we try to grasp an issue more clearly than before, we may succeed in getting an idea that dispels a great deal of previous confusion and displaces, to a considerable extent, earlier ideas. (That is not so easy, by the way, because existing ideas entrench themselves and we have a tendency to interpret new information as confirmatory, so that we dig ourselves deeper and deeper into what may be a quite unsuitable hole.) Over time scales such that forgetting is not a crucial factor, replication and population are not particularly relevant concepts to the success of an idea in the thinking of an individual person. What matters most is that at the real world level one idea has received more positive feedback than another and thus assumed a comparatively dominant position. Over a very long time scale, of course, every system eventually has to get rid of clutter in some way, so that erasure, forgetting, or some other kind of grim reaper has to come into the picture.

Looking at CAS overall, we see that fitness is a rather elusive concept when it is endogenous. If an exogenous criterion is supplied, as in a machine that is designed and programmed to win at chess, then of course the feedback loop involves a well-defined fitness. But when fitness is emergent, it is not so easy to define without a somewhat circular argument in which whatever wins is fit by definition, and whatever is fit is likely to win.

As everyone recognizes, fitness is even less well defined when it is acknowledged that the surroundings of the system are themselves undergoing change and often coevolving. In the latter case, fitness "landscapes," even to the extent that they could be defined for fixed surroundings, now give way to a picture of shifting and interdependent landscapes for the different adaptive components of the total system.

The greatest difficulty in discussing features of a system that are "adaptive" (or that render it "fit") is the distinction between what is adaptive and what has resulted from a process of adaptation. The latter may often be maladaptive. Let us discuss some common reasons for that.

The simplest reason is, of course, that a CAS engages, under the influence of selection pressures in the real world, in a search process over the abstract space of schemata that is necessarily imperfect. Even if fitness is well defined, a system that merely searches for local maxima by "hill climbing on a landscape" would most often get stuck on a molehill. To have a chance to find mountains nearby, the search process must include other features, such as noise (but not too much noise) or else pauses in climbing to allow for free exploration. Naturally, schemata that are more or less maladaptive are often selected.

Apparently maladaptive schemata often occur for another reason, namely that the system is not defined broadly enough to encompass all the important selection pressures that are operating on the schemata concerned. For example, in the scientific enterprise, it would be a mistake to ignore the pressures other than purely scientific ones that affect the viability of a schema, especially in the short run. Scientists often exhibit human frailty, and issues of jealousy, greed, and the misuse of power may play a role in the fate of theories; even observational data are occasionally falsified. Of course it is equally foolish to exaggerate the importance of these extra-scientific selection pressures and to ignore the powerful correcting effect that comparison with nature keeps supplying.

The prevalence of prescientific theories, such as those associated with sympathetic magic, provide even more striking examples of the breadth of selection pressures. Suppose the members of a tribe believe in the efficacy of bringing rain by pouring out on the ground water obtained in a special place in the mountains. Clearly it is not carefully controlled comparison with results that sustains faith in the procedure, but selection pressures of very different kinds. For instance, the authority of powerful individuals or groups may be enhanced by the prevalence of belief in the ceremony, which may, in addition, be part of a whole set of customs that cement the bonds holding the society together.

More generally, it is significant that any CAS is a pattern-recognition device that seeks to find regularities in experience and compress them into schemata. Often it will find fake regularities where there is in fact only randomness. A great deal of superstitious belief can probably be attributed simply to that effect, which might be labeled the "selfish schema." (I have already mentioned how new data are often interpreted so as to strengthen an existing belief.)

Of course, a CAS will often err in the other sense and overlook regularities. Both types of error are presumably universal. In the realm of human beliefs, overlooking obvious regularities can usually be identified with denial. It is striking that in human beings both superstition and denial are typically associated with the alleviation of fear: in the former case fear of the random and uncontrollable and in the latter case fear of regularities that are all too evident, like the certainty of death.

Another example of the breadth of selection pressures comes up in studying the evolution of human languages. Here one should first of all distinguish several different CAS, at different levels and on different time scales. One is the evolution, over hundreds of thousands or millions of years, of the biological capacity to use languages of the modern type. Another is the evolution of those languages themselves, over thousands or tens of thousands of years. Yet another is the learning of a native language by a child. Consider the second of these three systems, concentrating for example on the evolution of grammar and phonology. One encounters, of course, the usual mixture of fundamental rules (in this case the "innate" constraints on grammar and phonology determined by biological evolution), frozen accidents or founder effects (in this case arbitrary choices made in ancestral languages that may have been transmitted to their descendants), and what is adaptive (in this case features that make for more effective communication). However, the selection pressures in linguistic evolution are not wholly linguistic. A great deal depends on whether a people speaking one language is more advanced culturally or stronger militarily than a people speaking another language. Such matters may easily have a greater effect on the fates of the two languages than which one is more convenient for communication.

Another common reason why maladaptive features arise from a process of adaptation is that time scales are mismatched. When circumstances change much more rapidly than the response time of the CAS, traits occur that may have been adaptive in the past but are so no longer. For instance, global climate change on a scale of a few decades will not permit the same kind of ecological adaptation that would be possible in the case of much slower change.

The human tendency to form groups that don't get along with one another, based on what are sometimes rather minute differences that an outsider would barely perceive, may be to a considerable extent an inherited tendency, even though it is fortunately subject to modification through culture. If a hereditary component is really involved, it may have been adaptive under the conditions that prevailed many tens of thousands of years ago. For example, it could have served to limit the size of the population in a given area to a number that the area could support. Nowadays, in a world of destructive weapons, the tendency seems quite maladaptive.

The phenomenon of imprinting provides an extreme case of the mismatch of time scales. A greylag goose that glimpsed Konrad Lorenz instead of its mother when it was first hatched was condemned to treat Lorenz as its mother ever after. The process of imprinting, which works fine in the more common case when the gosling sees its real mother, compromised forever the chances of a normal goose life for any gosling that saw Lorenz instead.

A milder phenomenon is that of windows of maturation. Béla Julesz emphasizes that certain abnormalities in vision have to be corrected early in childhood if they are to be corrected at all. In the case of learning deficits, it is important for public policy to know the extent to which they must be remedied during the first two years or so of life and the extent to which plasticity of the central nervous system permits them to be dealt with later by such programs as Head Start. (Of course the chances of success of Head Start are in any case compromised if the duration and intensity of the program are insufficient, as they often are.)

We must pay attention to time scales for other reasons as well. Fundamental rules on one scale of space and time may reveal themselves to be the results of frozen accidents on a larger scale. Thus the rules of terrestrial biology (such as the occurrence of DNA based on the nucleotides abbreviated A, C, G, and T) may turn out to represent just one possibility out of very many. On a cosmic scale of space and time, the earthly rules would then have the character of a frozen accident or founder effect. That is already widely believed to be the case for the occurrence of certain right-handed molecules in important biological contexts where the corresponding left-handed molecules do not occur. (Attempts to derive that asymmetry from the left-handedness of the weak interaction for matter, as opposed to antimatter, do not seem to have succeeded.)

Some of the most interesting questions about CAS have to do with their relations to one another. We know that such systems have a tendency to spawn others. Thus biological evolution gave rise to thinking, including human thought, and to mammalian immune systems; human thought gave rise to computer-based CAS; and so on. In addition, CAS are often subsystems of others, as an immune system forms part of an organism. Often, a CAS is a collectivity of adaptive agents, each a CAS in its own right, constructing schemata describing one another's behavior. One of the most important branches of the emerging science of CAS concerns the inclusion of one such system in another and the functioning of collectivities such as ecological communities or markets.

One class of composite CAS of particular interest consists of natural or computer-based systems with human beings "in the loop," as in the breeding of animals or plants (what Darwin called artificial selection as opposed to natural selection) or as in a computer system that creates pictures by presenting a human being with successive choices of alterations in an initial pattern.

Pure computer-based CAS can be used for adaptive computation, for modeling or simulating in a crude fashion some natural CAS, and for study as examples of CAS. In all three capacities, they illustrate that astonishingly great apparent complexity can emerge from simple rules, alone or accompanied by a stochastic process. It is always a fascinating and useful exercise to try to prune the rules, making them even simpler, while retaining the apparent complexity of the consequences. Such investigations will gradually lead to a mathematical science of rules and consequences, with theorems initially conjectured on the basis of examples and then proved.

Applications to natural or behavioral sciences require, at a minimum, not just those abstract propositions about rules and consequences but also additional information specifying situations simulating in some convincing way ones that arise in the science in question.

Still more information must be supplied if the computer model is to have any relevance to policy. Conditions prevailing on the planet Earth, including human institutions as well as features of the biosphere, have to be at least vaguely recognizable in the model. Even then, it is critical to use the results mainly as "prostheses for the imagination" in forecasting or in discussing policy options. Trying to fit policy matters into the Procrustean bed of some mathematical discipline can have most unfortunate consequences.

It is a major challenge to the Santa Fe Institute to try to construct bridges connecting these different levels of abstraction, while maintaining the distinctions among them.

When we ask general questions about the properties of CAS, as opposed to questions about specific subject matter such as computer science, immunology, economics, or policy matters, a useful way to proceed, in my opinion, is to refer to the parts of the CAS cycle,

I. coarse graining,

II. identification of perceived regularities,

III. compression into a schema,

IV. variation of schemata,

V. application of schemata to the real world,

VI. consequences in the real world exerting selection pressures that affect the competition among schemata,

as well as to four other sets of issues:

VII. comparisons of time and space scales,

VIII. inclusion of CAS in other CAS,

IX. the special case of humans in the loop (directed evolution, artificial selection), and

X. the special case of composite CAS consisting of many CAS (adaptive agents) constructing schemata describing one another's behavior.

Here, in outline form, is an illustrative list, arranged according to the categories named, of a few features of CAS, most of them already being studied by members of the Santa Fe Institute family, that seem to need further investigation:

I. Coarse Graining

1. Tradeoffs between coarseness for manageability of information and fineness for adequate picture of the environment.

II. Sorting Out of Regularities from Randomness

1. Comparison with distinctions in computer science between intrinsic program and input data.
2. Possibility of regarding the elimination of the random component as a kind of further coarse graining.
3. Origin of the regularities in the fundamental laws of nature and in shared causation by past accidents; branching historical trees and mutual information; branching historical trees and thermodynamic depth.
4. Even in an infinite data stream, it is impossible to recognize all regularities.
5. For an indefinitely long data stream, algorithms for distinguishing regularities belonging to a class.
6. Tendency of a CAS to err in both directions, mistaking regularity for randomness and vice versa.

III. Compression of Perceived Regularities into a Schema

1. If a CAS is studying another system, a set of rules describing that system is a schema; length of such a schema as effective complexity of the observed system.
2. Importance of potential complexity, the effective complexity that may be achieved by evolution of the observed system over a given period of time, weighted according to the probabilities of the different future histories; time best measured in units reflecting intervals between changes in the observed system (inverse of mutation rate).
3. Tradeoffs between maximum feasible compression and lesser degree that can permit savings in computing time and in time and difficulty of execution; connection with tradeoffs in communication theory—detailed information in data base versus detailed information in each message and language efficiency versus redundancy for error correction.
4. Oversimplification of schema sometimes adaptive for CAS at phenotypic (real world) level.
5. Hierarchy and chunking in the recognition of regularities.

IV. Variation of Schemata

1. In biological evolution, as in many other cases, variation always proceeds step by step from what already is available, even when major changes in organization occur; vestigial features and utilization of existing structures for new functions are characteristic; are there CAS in which schemata can change by huge jumps all at once?
2. Variable sensitivity of phenotypic manifestation to different changes in a schema; possibility in biological case of long sequences of schematic changes with little phenotypic change, followed by major phenotypic "punctuations;" generality of this phenomenon of "drift."
3. Clustering of schemata, as in subspecies and species in biology or quasispecies in theories of the origin of life or word order patterns in linguistics—generality of clustering.
4. Possibility, in certain kinds of CAS, of largely sequential rather than simultaneous variants.

V. Use of the Schema (Reexpansion and Application to Real World)

1. Methods of incorporation of (largely random) new data.
2. Description, prediction, prescribed behavior—relations among these functions.
3. Sensitivity of these operations to variations in new data.

VI. Selection Pressures in the Real World Feeding Back to Affect Competition of Schemata

1. Concept of CAS still valid for systems in which "death" can be approximately neglected and reproduction and population may be correspondingly unimportant.
2. Exogenous fitness well-defined, as in a machine to play checkers; when endogenous, a elusive concept: attempts to define it in various fields, along with seeking maxima on "landscapes."
3. Noise, pauses for exploration, or other mechanisms required for the system to avoid getting stuck at minor relative maxima; survey of mechanisms employed by different systems.
4. Procedures to use when selection pressures are not derivable from a fitness function, as in neural nets with (realistic) unsymmetrical coefficients.
5. Possible approaches to the case of coevolution, in which the fitness concept becomes even more difficult to use.
6. Situations in which maladaptive schemata occur because of mismatch of time scales.
7. Situations in which maladaptive schemata occur because the system is defined too narrowly.

8. Situations in which maladaptive schemata occur by chance in a CAS operating straightforwardly.

VII,VIII. Time Scales; CAS Included in Others or Spawned by Others

1. Problems involved in describing interactions among CAS related by inclusion or generation and operating simultaneously on different levels and on different time scales.

IX. CAS with Humans in the Loop

1. Information about the properties of sets of explicit and implicit human preferences revealed by such systems.

X. CAS Composed of Many Coadapting CAS

1. Importance of region between order and disorder for depth, effective complexity, etc.
2. Possible phase transition in that region.
3. Possibility of very great effective complexity in the transition region.
4. Possibility of efficient adaptation in the transition region.
5. Possibility of relation to self-organized criticality.
6. Possible derivations of scaling (power law) behavior in the transition region.
7. With all scales of time present, possibility of universal computation for the system in the transition region.

ACKNOWLEDGMENTS

It is a pleasure to acknowledge the great value of conversations with John Holland and with other members of the SFI family. My research has been supported by the U.S. Department of Energy under Contract No. DEAC-03-81ER40050, by the Alfred P. Sloan Foundation, and by the U.S. Air Force Office of Scientific Research under the University Resident Research Program for research performed at Phillips Laboratory (PL/OLAL).

REFERENCES

1. Gell-Mann, Murray. "Complexity and Complex Adaptive Systems." In *The Evolution of Human Languages*, edited by M. Gell-Mann and J. A. Hawkins. Santa Fe Institute Studies in the Sciences of Complexity, Proc. Vol. X. Reading, MA: Addison-Wesley, 1992
2. Gell-Mann, Murray. *The Quark and the Jaguar.* New York: W. H. Freeman, 1994 (in press).

DISCUSSION

COWAN: If the language is dynamic and adaptive, to what extent does the grammar permit you to describe such changes?

GELL-MANN: I don't know exactly how to answer that, but I can answer a related question that may be of interest.

There are a lot of things called grammatical universals, rules that are true of grammars of all known human languages. They are usually of the "if-then" form, and they often refer to things like word order. Others are very simple and are not of the "if-then" form. For example, all known human languages have pronouns, and every known human language has a genitive construction of some kind. Some rules are like the following. If a language has a special form for *three* objects, then you may be sure it also has a special form for *two* objects. There's no language with a singular, a form for three objects, and a plural, without a form for two objects as well.

These grammatical universals are supplemented by grammatical near-universals, or statistical universals, which start out "For all known human languages except one or two, the following rule holds. . . ." Now, some of the people who study universals are very pure, rigid, dogmatic Chomskyans, who concentrate on innate, preprogrammed, biologically evolved rules and insist that grammatical universals must reflect only those rules. Other linguists study the utility of some of these rules in communication, and they say many of them could simply keep arising in the course of linguistic evolution. Other universals may be the results of frozen accidents. If modern human languages go back only something of the order of 10^5 years, then they may have features that are simply inherited from a unique acestral tongue.

All of these three mechanisms may be present, as well as mixtures of them. But the most important property of linguistic evolution is that language isn't a closed system. A society has lots of characteristics associated with it besides language, and the death of a language may be caused by events that have nothing to do with language as all. Linguistic evolution is a complicated business, and people who try to simplify it, by looking at only one aspect, are doing the subject an injustice.

WALDROP: In terms of taking experience and compressing that into a schema as opposed to a look-up table: look-up tables are fast, whereas expanding a schema can be slow, depending on the processing power of the machine you're doing it on.

GELL-MANN: Correct. I refer you to my abstract where I wrote the following (this is under "compression of perceived regularities"): "2. Tradeoffs between maximum feasible compression and lesser degree of compression which can permit savings in computing time and in time and difficulty of execution. Connection with tradeoffs in communication theory, detailed information in the data base, versus detailed information in each message. Language efficiency versus redundancy for error correction." Here I've consulted with Charlie Bennett, and John Holland, who really know about these matters, and they tell me that such tradeoffs are of very great importance. So I completely agree with you.

KAUFFMAN: Concerning the fundamental distinction you're making between direct adaptation and complex adaptive systems. Let me take what might be a form of the strong artificial life claim, that anything you can do with neurons and computer chips I can do with molecules. Consider *E. coli*, which has something called the lac operon in it. The function of the lac operon is that when lactose comes into *E. coli*, it binds to a receptor molecule. Normally the receptor molecule sits on the gene—called the operator—and stops transcription of the lactose gene. But when lactose is in the cell, the lactose binds to the repressor molecule and pulls it off the operator thereby allowing transcription so that *E. coli* now metabolizes lactose. The point here is when you're trying to draw the distinction between something called "direct adaptation" of *E. coli* becoming resistant to a toxin, which it might do for example by evolving a new protein, or evolving new regulatory circuitry to switch on in the presence of that toxin, which commonly happens, then if *E. coli* can do it, what's the distinction between *E. coli* doing direct adaptation to toxins, and the distinction you want.

GELL-MANN: If it happens genetically, it is not direct adaptation.

KAUFFMAN: So why isn't that a complex adaptive system?

GELL-MANN: It is. I said genetic adaptation *was*.

KAUFFMAN: I thought you said that direct adaptation is not complex.

GELL-MANN: I said Hinshelwood's kind of direct adaptation was not. If you are discussing genetic adaptation which I do not call direct, then you are dealing with complex adaptation.

KAUFFMAN: Oh. Well, what's excluded?

GELL-MANN: What is excluded is a case where you have simple cybernetic feedback without compression, without what, in biology, usually goes through the genes. In biology, compressed information is usually genetic.

COWAN: Hinshelwood did it by a chemical reaction in one generation.

KAUFFMAN: But suppose that that worked, suppose that you pulled out a modification in a protein molecule so that it is now a new protein. A Lamarckian-modified molecule would not have compressed information.

GELL-MANN: That's all right. You can imagine a biological process that involves compression of information and that *does* go on in the same generation, without genes, and that would still be a complex adaptive system. It would be a new, hitherto-undiscovered one, or maybe one that's been discovered already but is obscure.

KAUFFMAN: So your distinction is between whether or not the process that does the recognizing and reacting somehow compresses its description. . .

GELL-MANN: I distinguish between compressing a lot of experience into a small message and a look-up table. Most of the cybernetic devices that we discussed 50 years ago did not have the faculty of compression.

ANDERSON: Couldn't the degree of compression be a criterion, rather than a measure of complexity? In the case of scientific theories it very often is, if one takes the Bayesian attitude that a simple theory is intrinsically better than a complex theory.

GELL-MANN: Yes, I agree. I think you could use it as a quantitative measure, also. But I don't know exactly how. I don't know whether to take ratio of numbers of bits, or what. But there probably is some quantitative measure that we use.

PINES: A comment: As you remarked, the scientific enterprise is a good example of a complex adaptive system, and so are we all. And in a sense, I think it's interesting to view this workshop as a complex adaptive system in which, in a sense, the selection pressures have to do with which, if any, of those guiding principles have general applicability. Then I can imagine several ways of trying to test this. One might be that we arrive at a consensus that some of these work, some don't. But I'm not sure that's enough. I wonder if we shouldn't—as we look at each of these principles—ask the question, "Can I compare it to an experiment, or series of experiments? Can I compare it to a series of observations?" Or finally, "Can I carry out a set of computer simulations to test the principles involved?" I make this comment in the hope that as we go along in the meeting, we'll address these issues, and each of the principles, from this point of view: How does it work? Does it really work when applied to practice? And finally, a further question: Do any of the principles possess any predictive power in dealing with a particular simple system?

GELL-MANN: I don't think it's such a simple matter as just saying, "These are theories, and they should be tested by observation." I think it's more subtle. We structured the workshop so that during the first few days, some notions of general ideas of complex adaptive systems, and also nonadaptive complexity, would be presented. Then there would be a few days of discussion by people who are experts in fields such as immunology, economics, adaptive computation, and the origin of life—various fields where the rubber hits the road. It's, so to speak, our phenotypic arena. Although we don't ourselves do experiments, nevertheless, these are people who are in contact with experiment and whose theories *are* intended to predict correctly the results of observations. Then we come back and reexamine the discussions that we've had at the beginning and see how we want to modify our original ideas in the light of criticism by the other general theorists, but especially in the light of what we've heard from the people who are doing the work in the specific fields. Now that's not quite the same as what you said.

These proposals are not exactly theories. These are suggestions for how to organize the work, and the test is not whether they're, so to speak, "true" or not (although I may have lied here and there); the test is whether these are useful in organizing our thinking about all the things we're going to hear about in the general session, and especially in the individual sessions on the individual subjects. That's what we'll come back and discuss during the last day or so: how we want to modify our general ways of talking, in the light of their utility not so much in the lab as in the discussion. Those are my views of the selection pressures on the general ideas.

PINES: Now my question: why do you think so much more attention has been paid to trying to arrive at quantitative measures of complexity, and so little attention, relatively speaking, to quantitative measures of adaption?

GELL-MANN: I think that we have not had a lot of attention to either, in the sense of complexity that I've tried to illustrate here. What's happened is we've had a lot of attention to quantitative measures of complexity as defined by mathematicians, for mathematicians, in mathematical contexts. And that's no wonder. But the kind of complexity we're talking about here is still a bit ill-defined, it's quite subjective, as we've discussed. And likewise adaptation, evolution, and so on are tricky to work with. These are less clear-cut mathematical problems, and there has been less attention to them. However, in certain fields, like mathematical population biology, we find attention paid to clear-cut discussion of the issues.

FELDMAN: One of the answers to that is that each of the disciplines comes at the definition of adaptation in a different way, and in biology you don't often measure adaptation. They measure, as you said, fitness differences—relative fitnesses. In the experimental computer science that I've seen at Santa Fe Institute the criterion for adaptation is given to you before you begin, namely, you want to see the process which leads to the best version of this program, or the best instruction. You know what the criterion for adaptation is. Biologists, on the whole, don't do that.

GELL-MANN: But I think there isn't yet much of a general theory, apart from biology, apart from computer science, apart from other particular disciplines. I don't think there's been much work on pattern recognition compression, and especially variation and selection, as general phenomena rather than in connection with biology, thinking, computers, or whatever. What I'm proposing here is that it *might* be useful to think of them in that way.

JEN: I was confused by the distinction you were making in the context of speciation and evolution of individuality between physical and biological systems. It led me to think about what other sort of fundamental distinctions there are, what other contexts in which you think there are actually fundamental distinctions between physical and nonphysical systems (though I'm not even sure what a physical system is). Given the fact that you think that there are such distinctions, and given the fact that we are as a group, by and large, mostly people coming from the physical sciences looking at the nonphysical sciences, how should this affect our choice of what problems to work on, given the fact that we have so much enthusiasm now for looking at everything. You believe, for instance, that a process like evolution of human language is in fact tractable, or susceptible to these methods of analysis which is not at all *a priori* obvious, because of the fact that it is so different from what we know from physical science. What is the scope of things that we can look at, and what is the scope of things that we're really presumptuous to be thinking about at this point?

GELL-MANN: I spoke rapidly about differences, and of course I was eliding differences which in my writing I've tried to make more explicit.

There is no nonphysical system in the universe, according to the beliefs of everybody in this room. I don't think that anyone here would believe that there are fundamental "vital forces" which are outside of physics and chemistry, and govern life. I don't imagine there's anyone in this room who believes that there are fundamental "mental" processes that govern thinking and that are outside of biology and, therefore, outside of physics and chemistry. When I made the distinction between physical evolution and the kind we're talking about here I meant only that there are properties of complex adaptive systems—which I tried to describe—which are largely absent as far as we know in whole classes of physical evolution: evolution of galaxies, of stars, of planets, and so on. We have no evidence of compression, schemata, variation of schemata, selection; *maybe* they exhibited these phenomena, but if so we don't know about it.

I've used the issue of turbulence, for example, as a case. We know that turbulence in a complicated pipe with changing shape has little eddies in it, and the little eddies spawn smaller eddies, and the smaller eddies spawn smaller eddies, and certain eddies find their way through the pipe successfully and live to reproduce with lots of little tiny eddies. Other eddies don't make it through the pipe. Now, are we talking there about selection and evolution in the biological sense? Well, probably *not*, because we don't have any evidence that these eddies are doing the work of perceiving regularities, compressing them, and then constructing variant compressed schemata that undergo selection. Rather the selection of eddies seems to be taking place on the surface, with look-up tables. Of course, we don't *know* that, for sure.

So you are right to be skeptical, if that is the attitude you're expressing. Doyne Farmer has phrased it very well in saying that he would like some day to understand how, from the equations or the rule-based mathematics for a system, one could tell whether it was making a compressed model of its environment and of its own behavior. In other words, could you tell from a mathematical description whether the system is going beyond the physics and chemistry that everything shares to complex adaptive behavior. We don't know how to do that, and it would be an interesting challenge.

LANGTON: You made the statement that the difference between learning and culture and evolution is that evolution doesn't make big leaps.

GELL-MANN: No, no. *Biological* evolution, I said, tends not to make big jumps but works with what it has and proceeds by modifications of what is there—organs that are there, for example—for new functions. You can see that in societies, also. The British, for example, are very good at this. They have the Privy Council, which used to supply advice at the highest level to the ruler. The ruler doesn't rule anymore, but simply reigns; they've still got the Privy Council, however. So what do they do with it? Well, they make it an advisory committee on science, for example. Human thinking may have the possibility, occasionally, of operating in a different mode where you make a big jump. However, some investigators think that isn't so...

ANDERSON: I don't think it's so. I don't believe that the brain makes such big leaps. It always uses something it already has.

LANGTON: To flip the coin, I also think it's also true that evolution sometimes (rarely) does make very big leaps. For instance, the Cambrian explosion, the evolution of multicellularity, or the origin of eukaryotic cells...

GELL-MANN: Those look like very rapid processes viewed from the present, because they took place billions of years ago, but they took awhile.

LANGTON: I'm not sure I see such a big difference...

GELL-MANN: In my opinion there is a significant difference. I think that an engineer designing something—a hypothetical engineer, maybe not a real graduate of an engineering school—could make a more rapid jump than a gene. But I could be wrong. We should try to find some quantitative measure of jumping and see whether it's true.

FONTANA: The basic difference that I believe to have caught in the first two talks...your talk was centered around the genotype-phenotype distinction. This is actually the essential distinction that was lacking, it seemed to me, in the considerations that Phil was making.

GELL-MANN: *Until* today I always said it was essential, and today I partially took it back, in the sense that there are cases where the two are *physically* the same.

FONTANA: I'd like to convince you to draw this distinction more radically, actually. Because that is exactly the distinction that physics is lacking (I mean a traditional way of thinking in physics). This distinction actually has its counterpart in mathematics, and in particular in a theory of mathematics called recursive function theory. There the distinction goes under the heading "Object and Function." A function, for a mathematician—prior to recursion theory—or for a set theoretician, was essentially a phenotype. So all functions were phenotypes. There were infinite arrays of facts, all of which were mostly accidental. Now if you take such an infinite array of facts, of input-output pairs, and you can express it in terms of a rule by capturing a pattern in this infinite series of facts, then you have constructed a computable function. So recursion theory strikes me as being the most basic, and the first mathematical theory about compression at all, because it tells you what you can express with finite means. It tells you which phenotypes have a genotype, so to speak, in an abstract sense, in a very general sense.

In this respect I would say that it is not true that, for example, if you'd like to include Tom Ray's work in your definition of complex adaptive systems, I think you can still do it because I believe that in Tom's work there *is* a clear-cut distinction between a genotype and phenotype because the *program* is a genotype, but a program is also a function. It's a series of input-output pairs that can be captured by a pattern which is the program, but nevertheless the program is just a specification of actions so it has a phenotype, obviously.

GELL-MANN: But in a sense it's the same thing.

FONTANA: No, it's not the same thing. If I write down the function "x^2," that's not the same thing as the function that maps any natural number x^2.

GELL-MANN: What you're saying is that the distinction has been compressed very thin, but it's not absolutely zero.

FONTANA: I think the distinction becomes very important particularly when we live in a finite world where a function, or an object that expresses a function, never sees the entire domain over which it is defined, but only a tiny part. So you can have a function that is the identity only on a very specific domain but not in general. It depends on what other objects are there. This leads me to a second brief comment. . . .

ANDERSON: First this comment; you mentioned my talk. I think that this represents a misunderstanding of my assigned role in giving this talk and I do not like being used as a straw man in this way. Essentially you're saying it's the difference between two historical eras, which I would certainly agree with. I was describing all the historical approaches to complex adaptive systems and carefully avoiding any discussion of CAS's themselves

FONTANA: Let me add a brief remark. The interesting fact about recursion theory, if we would like to take this as a model of an abstract genotype-phenotype distinction, is that it is constructive. That's why I like it so much. It tells you how you build new things—in your terms, schemas—out of available ones in a totally nonrandom fashion, without mutation or recombination and so forth. Clearly, mutation and recombination are events that are most important for an adaptive system, and a complex adaptive system seems to me also to have a constructive part where you can't buy the actual numbers individually—you get them only as a package and you get the entire structure implied by them. So, if you have elements in complex adaptive systems that are objects and functions at the same time, that are genotypes and phenotypes, then by virtue of their being functions that can act on these phenotypes, you get constructive effects that are not taken into account by purely focusing on random mutation and recombination.

GELL-MANN: What you're saying, if I may just put it in very simple terms so I can follow you, is that while the distinction has practically disappeared in Tom Ray's work, and also in the RNA theory of the origin of life—because we're dealing with the same agents, really, that are the chemical agents, and are the bearers of the information—nevertheless, it's worth still making the distinction between these two *roles* and in that way we can preserve the genotype-phenotype distinction in a useful way. Because in all the cases where it does exist, it is a very important distinction, preserving it in its degenerate form is a very good idea.

FONTANA: Exactly. I would say it occurs already with molecules, where you would say there is no such distinction. I would still make a distinction between the structure of a protein and the set of chemical reactions it can undergo, and these are just two examples. You can think the one as being a representation of the others, but they're two different functions, one is extensional and one is intensional.

BROWN: Early on in your talk—you went over this very rapidly—I think I heard you say that complex systems can exist, and actually can evolve and develop if the environment, essentially, is a constant time series and doesn't change. I wonder if that is really correct...

GELL-MANN: By time series with constant properties I mean that it does have certain reactions to what's done to it and reacts back on the system, but it does it in a constant way, so to speak. So we're not including secular variation like changing of the earth's atmosphere, and we're not including coevolution.

BROWN: I want to challenge you on that and ask whether these major extrinsic changes may not be very important in the development of these kinds of systems. If we get to the question that Chris Langton asked about the leaps, for example, in biology and in ideas—I wonder if many times those leaps aren't triggered and in some fundamental way caused by a change in the environment. One way to get off the hill, to use the metaphor, is to change the landscape so the peak of the hill no longer is the top of the mountain. That involves potentially a major environmental change.

GELL-MANN: I think you're right in many cases, and in fact I have a whole lecture—which I will not inflict on you—on getting creative ideas, which depends precisely on that point: that one can search for artificial stimuli to get one out of traps, out of idea traps, to other regions where the better ideas may lie, through a changed environment. But what I wanted to say was something slightly different. I wanted to say that you can still maintain the definition of a complex adaptive system and study some of its properties by idealizing, by ignoring the change in the time series represented by the environment, and by ignoring coevolution. You will not get many of the most important properties of real ones, but you will still have something that would fall into the rubric of complex adaptive systems and would adapt to its environment. And that's worth studying, I believe, as a simple case that is still part of the general subject. But you're right; in most practical cases the changes are very important.

ANDERSON: The fitness-nonfitness distinction is already available at the rather primitive level of spin glasses. You can deal with them as neural nets, and deal with them—as John Hopfield did—with a fitness landscape. And in fact all of the mathematics works perfectly well with asymmetric coefficients.

GELL-MANN: He was able to translate what was done into the language of some kind of physics and that was very nice. But it was also a step backwards because the people working on psychological models were already using the correct, nonsymmetric coefficients.

ANDERSON: But one can now deal with this mathematically and formally without assuming a fitness (Lyapunov) function.

BUSS: I have a comment on these definitional issues and then a question for the speakers. The suggestion was made by Murray that one of the strengths of science is that there's some mild selection pressure on being correct. I would contend that another strength of science is that the definition of what you are working on is not a problem internal to the science itself. I think that if you look at fields where definition of your activity is in fact internal to the discipline—for example, philosophy—there is not quite the same illusion of progress as you have in other disciplines.

GELL-MANN: We've seen a lot of that in recent years in the penetration of what are fundamentally political disputes into the philosophical bases of various subjects. In archeology, for example, it's been a terrible problem, now receding. You're right. That kind of debate in a scientific field can be devastating to scientific progress.

That was why I mentioned three different ways in which a society can "adapt," and how futile disputes about which of those is *really* adaptation, which of those is *really* evolution, can cause the subject to grind to a halt. Is that the sort of thing you mean?

BUSS: I think it speaks to how much discussion there need be about what complexity is.

Now my question. I was struck by the difference between George's introduction, which focused on levels of complexity, the hierarchy of complexity—how do you get chemistry out of physics, how do you get biology out of chemistry, how do you get social systems out of biology—from the two largely methodological talks that followed. It wasn't obvious to me that the material covered in those two presentations in any sense went to the heart of the emergence of new organizational levels that George was focusing us on.

GELL-MANN: I think that's an excellent question. I don't have much that's intelligent to say about it, but that's not because I don't think it's a very important subject.

There are, presumably, thresholds of complexity allowing certain kinds of systems to function. And there is presumably some threshold of complexity for a complex adaptive system which is why we're allowed to call it a complex adaptive system. Or, if not a threshold, at least a mathematical relationship, so that if we define the degree of compression, the degree of variation or selection, and so on, we can relate the complexity of the system to those degrees. In other words, either a sharp threshold or a gentle one. Likewise, there's presumably some sort of threshold for self-awareness, which would be very exciting to understand. And again, that may be sharp or it may be gentle. At a place like this, one is presumably allowed to mention consciousness—although there are a number of campuses where saying the "C word" would lead to ostracism.

ANDERSON: I would agree with what Murray says very much; it's not that we don't agree, but we know less about that. Really, you build hierarchies—of course you build hierarchies.

GELL-MANN: Now on the other subject of what is fundamental, and how fundamental science gives rise to sciences on other levels; that I have written about a lot (but not here), and it's an important subject too—how particle physics and cosmology give rise to chemistry, giving rise to biology, and so on and so on, with the addition of new information at each stage.

PINES: I want to go back to one of Murray's points, which has been discussed a bit, that the difference between, as he suggested, a biological system— which goes along pretty much as it has been, within a certain set of regularities, small changes, and so forth—as compared to a thought, human thought, where one can suddenly get great leaps, or so one thinks.

GELL-MANN: I said I thought that there was a quantitative difference between the two, but people have disputed it.

PINES: I think there is, and I would suggest that it's the difference between learning and innovation. [In] learning, in some sense, you're adapting to be able to recognize an existing set of patterns, which are sort of spelled out for you by your environment, whereas with innovation you may suddenly view things differently. The complex adaptive system may be capable of saying, "Hey, there may be a quite different set of patterns out there other than those which the environment is providing to me." And I wonder if, in some poetic sense at least, this is also connected with the notion of emergent behavior. Namely, that you're going along with a system which seems to be, you know, bouncing up and down in some regular, or irregular, kind of way, and suddenly you find—subject to a given set of stimuli— that it quite changes its behavior pattern.

GELL-MANN: Let me argue on the other side, for a moment, since you've taken my side. Chris Langton isn't here, today; I'll argue on his side.

I've thought some, and read a lot, and written a little bit, about innovation in human thinking—in particular in connection with science, but also art and other subjects—and talked with a lot of practitioners of all sorts of subjects (in the arts, engineering, and the sciences). The methods seem to be very similar, and one thing that stands out is that usually the leap (if there is one), begins in a negative way, by getting rid of some unnecessary prohibition that was adopted a long time before along with a useful idea. Getting rid of that prohibition allows more freedom than was thought to exist, and it can lead quite rapidly to progress in solving or formulating a problem. Perhaps the wrong *Verbot* holds back to normal form of progress by small tips, so that when the brain is removed, a longer tip can be taken.

ANDERSON: Let's say that our answers to Leo kind of implied that evolution can make big jumps, because we were saying that there are quite different levels on the hierarchy and that they had to happen sometime. We have to have consciousness for the first time, in some sense.

GELL-MANN: Well maybe. That depends on how gentle the threshold is. We don't know how sharp these thresholds are.

PINES: Jim Brown makes the point that it may have a lot to do, in evolution at least, with a rapid change in the environment. And of course you can ask the same question about human innovation.

LLOYD: Saltation (the theory of evolution by leaps and bounds), rather than gradualism, is a viable theory of evolution, at any rate.

GELL-MANN: There may be some planets on which it usually goes that way. On this planet it doesn't seem to be true.

FELDMAN: This issue was addressed extensively by Saul Wright in the '20s and '30s, when he developed what he called the "shifting balance" theory. And it was designed to handle this sort of thing. You can have an evolutionary system— a biological evolutionary system—going along, changing—in the Darwinian sense, gradually—but with a number of different attracting points. Environmental change has the property of cutting down the population's size in a nonrandom way, so that each segment is now put into the domain of attraction of a different attractor. That's the shifting balance theory. But the long run of that is that you get an array of ultimate phenotypes, caused by these genotypes, and it might have been a stimulus from an environmental change, but it wasn't a large biological change at the level of the genotype that was necessary for this.

GELL-MANN: I think that much discussions of this topic from now on should refer to Tom Ray's program, and also to other programs that people are developing, in some of which there is a physical distinction between phenotype and genotype. For example at UCLA, they're working on such computer models of evolution, trying to test by computer modeling the idea of Hamilton and others, the idea that the value of sex—the role of the male—may have something to do with resistance to short-generation parasites. But to take Tom Ray's only, he has some very nice examples of cases where the situation is stable for a very very long time— apparently—but changes may be taking place that don't matter much for survival, and then all of a sudden, after a long period, huge changes in populations take place, as in the model of punctuated equilibrium.

FELDMAN: The key issue is, is there a biological organism that does that?

GELL-MANN: Yes that is the key issue. We have other cases where organisms are very similar to what they used to be billions of years ago, like extremophiles in hot deep acidic waters, with sulfur. Now we don't know whether those are *genetically* like the very ancient extremophiles, or whether they've undergone a substantial amount of genetic change while remaining phenotypically very similar. It could be either way, and either way would be theoretically satisfactory. We just don't know. But in any case it's a fact that there is a case where the environment has changed very little, and where the phenotypic response to that environment has stayed more or less constant. It's like one of John Holland's cases where the problem is so simple that it's *solved* by the computer. I mean, playing tic-tac-toe, for example. It's all very well to design a checkers automaton, or a chess automaton, but a tic-tac-toe automaton converges quite rapidly to something that plays tic-tac-toe perfectly. It may change genotypically after that, but it's not going to change phenotypically because it already plays tic-tac-toe perfectly.

BUSS: I'd like to disassociate the comments that I've made with respect to hierarchy of complexity from this issue of whether biological evolution can make leaps. There are certain classes of leaps that are simply undeniable, and likely do involve something more than our conventional apparatus. For example, when we go from a prokaryotic cell to a eukaryotic cell, when we're combining two self-replicating entities to make a third class of self-replicating entities—that is an organizational shift that has happened biologically. It is clearly on a different grade than whether in fact you make a long bristle fast. There's a long history of the relative magnitude of changes within a given organizational grade, and the rate of their appearance, and there are a wide number of scenarios that can predict the rate of your choice. But I would like to say that that's a different set of problems than the set of problems of how you get to new organizational classes in evolution, and those are necessarily rapid.

KAUFFMAN: I'm struck, Phil, by the fact that you're drawing a distinction between throwing away irrelevant degrees of freedom, a system losing entropy and contracting down to a region of its phase volume—and Murray's statement about throwing away clutter, and compression. If you have a reversible dynamical system, it's reversible, and you can't in that sense throw away degrees of freedom. It seems to me that there's a connection between the point that you've made and Murray's statements about compression. Then there's the question of what you mean by the relevant degrees of freedom in keeping them. The fundamental question is: relevant for whom? Attempting to describe that is inevitably going to get us into a notion of agency. For whom is it good and for what purposes?

For example, if you've got bugs—Darwinian evolution with self-reproducing things—then I know how to answer the question "for whom?" Just, is it toxin, or food, for this bug? But more generally, one needs some sort of notion of relevance for whom, for what: what is becoming? And I think that it's fundamental, but I don't think it's well-posed.

GELL-MANN: I don't think that deriving the selection forces from a potential—or even treating them necessarily as forces rather than a statistical distribution of forces—is always possible. In specific cases it may be, but as a general rule I think it's not, and therefore I wouldn't use a potential function as a fitness. I would just study the selection pressures as such.

KAUFFMAN: You're restating that it's a dynamical system.

GELL-MANN: Yes. The second thing I would say is that in connection with entropy, that it's by now well known, as a result of the work of Landauer, and Bennett, and Zurek, and Lloyd (and even me, to some extent), if you're going to consider entropy from outside the closed system, that's one thing. Then it just increases in the usual way. But if you're going to consider it from inside, with an observer, then every time that observer learns something, on the average over which alternative occurs and is learned, the effective entropy of the system *decreases*, and keeps on decreasing, as more and more stuff is learned, provided you define entropy in the old way. And it's useful, therefore, to have a newer definition of entropy in which you add in the algorithmic complexity of the record of what's been learned. In that case, on the average, it continues to increase just as before. And that helps to clear up one or two points of confusion to which you alluded.

In order to keep the second law of thermodynamics going, from the point of view of an observer inside the system, you have to modify the definition of entropy in this way.

ANDERSON: These are points you can argue indefinitely, and my answer is different from Murray's and much more practical: namely, there are no closed systems in the universe, and therefore what would happen in a closed system is irrelevant. We're always radiating to somewhere and you're always adding in food, and this process of turning one into the other is what's happening. The second different answer I would make is that the actual entropy in the information that you're using is so negligible compared to what you're using to function—what you must use to function, to function at some reasonable rate—that you don't need to keep count of that.

GELL-MANN: It is certainly negligible, compared with the usual entropy that we talk about in chemical processes. However, it's not negligible if you're concentrating on that issue.

ANDERSON: Yes. One should not concentrate on that issue, is my answer. But we have not really answered Stu's question which is a fascinating and provocative one—is there an analogy between Liouville's theorem in phase space and something like it in information space?

WALDROP: If I understand what John Holland has been telling us, the real distinction between learning as in the mind, and evolution as in a biological system, is the difference between implicit and explicit models. An implicit model is something like a rule, "If this is the situation, then do that." It's encoded a model of the world into a set of rules that are useful for the organism to do. The prototypic example is a bacterium swimming upstream in a glucose gradient. It is in effect executing a rule, "If the gradient is such-and-such, swim upstream." The bacterium has no brain, so it can't really know much about its world, but it does function as if it did. And this is a useful thing to do.

The explicit model is something much more like consciousness, where we do have an explicit model of, say, the physics of building a building so that we can reason about it, model it, and come to conclusions about it.

GELL-MANN: You're just talking again about the distinction between self-awareness and the lack of self-awareness, right?. At this point, we don't know all that much about what self-awareness is. It's true that human beings are supposed to have an unusual degree of self-awareness, and therefore human learning would be characterized, in many cases, by the properties of self-awareness. And we don't believe that's the case with biological evolution. Therefore that will be a difference. But I don't understand how that has to do with the question of whether there are significant jumps, or not. The fact that you mentioned, which is certainly true—that in human learning it's often related to self-awareness—doesn't change the dispute. People were asking whether in biological evolution there may not *also* be big leaps, and Leo Buss clarified very nicely the fact that there had been *certain* big leaps, of an organizational character. But then one can still ask: "Omitting those, does biological evolution proceed by big leaps, or by little ones?"

WALDROP: I'm saying that that might not be the important difference between learning and biological evolution, that both of them may in fact go by big leaps and small steps.

EPSTEIN: I've been having a crazy thought for a while. I don't know whether I'm the only one or not. But I've been thinking we've been talking about the scientific enterprise as an example of a complex adaptive system, with linguistic evolution as another example. And I've been thinking that maybe artistic evolution, and musical evolution, pose interesting problems for the construction of a good definition of a complex adaptive system. And what I'm thinking of is this: that in a certain strain of musical development, particularly western European classical music history since the sixteenth century forward, there's a very distinct line of development based on the emergence of the tonal system, and the gradual relaxation of that system with the Romantics, with Wagner, with extreme chromaticism, and finally the atonal movement. There seems to be directionality, the competition of schemata (in the form of different composers, and the search for different organizing principles); definite selection of some sort going on, in the dominant schools of musical thought, but no noncircular definition of fitness that I can come up with. And I'm thinking, "Here's a nice example of a complex adaptive system, that seems to evolve, where fitness is a strange notion." And I'm wondering what we do with things like that, and whether the problem of sort of the endogenous. . .that fitness itself is an emergent thing that the systems have to somehow come to discover. And I wonder whether this isn't an interesting model. . .

GELL-MANN: The selection pressures there have to do in part with peer evaluation, in part with audience evaluation, in part with historical evaluation in part with subjective evaluation by people of their own work, and so on. That this would all be summarized in a potential—which is what fitness ideas would say—seems somewhat unlikely. I don't believe that we should look everywhere for fitness. It works here and there, and I don't expect it to work everywhere. But the selection pressures certainly exist everywhere.

Something I find interesting about the kind of development you describe is the following: Archeologists these days are terribly reluctant to engage in what they imagine are "value judgments" about the past by talking about periods in terms of florescience and decline in the arts.

Nevertheless, it's my belief that one can identify certain properties of many periods and many kinds of art—phenomena that occur over and over again. Namely, people formulate, in a given art at a given time, certain requirements that hadn't been very important before. Then there's a period of challenge when people are trying to meet those requirements in all sorts of interesting ways. That's an archaic period. Then the requirements are not. The artists are able to do whatever it is, and the art *flourishes*. Then comes a period when they begin to go off into variant complications. And these are often called "archaic art," "high art," and "rococo art." I think it could be a scientific fact that there exists this kind of sequence.

EPSTEIN: I think in the case of music history that this happens; that in the case of Bach, for example, the art of the fugue is basically an attempt to construct large musical structures based on very rigorous limitations on thematic material. There's this tiny snippet, this little string of material, you're going to be permitted to use, and you construct this enormous thing. You run the string backwards at great, long expanse; you dilate it; you shrink it; you turn it upside down; you run it backwards. But the whole thing is very self-similar in that, anywhere you look, it's some variant of this thing. And then that impulse is in fact relaxed. And then it returns in twentieth century music, with serial composition that again says "Thou shalt use only this tone row, this string of material." So I agree.

BROWN: Isn't that just frequency-dependent selection played out over a long time course? Once you have a dominant thing, the rare allele gets favored and then, when that becomes dominant, the other gets favored. . .

EPSTEIN: Yes, except that it's part and parcel of a larger process, which is the gradual destruction of tonal music over four centuries, that has definite direction to it. I mean, you can document how this took place.

ANDERSON: I've already mentioned, in discussing George's comments, that there are many examples of evolution where the institution has its own fitness function that is irrelevant to the ostensible value and function. Modern music is clearly a selfish meme, because the populace liked Bach, the populace does not like John Cage or Schoenberg or Milton Babbitt. The populace invented a totally new atonal music, a very complicated music (often using the ideas of academic music) and abandoned some of the old rules, and this music is rapidly taking over from this other music—formal music—which is very much controlled by a selfish meme that has only institutional value.

Marcus W. Feldman,† **Luigi L. Cavalli-Sforza,**‡ **and Lev A. Zhivotovsky***

†Department of Biological Sciences, Stanford University, Stanford, CA 94305.
‡Department of Genetics, Stanford University Medical School, Stanford, CA 94305.
*Institute of General Genetics, Russian Academy of Sciences, Gubkin St., GSP-1, B-333, Moscow 117809, Russia.

On the Complexity of Cultural Transmission and Evolution

Abstract: The dynamical systems that underlie the evolution of genetic systems have been well studied in population genetics. The rules of transmission that might characterize cultural systems have had far less quantitative analysis and as a result the dynamics of cultural evolutionary systems are less well understood. We have attempted to develop a formal theory for the transmission and evolution of cultural traits. This theory has paralleled the development of population genetic theory in its focus on the reasons for the maintenance of cultural variation of change in the amount of such variation. Evolution can only occur in the presence of variation and different modes of transmission result in different levels of intrapopulation cultural variability.

When a trait is determined partially by genes and partially by culture, the mathematical expression of the transmission rule becomes much more complex. In the present paper, we explore what kinds of transmission rules might, in the process of gene-culture coevolution, lead, for example, to eventual independence of the levels of genetic and cultural variability. In this framework of gene-culture coevolution it becomes natural to investigate the properties of the transmission and natural selection that would lead to the evolution of a purely cultural transmission system from an initial state in which the trait under study was genetically transmitted. This ties directly to the issue of the evolution of learning.

INTRODUCTION

Cultural anthropologists disagree vehemently on what aspects of society the term "culture" should subsume. Indeed, more than one hundred definitions exist in the literature, all of which have been criticized. For our purposes the definition in Webster's Dictionary (Third International Edition) is adequate: "The total pattern of human behavior and its products embodied in thought, speech, action and artifacts and dependent upon man's capacity for learning and transmitting knowledge." The processes of information storage and retrieval that we recognized as learning, as well as the process of transmission, are presumably most highly developed in humans. But, as discussed at length by Bonner,[5] many aspects of animal behavior might be better understood within a framework that placed less emphasis on strictly biological (i.e., genetic) transmission, and more on the roles of learning and cultural transmission.

The "total pattern of human behavior" referred to above has no universally accepted formal particles and no hierarchy of atomization. Nevertheless, in attempting to build a productive theory of cultural change over time, that is, a theory of cultural evolution, it is natural to focus on specific traits that may vary among individual members of society, that may change over time, and that may be transmitted among individuals. Careful analysis of the consequences of assumptions about the rules of change and transmission may increase our understanding of the evolution of human societies, from the level of families to that of institutions and perhaps even to forms of government.

The theory of cultural transmission and evolution that we have developed over the past twenty years has been criticized by cultural anthropologists on the grounds that it focused on the "atoms" of culture, i.e., traits and their variants, and not on the "gestalt" of a culture. In countering this objection, it should be recalled that, following Darwin's[12] momentous suggestion that biological variation among individuals was the precursor to speciation, the *formal* theory of biological evolution had to await the discovery of rules of biological transmission, namely, Mendel's. In fact, prior to the rediscovery of Mendel's rules, the trajectory of biological variation over time was viewed in Galton's terms that actually involved continuous reduction of phenotypic variance over time. With particulate transmission and the Hardy-Weinberg law, maintenance of the biological variation became a theoretically reasonable foundation upon which subsequent mathematical theories of Fisher, Wright, and Haldane could stand.

Thus, we claim that "the total pattern of human behavior" in the above definition is not operationally useful, and we must focus on observable and measurable components. Our studies attempt to classify and quantify modes of cultural transmission, that is, transmission of these components, in order to find the relationship between these modes of transmission and evolutionary processes of culture.

BASIC ANALOGIES

Our theory of cultural evolution is based on analogies to biological evolution. These analogies have as their source the very well developed theory of population genetics. The analogy in culture to mutation in biology is invention and innovation. Of course, biological mutations are known to be *random* changes in the structure of DNA, while a cultural innovation might be intentionally designed to respond to a specific need. In his book *Natural Inheritance*, Galton[25] explained mutations, which at the time were called "sports," as analogous to inventions and, in particular, he cited changes in the brougham and the invention of the new electric battery. An innovation in a cultural artifact may, of course, simply be the result of a random error made in reproducing it. An example is given in Figure 1. When genetic variation exists in a population, its transmission from parents to offspring obeys the rules discovered by Mendel. Violations of these rules are well documented and are called *segregation distortion.* Although it is essentially a property of transmission, the effect of segregation distortion is natural selection at the level of gametes. We have no correspondingly precise rules for the transmission of cultural variants among the individuals in a population. For specific traits it is possible, however, to approximate quantitative rules. For example, we found[9] that, in a sample of Stanford undergraduates, political and religious affiliations are strongly transmitted from their parents. In a study of African Pygmies, hunter gatherers of the tropical forest, Hewlett and Cavalli-Sforza[27] found that almost 80% of specifically identified skills were learned from parents.

Natural selection occurs among biological variants if they have different mortality and/or fertility rates, that is, different potentials to pass their genetic information on to the next generation. This is Darwinian selection, and usually it is couched in terms of competition for food, space, mates, etc. Cultural selection may include a component of this kind, where one cultural variant might, for example, leave more biological offspring that carry that variant than another. But it may also involve a component of transmission; an idea may be more attractive than its alternatives, and obtain adherents without the occurrence of physical deaths or additional births. The analogy here to genetic segregation distortion is obvious. In the majority of cases of segregation distortion that have been studied in detail, the gene that has the advantage at the level of transmission also has a disadvantage at the level of either reproductive ability in adults or survival to adulthood. Such tradeoffs may be much more frequent in evolutionary biology than generally acknowledged; they are certainly difficult to study empirically and theoretically.[11]

The interaction between transmission and cultural selection becomes very complex in such situations as frequency-dependent attractiveness: an idea might be more attractive to some by virtue of its larger number of adherents, rather than through any intrinsic property. This is probably the case with many fashions.

Our final analogy involves the effect of small population size. In population genetics, sampling effects due to small population size are termed *random genetic drift*. Within this idea are subsumed the notions of founder effects, frozen accidents, and contingent evolution. Small populations are much more likely to be affected by

MS	1		10	20	30	40	50			60	70
Sg	FORE	TH 'E	NEIDFAERAE	NAENIG	UUIURTHIT	THONCSNOTTURRA	THAN	HIM	THARF	SIE	T–O
Ba	FORE	THAE	NEIDFAERAE	NAENIG	UUIURTHIT	THONCSNOTTURRA	THAN	HIM	THARF	SIE	T–O
Adl	FORE	TH–E	NEIDFAERAE	NAENIG	UUIURTHIT	THONCSNOTTURRA	THAN	HIM	THARF	SIE	T–O
K11	FORE	TH–E	NEIDFAER–E	NAENIG	UUIURTHIT	T–ONCSNOTTURRA	THAN	HIM	THARF	SIE	T–O
Mul	FORE	TH–E	NEYDFAER–E	NAENIG	VÜIÜRTHIT	THONCSNOTTURRA	THAN	HIM	THARS	SIE	T–O
K12	FORE	TH–E	NEIDFAOR–E	NAENIG	UUIURTHIT	THONESNOTTURRA	THAN	HI–	CHRAF	SIE	THO
Hk	FORE	TH–E	NEIDFACR–E	NAENIG	UUIURTHIT	THONCSNOTTURRA	THAN	HIM	THRAF	SIE	T–O

MS	80	90	100	110	120	130	
Sg	YMBHYCGGANNAE	AER	HIS	HINIONGAE	HUAET	HIS	GAS–	TAE	GODA–ES	A&–HTHA
Ba	YMBHYCGGANNE–	AER	HIS	HINIONGAE	HAUAET	HIS	GAS–	TAE	GODA–ES	A&–HTHA
Adl	YMBHYCGGANNAE	AER	HIS	HINIONGAE	HUAET	HIS	GAS–	T–E	GODA–ES	AETHTHA
K11	YMB–ICGGANNAE	AER	HIS	HINIONGAE	HUAET	HIS	GAS–	T–E	GODA–ES	AETHTHA
Mul	YMBHYCGGANNAE	AER	HIS	HINIONGAE	HUAET	HIS	GAS–	T–E	GODA–ES	AETHTHA
K12	YM–HICGGANNAE	AER	HIS	HYNIONGAE	HUAEX	HIS	GAS–	T–E	GODELES	AETHTHA
Hk	YM–HYCGGANNAE	AER	HIS	HINIONGAE	HUAEX	HIS	GAS–	T–E	GODELES	AETHTHA

MS	140	150	160	170	
Sg	YFLAES	AEFTER	DEOTHDAEGE	DOEMID	UUEORTHAE
Ba	YFLAES	AEFTER	DEOTHDAEGE	DOEMID	UUEORTHAE
Adl	YFLAES	AESTER	DEOTHDAEGE	DOEMUD	UUEORTH–E
K11	YFLAES	AESTER	DEOTHDAEGE	DOEMNL	UUEORTH–E
Mul	YFLAES	AESTER	DEOTHDAEGE	DOEMNL	VUEORTH–E
K12	YFLAES	AESTER	DEOTHDAEGE	DOEMIT	UUEORTH–E
Hk	YFLAES	AESTER	DEOTHDAEGE	DOEMIT	UUEORTHAE

FIGURE 1 "Bede's Death Song" (Bede died in 735 A.D.) is a famous Anglo Saxon poem. It shows slight differences in all manuscripts of the epistle of Cuthbert on Bede's death in which the song was found. Bede's song is only a small part of the epistle; it is given above in all versions from seven manuscripts found in continental Europe. The oldest "continental" version (Sg, from St. Gallen) is given in full. Hk is a summary of five Austrian manuscripts very similar to one another. A dash indicates absence of a letter found in other manuscripts. The display given here is equivalent to that of sequence differences in DNA. The words of Bede's song translate in English as: "Before the inevitable journey no man shall grow more discerning of thought than his need is, by contemplating before his going hence what, good or evil, will be adjudged to his soul after his death-day."

these stochastic effects than are large populations. Cultural drift may also occur. Rare innovations may disappear or spread as a stochastic sampling process. A trait that originates with a few individuals may spread to many via forced conversions, peaceful advertisement, mass media, etc. For such traits, the *effective* population may be very small which results in cultural inbreeding.[8,17]

During the past twenty years we have seen significant advances in both the quantitative aspects of the theory of cultural evolution, and more literary approaches. We do not intend to compare these approaches here; two recent reviews are recommended to the interested reader, one by Durham[13] and the other by Laland.[30] The former takes a more anthropological stance, while the latter relates the work that we initiated in 1973[6,7] to the psychological sciences.

In what follows, we first review the various types of cultural transmission that have been incorporated into our evolutionary models and the rates of evolution that these modes of transmission produce. We then discuss how these are relevant in the context of geographical studies of cultural variation. Next we extend the transmission system to allow for genetic variation in rates of cultural transmission and discuss how these genetic effects may interact with Darwinian selection and assortative mating. We conclude with two recent examples of the application of our ideas to specific cases of anthropological interest.

MODES OF TRANSMISSION AND RATES OF EVOLUTION

In epidemiology, transmission of a disease from a parent to child is said to be *vertical*. We have adapted this terminology for cultural traits that are transmitted from parent to child. In a number of studies, summarized by Cavalli-Sforza and Feldman,[8] we have examined mathematical properties of dynamical models for traits evolving under vertical transmission. In general, this kind of transmission is the most similar to biological in its consequences for evolution. It is highly conservative and tends to maintain the status quo. In a recent study, Tanaka[33] has shown that the grooming technique in free-ranging Japanese macaques follows maternal lineages, most probably by imitiation. Such observational field studies of transmission in nonhumans are, however, not numerous.

Transmission between unrelated members of the same generation is called *horizontal*, while *oblique* refers to transmission between members of different generations who are not vertically related. These are closest to the transmission of contagious diseases and can produce rapid evolution.

Horizontal and vertical transmission, if they are strong enough, balance adverse natural selection against a cultural trait. This brings us back to the fuzzy boundary between cultural selection and transmission. Thus, a custom that caused its adherents to lose fitness may be maintained, as long as the population does not go extinct, by powerful transmission of that custom from adherents to naive individuals. Transmission is playing a selective role in this case. Of course, the strength of transmission required to prevent elimination of a deleterious trait depends on how deleterious the trait is.

If very few individuals of one generation are able to transmit their form of a trait to large numbers of a succeeding generation, then the rate of evolution can be extremely rapid. This rate can be quantified[17] in terms of an approximating diffusion process. Qualitatively the result is analogous to evolution in a biological population sufficiently small that most members are related: successful innovations are able to spread very quickly by this mode of transmission.

In Table 1 we exhibit the transmission rule for a dichotomous vertically transmitted trait whose alternative forms are labelled $+$ and $-$. The dynamic that describes the evolution of the frequency of $+$ in the population is quadratic. It is clear that, if a mating between parents who are $+$ produces mostly $-$ offspring, and vice versa, then a cycle may be set up which we have called "the generation gap."[16]

GEOGRAPHY AND CULTURAL VARIATION

The spatial diffusion of cultural innovations has long interested sociologists[31,32] and has also begun to catch the attention of economists. Detailed quantitative studies, however, are not numerous. In evolutionary genetics, a widely accepted paradigm for the spread of an advantageous mutant is due to Fisher[24] who wrote the frequency $p(x,t)$ at position x on a line, and time t in terms of a local diffusion coefficient M as

$$\frac{\partial p}{\partial t} = M\frac{\partial^2 p}{\partial x^2}. \tag{1}$$

TABLE 1 The Transmission Rule for a Dichotomous Vertically Transmitted Trait Whose Alternative Forms are Labelled $+$ and $-$

Mating			Offspring Phenotype	
Father	$\times$	Mother	$+$	$-$
$+$	$\times$	$+$	b_3	$1 - b_3$
$+$	$\times$	$-$	b_2	$1 - b_2$
$-$	$\times$	$+$	b_1	$1 - b_1$
$-$	$\times$	$-$	b_0	$1 - b_0$

Subsequent models for the spread of a cultural variant, and indeed for the maintenance of cultural diversity, have used Eq. (1) as a point of departure. Ammerman and Cavalli-Sforza,[1] for example, explain modern patterns of gene frequency variation in Europe as a consequence of the diffusion of early farmers from the fertile crescent, where wheat agriculture originated some ten thousand years ago. This diffusion of the human carriers of the cultural variant was called *demic diffusion* to emphasize that there was movement of people whose cultural baggage accompanied them, resulting in cultural similarities between geographically very distant groups. In the absence of knowledge of genetic data, linguistic affiliation may serve as an indirect measure of demic diffusion.[10]

Cultural diffusion may occur in the absence of demic diffusion when innovations are observed, adopted, and transmitted horizontally among neighbors. The efficiency of this mode of geographic spread probably is a function of the number of contacts but may also have an ecological or environmental component. A method of making pots may be learned from a neighbor but its adoption may require access to specific clays, dyes, etc.

A final explanation for patterns of cultural variation is analogous to ecological convergence. Desert plants from geographically distant locations often use the same physiological mechanism to sequester water. Analogously, humans in similar ecological habitats may develop convergent cultural practices to deal with local flora, fauna, and weather.

Guglielmino et al.,[26] in a recent study of cultural variation in parts of sub-Saharan Africa, have found that kinship and family-related traits are patterned more by language and cultural history than by ecology or neighbors. They suggest that this reflects demic diffusion in conjunction with vertical transmission. Traits subject to cultural diffusion, such as the division of labor between the sexes, seem to cluster geographically and are, they suggest, mainly subject to horizontal transmission.

A class of geographically variable cultural traits that is especially interesting to evolutionists is that for which there is correlated genetic variation. If the cultural trait involves food production or processing, and if there is genetic variability for allergy to the food, then some correlation is expected. We have made a particular study of the joint evolution of milk use as a cultural trait, and the gene for the ability to absorb lactose. Those indigenous peoples without a history of pastoralism generally are unable to digest milk; their frequency of the lactose-absorbing allele is low. In contrast, those tribes that have practiced pastoralism through the millenia appear to have a relatively high frequency of lactose absorption.

The class of models that we have developed introduces a fitness advantage to the use of milk that is stronger among lactose absorbers. It was originally our hope that a reasonable value for this advantage might explain an increase in the frequency of the absorbing allele to levels that characterize northern European populations. It turns out, however, that our models[2,18] have not been very successful, in that the selective advantage required is unreasonably large. Work continues on this problem and others like it that involve geographical variation in both genes and cultural traits.

GENE-CULTURE COTRANSMISSION

The first evolutionary treatment of a trait influenced simultaneously by genes and culture was that by Cavalli-Sforza and Feldman.[7] The context here was a continuously varying phenotype in which the cultural transmission from parent to child had a genetic component modeled after Fisher.[22] It was shown that vertical cultural transmission could significantly affect correlations between relatives that would traditionally have been attributed to genetic transmission.

Feldman and Cavalli-Sforza[16] chose a simple dichotomous phenotype whose transmission from parent to offspring was influenced by genes. The focus here was the dynamical behavior of such a system, and how it differed from that of either purely phenotypic or purely genotypic transmission. One fundamental finding was that heterozygote superiority in the transmission of a selectively advantageous trait was not sufficient for the maintenance of genetic variability. This is of interest because, in classical evolutionary genetics, heterozygote advantage is sufficient to maintain a genetic polymorphism of at least two alleles.

As an example of how such a model can be constructed, suppose that the phenotype takes two values, which we call *bar* and *not bar*. Suppose also that the genetic system is a single diploid diallelic locus with alleles A_1 and A_2. In Table 2 we show a transmission regime in terms of five parameters, α, β, η, σ, and τ. For those not familiar with standard population genetics, β is a baseline probability of transmission, α is Fisher's[22] additive allelic contribution to this probability, and σ is a measure of the genetic dominance of A_1 over A_2; $\sigma = 1$ means A_1 is dominant to A_2, and $\sigma = 0$ means A_1 is recessive to A_2. Vertical cultural transmission is

TABLE 2 A Bilinear Transmission Scheme

| Father | × | Mother | Probability that offspring is phenotype Bar | | |
			AA	Aa	aa
Bar	×	Bar	$2\eta + 2\alpha + \beta$	$2\eta + \sigma\alpha + \beta$	$2\eta + \beta$
Bar	×	Not Bar	$\tau\eta + 2\alpha + \beta$	$\tau\eta + \sigma\alpha + \beta$	$\tau\eta + \beta$
Not Bar	×	Bar	$\tau\eta + 2\alpha + \beta$	$\tau\eta + \sigma\alpha + \beta$	$\tau\eta + \beta$
Not Bar	×	Not Bar	$2\alpha + \beta$	$\sigma\alpha + \beta$	β

measured by η with a "cultural dominance" component τ such that if $\tau = 2$, then one bar parent is as good as two in terms of transmission, while $\tau = 0$ means that one bar parent is as good as none.

For human behavioral traits, one further parameter is essential, namely the level of assortative mating for the phenotypic dichotomy, m. Assortative mating is just the propensity of individuals with like phenotypes to mate more often than randomly.

In this model, there are six phenogenotypes $\overline{A_1A_1}$, $\overline{A_1A_2}$, $\overline{A_2A_2}$, A_1A_1, A_1A_2, and A_2A_2 whose frequencies we write as $u_1, u_2, u_3, v_1, v_2,$ and v_3, which sum to unity. The frequency of the bar phenotype is $k = u_1 + u_2 + u_3$; the frequency of A_1 is $p = u_1 + v_1 + (u_2 + v_2)/2 = 1 - q$; the frequency of A_1 in bar individuals is $p_1 = (u_1 + u_2/2)/k = 1 - q_1$, and in the not bar individuals it is $p_2 = (v_1 - v_2/2)/(1 - k)$.

It is easy to see that the overall allele frequency of A_1 does not change over time: $p' = p$ where the prime denotes frequencies in the next generation. In general, however, both p_1 and p_2 do change over time as does k. For example, the recursion for k is

$$k' = \beta + 2\alpha \left[p + (\sigma - 1)pq(1 - F)\right] + 2\eta \left[(1 - m)k^2 + mk + \tau(1 - m)k(1 - k)\right] \quad (2)$$

where

$$F = mk(1 - k)(p_1 - p_2)^2/pq. \quad (3)$$

Thus, if there is no genetic dominance, $\sigma = 1$, and F is removed from Eq. (2) with the result that genes and phenotype become separated during the evolutionary process. On the other hand, it is possible to show that among offspring, the frequency of heterozygotes $u_2' + v_2'$ is reduced from the expectation $2pq$ by the factor $(1 - F)$. This actually explains the result referred to earlier about the insufficiency of heterozygote advantage in transmission for the maintenance of a genetic polymorphism.

We have recently completed a large analytical and numerical study of this transmission structure. Our objective was to generate familial data numerically and then to see to what extent standard statistics, i.e., correlations between relatives, would reflect the actual transmission parameters that produced the data.[21] Parent-offspring correlations are the most likely to produce confusion as to the role of genetic versus cultural vertical transmission: high parent-offspring correlation may be due to either strong genetic or strong vertical cultural transmission. As expected, all correlations between relatives are strongly affected by the level of mate preference, m.

GENE-CULTURE COEVOLUTION

The framework described above for a simple two-valued phenotype with one diallelic locus ignores the possibility of natural selection. A first cut at natural selection would limit its effect to phenotypes and, further, only to the phenotypes of offspring through differences in their survival probability.

The full exposition of a general model of this kind illustrates just how complex a biobehavioral "adaptive" system really is, even one that is fully deterministic. We begin with genotypes defined at an arbitrary number of loci and labelled $G_1, G_2, \ldots, G_L$. The phenotypes are $F_1, F_2, \ldots, F_M$ so that an individual of phenotype α and genotype i is written $FG_{\alpha i}$. A mother of phenogenotype $FG_{\alpha i}$ and a father of phenogenotype $FG_{\beta j}$ produce an offspring of phenogenotype $FG_{\gamma k}$ with probability $T_{\alpha\beta\gamma}^{ijk}$. Under most imaginable scenarios we would have

$$T_{\alpha\beta\gamma}^{ijk} = T_g\{ij \to k\}T_{ijk}\{\alpha\beta \to \gamma\}, \tag{3}$$

where $Tg\{\cdot\}$ is the usual genetic operator for multilocus transmission from parents $\{i, j\}$ to offspring k, and $T_{ijk}\{\alpha\beta \to \gamma\}$ reflects the kind of transmission illustrated in Table 2 where parents' genotypes and phenotypes may affect the offspring's phenotypes. In fact, in Table 2 we illustrate a case where $T_{ijk}\{\alpha\beta \to \gamma\}$ is actually $T_{..k}\{\alpha\beta \to \gamma\}$, because only the *offspring's* genotype affects its transmission probabilities. The more general case might include genetic effects on a mother's (i) or father's (j) ability to transmit phenotypes.

Write $P_\alpha(i)$ as the frequency of $FG_{\alpha i}$, and suppose that the survival probability of an offspring $FG_{\gamma k}$ with parents $FG_{\alpha i}$, $FG_{\beta j}$ is proportional to $w_{ijk}^{\alpha\beta\gamma}$. Finally, suppose that the ML phenogenotypes are divided into groups $R_1, R_2, \ldots, R_a$. Mating is assortative with coefficient m if a fraction $(1 - m)$ of all individuals mate randomly among and within these groups, while the rest mate randomly *within* these groups. The complete dynamical system for the ML phenogenotypes is then

$$\overline{w}P_\gamma' = (1 - m)\sum_{ij}\sum_{\alpha\beta}\hat{w}_{ijk}^{\alpha\beta\gamma}Tg\{ij \to k\}P_\alpha(i)P_\beta(j)$$

$$+ \sum_{\nu=1}^{a}\hat{P}_\nu\sum_{\alpha i \in R_\nu}\hat{w}_{ijk}^{\alpha\beta\gamma}Tg\{ij \to k\}P_\alpha^\nu(i)P_\beta^\nu(j) \tag{4}$$

where $\hat{P}_\nu = \sum_{\alpha i \in R_\nu} P_\alpha(i)$ is the sum of all frequencies in R_ν and $P_\alpha^\nu(i) = P_\alpha(i)/\hat{P}_\nu$ is the relative frequency of $FG_{\alpha i}$ in this group. The modified fitnesses $\tilde{w}_{ijk}^{\alpha\beta\gamma}$ are defined by

$$\tilde{w}_{ijk}^{\alpha\beta\gamma} = \frac{1}{2}\left[w_{ijk}^{\alpha\beta\gamma}T_{ijk}\{\alpha\beta \to \gamma\} + w_{jik}^{\beta\alpha\gamma}T_{ijk}\{\beta\alpha \to \gamma\}\right]$$

with the normalizer $\overline{w}$ defined as the sum over (k, γ) of the right sides of Eq. (4).

All of the models in the literature on gene-culture coevolution with vertical cultural transmission are special cases of the structure in Eq. (4). A detailed analysis of models of this form has recently been attempted by Feldman and Zhivotovsky,[20] but much remains to be done. One general point worth mentioning concerns our framework for gene-culture association measures. These are inspired by the notion of linkage disequilibrium in population genetics. Suppose that the frequency of phenotype F_α is $\sum_i P_\alpha(i) = P_\alpha$ and of genotype G_i is $P_i = \sum_\alpha P_\alpha(i)$. Then the *phenogenotypic association* is

$$\hat{A}_\alpha(i) = P_\alpha(i) - P_\alpha P(i) \tag{5}$$

after selection. If $p(s)$ is the frequency of a gamete g_s such that $G_i = g_s/g_t$, for example, then a phenogametic association measure is

$$\mathcal{A}_\alpha(s) = p_\alpha(s) - P_\alpha p(s), \tag{6}$$

where $p_\alpha(s)$ is the fraction of gametes g_s that are produced by phenotype F_α. It turns out that changes in these association measures are crucial in determining the evolutionary trajectories of the phenogenotypes under selection. For example, we find that when selection occurs on phenotypes, and the role of genetics is restricted to its effect on transmission rates, the rate of change of genotype frequencies is orders of magnitude slower than if the selection occurred directly on genotypes. This time lag was first observed in the simplest model of phenogenotypic evolution by Feldman and Cavalli-Sforza.[16]

TWO EXAMPLES

(1) DEAFNESS AND SIGN LANGUAGE

Aoki and Feldman[3,4,19] proposed a model for the transmission of sign language in the presence of hereditary deafness. Consider alleles A and a with genotype aa producing deafness, while AA and Aa individuals hear normally. We suppose mating to be assortative for homozygotes aa, as in the cited articles. In Aoki and Feldman's scheme of sign language transmission, deaf progeny can be taught the sign language only by their parents. Thus there are two "cultural" phenotypes. The first group, F_1, consists of those aa individuals who possess the sign language, while the second group (denoted as F_0) includes individuals AA, Aa and some of genotype aa who do not possess the sign language.

Let the probabilities of learning the sign language be c if both parents are signers, and c_f or c_m if only one of the parents (father or mother, respectively) are signers. Hence, the cultural transmission matrix is

$$
\begin{aligned}
&T_{aa}\{11 \rightarrow 1\} = c, &&T_{aa}\{11 \rightarrow 0\} = 1 - c, \\
&T_{aa}\{10 \rightarrow 1\} = c_m, &&T_{aa}\{10 \rightarrow 0\} = 1 - c_m, \\
&T_{aa}\{01 \rightarrow 1\} = c_f, &&T_{aa}\{01 \rightarrow 0\} = 1 - c_f, \\
&T_{aa}\{00 \rightarrow 1\} = 0, &&T_{aa}\{00 \rightarrow 0\} = 1, \\
&T_{Aa}\{** \rightarrow 1\} = 0, &&T_{Aa}\{** \rightarrow 0\} = 1, \\
&T_{AA}\{** \rightarrow 1\} = 0, &&T_{AA}\{** \rightarrow 0\} = 1,
\end{aligned}
$$

where the "$*$" means an arbitrary phenotype. If selection is due only to the genotypic differences (as it could well be in the case of deafness), we have $w_{ijk}^{\alpha\beta\gamma} = w_{ijk}$, where $i, j, k = 1, 2, 3$ for genotypes AA, Aa, and aa respectively. Assortative mating for deafness occurs at rate m. Then it can be shown that sign language will persist in the population if

$$
(c_m + c_f) \left[m w_{333} + (1 - m) q \frac{w_{323} c_m + w_{233} c_f}{c_m + c_f} \right] > w_{111}.
$$

(2) PREJUDICES FOR OFFSPRING GENDER

When the probability that an individual is male is determined by its genotype at one autosomal locus, then a famous result by Fisher[23] is that the fractions of males and females in the population tend to equalize. That is, the sex ratio should evolve towards evenness. General proofs can be found in papers by Eshel and Feldman[15] and Karlin and Lessard.[28] Now in many parts of the world, there is a strongly

TABLE 3 The Transmission of Cultural Phenotypes from Parent to Offspring

| Mating | | | Offspring Phenotype | | ASR | |
| | | | | | Fixed | Variable |
Father	$\times$	Mother	unbiased	biased	Adjustment	Adjustment
unbiased	$\times$	unbiased	b_3	$1 - b_3$	m_{ij}	m_{ij}
unbiased	$\times$	biased	b_2	$1 - b_2$	$m_{ij} + d$	$(m_{ij} + r)/2$
biased	$\times$	unbiased	b_1	$1 - b_1$	$m_{ij} + d$	$(m_{ij} + r)/2$
biased	$\times$	biased	b_0	$1 - b_0$	$m_{ij} + 2d$	r

held preference towards having sons. This preference often coexists with patrilocal marriage and is particularly pronounced in rural economies where security for the elderly is in the hands of male offspring.

Two sex ratios should be distinguished; the primary sex ratio, PSR, which is measured at birth, and the adult sex ratio, ASR, which takes account of the fact that one of the sexes, usually the female, has been subject to excess mortality during the zygote-to-adult phase of the life cycle. Female infanticide is the most widely publicized manifestation of this prejudice against one sex.

Suppose that the cultural dichotomy consists of being biased against the female sex, or unbiased. In the first three columns of Table 3 we show the transmission of this dichotomy. Now suppose that primary sex ratio is determined by one locus with two alleles A_1, A_2 with $m_{ij}(i, j = 1, 2)$, the probability that the offspring of genotype $A_i A_j$ are male at birth when both parents are unbiased.

Kumm et al.[29] propose two kinds of models to take account of the effect of bias on the ASR. These are called the Fixed Adjustment and the Variable Adjustment models. They have distinct probabilities for $A_i A_j$ to be male, depending on the parental phenotypes. Thus, in the Fixed Adjustment model, if both parents were biased, then fraction $m_{12} + 2d$ of $A_1 A_2$ offspring would be male. In the Variable Adjustment model, biased individuals attempt to bring the sex ratio of their offspring closer to a desired value, r, according to the last column of Table 3. In this second model, if $m_{ij} < r$, biased parents have to kill daughters to move the ASR closer to r. Thus, biased parents in this model may have lower reproductive fitness than their unbiased counterparts. This is a major difference between the two models.

It is of considerable interest, not only to scholars of genetically determined sex ratios, but also to students of the cultural biases that result in infanticide, that the results of these two models are different. For the Fixed Adjustment model, if the culturally transmitted bias is against female offspring, then new alleles producing a higher fraction of female offspring in the PSR will increase. On the other hand, in the Variable Adjustment model, if the bias in favor of males can overcome a significant fitness cost, the result will be selection for a male-biased PSR but, if the bias produces a sufficiently small effect on the ASR, then again, as in the Fixed Adjustment model, there will be selection for more females in the PSR.

These two examples serve, we hope, to demonstrate the utility of our approach to the interaction of genetics and culture in transmission and evolution of complex traits. The subject is in its infancy and new tools are needed to handle some of the dynamical systems that arise naturally in this context.

ACKNOWLEDGMENTS

Research supported in part by NIH grants GM28016 and GM10452, and a grant from the John D. and Catherine T. MacArthur Foundation.

REFERENCES

1. Ammerman, A. J., and L. L. Cavalli-Sforza. *The Neolithic Transition and the Genetics of Populations in Europe.* Princeton, NJ: Princeton University Press, 1984.
2. Aoki, K. "A Stochastic Model of Gene-Culture Coevolution Suggested by the 'Culture Historical Hypothesis' for the Evolution of Adult Lactose Absorption in Humans." *Proc. Natl. Acad. Sci. USA* **83** (1986): 2929–2933.
3. Aoki, K., and M. W. Feldman. "Recessive Hereditary Deafness, Assortative Mating and Persistence of Sign Language." *Theor. Pop. Biol.* **39** (1991): 358–372.
4. Aoki, K., and M. W. Feldman. "Cultural Transmission of a Sign Language when Deafness is Caused by Recessive Alleles at Two Independent Loci." *Theor. Pop. Biol.* (1993): in press.
5. Bonner, J. T. *The Evolution of Culture in Animals.* Princeton, NJ: Princeton University Press, 1980.
6. Cavalli-Sforza, L. L., and M. W. Feldman. "Models for Cultural Inheritance, I: Group Mean and Within-Group Variation." *Theor. Pop. Biol.* **4** (1973): 42–55.
7. Cavalli-Sforza, L. L., and M. W. Feldman. "Cultural Versus Biological Inheritance: Phenotypic Transmission from Parent to Children (a Theory of the Effect of Parental Phenotypes on Children's Phenotype)." *Am. J. Hum. Genetics* **25** (1973): 618–637.
8. Cavalli-Sforza, L. L., and M. W. Feldman. *Cultural Transmission and Evolution: A Quantitative Approach.* Princeton, NJ: Princeton University Press, 1981.
9. Cavalli-Sforza, L. L., M. W. Feldman, K. H. Chen, and S. Dornbusch. "Theory and Observation in Cultural Transmission." *Science* **218** (1982): 19–27.
10. Cavalli-Sforza, L. L., A. Piazza, and P. Menozzi. *History and Geography of Human Genes.* Princeton, NJ: Princeton University Press, in press.
11. Curtsinger, J. W., and M. W. Feldman. "Experimental and Theoretical Analysis of the "Sex-Ratio" Polymorphism in *Drosophila pseudoobscura.*" *Genetics* **94**: 445–466.
12. Darwin, C. *On the Origin of Species.* London: Watts and Company, 1859.
13. Durham, W. H. "Advances in Evolutionary Culture Theory." *Ann. Rev. Anthropol.* **19** (1990): 187–210.
14. Eshel, I., and M. W. Feldman. "On Evolutionary Genetic Stability of the Sex Ratio." *Theor. Pop. Biol.* **21** (1982): 430–439.
15. Eshel, I. and M. W. Feldman. "On the Evolution of Sex Determination and the Sex Ratio in Haplodiploid Populations." *Theor. Pop. Biol.* **21** (1982): 440–450.

16. Feldman, M. W., and L. L. Cavalli-Sforza. "Cultural and Biological Processes, Selection for a Trait Under Complex Transmission." *Theor. Pop. Biol.* **9** (1976): 238–259.

17. Feldman, M. W., and L. L. Cavalli-Sforza. "Random Sampling Drift Under Non-Mendelian Transmission." In the Proceedings of the 41st Session of the International Statistical Institute, held in New Delhi, *Bull. Intl. Stat. Inst.* *XLVII* **2** (1977): 151–164.

18. Feldman, M. W., and L. L. Cavalli-Sforza. "On the Theory of Evolution Under Genetic and Cultural Transmission with Application to the Lactose Absorption Problem." In *Mathematical Evolutionary Theory*, edited by M. W. Feldman, 145–173. Princeton, NJ: Princeton University Press, 1989.

19. Feldman, M. W., and K. Aoki. "Assortative Mating and Grandparental Transmission Facilitate the Persistence of a Sign Language." *Theor. Pop. Biol.* **42** (1992): 107–116.

20. Feldman, M. W., and Lev A. Zhivotovsky. "Gene-Culture Coevolution: Toward a General Theory of Vertical Transmission." *Proc. Natl. Acad. Sci. USA* **89**: 11,935–11,938.

21. Feldman, M. W., F. B. Christiansen, and S. P. Otto. "'Statistics of Discrete-Valued Traits Under Vertical Transmission." Unpublished manuscript, 1993.

22. Fisher, R. A. "The Correlation Between Relatives on the Supposition of Mendelian Inheritance." *Trans. Roy. Soc. Edinburgh* **52** (1918): 399–433.

23. Fisher, R. A. *The Genetical Theory of Natural Selection*, 2nd ed. New York: Dover, 1930, 1958.

24. Fisher, R. A. 1937. "The Wave of Advance of Advantageous Genes." *Ann. Eugen.* **7** (1937): 355–360.

25. Galton, F. *Natural Inheritance.* London and New York: Macmillan, 1889.

26. Guglielmino, C. R., B. S. Hewlett, C. Viganotti, and L. L. Cavalli-Sforza. "Mechanisms of Sociocultural Transmission and Models of Cultural Change." Unpublished manuscript, 1993.

27. Hewlett, B. S., and L. L. Cavalli-Sforza. "Cultural Transmission Among Aka Pygmies." *Am. Anthropol.* **88** (1986): 922–934.

28. Karlin, S., and S. Lessard. *Theoretical Studies on Sex Ratio Evolution.* Princeton, NJ: Princeton University Press, 1986.

29. Kumm, J., K. N. Laland, and M. W. Feldman. "Culturally Transmitted Prejudices for Offspring Gender and the Evolution of Primary Sex Ratios." Unpublished manuscript, 1993.

30. Laland, K. N. "The Mathematical Modelling of Human Culture and Its Implications for Psychology and the Human Sciences." *Brit. J. Psych.* **84** (1993): in press.

31. Rogers, E. M. *Diffusion of Innovations.* Galt, Ontario: The Free Press of Glencoe, Collier-Macmillan Canada, 1962.

32. Rogers, E. M., and F. F. Schoemaker. *Communication of Innovations: A Crosscultural Approach.* New York: The Free Press, a division of Macmillan, 1971.

33. Tanaka, I. "Dominance-Linked Variability in Louse-Egg-Handling Technique During Grooming Among Free-Ranging Japanese Macaques." Unpublished manuscript, 1993.

DISCUSSION

GELL-MANN: In the case where the very simplest problem—with two alleles, two phenotypes, and so on—gives sixty parameters and a vast amount of work for computers to do, I hate to ask the following question. But I've asked it before and I'd still like to know the answer.

Is there any future in population genetics, or in combined cultural and genetic transmission, for using computers to simulate a situation with *very many* genes, traits, kinds of population, and so on, and trying to search for properties that arise when you have numerous such entities, instead of just a few? Has anyone ventured into this very difficult domain?

FELDMAN: There's nobody that I know doing that work besides us.

GELL-MANN: But you have done that?

FELDMAN: We are in the process of doing that with a large number of genes and so far we don't have too many results.

GELL-MANN: No, I wouldn't expect that initially one would have. But it seems like a thing that somebody really ought to do. There might be entirely new domains of results in that limit.

FELDMAN: What has happened historically, in the last fifteen years, there's been a push to work on Gaussian fitnesses, so one puts the distribution of the fitness like that over the phenotype space, and then makes assumptions about how the genes influence that trait value. That's getting towards the thing, but it's not exactly, because there's a lot of phenotypes of interest that are not a Gaussian.

GELL-MANN: So this assumes a certain kind of limit theorem, whereas you might be dealing with a different kind of limit theorem.

FELDMAN: Absolutely. But there is some sound interest in work in that area.

FONTANA: You are describing a phenomenological theory, due to the fact that you talk about genotypes and phenotypes, or geno-phenotypes, and if one tends to think microscopically, then one might ask, "Is the map from a genotype to phenotype?" And then the first consideration that comes to one's mind is that there are obviously many more genotypes than there are phenotypes. This somehow relates to the comment made by Murray about equivalence classes, in other words, there are processes in nature that parse the space of genotypes into equivalence classes, such that we say, "they have the same phenotype," right? Different genotypes have the same phenotype. There are two extreme cases: either every genotype maps in just the same phenotype (that would be a pretty boring world); on the other hand, if every genotype would be a distinguishable phenotype, that would be pretty chaotic, and probably would not organize at all. So somewhere in between something happens. Do you have any notion about how this would enter your equations?

FELDMAN: Well, there's work that we are trying to do right now; it's actually rather difficult. There's a new idea—relatively new, 1976 I believe, was the first description—of what are called "quantitative trait loci," where one attempts to find out whether you can map a certain section of a chromosome to a certain percentage of a contribution to a phenotype. And the best known examples to date have to do with plants like the tomato in which there is a sugar contribution which you can map to a fairly narrow band in the chromosome. Unfortunately, I've just been to North Carolina and talking with corn breeders. They find that if they're interested in a trait, such as yield, or first tiller height (time to tiller reaching a certain height), they find every chromosome has contributions, and the genes are scattered all over. Empirically, it looks to me like a lot of the traits of interest are going to have contributions from a lot of genes. Now what we'd like to do is to have some kind of theory that would relate that percentage contribution to the phenotype itself. And that's the sort of thing that we're working on. It's a goal. But plyotropy is a very, very difficult problem, and every population geneticist always keeps it at the back of his mind, and says "I don't want to look on that right yet."

COWAN: Early in your talk, you referred to various kinds of cultural transmission, and one of them was "one to many"—I think maybe you used the term "broadcast" in your book.

FELDMAN: That's right.

COWAN: Electronic broadcast is "one to many" and if you have many channels of electronic broadcasting, you wouldn't want to call it cultural inbreeding any more; it may be in quite different. Do you see anything in modern communication technology which is like your "one to many" which has produced the kind of fast cultural transmission that you describe?

FELDMAN: The usual traits that people trot out for that are hemlines, things which are the subject of fashion. That is the definition of fashion: that everybody adopts them. I think that there are more important examples that involve coercion. So as we cited, if you have a dominant culture, or group that is dominating the culture, then things can change rapidly, like many words in Old English changed rapidly under the Normans. And the culture of celibacy in the church probably had a strong relationship to primogeniture considerations. There are things that you can't separate. If you have one form of transmission going on, as Murray said, it may have what you might call "epistatic effects" on a lot of other traits that are tied in some way to that. And it's very difficult to know which one of these is going to work quickest.

Some things we know about. We know that linguistic changes can work very quickly. Like, the reason that I say the *a's* different from the way you say your *a's* is because probably, not *my* ancestors, but the ancestors of the people who taught me how to speak English, came from central London, which is a different body of people that came to this country, and a different body that went to Pitcairn, or to Tristan da Cunha. Those are things that I think are really interesting to study, and Murray indicated that those are either dialectical or linguistic differences that I think you can put a handle on and eventually get quantifying.

W. Brian Arthur
Stanford University, Stanford, CA 94305-6084

On the Evolution of Complexity

Abstract: It is often taken for granted that as systems evolve over time they tend to become more complex. But little is understood about the mechanisms that might cause evolution to favor increases in complexity over time. This chapter proposes three means by which complexity tends to grow as systems evolve. In coevolutionary systems it may grow by increases in "species" diversity: under certain circumstances new species may provide further niches that call forth further new species in a steady upward spiral. In single systems it may grow by increases in structural sophistication: the system steadily cumulates increasing numbers of subsystems or subfunctions or subparts to break through performance limitations, or to enhance its range of operation, or to handle exceptional circumstances. Or, it may suddenly increase by "capturing software": the system captures simpler elements and learns to "program" these as "software" to be used to its own ends. Growth in complexity in all three mechanisms is intermittent and epochal. And in the first two is reversible, so that collapses in complexity may occur randomly from time to time.

Illustrative examples are drawn not just from biology, but from economics, adaptive computation, artificial life, and evolutionary game theory.

Complexity: Metaphors, Models, and Reality
Eds. G. Cowan, D. Pines, and D. Meltzer, SFI Studies in the
Sciences of Complexity, Proc. Vol. XIX, Addison-Wesley, 1994

INTRODUCTION

It is a commonly accepted belief—a folk theorem, almost—that as systems evolve over time they tend to become more complex. But what is the evidence for this? Does evolution, in fact, favor increases in complexity and, if so, why? By what mechanisms might evolution increase complexity over time? And can the process go in the other direction, too, so that complexity diminishes from time to time? In this chapter I will discuss these questions and, in particular, three different ways in which evolution tends to increase complexity in general systems.

In the biological literature, there has been considerable debate on the connection between evolution and complexity.[1,11] But much of this discussion has been hampered by the fact that evolutionary innovations typically come in the form of smooth changes or continuous, plastic modifications: in the size of organism,[1] in the morphology of body parts,[13] or in animal behavior,[1] so that increases in "complexity" are difficult both to define and discern. As a result, while most biologists believe that complexity does indeed increase with evolution, and particular mechanisms are often cited, the question remains muddied by problems of definition and observation, so that some biologists have expressed doubts about any linkage between evolution and complexity at all.[11] Fortunately, of late we are beginning to cumulate experience in evolutionary contexts that are not necessarily biological. These contexts include those of competition among technologies and firms in the economy, of self-replicating computer programs, of adaptive computation, of artificial life systems, and of computer-based "ecologies" of competing game strategies. Used as alternatives to biological examples, these have two advantages. Their alterations and innovations are very often discrete and well marked, so that in these contexts we can define and observe increases in complexity more easily. And many are computer based. Thus, they can provide "laboratories" for the real-time measurement and replication of changes in complexity in the course of evolution.

In discussing complexity and evolution in this chapter, I will draw examples from the economy and from several of the other contexts mentioned above, as well as from biology. I will be interested in "complexity" seen simply as complication. Exactly what "complication" means will vary from context to context; but it will become clear, I hope, in the mechanisms as they are discussed. And I will use the term "evolution" often in its phylogenetic sense, as development in a system with a clear lineage of inherited structures that may change over time. Thus, we can talk about the evolution of a language, or of a technology, without having to assume that these necessarily reproduce in a population of languages or technologies.

GROWTH IN COEVOLUTIONARY DIVERSITY

The first mechanism whereby complexity increases as evolution takes place, I will call *growth in coevolutionary diversity*. It applies in systems where the individuals or entities or species or organisms coexist together in an interacting population, with some forming substrates or niches that allow the existence of others. We may, therefore, think of such coevolving systems as organized into loose hierarchies or "food webs" of dependence, with individuals further down a hierarchy depending for their existence on the existence of more fundamental ones nearer the base of the hierarchy.

When the individuals (and their multiple possibilities in interaction) in such systems create a variety of niches that are not closed off to further newly generated individuals, diversity tends to grow in a self-reinforcing way. New individuals that enter the population may provide new substrates, new niches. This provides new possibilities to be filled or exploited by further new entities. The appearance of these, in turn, may provide further new niches and substrates. And so on. By this means, complexity in the form of greater diversity and a more intricate web of interactions, tends to bootstrap itself upward over time. Growth in coevolutionary diversity may be slow and halting at first, as when the new individuals merely replace uncompetitive, preexisting ones. But over time, with entities providing niches and niches making possible new entities, it may feed upon itself; so that diversity itself provides the fuel for further diversity.

Growth in coevolutionary diversity can be seen in the economy in the way specialized products and processes within the computer industry have proliferated in the last two decades. As modern microprocessors came into existence, they created niches for devices such as memory systems, screen monitors, and bus interfaces that could be connected with them to form useful hardware—computing devices. These, in turn, created a need, or niche, for new operating system software and programming languages, and for software applications. The existence of such hardware and software, in turn, made possible desktop publishing, computer-aided design and manufacturing, electronic mail, shared computer networks, and so on. This created niches for laser printers, engineering-design software and hardware, network servers, modems, and transmission systems. These new devices, in turn, called forth further new microprocessors and system software to drive them. And so, in about two decades, the computer industry has undergone an explosive increase in diversity: from a small number of devices and software to a very large number, as new devices make possible further new devices, and new software products make possible new functions for computers, and these, in turn, call forth further new devices and new software.

Of course, we should not forget that as new computer products and functions for computers appear, they are often replacing something else in the economy. Computer-aided design may eventually replace standard drawing board and T-square design. And so the increase in diversity in one part of a system may be partially offset by loss of diversity elsewhere. Occasionally, in a coevolving system, this replacement of an existing function can cause a *reversal* in the growth of coevolutionary diversity. This happens when the new entity replaces a more fundamental one in the system and the niches dependent on this disappear. In the economy of the last century, for example, there was a steady increase in the numbers of specialized, interconnected "niche firms" in the horse-drawn transportation industry; so that by the end of the century very many different types of coach builders, harness makers, smithy shops, and horse breeders coexisted. The appearance of the automobile caused all this to collapse, to be replaced, in turn, by a slow-growing network of interconnected niche manufacturers dependent on gasoline technology, oil exploration and refining, and the internal combustion engine. Thus, complexity—diversity in this case—may, indeed, tend to grow in coevolving systems, but it may also fluctuate greatly over time.

Growth in diversity can be observed in several artificial evolution contexts: for example, Tom Ray's Tierra system,[14] John Holland's ECHO system,[6] and Stuart Kauffman's various chemical evolution systems.[7] To take the Tierra example, Ray sets up an artificial world in which computer programs compete for processor time and memory space in a virtual computer. He begins with a single "organism" in the form of a set of self-replicating machine language instructions that can occasionally mutate. This forms a niche or substrate for the appearance of parasitic organisms that use part of its code to replicate—that "feed" on its instructions. Further organisms appear that are immune to the parasites. The parasites in turn form a substrate for hyper parasites that feed on them. Hyper-hyper parasites appear. And so on. New "organisms" continually appear and disappear, in a rich ecosystem of symbiotic and competing machine-language programs that shows a continual net growth of diversity. In several days of running this system, Ray found no endpoint to the growth of diversity. Starting from a single genotype, over 29,000 different self-replicating genotypes in 300 size classes (equivalent to species in this system) accumulated in this coevolving computer ecology.

At this point I want to note several things that apply to this mechanism.

First, the appearance of new entities may, in some cases, depend not so much on the existence of previous entities as on their possibilities in interaction. For example, in the economy, a new technology such as the computer laser printer mentioned above is possible only if lasers, xerography, and computers are previously available as technologies. In these cases, symbiotic clusters of entities—sets of entities whose collective activity or existence is important—may form many of the niches. We could predict that where collective existence is important in forming niches, growth in coevolutionary diversity would be slow at first—with few entities there would be few possibilities in combination and, hence, few niches. But as more single entities enter, we would see a very rapid increase in niche possibilities, as the number of possible niche clusters that can be created undergoes a combinatorial explosion.

Second, collapses will be large if replacement by a new entity happens near the base of the dependency hierarchy; small if near the endpoints. Therefore, the way in which expansion and collapse of diversity actually work themselves out in a coevolutionary system is conditioned heavily on the way dependencies are structured.

Third, two positive feedbacks—circular causalities—are inherent in this mechanism. The generation of new entities may enhance the generation of new entities, simply because there is new "genetic material" in the system available for further "adaptive radiation." And the appearance of new entities provides niches for the appearance of further, new entities. In turn, these mean that where few new entities are being created, few new entities can appear; thus, few new niches will be created. And so the system will be largely quiescent. And where new entities are appearing rapidly, there will be a rapid increase in new niches, causing further generation of entities and further new niches. The system may then undergo a "Cambrian explosion." Hence, we would expect that such systems might lie dormant in long periods of relative quiescence but burst occasionally into periods of rapid increase in complexity. That is, we would expect them to experience punctuated equilibria.

This mechanism, whereby complexity increases via the generation of new niches, is familiar to most of us who study complex systems. Certainly Stuart Kauffman has written extensively on various examples of self-reinforcing diversity. Yet strangely it is hard to find discussion of it in the traditional biological literature. Bonner's 1988 book, *The Evolution of Complexity*, does not mention it, for example, although it devotes a chapter to a discussion of complexity as diversity. Waddington[16] comes somewhat closer when he suggests that niches become more complex as organismal diversity increases. The more complex niches, he suggests, are then filled by more complex organisms, which in turn increases niche complexity. But he seems to have in mind an upward spiral of internal structural complexity, and not of ecological diversity. An intriguing mention of this mechanism—or something tantalizingly close to it—comes from Darwin's notebooks,[3] p. 422.[1]

> "The enormous number of animals in the world depends, of their varied structure and complexity...hence as the forms became complicated, they opened fresh means of adding to their complexity."

But once again this could be read as having to do with internal structural complexity, rather than ecological diversity.

STRUCTURAL DEEPENING

A second mechanism causing complexity to increase over time I will call structural deepening. This applies to single entities—systems, organisms, species, individuals— that evolve against a background that can be regarded as their "environment." Normally, competition exerts strong pressure for such systems to operate at their limits of performance. But they can break out of these limits by adding functions or subsystems that allow them to (a) operate in a wider or more extreme range, (b) sense and react to exceptional circumstances, (c) service other systems so that they operate better, and (d) enhance their reliability. In doing so, they add to their

[1]I am grateful to Dan McShea for pointing out this quotation to me.

"structural depth" or design sophistication. Of course, such functions or subsystems, once added, may operate at their limits of performance. Once again they can break through these limits by adding sub-subsystems according to (a)–(d) above. By this process, over time the original system becomes encrusted with deeper functions and subfunctions. It may improve greatly in its performance and in the range of environment it can operate in. But in doing so, it becomes increasingly complex.

The history of the evolution of technology provides many examples of structural deepening. The original gas-turbine (or jet) aero engine, designed independently by Frank Whittle and Hans von Ohain in the 1930s, for example, was simple.[2] It compressed intake air, ignited fuel in it, released the exploding mixture through a turbine that drove the compressor, and then exhausted the air mass at high velocity to provide thrust. Whittle's original prototype had one moving part, the compressor-turbine combination. But over the years, competitive pressures felt by commercial and military interests led to constant demands for improvement. This forced designers to overcome limits imposed by extreme stresses and temperatures, and to handle exceptional situations, sometimes by using better materials, but more often by adding subsystems.

And so, over time, higher air-compression ratios were achieved by using not one, but an assembly—a system—of many compressors. Efficiency was enhanced by a variable position guide-vane control system that admitted more air at high altitudes and velocities and lowered the possibility of the engine stalling. A bleed-valve control system was added to permit air to be bled from critical points in the compressor when pressures reached certain levels. This also reduced the tendency of the engine to stall. A secondary airflow system was added to cool the red-hot turbine blades and pressurize sump cavities to prevent lubrication leakage. Turbine blades were also cooled by a system that circulated air inside them. To provide additional thrust in military aircombat conditions, afterburner assemblies were added. To handle the possibility of engine fires, sophisticated fire-detection systems were added. To prevent the build up of ice in the intake region, deicing assemblies were added. Specialized fuel systems, lubrication systems, variable exhaust-nozzle systems, and engine-starting systems were added.

But all these required further subsystems, to monitor and control *them* and to enhance their performance when they ran into limitations. These subsystems, in turn, required sub-subsystems to enhance their performance. A modern, aero gas turbine engine is 30 to 50 times more powerful than Whittle's and a great deal more sophisticated. But Whittle's original simple system is now encrusted with subsystem upon subsystem in an enormously complicated array of interconnected modules and parts. Modern jet engines have upwards of 22,000 parts.[2]

And so, in this mechanism, the steady pressure of competition causes complexity to increase as functions and modifications are added to a system to break through limitations, to handle exceptional circumstances, or to adapt to an environment itself more complex. It should be evident to the reader after a little thought

[2] Personal communication from Michael Bailey, General Electric Aircraft Engines.

that this increase of structural sophistication applies not just to technologies, but also to biological organisms, legal systems, tax codes, scientific theories, and even to successive releases of software programs.

One laboratory for observing real-time structural deepening is John Holland's genetic algorithm.[5] In the course of searching through a space of feasible candidate "solutions" using the genetic algorithm, a rough ballpark solution—in Holland's jargon, a coarse schema—appears at first. This may perform only somewhat better than its rivals. But as the search continues, superior solutions begin to appear. These have deeper structures (finer subschemas) that allow them to refine the original solution, handle exceptional situations, or overcome some limitation of the original solution. The eventual solution-formulation (or schemata combination) arrived at may be structurally "deep" and complicated. Reversals in structural depth can be observed in the progress of solutions provided by the genetic algorithm. This happens when a coarse schema that has dominated for some time and has been considerably elaborated upon is replaced by a newly "discovered," improved coarse schema. The hierarchy of subschemas dependent on the original coarse schema then collapses. The search for good solutions now begins to concentrate upon the new schema, which in its turn begins to be elaborated upon. This may happen several times in the course of the algorithmic search.

John Koza's genetic programming algorithm, in which algebraic expressions evolve with the purpose of solving a given mathematical problem, provides a similar laboratory.[8] In Koza's setup, we typically see the algorithmic parse trees that describe the expressions grow more and more branches as increasing "depth" becomes built into the currently best-performing algebraic expression.

In Figure 1 I show the growth of structure as the search for good "solutions" progresses in one of Koza's examples. As we can see, once again this mechanism is not unidirectional. Reversals in structural depth and sophistication occur when new symbolic expressions come along that allow the replacement of ones near the "root base" of the original system. On the whole, depth increases, but with intermittent reversals into relatively simpler structures along the way.

Collapse near the base of a system can be seen in a very different context, the history of science, when new theories suddenly replace old, elaborate ones. An example is the collapse of Ptolemaic astronomy caused by the Kepler-Newton version of the Copernican theory. This novel system, that explained planetary orbits using only a few simple laws, struck at the root base of the hugely complicated Ptolemaic system; and it had such superior explanatory power that the Ptolemaic system never recovered. Similarly, Whittle's jet engine, with its extraordinarily simple propulsion principle, largely replaced the piston aero engine of the 1930s, which had become incurably complicated in attempts to overcome the limitations in operating internal combustion engines at high speed in the very thin air of higher altitudes.[4] And so in evolving systems, bursts of simplicity often cut through growing complexity and establish a new basis upon which complication can again grow. In this back-and-forth dance between complexity and simplicity, complication usually gains a net edge over time.

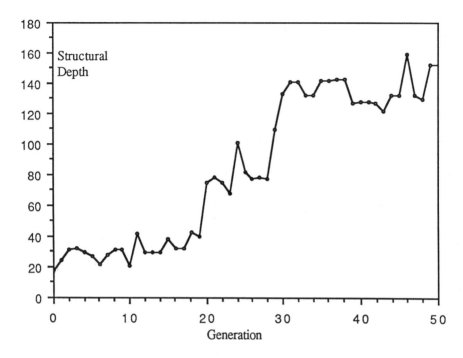

FIGURE 1   Structural depth (number of parts in parse tree) of the currently best expression plotted against number of generations of search in the problem of finding a Fourier series expression to match a given function (from Koza,[8] p. 502).

So far I have described two apparently separate mechanisms. In the first, ecosystems—collections of many individuals—become more complex, more diverse, in the course of evolution; in the second, individuals within ecosystems become more complex, structurally deeper, in the course of evolution. In many systems, of course, these mechanisms operate simultaneously, and they may interact, alternate, and even compete.

This can be seen clearly in Kristian Lindgren's study of strategies that evolve in a game-theoretic setting.[10] Lindgren sets up a computerized model populated by strategies that meet randomly and cumulate profit by playing, one-on-one, a finite version of the iterated prisoners' dilemma. The competing strategies are described as coded bit-strings, where the bits represent memory of previous plays that the strategies can take account of. The strategies can occasionally mutate. Successful ones proliferate in this coevolutionary environment; unsuccessful ones die out. In Lindgren's world, it can clearly be seen that the diversity of strategies increases as new coevolving strategies provide niches that can be exploited by fresh, new strategies, exactly as in the first mechanism I have discussed. But the strategies themselves also become increasing "deep"—their code string or memory lengthens—as competition rewards increasingly subtle strategies, as in the second mechanism. In fact, the two mechanisms interact in that the arrival of a new, successful, deeper strategy eliminates many of the previous, simpler strategies. Diversity collapses, and with it many of the niches it provides. There follows a phase in which the newer, deeper strategies mutate and proliferate, so that diversity increases again. And so new depth can both destroy old diversity and feed a new round of increased diversity among newer, deeper strategies. In this way, the growth of coevolutionary diversity alternates in a sporadic way with the growth of structural depth in the strategies. This process has obvious parallels in the history of biological evolution. Some biologists suggest, for example, that increased "depth" in the form of the appearance of multicellular, eukaryotic organisms fueled the Cambrian explosion of diversity 600 million years ago.[15]

CAPTURING SOFTWARE

The third mechanism in the growth of complexity that I will propose is completely different from the first two. Actually it has more to do with the rapid emergence of complexity than with any slow growth. It is a phenomenon I will call capturing software. This is the taking over and "tasking" of simpler elements by an outside system for its own (usually informational) purposes. Typically the outside system "discovers" the simpler elements and finds it can use them for some elementary purposes. The elements turn out to have a set of rules that govern how they may be combined and used—an "interactive grammar." This grammar typically allows many combinations of the simple elements; and as the outside system begins to learn this grammar, it also learns to take advantage of the elements in combination. At full fruition, the outside system learns to use this interactive grammar to "program" the simple elements and use them in complicated combinations for its own multi-purpose ends.

This mechanism may sound somewhat strange and unfamiliar; so let me clarify it by some examples. A very simple one would be electronics, taken as a technology. As humans, we have learned over the last couple of centuries to "task" electrons to carry out such activities as transmitting sound and vision, controlling sophisticated machinery, and computing. Originally, in the days of Faraday and Franklin, the workings of electrons and of static electricity were poorly understood. And so, uses were few. But in the last century and in the early decades of this one, we began to learn the "grammar" of electricity—the set of operational rules involving induction, capacitance, and impedance that govern the movements of electrons and amplification of their flow. And so we slowly learned to "capture" and "program" electrons for our own use. In this case the simple elements referred to above are electrons. The outside system is ourselves, the human users. The grammar is the laws of electromagnetism. And the programmable outputs are the various technical uses to which electronics are put. At the output level, there is swift "adaptation." The various technological purposes in which we use electrons as a "programmable software" shift and expand rapidly. But at the grammar and carrier level, in this case, adaptation is absent. The behavior of electricity and of electrons is fixed by physical laws that are, within the human time frame at least, immutable.

Sometimes with capturing software, the interactive grammar is not laid down unalterably, but can itself change and evolve in the process of "capturing" the software. An example is the way in which human language evolved. Early humans learned perhaps several hundred thousand years ago that crude, emitted sounds could be used for communicating warnings, pleasure, or simple needs. Very slowly, and comparatively recently on an evolutionary time scale, they began to generate some elementary rules—a grammar—to organize these into simple concatenated expressions. Eventually, over many thousands of years, these sounds or phonemes plus grammar evolved into a complex interactive system—a language. This could be "programmed" to form statements, queries, and commands that conveyed a high degree of nuance and subtlety.

In this example, the simple, carrier elements are the sounds or phonemes of human speech. The outside system is the human community that "captures" and makes them into a software, a language. And the grammar is the syntactical system that develops to ensure consistency and commonality of meaning. Of course, there is no single, natural syntactical grammar for human language. A grammar must emerge by the slow evolution of a social convention, with constraints exercised by the need for linguistic efficiency and consistency and by the way linguistic activities are organized in the human brain.[9] (Of course, both the human vocal anatomy and brain also changed as a response to the evolution of language.) The overall language that results from this evolutionary process is a programmable software whose potential output we may think of as the set of all meaningful sentences or statements the language can express.

Adaptation in this case can occur at all levels. At the program output level, adaptation is instantaneous. We can think of a sentence uttered as a one-off, extremely rapid adaptation of software output to the purpose of what the sentence is intended to communicate. At the grammar level, adaptation implies change in the language itself. This commonly takes the form of drift, and it happens slowly and continuously. This is because any abrupt alteration or large deviation in grammar would invalidate current "output programs." At the phoneme or simple element level, adaptation—or change and drift—is slowest of all. Slight changes at this carrier level, if not subtle and continuous, might upset all that is expressed in the system. Slow drift may occur, as when vowels shift over the course of a generation or two; but there is a powerful mechanism acting to keep the carrier elements locked-in to a constant way of behaving.

A particularly telling example of capturing software is the way in which sophisticated derivatives have arisen and are used in recent years in financial markets. In this case the outside system is the financial community. It begins by the simple trading of something of value—soybeans, securities, foreign currencies, municipal bonds, Third World debt, packages of mortgages, Eurodollars—anything to which title can be held. Such items may fluctuate in value and can be swapped and traded. They are called underlyings in financial jargon, and they form the simple, carrier elements of the system I want to consider now.

In early days of such markets, typically an underlying is simply held and traded for its intrinsic value. But over time, a grammar forms. Traders find they can: (a) usefully arrange *options* associated with contingent events that affect the underlying; (b) put several underlyings together to create an associated *index*, as with a stock index; (c) issue *futures* contracts to deliver or obtain an underlying at some time, say, 60 days or one year, in the future; and (d) issue *securities* backed by the underlying. But notice that such "derivatives" as contingent-event options, indices, futures, and securities are themselves elements of value. Thus, they, too, can become *underlyings*, with their own traded values. Once again the market could apply (a), (b), (c), or (d) to these new underlyings. We may then have options on securities, index-futures, options on futures, securities indices, and so on, with trades and swaps of all these.

With such a grammar in place, derivatives experts "program" these elements into a package that provides a desired combination of financing, cash-flow, and risk exposure for clients with highly particular, sophisticated financial needs. Of course, financial markets did not invent such programming all at once. It evolved in several markets semi-independently, as a carrier element was used, simply at first and then in conjunction with the natural grammar of finance.

From the examples I have given, it may seem that the system that uses and captures simple elements to its own uses is always a human one. But, of course, this is not the case. Let me point out two examples in the biological sphere. One is the formation of neural systems. As certain organisms evolved, they began to "task" specialized cells for the simple purposes of sensing and modulating reactions to outside stimuli. These specialized cells, in turn, developed their own interactive grammar; and the overall organism used this to "program" this interconnected neural system to its own purposes. Similarly, the ancestors of the cells found in the immune systems of higher organisms were used originally for simple purposes. Over time, these, too, developed useful rules of interaction—an interactive grammar— thereby eventually becoming a highly programmable system that could protect against outside antigens.

Biological life itself can be thought of in this way. Here the situation is much more complicated than in the previous examples. Biological organisms are built from modules—cells mainly—that in turn are built from relatively small and few (about 50 or so), fairly simple molecules.[12] These molecules are universal across all terrestrial life and are the carriers of biological construction. They are combined into appropriate structures using a grammar consisting of a relatively small number of metabolic chemical pathways. This metabolic grammar, in turn, is modulated or programmed by enzymes. The enzymes doing the programming of course have no conscious purpose. In fact they themselves are the carriers in a second programmed system. They are governed by a complicated gene-expression "grammar," which switches on or inhibits their production from the genes or DNA that code for them, according to feedback received from the state of the organism they exist in. And so we have one captured software system, the programming of the simple metabolic pathways via proteins or enzymes to form and maintain biological structures, modulated by another captured software system, the programming of proteins or enzymes via nucleic acids and the current state of the organism.

In this case the entire system is closed—there is no outside system programming the biological one to its own purposes. In the short term each organism programs itself according to its current development and current needs. In the long term the overall system—the resulting biospheric pattern of organisms that survive, interact, and coevolve—together with environmental and climatic influences, becomes the programmer, laying down its code in the form of the collection of gene sequences that survive and exist at any time. Of course, without an outside system, we can not say these programmable systems were ever "captured." Instead they emerged and bootstrapped themselves, developing carriers, grammar, and software as they went. Viewed this way, the origin of life is very much the emergence of a software system carried by a physical system—the emergence of a programmable system learning to program itself.

Capturing software in all the cases discussed here is an enormously successful evolutionary strategy. It allows the system to adapt extremely rapidly by merely reprogramming the captured system to form a different output. But because changes in grammars and in carriers would upset existing "programs," we would expect them to be locked in and to change slowly if at all. This explains why a genetic sequence can change easily, but the genetic code can not; why new organisms can appear, but the cell and metabolic chemistry remain relatively fixed; why new financial derivatives are constantly seen, but the securities-and-exchange rules stay relatively constant.[3]

CONCLUSION

In this chapter, I have suggested three ways in which complexity tends to grow as evolution takes place. It may grow by increases in diversity that are self-reinforcing; or by increases in structural sophistication that break through performance limitations; or by systems "capturing" simpler elements and learning to "program" these as "software" to be used to their own ends. Of course, we would not expect such growth in complexity to be steady. On the contrary, in all three mechanisms we

---

[3] Carriers do change, of course, if they can be substituted for one another easily. For example, options can be built on any underlying; and so, in this case, carriers can and do change rapidly. But the essential property of underlyings—that of being an object that carries uncertain value— remains necessary in all cases and does not change.

would predict it to be intermittent and epochal. And we would not expect it to be unidirectional. The first two mechanisms are certainly reversible, so we would expect collapses in complexity to occur randomly from time to time.

As we study evolution more deeply, we find that biology provides by no means all of the examples of interest. Any system with a lineage of inherited, alterable structures pressured to improve their performance shows evolutionary phenomena. And so, it is likely that increasingly we will find connections between complexity and evolution by drawing examples not just from biology, but from the domains of economics, adaptive computation, artificial life, and game theory. Interestingly, the mechanisms described in this chapter apply to examples in all of these evolutionary settings.

ACKNOWLEDGMENTS

This paper was originally presented at the Santa Fe Institute's Integrative Themes Workshop in July, 1992. I thank Dan McShea, Brian Goodwin, and the workshop participants for useful comments. I am grateful to Harold Morowitz in particular for several conversations on the themes of this essay.

REFERENCES

1. Bonner, J. T. *The Evolution of Complexity.* Princeton: Princeton University Press, 1988.
2. Constant, E. W. *Origins of the Turbojet Revolution.* Baltimore: Johns Hopkins University Press, 1980.
3. Darwin C. *From Charles Darwin's Notebooks*, edited by P. H. Barrett et al., 422. Ithaca: Cornell University Press, 1987.
4. Heron, S. D. *History of the Aircraft Piston Engine.* Detroit: Ethyl Corp., 1961.
5. Holland, J. *Adaptation in Natural and Artificial Systems*, 2nd ed. Cambridge: MIT Press, 1992.
6. Holland, J. "Echoing Emergence: Objectives, Rough Definitions, and Speculation for Echo-Class Models." Mimeograph, University of Michigan, 1993.
7. Kauffman, S. "The Sciences of Complexity and Origins of Order." Working Paper 91-04-021, Santa Fe Institute, 1991.
8. Koza, J. *Genetic Programming.* Cambridge: MIT Press, 1992.
9. Lieberman, P. *The Biology and Evolution of Human Language.* Cambridge: Harvard University Press, 1984.

10. Lindgren, K. "Evolutionary Phenomena in Simple Dynamics." In *Artificial Life II*, edited by C. Langton, C. Taylor, J. D. Farmer, and S. Rasmussen. Santa Fe Institute Studies in the Sciences of Complexity, Vol. X, 295–312. Reading, MA: Addison-Wesley, 1991.
11. McShea, D. "Complexity and Evolution: What Everybody Knows." *Bio. & Phil.* **6** (1991): 303–324.
12. Morowitz, H. *Beginnings of Cellular Life.* New Haven: Yale University Press, 1992.
13. Müller, G. B. "Developmental Mechanisms at the Origin of Morphological Novelty: A Side-Effect Hypothesis." In *Evolutionary Innovations*, edited by Matthew Nitecki, 99–130. Chicago: University of Chicago Press, 1990.
14. Ray, T. S. "An Approach to the Synthesis of Life." In *Artificial Life II,* edited by C. Langton, C. Taylor, J. D. Farmer, and S. Rasmussen. Santa Fe Institute Studies in the Sciences of Complexity, Vol. X, 371–408. Reading, MA: Addison-Wesley, 1991.
15. Stanley, S. M. "An Ecological Theory for the Sudden Origin of Multicellular Life in the Late Precambrian." *Proc. Nat. Acad. Sci.* **70** (1979): 1486–1489.
16. Waddington, C. H. "Paradigm for an Evolutionary Process." In *Towards a Theoretical Biology*, edited by C. H. Waddington, Vol. 2, 106–128. New York: Aldine, 1969.

DISCUSSION

WALDROP: I wasn't clear how you distinguish co-organization from tool use.

ARTHUR: Co-organization happens when similar cells or entities get together, then specialize to carry out some task. To some degree that did happen with the neural systems and with the immune system. I see that as a slightly different theme. But what I'm talking about here is a system's ability to co-opt an entirely different system, surround it, bring it in, and start to program it for its own purposes. For that you need elementary interactive devices. They need to have a grammar that you can use, and you need to be able to program them. I believe we see this again and again in emergence.

EPSTEIN: I want to bring to your attention an article that you would enjoy. It's a very nice result by a mathematician named Joseph So published in *Mathematical Biosciences* in 1978, that goes to the issue of collapse and this "root" thing you mentioned. He has a strict chain of specialists—Volterra-type specialists— and if you arrange for stable interior equilibrium of n species, then you can dial down the Malthusian parameter of the bottom species, and you'll get sequentially the extinction of species from the top down. So you can blip them out, by dialing down the bottom growth rate. But, to put it in your terms, you can speciate by dialing up that growth rate which is the throughput of the bottom industry. You have a slightly more interesting architecture: it's a tree, not a chain, but you'll get the same result. It gives the eerie feeling that we're always living in species flatland, and then, if you dial up some bottom parameter, you'll get more blips, and then there's other species out there waiting to happen. I think it's very much in line with what you're saying, and it clearly shows this collapse property.

Now a question: the basic mechanism of collapse that you posit is innovation, and it seems to be that the question begged is, "If I'm a member of this tree, the obvious defense strategy for the tree is to prevent innovation," since the real threat to everyone's existence is the entry of some innovation, like the steam engine or the automobile. The question is: what mechanisms do these ecologies evolve to prevent invasion by innovations? What are the barriers to entry? Obviously, the most fit of these systems will be the most dead in your sense; that is, they will want to devise mechanisms that prevent their collapse. But that means, prevent innovation. So the most living systems, the most "fit" ones in your terms, will also produce the deadest economy in the usual interpretation. How do they prevent innovation?

ARTHUR: I'd have to think about it biologically.

EPSTEIN: Or economically.

ARTHUR: Economically, I can answer you. That happens all the time.

EPSTEIN: So the economy *wants* to be dead.

ARTHUR: No, there are vested interests that want to be dead. Why do you think it's taken umpteen years to get a light bulb that lasts more than a couple of hundred hours? I don't know.

EPSTEIN: But the point is that it's struggling to be alive; it wants to be dead, and it's struggling to be alive. It's *hard* to innovate.

ARTHUR: I do think there's a tradeoff between vested interests, and interests that say, "We can make a large profit if we're the first people with the jet engine."

FONTANA: You drew close analogies between economics and biology, and convinced us all that the economy is alive, or you talked it alive, but we know that, from a theoretical point of view at least, theoretical biology is in bad shape. So the question to you is "What do you really require a theory of economics to fulfil? What kinds of laws do you think we can discover—are we bound to make models of very specific examples of small sectors in the economy or is there any way to attempt to make a theory about the economy like our attempts for a theory of life?"

ARTHUR: There *is* a theory of the economy. It's called neoclassical economics. The problem with all theory is that it tends to portray the system you are
looking at—say, the economy, or the biosphere—in terms of the dominant zeitgeist
metaphor of the time. For Adam Smith's time, in 1776, for 50 years—or longer—
the dominant metaphor was systems in stasis, systems that operate in a kind of
clockwork fashion, that are highly deterministic, and are in some sort of equilibrium balance. That was the notion of the Enlightenment, partially inherited from
Newton and others.

    We're entering the a different zeitgeist at this stage in the twentieth century,
where we're more interested in things that are in process, and pattern change.
So I'm hoping that new theories of economics can reflect that the economy is in
process, that the economy is always developing. The development never necessarily
stops. The question is how do we talk about that? How do we think about it? What
are the mechanisms? And above all, does that give us a feeling that we're closer to
reality.

GELL-MANN: Much of the work on laws was rule-based mathematics, which
isn't all that old for this kind of application. Maybe we haven't yet gotten used to
formulating laws, theorems—all the apparatus of academic, theoretical science—for
rule-based mathematics so that we're satisfied with the degree of region achieved.
With a differential equation, everybody knows what it's like—to propose a differential equation, to do all the associated proofs, and to check it against observation.
With rule-based mathematics, we probably haven't got a procedure yet, and that's
something we could work on.

ARTHUR: I believe that we are starting to check some of our theories
observationally, and we're finding that the kinds of ideas we're justifying here can
be checked and do have scientific credibility in the hard sense; that is, they can be
empirically validated.

HÜBLER: Some comments on the jet engine. I think a jet engine has an
approximate lifetime of twenty years. Now this makes it, according to the study that
we did on adaptive complex systems, an example of maladaptation for two main
reasons. The first is the fitness function; these are complicated fitness functions
because you have several fitness functions at the same time which are competing.
First of all, the engine should be simple. Second, you have employment issues of
local government, which might support a certain company just in order to keep the
manpower there. Third, you have the performance issues. Now, you have at least
three different fitness functions, and the time scale at which the weight is changing
of this fitness function is typically on the order of four years—which is the election
period—which is much shorter than the lifetime of a jet engine; and that you need
to have the lifetime in order to figure out whether the jet engine was something
good or bad.

The second reason is that the state space is changing at a faster rate. The state space is basically the engineering possibilities—what is developed in research during that period of time—and again this is changing very fast compared to the twenty years which you need to figure out whether the jet engine was good or bad. Okay, so I would think that from this point of view, it is very difficult to adapt an airplane engine right now because the rate at which the environment and the fitness functions are changing is much faster than the time scale you need to figure out whether a certain engine was good or bad. And, therefore, I think it could be easily the case that the jet engine was a bad idea, and that step now back to the propeller engine is something very natural, because of this difficulty of the adaptation process. It was just misdevelopment.

ARTHUR: It is that there are pressures to improve the performance of any technological device, jet engines among them. Those pressures can be responded to very often by adding subsystems, or subparts, or new functions. Those new functions lead to extra complication. That's the mechanism I'm talking about. I agree it's subject to marketing whimsy. But that's the simple mechanism I'm talking about. But it doesn't always go in one direction because occasionally you can get subsystem collapse.

Stuart A. Kauffman
Santa Fe Institute, 1660 Old Pecos Trail, Suite A, Santa Fe, NM 87501

Whispers From Carnot: The Origins of Order and Principles of Adaptation in Complex Nonequilibrium Systems

Abstract: Living systems do not merely process information, they carry out activities by which they reproduce and evolve. Prior to the origin of life, some form of prebiotic chemical evolution must have taken place on the early Earth. This evolution, in general, was displaced from equilibrium, and presumably built up the diversity of organic compounds from which life was able to crystallize. We need new bodies of theory to begin to understand such nonequilibrium systems, open in at least two senses: First, they are open to matter and energy in the familiar thermodynamic sense. Second, these systems are open in the sense that the molecular constituents of the system themselves change via their own unfolding dynamics. The autocatalytic polymer models, which crystalize at a phase transition in chemical reaction graphs, that I have worked on with Doyne Farmer and Norm Packard, and as Doyne, Norm, Walter Fontana, and Rick Bagley have extended them, are new members of a Universality class of models to examine these issues. Fontana's AlChemy, and the random grammar models I am now using, appear to be useful formalizations in this arena.

Our joint major concern at this meeting has to do with complex adapting systems (CAS). Darwin taught us how to think about such adaptation

Complexity: Metaphors, Models, and Reality
Eds. G. Cowan, D. Pines, and D. Meltzer, SFI Studies in the
Sciences of Complexity, Proc. Vol. XIX, Addison-Wesley, 1994

for evolving entities capable of self-replication and heritable variation. In a deep sense, Darwin taught us that in such open, far from equilibrium systems, it is kinetics which matters most. Those entities which reproduce most rapidly, which capture the most flux of matter and material to make more of themselves, are the entities which shall predominate. But Darwin captured only a piece of the problem. Prior to life, prior to self-reproducing entities, during the evolution of the open-far-from-equilibrium chemical systems that begot life, what does adaptation and heritable variation mean? What principles might govern the unfolding dynamics? The most difficult task I will undertake is to try to articulate what such principles might look like. Like Doyne, I suspect that some very general laws may govern such open, self-constructing systems.

In addition to trying to tread turf which I do not understand at all, but which may be among our most important issues, I will discuss the emergence of autocatalytic systems, as a test bed for us to think about ways such systems, parallel processing collective molecular machines, coevolve with one another, build models of their worlds, and must carry out practical action. By considering these examples, I hope to address the relationship between basins of attraction and attractors in parallel-processing dynamical systems and "schemas" in John Holland's sense. In addition, I will address the relation between basins of attraction, and more generally, the dynamics of such systems, and the notions of compression and evolvability.

These topics will lead me to more familiar territory. I will discuss two tentative general principles concerning CAS: (1) Adaptation to the edge of chaos in parallel-processing Boolean networks and cellular automata, as Chris Langton, Norm Packard, and I have been urging. (2) Coevolution to the edge of chaos, as I have been urging. Evidence supporting these putative principles is scant, but important. I will outline that data, and sketch my own plans for trying to nail these issues. If either were true in general, we would have grounds to be very pleased indeed.

Finally, I will want to discuss the bearing of these two putative principles on the basic problem of bounded rationality. The theme I am investigating is that we are "optimally" myopic strategists, hence coevolve with one another on coupled deforming landscapes. We are myopic, I want to show, not because of costs of computation, but because if we are individually too clever, we tend to transform the world in which we are adapting into a yet more chaotic world in which we fare less well. Results by myself, by Jean-Michel Grandmont, and by Alfred Hübler point in this direction. I suspect that we tune the games we play with one another, as in the coevolution to the edge of chaos story, and we tune the amount of data from the past that we use, and forecasting into the future that we do, such that the risks of being wildly wrong balance the rewards of foresight. I want to believe that entities that coevolve by tuning the amount of foresight they employ will attain some analogue of the edge of chaos where all agents, on average, do the best that they can, but a power-law distribution of avalanches of bankruptcies propagate through the system. I also want to believe that such a self-consistent solution among economic agents leads to a nonequilibrium solution concept. Markets will not clear, but come close to clearing. And, to end, note that firms, unlike *E. coli*, are not self-reproducing. But, rather like prebiotic chemical evolution, an evolution occurs such that the richly endowed firms, abundant in resources, predominate. Do similar evolutionary principles apply to the prebiotic world and the economic world?

I. INTRODUCTION

The title of this chapter, "Whispers from Carnot," foretells its direction. If there is a core interest at the Santa Fe Institute, it is the search for general principles governing adaptation in complex systems. In the deepest sense, we seek universal principles governing the emergence of nonequilibrium, self-organizing, living, evolving systems on the earth and elsewhere in the Universe. It is life, after all, which has yielded our only natural examples of complex adaptive systems (CAS). Chemical evolution on the prebiotic earth begat protolife, which in time begat the evolution of organisms, ecological, economic, and political systems. In senses to be explored below, the emergence and coevolution of living systems was driven by properties of self-organization in nonequilibrium systems wedded to Darwin's themes of variation and natural selection. These together have given rise to complex adaptive living entities which ultimately were able to form adaptive internal models of the worlds they mutually create.

The bold promise of the title, however, would seem foredoomed. Good arguments suggest that there can be no general body of laws governing nonequilibrium, self-organizing systems. We know that Turing machines are examples of systems capable of universal computation. Universal computation implies that *any* well-specified algorithm can be carried out by some machine. That computing machine can, within the limitations of a finite universe, be made of real physical materials, and can carry out the algorithm if the system is displaced from equilibrium, open to matter and energy. Thus, physical systems, displaced from equilibrium and appropriately constructed, can carry out any well-specified sequence of behavior. But some computations cannot be described in a more compact form than to carry out the computation and observe its unfolding. We could not, in principle, have general laws (shorter, compact descriptions) covering such behaviors. In short, the theory of computation, coupled to the physical realizability of such algorithms, assures us that there could not, in principle, be general laws for the behavior of all possible nonequilibrium systems.

This argument, however, makes the critical assumption that the nonequilibrium system be "appropriately constructed" in some essentially arbitrary way so as to carry out the algorithm. The system does not construct itself; we construct it for our purposes. But no agent constructed living systems for an exogenous design purpose. Living systems unequivocally demonstrate that certain forms of matter and energy, displaced from thermodynamic equilibrium, emerged spontaneously and coevolved to form the biosphere. The emergence and evolution of life must be a natural expression of a properly defined class of matter and energy. While the arguments above based on the theory of computation do appear to imply that no general laws could govern all possible nonequilibrium systems, the spontaneous emergence of life suggests that we consider the possibility that universal laws may govern the unfolding of *self-constructing, self-organizing, far from equilibrium systems.*

In this chapter, I propose to investigate several candidate general principles governing the origins and evolution of such CAS. None is established. At this stage it may suffice that we can even begin to frame candidate general principles.

In this chapter I will discuss properties of self-organization in nonequilibrium systems which I believe to underlie the origin of life itself, the origin of the order in ontogeny, the emergence of adaptive complex systems which build models of their worlds, the generation of biodiversity, and the drive to hierarchical complexity. This new view requires fundamental revisions of biology. Since Darwin, we have come to think of organisms as tinkered together contraptions, and selection as the sole source of order. Yet Darwin could not have begun to suspect the power of self-organization. We must seek our principles of adaptation in complex systems anew, as a proper marriage of self-organization and selection. In this chapter I discuss a modest number of such *candidate* general principles.

1. Life, self-reproducing chemical systems capable of adaptive evolution, is an expected emergent collective property of sufficiently complex nonequilibrium chemical systems.
2. The far-from-equilibrium evolution of prebiotic and biosphere chemical diversity is catalytically self-extending in a supracritical process.
3. Free living self-reproducing systems, living cells, achieve a slightly subcritical internal molecular diversity on a subcritical-supracritical axis.
4. Coevolution among free living self-reproducing systems drives the biosphere as a whole to be strongly supracritical. The total nonequilibrium system may maximize a measure of total nonequilibrium action.
5. Complex adaptive parallel-processing systems, from protoorganisms to the genomic regulatory networks in contemporary cells, individually adapt their internal logic and dynamics to the ordered regime near the edge of chaos.
6. Coevolving adaptive systems, organisms, economic agents, and others, collectively adapt their internal structure and games they mutually play to the edge of chaos.

7. CAS, in order to predict optimally, build optimally complex and thus optimally boundedly rational internal models of their worlds. These models may coevolve to the edge of chaos.
8. CAS are driven to build hierarchical complexity due to proliferating task differentiation, the consequent emergence of mutualisms, and thus the emergence of advantages of trade. Symbiosis, or its analogs, holding the mutualists in close collaboration, arises at all levels from self-reproducing molecular systems upwards.

This chapter is organized as follows: Section II discusses life as a phase transition between subcritical and supracritical behavior at a critical diversity in chemical reactions systems. Section III examines the hypothesis the prebiotic chemistry and the current biosphere are catalytically self-extending supracritical systems. The biosphere appears supracritical, cells appear just subcritical. Economic interactions, including the advantages of trade, arise even among molecular reproducing systems and drive hierarchical integration. Section IV examines the unexpected and powerful "antichaos" in parallel-processing networks and the hypothesis that natural selection culminates in networks poised in the ordered regime near the edge of chaos. Section V explores the novel possibility that coevolutionary systems, as if by an invisible hand, coevolve to the edge of chaos. Section VI discusses a new general approach to the problem of bounded rationality and behavior coordination. Agents build internal models of one another. Those models coevolve to the edge of chaos. The concluding section seeks a unifying framework, a marriage of Charles Darwin and Adam Smith, for nonequilibrium, coevolving, self-reproducing systems.

II. THE ORIGIN OF LIFE AS A PHASE TRANSITION AT A CRITICAL CHEMICAL DIVERSITY

There is a received view about life's origin. DNA and RNA molecules command attention as the best candidates for the first living molecules because of the self-templating character of the polynucleotide double helix.[64,65] Thus, efforts for the past several decades have focused on two areas: First, work has been carried out demonstrating the prebiotic synthesis of critical biomolecules such as amino acids, sugars, and even nucleosides and nucleotides.[62,66,67] Second, considerable work has attempted to achieve template replication of arbitrary RNA sequences, in which the template strand uses Watson-Crick base pairing to line up the complementary free nucleotides, join them together to create the complementary strand, melt back to single strands, then recycle.[38] This work has not yet been successful, but may well be in the future.

While template replication has failed to date, it has been possible to create small *autocatalytic* sets of specific polynucleotides. In one case a hexamer binds two trimers by Watson-Crick base pairing, then ligates the two to form a hexamer identical to the initial hexamer.[86] Here, then, the hexamer acts as a specific *ligase* to join two trimer precursors which then create the hexamer itself. Similar work has been carried out using a tetramer as the ligase and two dimers as the substrates. And more recently, modified nucleotides able to function in this way have been explored.

The new view of the origin of life is based on a generalization of small autocatalytic sets. I believe good theory exists to support the possibility that sufficiently complex sets of catalytic polymers will almost inevitably contain *collectively autocatalytic sets.*[23,40,43,73] In such a collectively autocatalytic set of monomers and polymers, no molecule need catalyze its own formation; rather, each molecule has its formation catalyzed by some molecule in the set such that the set is collectively autocatalytic. The set constructs itself from exogenously supplied monomers or other building blocks.

This new theory is based on the discovery of a phase transition in chemical reaction graphs.[43] To be concrete, consider a set of amino acids and peptides, or a set of nucleotides and oligonucleotides, where the longest polymer is of length M. As M increases, the number of types of polymers in the system, from monomers to those length M, increases exponentially. The next step considers all the possible cleavage and ligation reactions among all the types of molecules, ranging from monomers to M in length. Since a polymer of length M can be made in $M - 1$ ways from smaller fragments, it is obvious that there are *more reactions among the polymers than there are polymers.* More formally, a reaction graph considers each types of molecule in the system as a point. A ligation reaction can be represented by two arrows leaving the points corresponding to the two precursors and entering a square "reaction box." An arrow leaves the reaction box and ends on the larger polymer which is the product achieved by ligating the two smaller polymers. Since reactions are reversible, arrows in the opposite direction depict the corresponding cleavage reaction. For simplicity, picture arrows directed toward the ligation product in the following. The set of all such triads of arrows and all reaction boxes by which the set of monomers and polymers in the system are connected comprises the *reaction graph.*

The next step notes that the polymers in the system are themselves candidates to catalyze the reactions in the reaction graph. An autocatalytic set would consist of a set of polymers mutually catalyzing their own formation from a maintained founder set of "food" molecules, say, monomers. To build the theory further, we need some idea of which polymers catalyze which reactions. A first simplest model assumes that any polymer has a fixed probability, say, one in a billion, to be able to act as an enzyme to catalyze any specific reaction. But the implication of this hypothesis is this: *At a critical diversity of monomers and polymers, the system will contain collectively autocatalytic sets.*

The emergence of autocatalytic sets as expected objects is due to a phase transition in the reaction graph. When a sufficient fraction of the reactions are catalyzed, a connected web of catalyzed transformations "crystallizes." This web is typically collectively autocatalytic. A simpler image tunes intuition. Consider

10,000 buttons on the floor. Begin to connect them at random with red threads. Every now and then, pick up a button and see how many other buttons are raised with it. Such a connected set of buttons is called a "component" in a "random graph." A random graph is a set of points, or nodes, connected at random with a set of edges, or lines. Erdos and Renyi[22] showed that at a critical ratio of edges to nodes, $E/N = 0.5$, a *giant component* crystallizes. Most, but not all nodes are connected directly or indirectly in the giant component. As the number of nodes increases to infinity, this becomes a first-order phase transition. In a similar way, in reaction graphs, when a sufficient fraction of the reactions are catalyzed, a giant component crystallizes. But this is inevitable: As the maximum length polymer, M, increases, the number of types of polymers increases exponentially, but the number of types of reactions among them increases even faster. Indeed, for ligation and cleavage reactions the ratio of reactions to polymers is proportional to M. Thus, for any fixed probability that an arbitrary polymer can act as a catalyst to catalyze an arbitrary reaction, eventually there are so many possible reactions per polymer that at least one reaction per polymer is catalyzed by some polymer. At that point, the analogous phase transition has been passed. Autocatalytic sets crystallize.[43,48]

Several years of numerical and theoretical work now substantiate these ideas (see Figure 1).[23,43] Under a variety of hypotheses about the distribution of catalytic activity among the polymers in the system, autocatalytic sets still form. For example, Bagley and I[48] examined the plausible case in which the candidate RNA catalyst must base pair match the 3′ and 5′ terminal several nucleotides of the two RNA substrates to be ligated, and then still only has a small probability of being able to function as an enzyme to catalyze that ligation and the reverse cleavage reaction. Here autocatalytic sets still form. Further, under a variety of plausible models of detailed chemical and catalyst kinetics, and about thermodynamic conditions, it is now clear that polymer systems surpassing a critical diversity, hence phase transition, are expected to contain collectively autocatalytic sets which, in computer experiments, do reproduce, and do show the capacity to survive fluctuations in their food environment.[3,4] More surprisingly perhaps, such collectively autocatalytic sets also show the capacity to evolve to neighboring autocatalytic sets. This evolution occurs via fluctuations in molecular concentrations due to spontaneous, uncatalyzed reactions by which molecular neighbors of the autocatalytic set are formed and may be grafted into the set.[4]

These results suggest that life may emerge as an inevitable consequence of polymer chemistry. At a critical diversity a phase transition is surpassed. Collectively reproducing metabolisms, able to grow, survive environmental changes, and able to evolve in the *absence of a genome*, are expected. If so, the routes to life in the Universe may be broad boulevards, not back alleys of thermodynamic improbability.

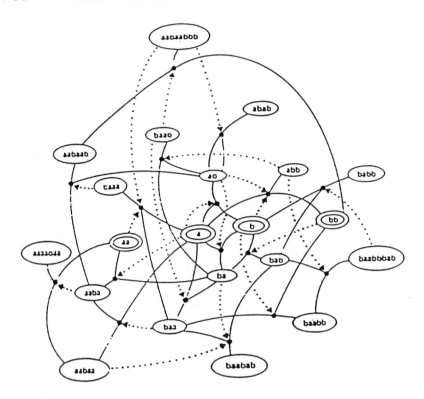

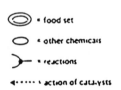

FIGURE 1   A typical example of a small autocatalytic set. The reactions are represented by points connecting cleavage products with the corresponding larger ligated polymer. Dotted lines indicate catalysis and point from the catalyst to the reaction being catalyzed. Monomers and dimers of a and b constitute the maintained food set (double ellipses).

These results lead to a candidate principle. Life—self-reproducing chemical systems capable of evolution—is an expected collective emergent property of critically complex, far-from-equilibrium chemical systems.

This theory is no mere academic model. It is open to experimental test in the very near future. It is now possible to use genetic cloning techniques to generate vast libraries of partially stochastic DNA, RNA, and proteins. For example, libraries of up to 10^{13} RNA sequences are being screen with positive results for the capacity to bind arbitrary ligands,[21] and to catalyze arbitrary reactions. The capacities of such libraries acting upon themselves to collectively surpass the diversity threshold and exhibit supracritical behavior and the formation of collectively autocatalytic sets can now be examined. Nor are the implications limited to our understanding of the origin of life. Applied molecular evolution, based on generating such large libraries of novel biopolymers, promises new drugs, vaccines, enzymes, cis-acting DNA and RNA regulatory sites, biosensors, and, via autocatalytic sets, selectable chemical robots.[7,47]

III. CATALYTICALLY SELF-EXTENDING MOLECULAR EVOLUTION ON THE PREBIOTIC EARTH

Life emerged on the Earth about 3.8–3.45 billion years ago, a scant 100,000,000 or so after the crust cooled sufficiently for liquid water to form and remain on the surface. Thus, the emergence of life could not have been an utterly remote possibility. While the detailed chemistry of the atmosphere and crust of the early Earth are not established, it is generally supposed that some form of chemical evolution occurred which gradually built up larger organic compounds from the small precursors—methane, hydrogen, water, and others—present.

In the current section, I wish to explore the new possibility that prebiotic chemistry was a catalytically self-extending supracritical system, that the current biosphere is now a supracritical system, but that cells are constrained to be marginally subcritical, and that economic interactions, including the advantages of trade, arise among self-reproducing molecular systems and drive hierarchical integration.

SUBCRITICAL AND SUPRACRITICAL BEHAVIOR

To discuss the new possibility more fully, I briefly characterize supracritical and subcritical behavior.[23,43] Rather than focusing on presumed prebiotic chemistry, I shall first consider contemporary proteins as candidate enzymes, and organic molecules as candidate substrates. I shall suggest that at a sufficient sustained diversity and concentration of candidate enzyme proteins and candidate substrate organic molecules, the proteins should catalyze an exploding diversity of organic molecules. This explosion is supracritical behavior. Conversely, at sufficiently low diversities of candidate enzymes and substrates, the candidate enzymes should catalyze few or no reactions forming new organic molecules. This limited response is subcritical behavior.

Supracritical behavior will occur if a sufficient number of candidate enzymes interact at high enough concentrations with a sufficient number of candidate substrates that a connected web of reactions is catalyzed. To estimate whether a system will be supracritical, therefore, we need to estimate the probability that candidate enzymes catalyze reactions, and the number of candidate reactions.

It is now known that antibodies raised against the stable analog of the transition state of a reaction can often act as catalysts for the reaction itself.[69,70,83,84]

The available data allow us to estimate that the probability that an arbitrary antibody catalyzes a reaction is about one in a million. Based on this, it is plausible to estimate that the probability that an arbitrary polypeptide catalyzes a given reaction is between one in a million and one in a billion. Assume the more pessimistic figure.

Start with a set of 1000 organic compounds, each with a modest number of carbon atoms and reactive groups. We need to estimate the total number of organic reactions that can occur among these 1000 compounds, counting single substrate single product reactions, two substrate one product reactions, one substrate two product reactions, and two substrate two product reactions. The expected number of such reactions is unknown, but it is reasonable to estimate crudely that any pair of these organic molecules can undergo at least one possible reaction. Thus, the number of reactions scales as the *square* of the diversity of organic molecules. One-thousand organic molecules afford about 1,000,000 potential reactions.

Reacting these 1000 organic molecules with a sufficient diversity of candidate enzymes should yield supracritical behavior. As a concrete example, consider use of the standing diversity of types of antibody molecule in a human, about 100,000,000. Then, by our assumptions above, the number of reactions which should have catalysts is $10^6 \times 10^8/10^9 = 10^5$. In other words, about 100,000 of the possible million reactions should find catalysts among the population of antibody molecules. Thus, on the order of 100,000 organic molecules should be formed by catalyzed reactions from the initial diversity of 1000 organic molecules. Most of these 100,000 will be new kinds of molecular species, not present among the initial 1000 substrates. The 100,000 types of organic molecules now afford the square of that diversity, hence 10^{10} possible reactions. Thus, on this next cycle, the expected number of reactions catalyzed will be $10^{10} \times 10^8/10^9 = 10^9$. Thus, the diversity of organic molecules should explode to a billion! Over successive cycles, the diversity will explode still further, until limited by the ever lower concentrations of the myriads of types of molecules formed from the 1000 organic molecules which constitute what I will call a sustained "founder set." This is *supracritical behavior*. In principle, except for the limitation due to decreasing concentrations, and other thermodynamic conditions a nonequilibrium supracritical system driven by adequate supply of founder molecules would increase in diversity indefinitely.[23,43,48]

In contrast, if very few candidate enzymes are reacted with very few candidate substrates, few or no new reactions will be catalyzed. This is subcritical behavior. Suppose we exposed ten organic molecules to a randomly chosen antibody molecule. The expected number of the 100 reactions that will be catalyzed is $100 \times 1/10^9 = 10^{-7}$. Thus, virtually certainly, no reaction will be catalyzed; hence, no new organic molecules are formed.

A phase transition between subcritical and supracritical behavior occurs as either the diversity of organic molecules, hence of reactions is changed, or the diversity of candidate enzymes is changed. Above a critical curve in the corresponding two-dimensional space, substrate diversity on the abcisa, candidate enzyme diversity of the ordinate, behavior is supracritical, below it, behavior is subcritical. As the boundary is approached from the subcritical region, at first no reactions are catalyzed, then a few reactions are catalyzed forming a few new organic molecules, but over cycles the cascade of novel molecules formed dies out. Closer to the boundary, the diversity explodes faster, but remains bounded, not merely by concentration or thermodynamic considerations, but because new reactions find no catalysts.

POSSIBLE EMERGENCE OF CATALYTICALLY SELF-EXTENDING, SUPRACRITICAL CHEMICAL SYSTEMS ON THE PREBIOTIC EARTH

The predicted phase transition between subcritical and supracritical behavior raises the possibility that such catalytically self-extending reaction systems may have been of fundamental importance to the chemical evolution which occurred on the prebiotic earth. Here it is important to stress that small organic molecules, as well as polymers such as RNA or proteins, can catalyze reactions among organic molecules. A phase transition to catalytically self-extending supracritical behavior need not require complex polymers. Furthermore, as pointed out by Oxford chemist Jack Baldwin,[6] cyclic reactions themselves constitute formal catalysts. Thus the reaction sequence $A \rightarrow B \rightarrow C \rightarrow D \rightarrow B$ plus $B \rightarrow E$ yields the $B - C - D$ reaction cycle as a formal catalyst for the conversion for A to E. Based on Baldwin's suggestion, it may be the case that complex networks of linked cyclic and branched organic reaction pathways can yield supracritical behavior and autocatalytic sets by virtue of such formal catalysis.

What might the expected emergence of supracritical behavior and collectively autocatalytic sets portend for the chemical evolution which must have occurred on the prebiotic Earth? Given an initial low-diversity sustained founder set of small organic molecules, *spontaneous* reactions among them must tend to increase molecular diversity, yet be countered by the breakdown of larger molecules driven by thermal energy, high-energy photons, or other sources of disordering energy. As diversity increased in any localized region, gradually a few molecular species or reaction cycles might emerge which would *catalyze* one or a few of the reactions afforded by the organic molecules present, thus increasing the diversity on a *fast* time scale compared to the processes leading to spontaneous breakdown of larger more complex molecules. Clays, for example, can catalyze a variety of reactions and may have been important in such prebiotic chemical evolution. If, in any locale, the rate of build-up of catalyzed diversity were faster than the processes of disordered breakdown, the coupled system of spontaneous and catalyzed reactions and reaction cycles would persistently increase in diversity, thereby catalyzing ever larger numbers of reactions. It seems plausible that eventually the diversity threshold was passed in some locales, leading to a local explosion of diversity and the formation of collectively autocatalytic sets of organic molecules. Protoorganisms would have emerged.

CATALYTICALLY SELF-EXTENDING REACTION SYSTEMS MAY PREDICT THE SIZE DISTRIBUTION OF ORGANIC MOLECULES

An important feature of supracritical reaction systems is that, as the set of catalyzed reactions extends, a characteristic unimodal distribution arises in the diversity of molecules plotted against the number of atoms per molecule. This unimodal distribution also seems to characterize the organic molecules in the biosphere.

The unimodal distribution is based on two countervailing features of reaction graphs among a set of small and large organic molecules: (1) There are always more potential reactions forming small molecules than forming large molecules. (2) There are vastly more possible large molecules. In the models examined, the joint effect of these countervailing effects creates a unimodal distribution with an exponential tail.[23,43,48]

It is easy to see that there are more ways to form a small molecule than a large molecule by considering a space of all peptides up to length 20. Any peptide length 20 can be made by 19 alternative reactions, ligating two smaller peptides which, when joined, form that 20 amino acid sequence. Conversely, a specific peptide with four amino acids can be cleaved from the amino or carboxy terminus of *all* 2×20^{16} peptides having the desired tetramer at one or the other terminus. This combinatorial feature of chemistry is not altered when complex organic molecules are considered. There are more ways to form a small molecule, with rather few atoms, by cleavage from the set of possible larger molecules, than to form the larger molecules. A more formal statement of this property is that, unlike random graphs, reaction graphs are highly nonisotropic.

Consideration of peptide space also makes it obvious that there is an exponentially increasing diversity of polymers as length increases. For organic molecules, including heterocyclic compounds, the numbers of types of molecules as a function of the number of atoms per molecule has not yet even been counted successfully,[72] but explodes extremely rapidly as the atoms per molecule increase.

In Figure 2(a) I show an intermediate stage in the evolution of a model supracritical system based on model peptides. The interaction between the countervailing effects noted above yield a unimodal distribution when diversity of peptides at each length is plotted against peptide length. Replotting on a semi-log scale, Figure 2(b), demonstrates that the right tail falls off exponentially. In Figure 3, I show the estimated distribution of organic molecules in the biosphere as a function of carbon atoms per molecule, estimated by Morowitz,[63] by the simple, undoubtedly biased, but nevertheless informative procedure of counting pages in a chemical index. The striking feature is the unimodal character of the distribution and the rough exponential fall off in the right tail.

The distribution exhibited by Figure 3 is probably roughly correct. It may also be a biological universal. If carbon-based life were to evolve again on the Earth, Tyrannosaurus Rex and Oedipus Rex might not recur, but I find it hard to believe that a similar distribution would fail to recur. I take Morowitz's distribution as tentative evidence that the organic diversity of the biosphere reflects a catalytically self-extending supracritical system to this day. Molecular evolution in contemporary organisms permits novel organic molecules to enter naive cells which must cope with the novel input by evolution of enzymes to carry out detoxification, transformation, or some other reaction. Often these reactions will yield further novel molecules which reach other naive cells and demand the evolution of yet further responses.

These arguments lead to a clear hypothesis. The biosphere is supracritical. Is it? I next present reasons to suppose that free living cells are constrained by the supracritical-subcritical boundary to be just subcritical, while the coevolution of such cells has driven the biosphere to be strongly supracritical.

(a)

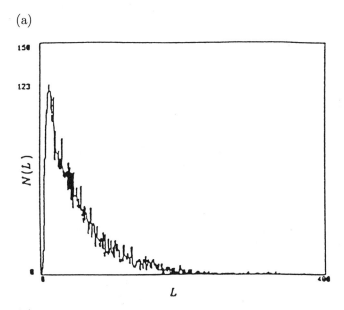

(b)

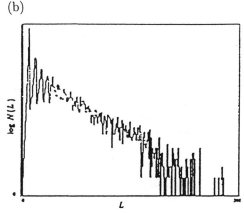

FIGURE 2   (a) Number of kind of organic molecules $N(M)$ as a function of number of atoms per molecule M. (b) Distribution of polymer species as a function of length. Length distribution for a system with only two amino acids used to form the monomers and dimers of the foot set, which causes a cycling or increased abundance at small multiples of the maximum polymer length maintained in the food set.[23]

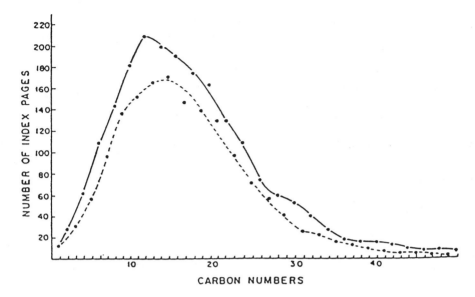

FIGURE 3   Distribution of number of known organic compounds (plotted in number of pages in chemical index volumes) as a function of number of carbon atoms per compound.[63]

INDIVIDUAL AUTOCATALYTIC SUBSYSTEMS SHOULD EVOLVE TO THE SUBCRITICAL-SUPRACRITICAL BOUNDARY

A very general argument suggests the possibility that individual autocatalytic systems, such as those enclosed in a membrane like true cells, should adjust their internal molecular diversity such that each is just subcritical.

Protoorganisms, if formed in supracritical chemical systems, must soon have developed means to protect themselves from the catastrophic disrupting effects of persistent molecular novelty. The simplest means to do so utilizes spatial compartmentation. For example, autocatalytic systems might become compartmentalized inside membrane or micelle boundaries. Novel autocatalytic systems have, in fact, recently been experimentally created consisting of a micelle formed by one or more kinds of amphipathic molecules, enclosing an interior containing other molecular species in an aqueous environment. To this system a third type of molecular species is added. This food molecule crosses into the interior of the micelle, is converted there to one of the kinds of molecules forming the micelle. The micelle grows, and ultimately buds in two.

It is instructive to consider the consequences of tuning the molecular diversity within such a protocell. If the internal diversity is strongly supracritical, the system will almost certainly soon disrupt any useful internal coordination. First, any novel type of molecule which gains entry will unleash a cascade of synthesis forming molecules never before seen by the system. These novel molecules are highly likely to destroy the organization of the protocell which allows it to take in food molecules over its boundary, to grow and bud. For example, many such novel molecules might intercalate into the micelle and totally disperse it in the aqueous medium. More generally, if the internal system is already supracritical then the molecular species within the system would themselves change indefinitely. But many novel molecules might be inhibitors of critical reactions catalyzed by other members of the set, hence lethal to the system. In short, on general grounds, an autocatalytic system which is strongly supracritical is in marked danger of persistently disrupting any useful internal coordination. Supracritical systems might not adapt well because they cannot *stop changing*.

Strongly subcritical systems might also evolve poorly. Imagine a micelle with a single internal molecular species. In the presence of one species of exogenous food molecule, suppose the micelle reproduces well. But were that food source cut off, and a new one supplied, the markedly subcritical micelle would be markedly unlikely to be able to catalyze a reaction involving the new potential food source. Strongly *subcritical* autocatalytic systems cannot evolve well because they cannot *start changing*.

These twin pressures suggest that heritable variation in internal molecular diversity might tend to approach an intermediate internal complexity which is just subcritical. Such systems might maximize metabolic diversity and capacity to adapt, yet remain subcritical hence not be likely to explode with molecular novelty at each new molecular insult.

THE BIOSPHERE IS PROBABLY SUPRACRITICAL, CELLS APPEAR JUST SUBCRITICAL

As noted, these ideas suggest that the biosphere as a whole should be supracritical, but any free living cell, or cell in a multicelled organism, should be somewhat subcritical. Surprisingly, it appears that both claims are likely to be true.

The biosphere is probably strongly supracritical. This claim rests on an estimate of the diversity of proteins, as candidate enzymes, in the biosphere, and the diversity of organic molecules in the biosphere.

A crude estimate of the gene-encoded protein diversity in a human, exclusive of the immune repertoire, is about 100,000. A estimate of the number of species extant now is about 100,000,000. If all were entirely distinct genetically, were genetically homozygous, and were of the diversity of a human, the total protein diversity of the biosphere would be about 10^{13}. Conversely, genes in species overlap strongly, yet species are not homozygous. Guess very conservatively that the total protein diversity of the biosphere is at least 10^8. Since 10,000,000 organic compounds are known, but many were synthesized by chemists, a crude estimate of the number of kinds of organic molecules produced naturally in the biosphere might be about 10^6.

Evolved proteins have evolved to perform specific tasks but are still candidates to catalyze novel reactions. Proteins are large molecules with many epitopes in addition to catalytic sites, binding sites, or other functionally selected features. Thus, any protein has nooks and crannies which might bind substrates and catalyze novel reactions. Based on the case of catalytic antibodies, a crude estimate of the probability that an arbitrarily chosen protein catalyzes an arbitrarily chosen reaction is 10^{-9}.

Given on these estimates, let us perform what I shall call "The Noah Experiment." Two of each species are collected, with due allowance for size differences between elephants and gnats. Equal volume samples of all are placed in a large blender, their proteins extracted, salt precipitated, spun down, lyophylized, then resuspended in measured aliquots. (The Noah Experiment requires proper biochemical details.) The resulting protein mixture is then exposed to the 10^6 kinds of naturally occurring organic molecules, most of which were extracted from the pairs of members of the species assembled, many from tropical regions. The 10^6 organic molecules afford 10^{12} kinds of reactions. The probable consequence is the "Noah Blowup." The expected number of reactions catalyzed on the first cycle is $10^{12} \times 10^8/10^9 = 10^{11}$. The organic molecular diversity should explode from a million to a hundred billion. In short, the biosphere is probably powerfully supracritical as a whole.

Is a human cell supracritical? Estimate the protein diversity at 100,000.[1] Estimate the organic molecule diversity of metabolism at 1000. Our proteins have evolved to *avoid* cross reactions among molecules in metabolism which are not advantageous. However, as noted, each protein has many epitopes beyond its tuned catalytic or binding site. Those epitopes, inevitably, are candidates to bind transition states and catalyze reactions.

Let a novel organic molecule, Q, manage to enter one of my cells. Q can form one of two substrates with each of the 1000 other organic molecules, hence a total of about 1000 new reactions are afforded. The number of these which would be expected to find catalysts is $10^5 \times 10^3/10^9 = 10^{-1}$. The human cell seems to be just subcritical!

Having seen that the human cell seems to be just subcritical, we might notice that any cell ought to be. Suppose one of your cells housed a supracritical diversity. Then if a millimolar concentration of molecules of type Q entered, they would, in general, generate several novel types of molecules, $R, S, T, \ldots$, which in turn would generate still more entirely novel molecules, in a supracritical avalanche. A thousand-fold increase in diversity would leave product molecules at an average micromolar concentration, sufficient to wreak havoc. A million-fold increase in diversity would leave products at an average nanomolar concentrations, still sufficient to cause harm. Some of these would have a strong probability of binding to and disrupting some critical cellular function. Thus, in retrospect, it seems fully plausible that cells can avoid this potential catastrophe by remaining sufficiently subcritical that almost no molecular insult starts such an avalanche of novelty. Alternatively, each cell must erect powerful barriers to prevent entry of novel molecules. At the whole body level, one notices, at least many vertebrates have developed such a defense. It is called the immune system. But even with the immune system, it seems sensible that cells remain subcritical.

If the human cell is just subcritical, then probably all cells are just subcritical. The encoded diversity in humans appears to be as high as any other type of cell. Bacteria have far fewer genes, encoding perhaps 3000 proteins, but probably have a higher metabolic diversity. The important tentative conclusion is this: If all contemporary cells are slightly subcritical, then it begins to appear that 3.45 billion years of evolution from the first simple cells have respected the subcritical-supracritical boundary. Cells may optimize their molecular diversity and metabolic capabilities by achieving a position somewhat below the boundary. This constraint of intracellular molecular diversity by the subcritical-supracritical boundary may be an ahistorical universal in biology.

ECONOMIC INTERACTIONS, ADVANTAGES OF TRADE, AND HIERARCHICAL INTEGRATION IN NONEQUILIBRIUM AUTOCATALYTIC SYSTEMS

An important feature of supracritical chemical systems containing autocatalytic subsystems is that fundamentally *economic* interactions, with advantages of trade, can emerge among the autocatalytic subsystems. Peter Schuster, his colleagues, and I have recently explored a simple model of two template-replicating RNA species, *A* and *B*. Each species can not only template replicate its two complementary strands by familiar Watson-Crick pairing, but each strand of each pair can help the other species replicate. However, *A* helps *B* at a cost to itself, since *A* strands are unavailable to template replicate while helping *B* strands replicate. Similarly, *B* helps *A*, but at a cost to itself. Thus, if each is helped by the other more than its own cost in affording help, *advantages of trade exist.* Under chemostat conditions, two regimes exist. In one regime, either *A* or *B* survives alone. In the second regime, *A* and *B* coexist. Due to the advantages of trade, in the coexistence regime the total concentration of *A* and *B* strands can be higher than if either survives alone. An optimal *A* to *B* ratio exists at which both sequences multiply optimally. At that ratio, monomer resources used to build the *A* and *B* strands are used up most rapidly.[75]

Physical analogs to utility and to price exist in this molecular system of two autocatalytic molecular species which can act as mutualists. The rate of replication corresponds to the economist's concept of an agent's utility. Agents act to optimize utility, molecular evolution acts to optimize the rate of replication of the molecular autocatalysts. The optimal ratio of the two mutualist species, such that both sequences help one another reproduce optimally, corresponds roughly to price. The parallel is clear if one considers two economic agents each endowed with two goods, but with different preferences for the two goods. The optimal exchange ratio of the two goods to maximize the utility of each agent, forms a price of one good in terms of the other. Since both A and B reproduce faster as mutualists, a selective advantage would be expected to favor the evolution of such mutualisms and of an optimum exchange ratio of their help for one another, or price.

Mutualism, as well as competition, can be expected to arise among autocatalytic subsystems of supracritical chemical systems. This should become true, if only because each individual autocatalytic system must presumably remain slightly subcritical as the biosphere is driven to become ever more supracritical. Molecular metabolic complementarities will inevitably arise among the subcritical free living cells. In turn, coexistence of mutualists affords a selective pressure towards symbiosis. Symbiosis assures attaining the advantages of trade in a sustained reliable trading partnership. Thus, *molecular complementarities and the emergence of advantages of trade help drive integration into hierarchically organized structures.* Integration of mitochondria and chloroplasts into the forming eukaryotic cell are obvious examples. Thus, molecular task differentiation and the emergence of advantages of trade, must be among the most primal sources driving the generation of hierarchical integrated complexity.

The physical grounding of economic interactions, due to the emergence of advantages of trade, suggests that Adam Smith's invisible hand may apply to the open nonequilibrium chemical evolution which occurred on the prebiotic and protobiotic earth. Self-constructing, autocatalytic, nonequilibrium molecular processes may, as if by an invisible hand, evolve mutualisms as well as competition in such a way so as to optimize the *proliferation* of those coupled processes. If Carnot is whispering anything, he may be whispering that general principles governing the evolution of open, self-constructing, nonequilibrium systems shall require us to understand the emergent economic behavior of coupled autocatalytic nonequilibrium processes. I will build towards the possible union of Adam Smith and Charles Darwin in the ensuing sections, and return to it in the final section.

IV. ADAPTATION TO A PREFERENCE POSITION ON THE ORDER-CHAOS AXIS

A candidate principle, to be held as a working hypotheses at this state, suggests that complex parallel-processing systems may evolve to a preferred position on an axis from order to chaos. The preferred position may lie in the ordered regime near the transition to chaos. This principle may bear on the organization, behavior, and evolution of parallel-processing systems, from autocatalytic molecular systems, to the hundred thousand genes comprising the parallel-processing genomic regulatory system harbored in the nucleus of each human cell. Evidence has accumulated over the past two decades that large classes of parallel-processing networks behave in three broad regimes: ordered, complex, and chaotic. The unexpected discovery of an ordered "antichaotic" regime in vastly disordered systems suggests that new principles of spontaneous self-organization may underlie much of evolution. Order is not the result of selection alone. Perhaps most intriguing is the complex regime located at the phase transition between order and chaos. It is an interesting speculation that systems in the complex regime can carry out and coordinate the most complex behavior, can adapt most readily, can build the most useful models of their environments. The bold working hypothesis is that CAS adapt to and on the edge of chaos. The more general question is whether CAS evolve to a preferred position on the order-chaos axis, perhaps within the ordered regime near the edge of chaos.

BOOLEAN NETWORKS

Among the simplest general mathematical models of such networks are Boolean networks.[11,12,39,40,41,42,48] In the simplest case, one considers a fixed set of binary variables, N. Each binary variable receives inputs from some subset of the N variables. Each binary variable is supplied a logical, or Boolean function, specifying its activity at the next discrete moment as a function of the 2^K combinations of activities of its K inputs. Once this is done, the "wiring diagram" structure of the system is specified, and its logic is specified.

In the simplest case, such Boolean networks are synchronous finite state machines. A state of the entire network consists in the current value of all the N binary variables; hence, there are 2^N states. All binary variables are governed by a central clock. At the same synchronized moment, each variable examines the activity values of its K inputs, looks up the proper response given by its Boolean function, and assumes the proper next activity, 1 or 0. Thus, at each time moment, the system undergoes a transition from a state to a state. Over time, the system passes along some trajectory in state space. Since the number of states is finite, eventually the system reenters a state previously encountered. Thereafter, since the system is deterministic, the system cycles around a recurrent set of states called a state cycle. Any network has at least one such state cycle, but may have many. The number of states on a state cycle can range from one, a steady state which transforms to itself, to all the states of the system. Every state lies on a trajectory which is either on a single state cycle, or flows to a single state cycles. The collection of trajectories flowing to one state cycle constitutes the basin of attraction of that state cycle. Thus, the basins of attraction partition the state space, and asymptotically, in the absence of noise, the system will arrive at and cycle around one state cycle attractor.

The simplest class of Boolean networks are not only synchronous, but autonomous. Networks receive no inputs from the environment. Obvious directions to relax these idealizations allow the binary elements to respond nonsynchronously, allow response to be error prone with a given probability of the "incorrect" response, and allow exogenous input. The response of a network to exogenous inputs is central to thinking about how such a network of processes might build an *internal model of its external world.* I will return to this below. First, however, intuitive understanding of these relaxations are aided by understanding the simplest synchronous autonomous case.

THE ORDERED, CHAOTIC, AND COMPLEX REGIMES IN RANDOM BOOLEAN NETWORKS

Autonomous Boolean networks generically lie in an ordered, a complex, or a chaotic regime. These results rest on creating well-specified ensembles of Boolean networks, sampling at random within such ensembles, and demonstrating or proving that most members of the ensemble exhibit the properties in question. Like spin-glasses, analysis here is a new kind of statistical mechanics over ensembles of systems. In general, ergodicity is broken. Two networks sampled at random from the same ensemble behave differently in detail. However, those features which are robustly present in the ensemble are the new observables we seek.

A simple way to study the ordered, complex, and chaotic regimes considers a finite two-dimensional lattice of binary sites, each coupled to its four neighbors.[24,25,80,81,89] To avoid edges, the lattice can be formed to a torus by joining left and right edges and top and bottom edges. Each binary variable is governed by a Boolean function on its four neighbors. Such a function specifies the response of the element for each of the $2^4 = 16$ combinations of activities of its four inputs.

At each clocked moment, the lattice passes from a state to a state. Color green those elements which have *changed activities* from on to off, or off to on. Thus green denotes those elements which are changing. Color red those elements which remain in the same activity over the state transition, on staying on, or off staying off. In the *ordered regime*, a movie of the lattice started from a random initial state begins with most elements colored green. Gradually an increasing fraction of the elements become red. Eventually, a connected red *"frozen sea"* of elements, each frozen active or frozen inactive, spans, or *percolates*, across the lattice,[14,24,25,40,41,42,48,89] leaving behind isolated green "islands" whose elements twinkle on and off. In contrast, in the chaotic regime, a green twinkling unfrozen sea percolates across the lattice, leaving behind one or more isolated red frozen islands. At the phase transition, in the complex regime between order and chaos, the green unfrozen sea is just breaking into unfrozen islands.

Two hallmarks characterize the chaotic regime: (1) Sensitivity to initial conditions, and (2) exponentially long state cycle attractors. Sensitivity to initial conditions can be studied in simulations by creating two identical Boolean lattices, and starting the two copies in initial states that differ in the activity of a single binary variable. Color purple the subsequent differences in the activity values of elements in the perturbed and unperturbed copy of the lattice. In the chaotic regime, a purple "stain" of *damage* typically propagates throughout much of the unfrozen green sea. This damage is sensitivity to initial conditions. In the chaotic regime, generically, damage spreads to a finite fraction of the sites bounded away from zero as the size of the lattice increases to infinity, and propagates with a fixed mean velocity.[80,81]

In the chaotic regime, the expected lengths of state cycle attractors increase exponentially as the number of sites, N, increases. The most careful analytic and numerical work with respect to these properties have not utilized Boolean lattices but have utilized Boolean networks which are fully random with respect both to the "wiring" among the binary elements as well as the Boolean function governing each element.[14,39,41,42,48] In the limiting case of networks in which each binary variable receives an input from all N binary variables and is assigned at random one of the possible Boolean functions on N inputs, the expected median state cycle length scales as the square root of the number of states. Thus, a small system with a mere 200 binary variables has state cycles on the order of $2^{100} = 10^{30}$ states in length. At a microsecond per state transition, such a system would require billions of times the history of the universe to traverse its attractor! Analytic mean field results for random Boolean networks with N binary variables and K inputs per variable show, for $K > 4$, that the expected state cycle lengths increase exponentially in N. The rate of exponential increase is an increasing function of K, reaching $0.5N$ for $K = N$.[43] Simulation results confirm that cycle lengths increase exponentially in N for $K > 4$ at rates closely approximated by the mean field theory.[30,48] Thus, both analytic and numerical results demonstrate that random Boolean networks with $K > 4$ lie in the chaotic regime.

Several hallmarks characterize the ordered regime: (1) networks lack sensitivity to initial conditions; (2) networks have short, stable state cycles; (3) in random Boolean networks, the number of alternative state cycle attractors is small; and (4) in random Boolean networks perturbations to each attractor can cause it to flow to only a few other attractors.

Perhaps the most stunning feature of random Boolean networks is the unexpected emergence of an ordered regime in networks of vast and apparent mind-boggling disorder. Boolean networks with $N = 100,000$ binary variables, rivaling the human genome in complexity, lie in the ordered regime if constructed subject to the simple constraint that each binary variable have $K = 2$ inputs chosen at random among the N.[2,11,24,25,29,39,40,41,42,48,77,87] Since the two inputs to each element, and the Boolean function governing its activity, are chosen at random, such a network of 100,000 elements is a scramble of "wires" and logic. Yet it exhibits very powerful "antichaotic" order.

Networks in the ordered regime do not show sensitivity to initial conditions. The most detailed work has been carried out with random Boolean networks rather than lattices. In the ordered regime, the "red frozen component" percolates across the network, leaving isolated "green twinkling unfrozen" islands. Each such island is functionally isolated from the others since no signal can propagate through the frozen component. Thus, in the ordered regime, binary elements which are structurally wired to one another directly or indirectly, can be functionally isolated from one another. Damage due to perturbations, unleashing purple stains spreading from the perturbed element, propagate only within the unfrozen island where the perturbation occurred, but do not propagate across the frozen sea. Thus, damage does not spread to a finite fraction of the binary elements as the number of elements, N, goes to infinity. In short, networks in the ordered regime are not highly sensitive to small alterations in initial conditions.

The lengths of state cycles are short in the ordered regime, and the state cycles are stable to perturbations. In $K = 2$ random Boolean networks, the lengths of state cycles scale as roughly the square root of the number of variables, $N^{1/2}$. Thus, a network with 100,000 binary variables, and a state space of $2^{100,000} = 10^{30,000}$ cycles among a mere 317 states! Here is truly a profound exemplar of spontaneous order. I sketch some biological implications below. Because networks in the ordered regime are not overly sensitive to small perturbations, it is not surprising that attractors are stable to most minimal perturbations. Thus attractors in the ordered regime exhibit "homeostasis." By contrast, attractors in the chaotic regime are unstable to most perturbations. Homeostasis is a near inevitable consequence of behavior in the ordered regime.

Each attractor is "near" only a few other attractors in the ordered regime. Here I mean that if a system on an attractor is minimally perturbed, it will either return to the attractor from which it was perturbed or enter a new basin of attraction and flow to a new attractor. Random Boolean networks in the ordered regime have the property that only a few basins of attraction have states which lie a small Hamming distance from any other attractor. Here, the Hamming distance between two states is just the number of elements in different activity values in the two states.

Work has begun on asynchronous, and on probabilistic Boolean networks.[24,25,82] In the ordered regime, once the "red" frozen component has fallen to its fixed state of permanently active or inactive elements, then it is intuitively clear that asynchronous behavior will not alter the major features of the system. Each of the frozen elements is typically held fixed in its active or inactive state by the each of the activities of a number of its inputs. Thus, the time order in which those inputs reconfirm their constant activity is immaterial. However, when the sequence of element updating is random, the unfrozen islands may not progress through simple state cycles, but can follow more complex trajectories. The behavior of networks deep in the ordered regime to a low level of "error" in the activity of any element is also intuitively clear. The percolating frozen component is typically immune to occasional misbehavior of elements in the frozen component, precisely because damage does not propagate. Such errors do propagate in the unfrozen islands, however, and will yield more complex trajectories. If the error rate, or probability of misbehavior, is high enough, even the frozen component will "melt." The system will behave in a chaotic fashion.

TUNING FROM THE CHAOTIC TO THE ORDERED REGIME VIA THE COMPLEX REGIME

Random Boolean networks can be tuned from the chaotic to the ordered regime by changing a number of construction parameters. The first to be investigated was K, the number of inputs per element.[39,40,41,42,48,87] Numerical evidence showed a chaotic regime for large values of K and the ordered regime for $K = 2$, or less. Analytic characterization of a percolating frozen component in such networks was demonstrated by showing that such systems cross a percolation threshold above which frozen structures form. The corresponding frozen structures are called forcing structures.[40,41,42]

A second analytic approach has been taken by B. Derrida and colleagues. This work is based on the annealed approximation, which imagines a succession of Boolean networks in which the binary elements retain their identity, but at each moment all the connections and Boolean functions in the network are randomized. Derrida and coworkers demonstrated a phase transition in the behavior of the annealed model as the number of inputs per variable, K, decreases to 2 or lower. The analysis considers whether arbitrary pairs of states, S_1 and S_2 lie on trajectories which *converge or diverge* in state space. Distance between two states is measured as the normalized Hamming distance. Here the Hamming distance is divided by the number of elements, N. In the ordered regime nearby states converge. In the chaotic regime nearby states diverge, an expression of sensitivity to initial conditions.

In Figure 4 I map the current normalized Hamming distance between two states on the abcissa to the distance between their successor states, graphed on the ordinate. The main diagonal corresponds to cases where the distance between initial states is the same as the distance between successor states. Flow neither converges nor diverges. In the chaotic regime, nearby states diverge; thus, the successor states are further apart, and the map lies above the main diagonal. Even in the chaotic regime very distant states tend to converge. Thus, the curve corresponding to chaotic behavior lies above the main diagonal for small initial distances and crosses to lie below the main diagonal for large initial distances. In contrast, in the ordered regime, both nearby and distant pairs of states flow on converging trajectories. The result is that the map lies below the main diagonal for all initial distances.

Derrida and Pomeau[11] demonstrated in the annealed model that for $K > 2$ the map lies in the chaotic regime. Nearby states diverge. But for $K = 2$, the curve is everywhere below the main diagonal. $K = 2$ networks lie in the ordered regime. At present, this is the best analytic demonstration of the phase transition from the chaotic to the ordered regime.

Derrida and Weisbuch,[13] and Derrida and Stauffer,[12] then introduced and explored another parameter called P. A Boolean function of K inputs specifies the response to each of the two combinations of activities of those inputs. The response

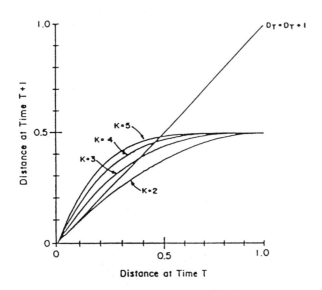

FIGURE 4   Recurrence relation showing the expected distance $D_t + 1$ between two states at time $T + 1$ after each is acted upon by the network at time T. Distance is normalized to the fraction of elements in different activity values in the two states being compared. That is, Hamming distance at time T is $H_t = N[1 - a_{12}(T)].D_t = H_t/N$. For $K = 2$, the recurrence curve is below the 4500 line, and hence distance between arbitrary initial states decreases toward zero over iterations. For $K > 2$, states which are initially very dose diverge to an asymptotic distance given by the the crossing of the corresponding K curve at the 45∞ line. Thus $K > 2$ networks exhibit sensitivity to initial conditions and chaos, not order. Based on annealed approximation.[11]

might be 1 for most of those combinations, might be 0 for most of those combinations, or might be equally divided between 1 and 0. P measures the deviation above 0.5 for the dominating response, 1 or 0. Thus, if .8 of the 2 input combinations yielded a 0, P would be .8. Derrida and his colleagues showed that if P is above a critical value, P_C, then the Boolean system is in the ordered regime. The critical value of P depends upon the structure of the network and the number of inputs per element. On square lattices, P_C is about 0.72.

Stauffer[80,81] has investigated the propagation of damage in the chaotic regime, where damage spreads at a finite velocity throughout a finite fraction of the unfrozen sea. At the phase transition, it would be expected, but remains to be shown, that Boolean networks would exhibit no natural length scale. Damage spreading from minor perturbations creates avalanches on all length scales, with a power-law distribution. Thus, at the phase transition, Boolean networks can exhibit very long transients following minor perturbations, and distant regions of the network can be in communication without perturbations triggering unending chaotic behavior.

Langton has independently explored the same phase transition in cellular automata rules, tuning a parameter, λ, which is essentially the same as the P parameter.[55,57] For cellular automata rules in which elements may have binary, or M-ary finite alternative values, lambda measures the frequency of response with a specific value, say 0, among the M. Exploring the space of cellular automata rules as λ is varied, Langton finds the ordered regime when the frequency of a 0 response is above a critical value. He finds the chaotic regime when the frequency of a 0 response becomes sufficiently less prominent Between the ordered and the chaotic regime, Langton finds a range of λ values where complex behavior is found. One measure of this complexity rests on use of a measure of mutual information, defined as the sum of the entropy of two sites minus their joint entropy. Mutual information is low if both sites are frozen, or if the flickering behavior of one site is uncorrelated with that of the second site. Thus mutual information is low deep in the ordered regime and deep in the chaotic regime. Langton's analysis showed that mutual information is highest in the complex regime located at the edge of chaos between the ordered and chaotic regime.

THE ORDER-CHAOS AXIS, CATEGORIZATION, AND INTERNAL MODELS OF A WORLD

A working hypothesis, indeed one of the candidate principles above, states that complex interacting adaptive systems evolve to a preferred position on the order-chaos axis, in the ordered regime near the edge of chaos. The hypothesis seems plausible on the face of it. Between the chaotic and the ordered regime lies the complex regime where the unfrozen percolating sea is just breaking into unfrozen islands. These islands are in fluctuating, tenuous functional contact with one another. It is intuitively plausible that the most complex coordinated behavior should be attainable in this complex regime. Consider systems deep in the ordered regime, and imagine requiring the system to coordinate an action requiring a series of outputs from elements in different isolated unfrozen islands. Since those islands are functionally isolated from one another, no behavior within the system can carry out such coordination. Conversely, consider achieving the same coordination in a system deep in the chaotic regime. Minor perturbations at distant loci will unleash avalanches of damage which propagate and discoordinate the activities required. Just in the ordered regime in the vicinity of the phase transition, it seems reasonable to suppose, the most complex coordinated behavior can arise. Further, as Langton in particular has emphasized, the long transients which arise in the complex regime, and the maximization of mutual information are strong hints that parallel-processing systems at the edge of chaos should be able to carry out the most complex computations.

Categorization is fundamental to the capacity of adaptive systems to respond to their world. Thus, it is natural to wonder how the properties of parallel-processing systems bear on the capacity of systems harboring them to categorize. Again, general considerations suggest that the edge of chaos may optimize the capacity for flexible, fine-grained categorization.

Ready insight into the relation between network dynamics and categorization can be gained by noting the near identity between "connectionist" parallel-processing Boolean networks and the somewhat more familiar connectionist neural network models.[74] Within the connectionist neural network paradigm, categorization is intimately related to basins of attraction and attractors. Indeed, the attractor constitutes the "class," and the basin constitutes the set of initial states, hence "observations" which are classed identically. For example, in one natural interpretation, one imagines an artificial retina which projects an input pattern onto the network as an initial state. The network flows to an attractor from that state. In the familiar Hopfield networks,[37] symmetrical synaptic weights are chosen and the point attractors are considered memories. Each memory is content-addressable by release of the system from any state within the basin of attraction of that memory.

The ordered, chaotic, and complex regimes carry obvious implications for network categorization. In the chaotic regime, nearby states—thus states that are very similar to one another—generically diverge, and often flow to distinct attractors. Since natural classification generically classes sets of similar things as "the same," it is obvious that networks in the chaotic regime cannot classify naturally. Furthermore, categories must be learned by the adapting system during its lifetime, or must evolve over evolutionary time. Thus, in true neural networks, learning is presumed to occur by alteration in synaptic weights. But in the chaotic regime small alterations in synaptic weights cause very marked alterations in the dynamical behavior of the chaotic system. Thus, chaotic systems cannot easily adapt by small incremental changes in structure. In the language of fitness landscapes described in the next section, chaotic systems evolve on very rugged, multipeaked fitness landscapes and rapidly become trapped on poor local optima.

These considerations suggest that natural classification, natural both in the sense that nearby states tend to be classed the same, and that useful categories can be learned by local walks in the parameter space of the system, are readily available in networks lying in the ordered regime. A further rather general argument suggests that networks in the ordered regime near the edge of chaos may be best able to categorize. If attractors drain basins of attraction, and the latter are the generalization class of states to be cocategorized, then a system able to interact subtly with its environment would seem to require trajectories which, on average, neither diverged nor converged too strongly. Useful categorization requires that useful clusters of initial states flow to the same attractor, but that the system be able to evolve to *distinguish* formerly cocategorized objects, hence now to send nearby initial states to different attractors. If convergence is too powerful, then it would seem that over-generalization would be a danger. Sculpting attractors and basins with finesse would seem easiest in the ordered regime near the phase transition between order and chaos, where trajectories tend to converge slightly on the average.

Organisms, to survive, must have always constructed more or less useful internal models of their environment. As a gedanken example, let us return, for a moment, to my hoped for collective autocatalytic sets. Once such an autocatalytic set exists and reproduces in a chemical environment, it is clear that some molecular inputs can be food, others can be toxins which poison catalysis and might be lethal to the system. Darwin supplied us with the criterion to discriminate what is of utility and what is not for a reproducing system. The "meaning" to a self-reproducing system of environmental variables is the implication for survival: food or toxin, friend or foe, for I live or die. To evolve to cope with such molecular perturbations, the system must come to have means to take in and utilize food molecules, and to ward off or transform toxic molecules. Either activity will typically require at least some molecules to touch, bind, and perhaps catalyze reactions. More generally, as in Holland's Classifier system,[36] a network of molecular activities ramifying throughout the parallel-processing autocatalytic system will occur. Those coping responses constitute the *categorization* by the autocatalytic set of some molecular inputs as food, others as toxin. The same coping response constitutes part of its model of its world. Useful categorization, both in terms of classification of nearby molecular input states, and the capacity to evolve such categories, would again seem to require that the parallel-processing autocatalytic molecular system behave in the ordered regime, perhaps near the edge of chaos. And it is worth noting that almost any such nonlinear dynamical system will have multiple basins of attraction in any fixed environment. The multiple basins imply alternative classifications of the same environment, and therefore the possibility of alternative responses to the same environment.

SELECTIVE EVOLUTION TO A PREFERRED POSITION ON THE ORDER-CHAOS AXIS

If the ordered regime near the phase transition between order and chaos, the complex regime, is optimal for the coordination of complex tasks, for categorization, and practical action, then it becomes urgent to ask whether natural selection causes CAS to evolve to the complex regime.

Packard was the first to pose this question, and obtain evidence in favor of it.[68] Packard tested the effects of selection on a population of cellular automata asked to perform a specific calculation. In particular, automata were asked to classify initial states with more than half the sites having 1 values by flowing to a terminal state of uniform 1 values, and classify those initial states with half or less having 0 values by flowing to a terminal state of uniform 0 values. Packard implemented a population genetic algorithm in which automata were assigned a fitness according to how well each performed the task, offspring were produced proportional to fitness, those offspring were subjected to mutation, and this cycle of processes was iterated for many model generations. Packard found that his automata flowed to a region of cellular automata rule space which corresponded to the phase transition regime. More recently, however, Mitchell et al.[61] have found evidence contradicting Packard's[68] results. Thus, the question is entirely open.

The hypothesis that natural selection will cause CAS to evolve to the vicinity of the edge of chaos requires a clear understanding of the *complexity of the task* the system is to accomplish. One natural framework in which to explore such issues is game theory. Here each agent has a set of possible actions, or strategies, agents "play" one another, and the payoff to each agent depends upon its own actions and those of the other player(s). For example, in a two-person game, a payoff matrix specifies the payoff to each player for each pair of actions of the two players.

In order to investigate whether Boolean networks challenged to carry out a complex task will evolve to the complex regime, my colleagues Sonke Johnsen, Emily Dickenson, and I have recently studied the evolution of Boolean networks playing the "mismatch" game. Here each network has a set of "input" sites and a set of "output" sites. The output sites of one network serve as temporary inputs to the input sites of the second network. In the mismatch game, the aim of each network is to have its own output sites be exactly the opposite of the output sites of the network it is playing. Thus, for six input-output sites, if network A has its output sites in state (000000), then network B will achieve a maximum score if a set time later its own output state is (111111). Payoff is proportional to mismatch. Were only two networks to play one another, a trivial solution to the game would have each fall to a fixed state, with A the Hamming opposite of B. If three or more networks play one another, this trivial solution is a poor one.

Evolution of networks playing the mismatch game occurs as follows: At each generation, each network plays the other networks in the population, and achieves a mean normalized score between 0.0 and 1.0 which corresponds to its fitness. Populations of networks evolve by mutation and selection. At each generation, each network produces, at random, a one-mutant variant network, obtained by mutating a single bit in a Boolean function, or by changing one input or output connection. Thus, networks can mutate logic and wiring diagrams among a fixed set of N binary variables. The wild type and its mutant network both play the remaining wild type networks. If the mutant is fitter than the wild type, it replaces the wild type version of that network in the population. In typical experiments, the number of binary variables, N was fixed at 20 or 30. Typically, several independent sets of three networks played the mismatch game with one another.

The central questions under investigation are two:

1. Can Boolean networks evolve by mutation and selection such that they play the mismatch game better than do unevolved networks?
2. If so, do Boolean networks approach a preferred position on the order-chaos axis near the phase transition between order and chaos? To test this, populations of Boolean networks with $K = 1$, $K = 2$, $K = 3$, and $K = 4$, were explored. $K = 1$ networks lie in the ordered regime. $K = 3$ and $K = 4$ networks lie in the chaotic regime. $K = 2$ networks lie near the phase transition. Thus, we are able

to test whether networks improve under mutation and selection, and whether they approach the phase transition.

The results, with affirmative answers to the two central questions, are shown in Figures 5(a)–(d). Networks improve at playing the mismatch game, and approach the phase transition between order and chaos during this evolution. In order to assess, for any network, where it behaves on the order-chaos axis, we sampled mean convergence or divergence along trajectories in the state space of the system, for different initial normalized Hamming distances between the two initial states. Thus, we constructed the homologue to the Derrida mapping shown in Figure 4.

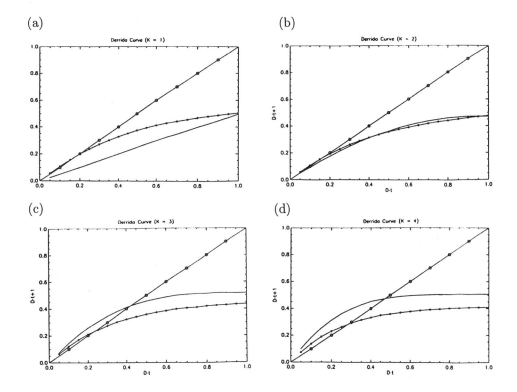

FIGURE 5   (a) Derrida curves for adapting population of Boolean networks, initially with $K = 1$, hence in the ordered regime, playing the mismatch game. Solid line shows mean Derrida curve in the adapting population at the onset of selection. Line marked with + shows the Derrida curve after 100,000 generations of adaptive evolution. Fitness, unshown, increased over the period of adaptation. (b) As in 5(a), except $K = 2$ in the initial population. (c) As in 5(a), except $K = 3$ in the initial population. (d) As in 5(a), except $K = 4$ in the initial population.

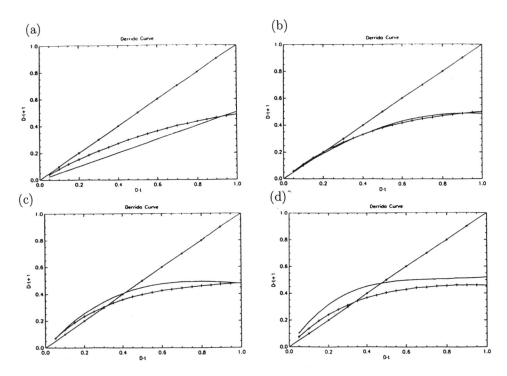

FIGURE 6   (a) Derrida curve for a population of Boolean networks, initially with $K = 1$, hence in the ordered regime, playing a random map game. Solid line is the initial Derrida curve at the onset of selection. Line marked with $+$ shows Derrida curve after 100,000 generations of adaptive evolution. Fitness, unshown, increased over the period of adaptation. (b) As in 6(a), except $K = 2$ in the initial population. (c) As in 6(a), except $K = 3$ in the initial population. (d) As in 6(a), except $K = 4$ in the initial population.

To do so, 1000 pairs of initial states at defined Hamming distances were sampled at random, their successor states computed with the network, and the Hamming distance between the successors assessed. Figures 5(a)–(d) show the mean "Derrida curves" for populations networks in the ordered and in the chaotic regimes at the outset of their evolutionary exploration, and also shows, for each such population, the mean Derrida curve after 100,000 generations of evolution. $K = 1$ networks begin in the ordered regime and approach the phase transition main diagonal. $K = 2$ networks begin and remain near the phase transition. $K = 3$ networks and $K = 4$ networks begin well into the chaotic regime and clearly approach the phase transition.

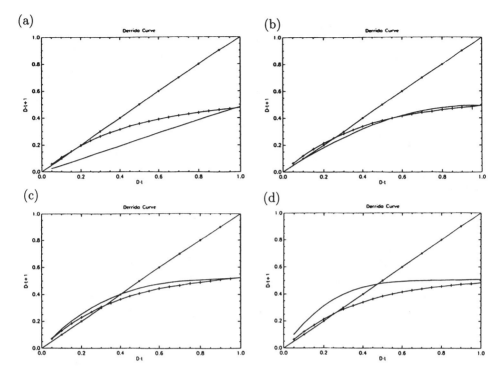

FIGURE 7   (a) As in 6(a), except a different random map game is being played. (b) As in 6(b), except a different random map game is being played. (c) As in 6(c), except a different random map game is being played. (d) As in 6(d), except a different random map game is being played.

In addition to the mismatch game, we have tested whether Boolean networks playing a "random map" game also evolve towards the edge of chaos. The "random map" game assigns as a "correct" output response of a network's six output variables for each of the $2 = 64$ states of the six input variables a randomly chosen six variable binary state. Thus, for each of the 64 input states a randomly chosen, but then fixed, output state among the possible output states 64 is chosen. Fitness is proportional to the Hamming distance from the correct response. Both players in a random map game attempt to "learn" the same correct random map. In Figures 6(a)–(d) and 7(a)–(d) I show that for two different random maps, Boolean networks from the ordered or from the chaotic regime, selected successfully to learn the random map, all evolve toward the edge of chaos.

In one set of control numerical experiments, we explored the evolution of Boolean networks with initial $K = 1$, or $K = 4$, playing a "game" where the payoff was 0.5, independent of the action of each network. Since fitter variants cannot arise in this control, we introduced random one-mutant variants with a low probability per generation. We also explored a second control where the "random map" was randomized after each generation. In neither case do networks evolve to the phase transition between order and chaos, Figures 8(a),(b) and 9(a),(b).

These results clearly show that this evolutionary process can, in fact, shift Boolean networks evolving to play the mismatch game and random map games from deep in the ordered regime and from the chaotic regime towards the phase transition, complex regime. The results do not, however, establish that the immediate vicinity of the phase transition is always optimal or always attained. In fact, the phase transition regime, even if optimal, is *not* always attained. Selection toward favorable networks in the ordered regime near the phase transition is counteracted by random drift in the ensemble of systems being explored. If the average, or generic, properties of that ensemble are too far from those sought by selection, the balance between drift and selection forces can leave the adapting population displaced from the optimal networks.[48] For example, in the numerical studies described above, a lower and an upper bound on K, the number of inputs to a binary variable, was set at 1 and 5. Similar results are found if the upper bound is increased to $K = 10$, but drift of the adapting population toward the increased mean value of K in the allowed range, 5.5 rather than 3.0, becomes a more powerful force resisting selective forces toward the phase transition. Networks playing the mismatch game evolve from the chaotic regime fairly far toward the phase transition. Those in the ordered regime evolve through the phase transition and slightly into the chaotic regime. As they pass the into the chaotic regime, fitness falls. Thus, selective forces are unable to overcome drift. All populations appear to end at the same slightly chaotic configuration. Drift is an even more powerful force counteracting selection among networks playing the random map game. When the maximal value of K in the ensemble is increased to 10, networks in the chaotic regime are shifted only slightly towards the phase transition. Networks initially in the ordered regime drift well past the phase transition, their fitness falls as they enter the chaotic regime, and the balance between selection and drift is struck modestly into the chaotic regime. Thus, for both the mismatch game and random map game, increasing K, hence the ensemble mean value of K, drives the adapting population to a new balance between selection and drift lying in the chaotic regime.

The results give qualitative support to the general hypothesis that selection operating on CAS to solve complex tasks pulls such systems toward a preferred position on the order-chaos axis near the edge of chaos. This pull can be countered by drift forces toward the average properties of the ensemble explored by the adaptive process. Further work is underway to establish the range of games, from simple to complex, for which networks in the phase transition regime on other positions on the order-chaos axis are optimal, and the conditions under which the optimal position is actually the attractor of an adaptive evolutionary dynamics balancing selection and drift. If, for demonstrably complex tasks, the ordered regime near the edge of chaos is typically both the target and the attractor of natural selection, we will have good grounds to support the bold hypothesis.

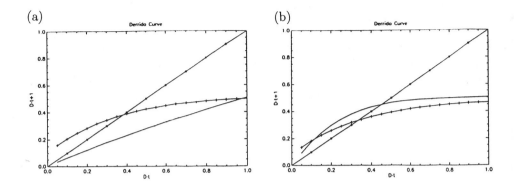

FIGURE 8   (a) Derrida curve for a control adapting population of Boolean networks, initially with $K = 1$, in which fitness is fixed at 0.5 regardless of network action. Solid line is initial Derrida cwe at the onset of selection. + marked line shows Derrida curve after 100,000 generations of adaptive evolution. Fitness, unshown, did not increase over the period of adaptation. (b) As in 8(a), except that $K = 4$ in the initial population.

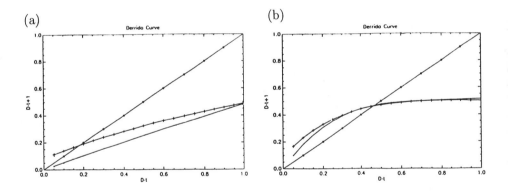

FIGURE 9   (a) Derrida curve for a control adapting population of Boolean networks, initially with $K = 1$, in which the random map is randomized at each generation. Solid line is initial Derrida curve at the onset of selection. + marked line shows Derrida curve after 100,000 generations of adaptive evolution. Fitness, unshown, did not increase over the period of adaptation. (b) As in 9(a), except that $K = 4$ in the initial population.

GENOMIC REGULATORY NETWORKS APPEAR TO BE IN THE ORDERED REGIME, PERHAPS NEAR THE EDGE OF CHAOS

One further body of evidence can be adduced to support the hypothesis that CAS may evolve to the ordered regime near edge of chaos. The data discussed next strongly suggest that parallel-processing genomic regulatory systems, in which thousands of genes and their product regulate one another's activities in the control of ontogeny, lie in the ordered regime, perhaps near the phase transition. If this is true across wide ranges of phyla, as I believe it is, then a billion years of evolution or more will have conformed to our candidate principle. More, we shall have the outline of a theory of the integrated behavior and evolution of the genome.

Random Boolean networks were initially invented as models of the parallel-processing genomic regulatory systems underlying cell differentiation and ontogeny. Examination of known regulated genes shows that most genes are directly regulated by few other genes, and that the Boolean functions utilized are strongly biased towards a class, called canalyzing functions, which ensures the emergence of the ordered regime. Thus, based on current genetic data and mathematical results, it is a plausible hypothesis that genomic systems lie in the ordered regime. If so, then the typical properties of such systems should characterize the behavior of genomic regulatory systems. To a remarkable extend, the parallels are found.[39,40,41,42,48]

The central interpretation that I shall make identifies a *cell type as an attractor*. According to this mapping, the human genome, with 100,000 genes, is capable of a hyperastronomical diversity of patterns of gene activity, but the parallel-processing molecular network flows along trajectories which end up asymptotically on attractors. The repertoire of alternative recurrent attractors then comprise the repertoire of alternative cell types of the organism. Given this mapping, then recurrent state cycles correspond to periodic or quasi-periodic behaviors of cells.

1. Small attractors correspond to expected cell cycle times. A genome with 100,000 genes might have state cycles with 317 states. Activating a gene requires from one to ten minutes. Cycles should require on the order of 317 to 3170 minutes to traverse. Cell cycle times lie in this range, and do, in fact, scale as a about a square root function of the estimated number of genes in a cell.[39,40,41,48]

2. The number of alternative attractors as a function of the number of binary elements, a rough square root function, predicts the number of cell types as a function of the number of genes in an organism. Humans, with about 100,000 genes should have 317 cell types,[39,48] and appear to have 254.[1]

3. Real cells exhibit homeostasis after most molecular perturbations. So too do attractors in the ordered regime.

4. The ordered regime is characterized by a frozen component of binary variables which percolates across the system; each variable is in the same state of activity on all the alternative attractors of the system. Those alternative attractors are due to the alternative behaviors of the functionally isolated unfrozen islands. Similarly, in all higher organisms studied a common core of genes, up to 70%

or more are transcribed to RNA and translated to protein in common in all cell types of the organism.

5. Human cell types differ because different subsets of the 100,000 genes are expressed in each cell type. Typically, cells differ in the activity of a small fraction of the genes. The predicted differences in model gene activity patterns between different attractors comes close to predicting the gene activity patterns in different cell types. For example, typical plant cells have a genomic diversity of about 20,000 and differ in the activities of about 1000 to 1500 genes. Theory predicts about this difference.[48]

6. "Damage" due to alteration in the activity of a single element in the ordered regime spreads to a small fraction of the elements. The expected sizes of such avalanches are bounded by the expected sizes of the unfrozen islands. Numerical results show that the number of elements which alter behavior parallels the number of genes which typically alter activity after hormonal or other stimuli. Thus, a cascade of about 155 genes activity changes are triggered in the salivary glands by ecdysone in the fruit fly, *Drosophila melanogaster*, while about 170 would be predicted for a genome of comparable complexity.[48]

7. Any attractor is the neighbor of only a few other attractors under the drive of small perturbations. Thus, under the drive of random perturbations, the system can move from one attractor to a few neighbors, and from them to a few more neighbors along branching pathways. Such transitions correspond to differentiation from one to another cell type, and predict that ontogeny from the fertilized egg, or zygote, should be organized around branching pathways to all the cell types of the organism. Since the Cambrian, presumably, all metazoans and metaphytens have organized ontogeny around just such branching pathways.[39,41,48]

In brief summary, Boolean networks in the ordered regime exhibit a range of ordered properties which find clear parallels in ontogeny. Thus, the spontaneous "antichaos," the order for free exhibited by such systems, must emerge as a candidate explanation of much of the order found in ontogeny. If one had to make a guess, it is quite plausible to suppose that genomic systems are in the ordered regime.

The hypothesis, elevated to a candidate principle in the Introduction, that complex systems evolve to a preferred position on the order-chaos axis, in the ordered regime near the edge of chaos, emerges as just that, a candidate principle. The data supporting it, from theoretical work to experimental work, are promising, perhaps very hopeful, but not yet convincing.

V. COEVOLUTION TO THE EDGE OF CHAOS

It is possible that a previously unexpected dynamical behavior is a general attractor of many coevolutionary processes. Coevolving agents may evolve their internal structure and their couplings to other agents, such that the entire system coevolves to a phase transition between order and chaos.

Coevolution is a process in which adaptive entities make adaptive hill-climbing moves towards "fitness peaks" on their own fitness landscapes, but in doing so, each entity may deform the fitness landscapes of its coevolving partners. The results can be persistent chaotic motion by all coevolving entities. Such persistent adaptation without improvement in a chaotic regime is sometimes called "Red Queen" behavior.[85] Alternatively, all agents can reach mutually consistent local peaks and cease changing. This ordered regime is sometimes called an Evolutionary Stable Strategy, or ESS.[60]

While chaotic and ordered regimes have long been recognized in models of coevolution, recent examination of a very general class of coevolutionary systems reveals that a phase transition regime lying between order and chaos exists.[50] Three features are known to characterize the phase transition regime. First, the entire ecosystem is marginally frozen. Second, the mean fitness of adaptive agents is maximized at the phase transition. Third, if the system of coevolving agents is not at the phase transition, it is to the selfish advantage of each agent to alter the statistical structure of its own fitness landscape toward the optimal structure corresponding to the phase transition region. These three established features suggest a fourth, as yet untested, consequence: A selective metadynamics acting on each agent for its own selfish benefit may, as if by an invisible hand, cause all agents to coevolve to the phase transition between order and chaos.

THE NK FAMILY OF FITNESS LANDSCAPES

I begin by introducing the general NK family of fitness landscapes used in this work.[44,45] Sewall Wright introduced the concept of fitness landscapes.[90] In the simplest case one can think of organisms with a haploid genome and ascribe a fitness to each possible genotype. To be concrete consider an organism with N genes, having two versions, or alleles, per gene. Then the set of possible genotypes is the 2^N possible combinations of alleles of the N genes. These genotypes can be located in a genotype space comprised by the N-dimensional Boolean hypercube. Each genotype is at a vertex of the hypercube, and located next to N one-mutant neighbors in the space. The fitness of each genotype, normalized to range from 0.0 to 1.0, yields a more or less mountainous fitness landscape over the space of genotypes.

Given a fitness landscape, adaptive evolution by mutation and selection is then a process whereby a population of individuals, each with a genotype located at a specific vertex, flows over the landscape. In fact, this process is complex. At a low mutation rate a population initiated at a single genotype will climb towards a local peak and hover as a cloud in its vicinity. At higher mutation rates, the population will begin to "melt" off the peak and flow along ridges of near equal fitness in the landscape. At still higher mutation rates, the population will flow down from the peaks and wander across fitness lowlands.[17,19,44,45,48,49] Thus, in analogy with statistical mechanics and systems with many potential wells and at finite temperatures, evolving populations can undergo phase transitions as mutation rate is tuned from low to high. Eigen and Schuster have termed such a melting transition the "error catastrophe."[18]

The NK family of fitness landscapes takes seriously the fact that the fitness contribution of the allele of one gene can depend on the alleles of other genes. Such intergenic couplings are called epistatic interactions. To capture the fact that, in general, we have no idea what such effects are, the NK model uses an idea made familiar from models of spin glasses. The model assigns the effects of epistatic interactions at random from some defined distribution.

The genome consists of N genes, each with A alleles. Here let $A = 2$. Each gene is assigned K other genes which serve as epistatic inputs to that gene. The K inputs might be flanking neighbors along a chromosome, or assigned at random. The fitness contribution of each gene as a function of the allele of that gene and the alleles of its K inputs, is assigned a random from the uniform interval between 0.0 and 1.0. Thus, the fitness contribution of each gene is specified for the 2^{K+1} possible combinations of its own allele state and the allele states of its K inputs. The fitness of a genotype is just the mean of the fitness contributions of the N genes.

The NK model yields a family of fitness landscapes ranging from smooth and single peaked to fully random and multipeaked. When $K = 0$, each site makes an independent fitness contribution. The landscape is Fujiyama-like, with a single high peak and smooth flanks sloping to it. When $K = N - 1$, fitness landscapes are completely random. Here the NK landscape model is identical to Derrida's Random Energy Model for spin glasses. In this limit the system has on the order of $2^N/(N + 1)$ local peaks, each accessible from a rather small neighborhood.

Three features of the NK family of landscapes are important for the following discussion of coevolution: (1) As K increases, the number of local peaks increases. (2) As K increases, the sides of fitness peaks become steeper. (3) As K increases, the heights of fitness peaks decrease. These three properties are all consequences of the increasing numbers of conflicting constraints, or frustration, as K increases. As K increases, these conflicting constraints create an increasingly multipeaked fitness landscape, but the peaks are lower because they are ever poorer compromises among the increasing numbers of conflicting constraints.

The NK family of landscapes is of general interest precisely because it is so general. It now appears that the statistical structure of cost landscapes in combinatorial optimization problems ranging from RNA folding landscapes to the Traveling Salesman Problem may map to the NK family.[79]

COEVOLUTION ON COUPLED NK LANDSCAPES

The NK family of landscapes affords a testbed to study the effects of landscape structure and the couplings between landscapes on coevolutionary dynamics. To couple two-fitness landscapes, Sonke Johnsen and I[50] supposed that each gene in each organism makes a fitness contribution which depends epistatically on K other genes within that organism, and upon C genes in each of the other types of organisms with which the first organism coevolves. Thus, we coupled landscapes. Adaptive moves by an organism on one landscape now deforms the fitness landscapes of its partners. Due to this coupling, model ecosystems in general lack a potential function with point attractors at fitness peaks, and become general dynamical systems. As such, these systems might exhibit chaotic, or ordered behavior.

Coupled NK landscapes pass from a chaotic regime through a phase transition to an ordered regime as the parameters of the coevolving system are altered. Indeed, the behaviors are remarkably like those of NK Boolean networks. Consider a *square lattice ecosystem*, in which each agent interacts with its four neighbors. At each "ecosystem generation," each agent in turn considers all its N one-mutant variants, and moves at random to a fitter variant if one exists. Color green any agent able to mutate to a fitter variant. Color red any agent which is currently at a local peak, hence unable to mutate to a fitter variant. In the chaotic regime, almost all agents move at each ecosystem generation. The lattice has a green percolating sea, with small red clusters in it. Deep in the ordered regime, all agents rapidly reach mutually consistent local optima and stop mutating. The entire system has a red frozen sea. The agents are in local pure strategy Nash equilibria: No agent can improve its fitness by changing as long as the other agents do not change. Here the phrase "local" refers to the fact that each agent can only search its genotype strategy space via local one-mutant search. At the phase transition, the red and green regions vie with one another for a long period, fluctuating and interpenetrating. Eventually, all agents stop mutating.

In Figure 10 I show the mean fitness of adaptive agents as ecosystems are tuned from the chaotic regime through the phase transition and into the ordered regime by increasing K, the level of epistatic interactions within genes of one organism. There is an optimal intermediate value of K where mean fitness is maximized. In Figure 11 I demonstrate that this value of K corresponds to the phase transition, where ecosystems are gradually freezing. In Figure 12 I demonstrate that, even when coevolving with other agents having nonoptimal K values, it is to the advantage of each agent to alter its own K value towards the optimal K value.

Figures 10, 11, and 12 carry powerful implications. Very general coupling of NK landscapes to form coevolving systems demonstrates a phase transition between an ordered and a chaotic regime. In the chaotic regime, fitness landscapes have high peaks, but landscapes deform so rapidly as other agents move that each agent's peaks recede faster than they can be climbed. Sustained fitness is low. In the ordered regime, agents climb and stay on mutually consistent

(a)

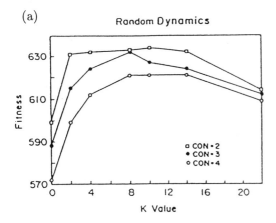

(b)

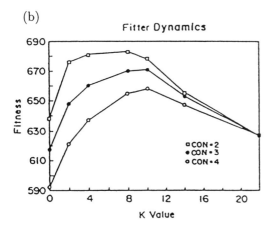

FIGURE 10 (a) Mean sustained fitness in 5×5 ecosystems as K varies from 0 to 22. In all cases, $N = 24$, $C = 1$. Corner species are connected to two others (top curve), interior species to four others, (bottom curve). Note that sustained fitness first increases, then decreases as K increases. Random dynamics was used. (b) As in 10(a), except here, fitter dynamics was used.

peaks, but the peaks are low due to the conflicting constraints implied by high epistatic interactions. Mean fitness is low. Optimal mean fitness occurs at the phase transition itself. In retrospect, this is to be expected. As K decreases towards the optimal value, fitness peaks become higher. Thus, as K decreases from the ordered regime, mean fitness will increase until the slopes towards peaks are so smooth and peaks so rare, that peaks recede faster than agents can climb them. But this transition is just the phase transition to the chaotic regime. It is less obvious that selfish moves by each agent to change the statistical structure of its own landscape by altering the level of epistatic interactions towards optimal landscape ruggedness will help that agent regardless of the actions of the other agents. Yet this appears to be true.

The most powerful implication of Figures 10, 11, and 12 requires further testing. All agents should modify the structure of their own fitness landscapes, each

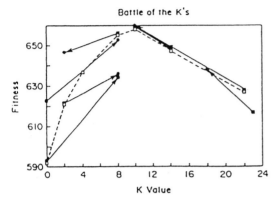

FIGURE 11   Fraction of 5×5 ecosystems which have not yet become frozen in an overall Nash equilibrium plotted against generation. Note for $K \leq 8$, none of the ecosystems attained a frozen Nash equilibrium in the time available. For $K \geq 10$, some or most systems freeze at Nash eqilibria and do so more rapidly as K increases.

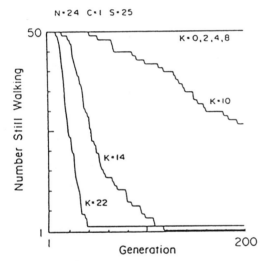

FIGURE 12   Selection force towards K_{opt} value of $K = 8$ to 10. Two experimental species located adjacent to the central species in the 5×5 ecosystem were constructed with a K value that was different from that of the remainder of the ecosystem. In all cases, deviation of the K values of the experimental pair (•) toward K_{opt} increased the sustained fitness of the experimental species relative to the unperturbed ecosystem (○) and relative to the control species (□) in the perturbed ecosystem.

to its own myopic advantage, and the entire system should coevolve to the edge of chaos. Here, if true, would be an invisible hand at work achieving coordination of coevolving players. These ideas require us to consider a new level of abstraction. Community models are familiar in ecology. Here organisms do not alter genotypes, but alter in population abundance. Coevolutionary models are familiar in which organisms alter genotypes and climb deforming fitness landscapes of fixed statistical structure. These models exhibit chaotic Red Queen behavior or ordered ESS behavior. At this new level of abstraction we consider a family of landscapes where organisms engage in the familiar dynamics of hill climbing, but also, presumably on a slower time scale, modify the kinds of hills they climb. The general, if tentative, possibility is that coevolving myopic agents can alter their internal structure and couplings to other agents, to evolve to a universal coevolutionary attractor at the edge of chaos. With Per Bak, I hope to test this possibility further.

(a)

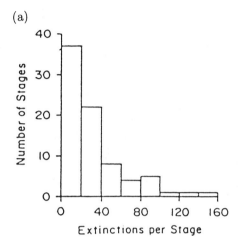

(b)

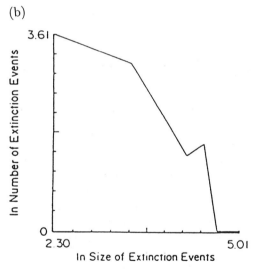

FIGURE 13 (a) Raup's data for the sizes of extinction events versus the number of events at that size. (b) Replot of Raup's data in natural logarithm form.

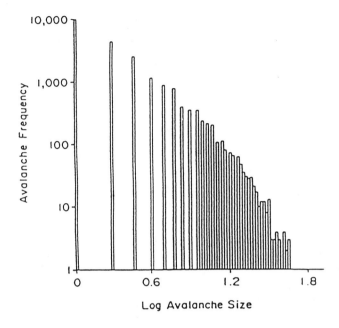

FIGURE 14 Logarithmic plot of the distribution sizes of extinction events in Ray's Tierra simulations. The size of an extinction event corresponds to the number of genotypes that go extinct over short intervals. (Courtesy of Tom Ray.)

A POWER-LAW DISTRIBUTION OF EXTINCTION EVENTS AT THE EDGE OF CHAOS

Two bodies of evidence tentatively support this general conclusion. Both concern the distribution of avalanches of coevolutionary change which propagate through model ecosystems at the phase transition between order and chaos. Boolean networks at the phase transition show a power-law distribution of avalanches of damage. Coupled NK landscapes at the phase transition also show a power-law distribution of coevolutionary changes propagating through the ecosystem after minor perturbations to the fitness landscape of one agent. Thus, in clear analogy to the power-law distributions of avalanches characteristic of self-organized criticality,[5] the selective metadynamics in model ecosystems may attain a self-organized critical state.

During such avalanches, fitness of individual agents should fall. Such agents, representing spaces, may therefore go extinct. Hence the size distribution of extinction events at the phase transition should be a power law. Were the ecosystem within the ordered regime, then the distribution would not be a power law. Too few large avalanches would occur. In Figure 13(a) I show the distribution of sizes of extinction events at the family level since the Cambrian.[50,71] In Figure 13(b)

I transform the data to log-log plots. In Figure 14 I show the distribution of extinction events arising in Tom Ray's Tierra model of coevolving "critters" in a computer's core. The biological data correspond to an ecosystem modestly in the ordered regime. Only 71 data points are available. Ray's data, with thousands of extinction events, are close to a power law which bends over in the large avalanche limit due to finite size effects. These data are insufficient. Thus, some biological extinctions may be due to large meteors. Nevertheless, one needs to explain the entire extinction distribution. The distribution as a whole may reflect endogenous rather than exogenous causes. While the data are insufficient, they are suggestive.

In summary, a new possibility confronts us. Myopic adaptive agents not only play games with one another, but can myopically alter the games they mutually play. Such agents might range from coevolving protoorganisms, to higher metazoans and metaphytens, to human economic agents. Indeed, it is not implausible that collective human agencies, such as firms in an economy, can be viewed as more or less myopic agents which coevolve with their economic partners. Many such coevolving systems are capable of chaotic dynamics, of ordered dynamics, and hence pass through a phase transition. In one broad family of such systems, the phase transition now appears likely to be the natural attractor of the coevolutionary process. The edge of chaos may be a universal attractor of coevolutionary processes.

VI. BOUNDEDLY RATIONAL EXPECTATIONS: COORDINATION OF MUTUAL BEHAVIOR AT THE EDGE OF CHAOS

No adequate "solution concept" is available to explain the coordination of behavior among many optimizing agents, each able to carry out complex planning processes. As I discuss briefly, familiar approaches in economic theory all seem to fail. In the present section I wish to consider a new solution concept which may apply to human economic agents, but also to simpler organisms, and to collective human organizations such as firms, as well. I stress that the idea are only suggestions and still untested. The core of the new idea is that agents persistently attempt to predict optimally the behaviors of other agents. To do so, each is driven to construct optimally complex models of the behaviors of the other agents. Inevitable disconfirmation of these models drives a coevolutionary process among the models adopted by the agents, as each agent replaces disconfirmed models with new hypotheses. This coevolutionary dynamics can lie in a chaotic regime, in an ordered regime, or at the phase transition between order and chaos. An inevitable metadynamics, I speculate, might drive the system to the edge of chaos. Here mutual behavior coordinates for intervals only to be disrupted by avalanches of changes in internal models driving avalanches of changes in decision rules and behavior. Mutual models, in short, would never settle down, but achieve a self-organized critical state. At this self-organized critical state, models would have an optimal bounded complexity. Thus, in this sense, agents construct optimally boundedly "rational" models.

ECONOMIC MODELS OF BEHAVIOR COORDINATION

Since Adam Smith's "invisible hand," economics has been the mother science seeking a natural explanation of the coordination of behavior among many optimizing agents. The founding, and indeed core, theory locates such coordination in the advantages of trade, and ultimately in price formation. Yet price formation itself has turned out to be difficult in model economies with more than one good, and consumption complementarities between the goods. In contrast, in a single good economy, price formation is better understood. Price changes while supply and demand curves cross until an equilibrium is attained. At that equilibrium price the market clears. No excess demand remains. But it is well known that, with two complementary goods, bread and butter, this simplicity vanishes. Change in the price of bread alters the demand for butter. In general, for any defined and fixed mechanism by which a vector of prices are to be adjusted for a given vector of excess demand, some model economy exists such that the equilibrium price at which all markets clear is unstable to fluctuations. Under that price adjustment mechanism, then, prices diverge from equilibrium, and markets fail to clear.

The most celebrated response to this dilemma is the Arrow-Debreu theory of general competitive equilibrium. As Arrow notes, general equilibrium theory, despite its flaws, remains the only general framework for economics. But the flaws are serious. Roughly speaking, general equilibrium demands two rather impossible premises: (1) complete markets for all dated contingent goods, and (2) infinitely rational economic agents.

Complete markets require that all possible types of goods are known at the start of the process. Formal models of technological evolution, based on an analogy to supracritical chemical reaction systems, suggest this is unlikely.[48] Goods are treated as words in a generative grammar. Production of goods is thought of as a set of grammatical transformation rules yielding new words. Economic takeoff occurs at a critical diversity of goods and services. It can be formally undecidable whether a given word will, or will not, ultimately be produced by iterative application of the grammar rules. In this model economic world, then, it can be formally undecidable what goods will ultimately be produced by the economic system. Similarly in the supracritical chemical world, it can be formally undecidable whether a given organic molecule will be produced from a given founder set. Thus, the requirement that all production processes, goods, and services be known at the outset seems implausible in the real world. In turn, a requirement for complete markets seems implausible.

Most economists are properly troubled by positing infinitely rational agents. Simon has suggested the concept of bounded rationality, and satisficing rather than optimizing. Experimental work in economics includes efforts to discover the ways human agents are less than infinitely rational. However, one would like a theoretical basis to understand whether and in what manner complex adaptive agents ought to be boundedly rational. I return to my own suggested mechanism below.

Game theory is another cornerstone effort to account for behavior coordination among optimizing agents. The most familiar solution concept in game theory is that of a Nash equilibrium. Such an equilibrium is defined as a strategy adopted by each agent having the property that, if the other agents do not change their own strategy, then no agent can improve its payoff by altering its own strategy.

While appealing, Nash equilibria suffer serious flaws as a general solution concept. First, payoff at Nash equilibria need not be optimal in the game. The Prisoner's Dilemma is the simplest and best known example. Second, in standard forms of game theory, agents achieve Nash equilibria by pure brilliance. Each reasons that the others will realize that such and such strategy is Nash, and each agent simply adopts the Nash strategy. Third, no general learning dynamics guarantees that agents might gradually alter strategies and all approach the Nash equilibrium. Fourth, most games have multiple Nash equilibria, and no dynamics to choose among them. Fifth, realistically complex games could have exponentially many Nash equilibria, hence the solution might typically be NP complete. A concrete example would be the coupled NK coevolving landscapes of the previous section under the further assumption that each agent could alter all alleles of all genes at any instant.

A more recent approach in economics is the body of work in rational expectations. An example of the kind of problem under investigation is the rationality of investing during periods of speculation in an asset. It can be rational for me to buy tulips because I believe all others will buy tulips. The framework of analysis uses learning rules together with rules coupling beliefs to the movement of the economic system, and shows that there are self-consistent solutions, hence the phrase rational expectations.

Yet it appears to some economists that the rational expectations framework is seriously flawed. Jean-Michel Grammont,[35] has considered simple model economies with such learning rules coupled to the behavior of the economy as a function of the beliefs adopted. His central point is this: As the set of data in a time series, for example of prices, grows larger, the probability increases that a Fourier analysis of that time series will have one or more unstable, amplifying modes. The entire spectrum of eigen-values and eigen-vectors yielded by the Fourier analysis are available for the agents to make use of in their planning. No mechanism within the theory exists which will lead all agents to coordinate on any particular mode. And in general, some or all agents will make use of the unstable modes in their predictions of the behavior of the economy, thus destabilizing any rational expectations solution. If Grammont is correct, then, as the data for analysis by agents increases in abundance, rational expectation solutions are generically not stable. Here Grammont stops. The economic system diverges from an equilibrium. But where does it go? Perhaps, as I suggest next, to the edge of chaos.

COEVOLUTION OF INTERNAL MODELS TO THE EDGE OF CHAOS

The framework I now describe is both new and entirely untested beyond the considerations below. Thus, it is a hope, not an established body of work. Nevertheless, I feel the framework is promising. It develops in several steps.

1. The central premise is that adaptive agents which interact with one another must each always try to predict the behavior of the other agents as well as possible.

2. Given *finite data*, say, a time series, models which can optimize the capacity to predict optimally must be of an optimum intermediate complexity. To take a concrete example, consider the daily price of wheat on a given commodity exchange over a 30-day period. Consider building a model of that time series based on use of Fourier modes. Then it is clearly possible to underfit the data, or overfit the data. For example, a straight line through the data typically underfits that data, and predicts the future very poorly. The straight line approximation ignores too much information in the data Conversely, use of several hundred Fourier modes to fit the 30 data points yields a curve which passes through each data point, but again predicts the future very poorly. The over-fit curve pays too much attention to noise in the data. Some *intermediate* number of Fourier modes, perhaps three or four, yields an approximation to the data which will predict optimally. Note that choice of Fourier modes here is arbitrary. Any other basis set of functions might have been used. Indeed, one might switch from Fourier modes to a different basis set at any point. The key and very general conclusion is that given *finite reliable data*, a model *of finite* bounded complexity optimizes the capacity to predict.

3. Adaptive agents make models of one another's behaviors as part of their mutual ongoing behavior. Since each agent has a finite model, it will eventually fail. At that point, the agent will substitute another optimally complex hypothesis, or model, about the behavior of the other agent from set of competing alternative hypotheses all of which fit the data reasonably well. When the first agent alters its model of the second agent, the first agent will typically change its own decision rules. In turn, the second agent's model will typically fail, if only because the first agent has changed its own decision rule. Thus, the second agent will be driven to adopt a new, optimally complex model of the first agent.

4. It follows that the adaptive agents are engaged in a *coevolutionary process*. Each moves in a space of modestly complex models or hypotheses, about the behavior of the other. The agents mutually flow in a space of models about one another's behavior.

5. Note that, as the first agent changes his model of the second agent, his own decision rule changes; therefore, past data about the first agent's behavior *is no longer reliable*. It follows that only a *finite* amount of data about the recent behavior of each agent is reliable. Thus, when models of one another are

coevolving, the very alteration of internal models by each agent delimits finite intervals over which reliable data are available.

6. Such a coevolutionary dynamics should exhibit an ordered regime, a chaotic regime, and a phase transition. In the ordered regime, each agent has a model of the other agent(s) such that all models are mutually consistent and do not require changing. This regime presumably corresponds to rational expectations solutions. In the chaotic regime, each agent alters its own model of the other agents frequently. Turbulent flow in the space of models, and resulting behavior, ensues.

7. An internal dynamics of *persistent optimal model building* by the agents should drive the set of agents to the edge of chaos.

a. If the dynamics are in the ordered regime, thus very stable and consistent, then each agent has a lot of reliable data about the other agent's behavior. Given more reliable data, each agent naturally attempts to improve its capacity to predict more precisely the other agent's behavior by constructing a *more complex model* of the other's action. This more complex model is necessarily more sensitive to small alterations in the other's behavior, hence is more easily disconfirmed. Thus, adopting more complex models to predict more precisely tends to drive the mutually coevolving system from the ordered regime toward the chaotic regime.

b. Conversely, in the chaotic regime, each agent has very limited reliable data about the other's behavior. In part this reflects the fact that, in the chaotic regime, each agent changes its own model of the other agent, hence its own decision rules governing its own behavior. To optimize the capacity to predict the behavior of the other agent, given the small amount of reliable data, each agent is driven to build a *less complex model* of the other's action. These less complex models predict less detail, hence are less sensitive to the behaviors of the other, hence less readily disconfirmed, and therefore less often changed. This drives the system from the chaotic regime toward the ordered regime.

c. The coevolutionary metadynamics should converge at the phase transition between order and chaos. Here, no internal model of the other agents would be valid indefinitely. The system would be poised at a self-organized critical state. Avalanches of changes in models of other agents would be triggered by small events arising among from the endogenous dynamics of the agents, or exogenous shocks.

EVIDENCE FOR AND IMPLICATIONS OF THIS CANDIDATE SOLUTION CONCEPT

While I find the framework suggested above appealing, I stress that I have as yet carried out no further work validating it as a body of theory. In particular, these ideas must be worked out in the context of the agents' purpose, or games the agents are playing. The structure of the game itself, say, the existence of a single Nash equilibrium, will modify the coevolutionary dynamics of the internal models. I hope, with Alfred Hübler, to test whether adaptive agents in lattice systems behave as expected based on this model. Elsewhere in this volume Arthur discusses simulations of stock market behavior using genetic algorithms to model economic agents. The parallels are clear. Holland discusses default hierarchies. Rules low in the hierarchy are very simple hypotheses. Rules high in the default hierarchy are very complex hypotheses. Simple and complex hypotheses are similar to simple and complex models. Adaptive agents in the stock market model should coevolve to internal models of intermediate complexity, and to the edge of chaos.

Evidence from technical traders supports a picture of agents altering models of one another. Technicians speak of decision rules which used to work, but now must be replaced with new rules which seem to fit recent market behavior better.

If the analysis sketched above can be carried out, it may supply a new and useful solution concept in economics and elsewhere. The theory offers a potential account of optimally bounded rationality, in the sense that it becomes counterproductive to each agent to build over-complex models of the other agents. As such, this approach should be applicable to all adaptive agents that interact, that gain or lose by the outcomes of the interaction, and that plan. In the central context of price formation, it would be fascinating, indeed, if this framework led to a theory of nonequilibrium price formation. Markets would not clear, but would always come close to clearing.

CONCLUSION: WHISPERS FROM CARNOT

Might we find a conceptual framework that unites the issues we have discussed? Carnot taught us that, with the proper abstractions, very general laws governing equilibrium systems could be found. But all the phenomena in which we are interested are nonequilibrium. Chemical evolution on the prebiotic earth, protobionts, current free living cells, organisms, ecosystems, and economic systems all demand a flow of matter and energy. Can we seek a new kind of second law governing nonequilibrium self-constructing systems?

We have hints that such a general law may be possible. It may require a marriage of Charles Darwin and Adam Smith, natural selection and the invisible hand. Natural selection is nothing but a specific kind of stability analysis with respect to self-reproducing entities. In equilibrium systems, in general, and autocatalytic systems in particular, kinetics dominates what occurs. We will see that which makes the most of its kind before that kind dissipates or becomes extinct. The heart of

Darwin is just this insight. Further, as Eigen and Schuster have shown,[18] natural selection can be given a crisp physical interpretation among replicating RNA molecules. The molecular quasi-species, or mutant cloud including and surrounding the fastest replicating molecular sequence, dominates.

I suggested above that the economic concepts of utility, the advantages of trade, and price can also be given a clear physical interpretation. Among replicating molecules, utility is the rate of reproduction, increased utility is an increased rate of reproduction. The advantages of trade occur when two replicators become mutualists and form a coevolving community. Coevolution for joint maximal rates of reproduction occurs at a fixed ratio of exchange of help versus cost to each, the analog of price. Coevolution to that ratio of exchange is price formation. Each replicator, acting for its own selfish benefit, as if by an invisible hand, unwittingly acts for the benefit of the small replicator society: the abundance of both replicators is maximized at the analog of price equilibrium.

Coevolution to the edge of chaos hints again at an invisible hand where each agent acts selfishly, yet acts unwittingly for the benefit of all. Again, a physical interpretation among coevolving autocatalytic systems can probably be given. Nonequilibrium, self-reproducing molecular systems may tune their internal structure and couplings to other agents such that all coevolve to the edge of chaos. It is natural selection acting on individuals for their selfish reproductive purposes that would attain that poised state. At the poised coevolutionary state, all entities, on average, sustain the highest average rate of reproduction, although avalanches of extinction events propagate through the system.

Natural selection and the invisible hand, the emergence of self-reproducing molecular systems which inevitably formed coevolving systems of mutualists and competitors, raise the clear possibility that life has arranged its coupled processes in a manner which optimizes some property of the global system. All efforts to characterize an optimized property in evolution or ecology have met with scant success. Nevertheless, we have new clues in the edge of chaos which arises at so many levels in the models discussed. It is as if the entire biosphere, the coupled living processes within and between organisms, have achieved a self-organized critical state which maximizes the exploration of novel molecular and other organizational possibilities.

REFERENCES

1. Alberts, A., D. Bray, J. Lewis, M. Raff, K. Roberts, and J. D. Watson. *Molecular Biology of the Cell.* New York: Garland, 1983.
2. Atlan H., F. Folgelman-Goulie, J. Salomon, and G. Weisbuch. "Random Boolean Networks." *Cybern. & Sys.* **12** (1981): 103.

3. Bagley, R. J., and J. D. Farmer. "Spontaneous Emergence of a Metabolism." In Artificial Life II, edited by C. G. Langton, C. Taylor, J. D. Farmer, and S. Rasmussen, 93–140. Santa Fe Institute Studies in the Sciences of Complexity, Proc. Vol. X. Redwood City, CA: Addison-Wesley, 1992.

4. Bagley, R. J., J. D. Farmer, and W. Fontana. "Evolution of a Metabolism." In *Artificial Life II*, edited by C. G. Langton, C. Taylor, J. D. Farmer, and S. Rasmussen. Santa Fe Institute Studies in the Sciences of Complexity, Proc. Vol. X, 141–158. Redwood City, CA: Addison Wesley, 1992.

5. Bak, P., C. Tang, and K. Weisenfeld. "Self-Organized Criticalcality." *Phys. Rev. A* **38(1)** (1988): 368–374.

6. Baldwin, J. Personal communication.

7. Ballivet, M., and S. A. Kauffman. *Adaptation on Rugged Landscapes.* 1985.

8. Deamer, D. W., and G. L. Barchfeld. "Encapsulation of Macromolecules by Lipid Vesicles Under Simulated Prebiotic Conditions." *J. Mol. Evol.* **18** (1982): 203–206.

9. Derrida, B. "Valleys and Overlaps in Kauffman's Model." *Phil. Mag. B.* **56(6)** (1987): 917–923.

10. Derrida, B., and H. Flyvberg. "Multivalley Structure in Kauffman's Model: Analogy with Spin Glasses." *J. Phys. A: Math. Gen.* **19** (1986): L1003–L1008.

11. Derrida, B., and Y. Pomeau. "Random Networks of Automata. A Simple Annealed Approximation." *Europhys. Lett.* **1(2)** (1986): 45–49.

12. Derrida, B., and D. Stauffer. "Phase-Transitions in Two-Dimensional Kauffman Cellular Automata." *Europhys. Lett.* **2(10)** (1986): 739–745.

13. Derrida, B., and G. Weisbuch. "Evolution of Overlaps Between Configurations in Random Boolean Networks." *J. Physique* **47** (1986): 1297–1303.

14. Derrida, B., and H. Flyvberg. "Distribution of Local Magnetization's in Random Networks of Automata." *J. Phys. A. Lett.* **20** (1987): L1107–L1112.

15. Devlin, J. J., L. C. Panganiban, and P. E. Devlin. "Random Peptide Libraries: A Source of Specific Protein Binding Molecules." *Science* **249** (1990): 404–406.

16. Dyson F. *Origins of Life.* London: Cambridge University Press, 1985.

17. Eigen, M. "The Physics of Molecular Evolution." *Chemica Scripta* **26B** (1986): 13–26.

18. Eigen, M., and P. Schuster. *The Hypercycle. A Principle of Natural Self-Organization.* New York: Springer-Verlag, 1979.

19. Eigen M., J. McCaskill, and P. Schuster. "The Molecular Quasi-Species." *J. Phys. Chem.* **92** (1988): 6881.

20. Eisen, H., and M. Ptashne. "The Bacteriophage Lambda." In *Cold Spring Harbor Symposium on Quantum Biology*, edited by A. D. Hershey, 239. New York: Cold Spring Harbor Laboratory, 1971.

21. Ellington, A. D., and J. W. Szostak. "*In Vitro* Selection of RNA Molecules that Bind Specific Ligands." *Nature* **346** (1990): 818.

22. Erdos and Renyi. 1956

23. Farmer, J. D., S. A. Kauffman, and N. H. Packard. "Autocatalytic Replication of Polymers." *Physica* **22D** (1986): 50–67.

24. Fogelman-Soulie, F. Ph.D. Thesis, Universite Scentifique et Medical de Grenoble. 1985.

25. Fogelman-Soulie, F. "Parallel and Sequential Computation on Boolean Networks." *Theor. Comp. Sci.* **40** (1985): 121.

26. Fontana, W. "Algorithmic Chemistry." In *Artificial Life II*, edited by C. Langton, C. Taylor, J. D. Farmer, and S. Rasmussen. Santa Fe Institute Studies in the Sciences of Complexity, Proc. Vol. X, 159–210. Redwood City, CA: Addison Wesley, 1992.

27. Fox, S. W. "Metabolic Microspheres: Origins and Evolution." *Naturwissenschaften* **67(8)** (1980): 373–383.

28. Fox, S. W., and H. Dose. *Molecular Evolution and the Origin of Life*. New York: Academic Press, 1977.

29. Gelfand, A. E., and C. C. Walker. "On the Character and Distance Between States in a Binary Switching Net." *Biol. Cybern.* **43** (1982): 79–86.

30. Gelfand, A. E., and C. C. Walker. *Ensemble Modeling*. New York: Dekker, 1984.

31. Gillespie, J. H. "A Simple Stochastic Gene Substitution Model." *Theor. Pop. Biol.* **23(2)** (1983): 202–215.

32. Gillespie, J. H. "Molecular Evolution over the Mutational Landscape." *Evolution* **38(5)** (1984): 1116–1129.

33. Glass, L., and S. A. Kauffman. "The Logical Analysis of Continuous Nonlinear Biochemical Control Networks." *J. Theor. Biol.* **39** (1973): 103–129.

34. Glass, L. "Boolean Networks and Continous Models for the Generation of Biological Rhythms." In *Dynamical Systems and Cellular Automata*, edited by J. Demongeot, E. Goles, and M. Tcheunte, 197–207. London: Academic Press, 1985.

35. Grammont, J.-M. Personal comments, 1992.

36. Holland, J. H. "Escaping Brittleness: The Possibilities of General Purpose Learning Algorithms Applied to Parallel Rule-Based Systems." In *Machine Learning II*, edited by R. S. Michalski, J. G. Carbonell, and T. M. Mitchell. Los Altos, CA: Morgan Kauffman, 1986.

37. Hopfield. "Neural Networks and Physical Systems with Emergent Collective Comoputational Abilities." *Proc. Natl. Acad. Sci.* **79** (1982): 2554.

38. Joyce, G. E., and L. E. Orgel. "Non-Enzymatic Template Directed Synthesis on RNA Random Copolymers. Poly-(C,G) Templates." *J. Mol. Biol.* **188** (1986): 433.

39. Kauffman, S. A. "Metabolic Stability and Epigenesis In Randomly Constructed Genetic Nets." *J. Theor. Blol.* **22** (1969): 437–467.

40. Kauffman, S. A. "Cellular Homeostasis, Epigenesis and Replication in Randomly Aggregated Macromolecular Systems." *J. Cybern.* **1** (1971): 71–96.

41. Kauffman, S. A. "The Large Scale Structure and Dynamics of Gene Control Circuits: An Ensemble Approach." *J. Theor. Biol.* **44** (1974): 167–190.

42. Kauffman, S. A. "Emergent Properties in Random Complex Automata." *Physica* **10D** (1984): 145–156.
43. Kauffman, S. A. "Autocatalytic Sets of Proteins." *J. Theor. Biol.* **119** (1986): 1–24.
44. Kauffman, S. A. "Adaptation on Rugged Fitness Landscapes." In *Lectures in the Sciences of Complexity*, edited by D. L. Stein. Santa Fe Institute Studies in the Sciences of Complexity, Lect. Vol. I, 527–619. Redwood City, CA: Addison Wesley, 1989.
45. Kauffman, S. A. "Princples of Adaptation in Complex Systems." In *Lectures in the Sciences of Complexity*, edited by D. L. Stein. Santa Fe Institute Studies in the Sciences of Complexity, Lect. Vol. I, 619–713. Redwood City, CA: Addison Wesley, 1989.
46. Kauffman, S. A. "Antichaos and Adaptation." *Sci. Amer.* **265(2)** (1991): 78–84.
47. Kauffman, S. A. "Applied Molecular Evolution." *J. Theor. Biol.* **157** (1992): 1–7.
48. Kauffman, S. A. *The Origins of Order: Self Organization and Selection in Evolution.* Oxford University Press, 1993.
49. Kauffman, S. A., and S. Levin. "Towards a genelal Theory of Aadaptive Walks on Rugged Landscapes." *J. Theor. Biol.* **128** (1987): 1145.
50. Kauffman, S. A., and S. Johnsen. "Coevolution to the Edge of Chaos: Coupled Landscapes, Poised States, and Coevolutionary Avalances." *Theor. Biol.* **149** (1990): 467.
51. Kauffman, S. A., and S. Johnsen. "Coevolution to the Edge of Chaos: Coupled Fitness Landscapes, Poised States, and Coevolutionary Avalanches." In *Artificial Life II*, edited by C. G. Langton, C. Taylor, J. D. Farmer, and S. Rasmussen. Santa Fe Institute Studies in the Sciences of Complexity, Proc. Vol. X, 325–370. Redwood City, CA: Addison Wesley, 1992.
52. Kauffman, S. A., and E. D. Weinberger. "The NK Model of Rugged Fitness Landscapes and Its Application to the Maturation of the Immune Response." In *Molecular Evolution on Rugged Landscapes: Proteins, RNA and the Immune System*, edited by A. S. Perelson and S. A. Kauffman. Santa Fe Institute Studies in the Sciences of Complexity, Proc. Vol. X, 135–177. Redwood City, CA: Addison Wesley, 1992.
53. Kauffman, S. A., E. Weinberger, and A. S. Perelson. "Maturation of the Immune Response via Adaptive Walks on Affinity Landscapes." In *Theoretical Immunology, Part One*, edited by A. S. Perelson. Santa Fe Institute Studies in the Sciences of Complexity, Proc. Vol. II, 349–83. Redwood City, CA: Addison Wesley, 1988.
54. Kirkpatrick, S., C. D. Gelatt, and H. P. Vecci. "Optimization by Simulated Annealing." *Science* **220** (1983): 671.
55. Langton, C. "Studying Artificial Life with Cellular Automata." *Physica* **22D** (1986): 120.

56. Langton, C. "Adaptation to the Edge of Chaos." Presented at the Second Artificial Life Conference, Santa Fe, New Mexico, 1990.

57. Langton, C. " Life at the Edge of Chaos." In *Artificial Life II*, edited by C. G. Langton, C. Taylor, J. D. Farmer, and S. Rasmussen. Santa Fe Institute Studies in the Sciences of Complexity, Proc. Vol. X, 41–92. Redwood City, CA: Addison Wesley, 1992.

58. Lewis, J. E., and L. Glass. "Steady States, Limit Cycles and Chaos in Models of Complex Biological Networks." *Intl. J. Bifur. & Chaos.* 1 (1991): 477–483.

59. Mandecki, W. "A Method for Construction of Long Randomized Open Reading Frames and Polypeptides." *Prot. Eng.* 3 (1990): 221–226.

60. Maynard Smith, J. "The Theory of Games and the Evolution of Animal Conflicts." *J. Theor. Biol.* 47 (1974): 209–221.

61. Mitchell, M., J. P. Crutchfield, and P. Hraber. "Dynamics, Computation, and the 'Edge of Chaos': A Re-Examination." This volume.

62. Miller S., and L. Orgel. *The Origins of Life on the Earth.* Englewood Cliffs, NJ: Prentice Hall, 1973.

63. Morowitz, H. *Energy Flow and Biology: Biological Organization as a Problem in Thermal Physics.* New York: Academic Press, 1968.

64. Orgel, L. E. "RNA Catalysis and the Origins of Life." *J. Theor. Biol.* 123 (1986): 127.

65. Orgel, L. E. "Evolution of the Genetic Appartus: A Review." *Cold Spring Harbor Sympsia on Quantitative Biology*, Vol. 52, 9-16. New York: Cold Spring Harbor Laboratory, 1987.

66. Oro, J., and A. P. Kimball. "Synthesis of Purines Under Possible Primitive Earth Conditions I. Adenine from Hydrogen Cynaide." *Arch. Biochem. Biophys.* 94 (1961): 221.

67. Oro, J., and A. P. Kimball. "Synthesis of Purines Under Possible Primitive Earth Conditions II. Purine Intermediates from Hydrogen Cynaide." *Arch. Biochem. Biophys.* 96 (1961): 293.

68. Packard, N. *Adaptation to the Edge of Chaos in Biological Modeling*, edited by S. Kelso and M. Schlesinger, 1988.

69. Pollack, S. J., J. W. Jacobs, and P. G. Schultz. "Selective Chemical Catalysis by an Antibody." *Science* 234 (1986): 1570–1573.

70. Pollack, S. J., and P. G. Schultz. "Antibody Catalysis by Transistion State Stabilization." In *Cold Spring Harbor Symposia on Quantitative Biology*, Vol. II, 97-104. New York: Cold Spring Harbor Laboratory, 1987.

71. Raup, D. M. "Biological Extinction in Earth History." *Science* 231 (1986): 1528–1533.

72. Read, R. C. "The Enumeration of Acrylic Chemical Compounds." In *Chemical Applications of Graph Theory*, edited by A. T. Balabanm. London: Academic Press, 1976.

73. Rossler, 0. "A system Theoretic Model of Biogenesis." *Z. Naturforsch* B266 (1971): 741–746.

74. Rumelhart, D. E., J. L. McClelland, and the PDF Research Group. *Parallel Distributed Processing: Explorations in the Microstructure of Cognition*. Vols. 1 and 2. Cambridge, MA: Bradford, 1986.

75. Schuster, P. et. al. Manuscript in preparation, 1993.

76. Scott J. K., and G. F. Smith. "Searching for Peptide Ligands with an Epitope Library." *Science* **249** (1990): 386.

77. Sherlock, R. A. "Analysis of the Behavior of Kauffman Binary Networks, I. State Space Description of Limit Cycle Lengths." *Bull. Math. Biol.* **41** (1979): 687–705.

78. Sherrington, D., and S. Kirkpatrick. "Solvable Model of a Spin Glass." *Phys. Rev. Lett.* **35** (1975): 1792.

79. Stadler, P. Personal comments, 1992.

80. Stauffer, D. "On Forcing Functions in Kauffman's Random Boolean Networks." *J. Stat. Phys.* **40, (3/4)** (1987): 789.

81. Stauffer, D. "Random Boolean Networks: Analogy with Percolation." *Phil. Mag. B* **56(6)** (1987): 901–916.

82. Thomas, R. "Kinetic Logic: A Boolean Analysis of the Dynamic Behavior of Control Circuits." In *Lecture Notes in Biomathematics*, edited by R. Thomas, Vol. 29, 107–142. New York: Springer-Verlag, 1979.

83. Tramontano, A., K. D. Janda, and R. A. Lerner. "Catalytic Antibodies." *Science* **234** (1986): 1566.

84. Tramontano, A., K. D. Janda, and R. A. Lerner. "Chemical Reactivity at an Antibody-Binding Site Elicited by Mechanistitic Design of a Synthetic Antigen." *Proc. Natl. Acad. Sci.* **83** (1986): 6736.

85. VanValen "A New Evolutionary Law." *Evol. Theor.* **1** (1973): 1–30.

86. Von Kiedrowski, G. "A Self-Replicating Hexadeoxynucleotide." *Agnew. Chem. Int. Ed. Eng.* (1986).

87. Walker, C. C., and A. E. Gelfand. "A System Theoretical Approach to the Management of Complex Organizations: Management by Exception, Priority, and Input Scan in a class of Fixed Structure Models." *Behav. Sci.* **24** (1979): 112–120.

88. Weinberger, E. D. "Correlated and Uncorrelated Fitness Landscapes and How to Tell the Difference." *Biol. Cybern.* (1990).

89. Weisbuch, G. "Dynamical Behavior of Discrete Models of Jerne's Network." In *Theories of Immune Networks*, edited by H. Atlan. Springer Lecture Notes in Biomathematics. New York: Springer-Verlag, 1989.

90. Wright, S. (1932). "The Roles of Mutation, In-Breedinig, Cross-Breeding and Selection in Evolution." *Proceedings of the Sixth International Congress on Genetics* **1** (1932): 356–366.

DISCUSSION

WALDROP: I'm beginning to understand your argument about bounded rationality but one thing still bothers me. If this argument is true, how come we have human beings able to make pretty sophisticated models of the world and each other, and bacteria making very simple models, and everything in between—all in the same ecosystem? Why shouldn't everything be driven to the same state?

KAUFFMAN: I don't have the answer to it. Brian, I think, was trying to poke us toward an answer, namely: What's missing from this entire story is, why in the world do things get more complicated? Why do things get more complicated? Why are they pushing the envelope, as Brian was, roughly, telling us? I think that's an extraordinarily deep question, and I don't find yet a simple, powerful, inevitable answer. Why is it good to get complicated so that things are always getting more complicated and pushing the envelope? I think that Brian has given us some clues.. It's division of labor, and I was pleased that Brian came up with the same one. I hadn't thought about hanging the subsystems on. I think that we need a deep argument that makes it inevitable that things get more complicated, because life has, technology has, cultures have; and even though we have these issues of grammar models that can proliferate in the diversity of kinds of things at *one* level, it's not obvious why at multiple levels things get more complicated. So I think you put your finger on a fundamental question.

ARTHUR: I've written another paper which comes to similar conclusions to the points you were making in the last part of your talk. I think actually you have your finger on something very important.

What you consistently argue is that everything you've been looking at is driven to this edge of chaos. When we were carrying out models of the stock market, we found a tradeoff. You can get very very sophisticated forecasting rules that appear in that system but, because they're sophisticated, it means that they're operating under very particular conditional rules: *If* the price has gone up in the last three periods, *and* if some extraneous index has fallen, *and* if some Fourier component of the prices in the last twenty periods has this property, then buy into the market. That's an incredibly complicated rule and, because it's complicated, it can't be validated very often because those conditions don't arise very often. Meanwhile the general rules are much more general and they don't make much profit, but they're getting validation every time. So what we began to see was that if a rule just gets too smart for its own good, it doesn't get much validation because it's not appropriate very often. If something is general and applies virtually every time, it turns out it's not that profitable, because it doesn't have this sophistication.

And so, in looking at what is coming out of our market simulation, you get a balance between very sophisticated rules that are not validated that often and other general rules that don't make that much profit but do quite well.

What you're telling us is something in addition, and that's that you have a balance between these two sorts of things in any ecology and you're sort of driven into the middle, the edge of chaos. You keep telling us that but I still think it's an open question.

KAUFFMAN: I completely agree that it's an open question. In the first place, I've been looking for this for four years and I found it two weeks ago. So to say that there's anything established about it is ridiculous. Meanwhile, the point that you're making is a reasonable one: you don't get much validation of very complicated rules because the circumstances don't come up. The point here is that very complicated rules are not very sensitive as well and, therefore, they can go out of date pretty rapidly.

One of the things I've noticed is that, if you listen to traders on Wall Street Week, they tell you "The following rule *used* to work, but more recently it looks like this other rule is a better decision rule." So there really is an ongoing evolution of decision rules among traders. I would think that part of what we could actually look for is ways to validate that, in chaotic times, people simplify their action rules—on pricing or whatever—and that in stable times, people do actually get fancier. We would need a lot of models to test something like this. In fact, I want to talk to you about whether or not one could use your market model to actually look at this. It's just a framework right now.

GELL-MANN: There's a very important issue in the fact that a complex adaptive system tries to compress regularities, and later on to use what it identifies as random features for extra input. I don't think we've devoted enough attention to that in some of these other discussions. The way it works in science is that these are parameters, boundary conditions—special circumstances you called them—and there's always a tradeoff between making theories that are extremely particular (where all these special conditions, boundary conditions and so on, are put in as given) and other cases where they're supplied as needed because they're external to the theory they're external to the model or schema. And I think that's a *very* fundamental point that we have to keep looking at in each discussion.

So when you say "better and better model," that isn't the same as what it means in science. In science we can have a better theory, even though it may have a lot of stuff as inputs: unexplained parameters, boundary conditions, special circumstances that come up in different cases. It's the high compression of what remains, of the identified regularities, that to a great extent identifies a good model. And that just ties in with what you are saying, Brian, about a tradeoff. You can, of course, get what looks superficially like a very successful picture if you build in all the extra boundary conditions as part of the theory. But that's not generally the best way to do it. Much better, in most cases, to have a general theory and put that stuff in as needed. I think that should be part of every single one of our discussions.

KAUFFMAN: I take your point very much about science, Murray. The intriguing thing to realize...

GELL-MANN: It's true everywhere, not just science.

KAUFFMAN: Well, hang on. I actually don't think it's true everywhere, and I think that the point is that when we are coevolving in our behavior with one another, where I have to try to understand what your decision rule is, and you have to understand what my decision rule is—I'm not sure that the framework you've just painted is the right one. We want to try to understand how independent agents come to coordinate their behavior. Now the invisible hand in economics is the first answer to that question. It's supply and demand balance at a given price. And so we are all led to try to optimize our own good as if by an invisible hand to coordinate things for the benefit of many. But it fails because in an economy with multiple goods, there are consumption complementarities: I like bread and butter together; when the price of bread changes it changes my demand for butter. And that means that if you look at price equilibrium and you look at fluctuation away from price equilibrium, the amplitude of that fluctuation can grow in time under any decision rule that you have about how people respond to excess demand. It's this that's driven the economists to their infinitely rational agents, and that's the next solution. It's fundamentally the Nash equilibrium idea: we will coordinate on a Nash equilibrium.

Now, there are several things wrong with the Nash equilibrium. First, there's no dynamics that gets you to a Nash equilibrium. We just calculate it and assume that the other guys will do it. And the same thing is true with all kinds of arguments in economics that require infinite rationality in dynamic programming problems. So there's no dynamics.

Second, Nash equilibria in general aren't necessarily good solutions, so there's no way of getting to them nor are they necessarily good.

The third approach to coordinating behavior is rational expectations, which Tom Sargent and others have pushed. I only partially understand it, but here the idea is the one that Brian told us: it's what the average investor thinks the average investor thinks, about thinking what the average investor thinks, (that John Maynard Keynes was taking us to), which is now the attempt to coordinate by rational expectations and, therefore, speculative bubbles. But that's not necessarily stationary, as the examples that Brian's looking at in his simulations, and as the *gedanken* example I'm giving you, is. Precisely because my law of motion depends upon what my projection of your law of motion is, and that changes because it's always only finitely accurate—because when you change your model of me, you change your own law of motion, and the past I have about you is irrelevant.

GELL-MANN: You repeated your talk very elegantly, but that wasn't the point of my remark. The point of my remark was that in every single one of these schema-building processes, it's critical to try to understand how the distinction is made between intrinsic program and input data. The input data are treated as given, random, not understandable; while the program is treated as something that is generalizable, compressible, and so forth. The optimization of that distinction is what Brian was talking about: the tradeoff. You run into trouble at each end. If you try always to include the input data as part of the intrinsic program, you're burdening the intrinsic program, you're not able to compress it, because most of it is incompressible. If you proceed the other way around—you just treat everything as an input datum—you don't have any theory. That's something we have to consider, I believe, in every single one of our discussions.

KAUFFMAN: The point that I was making was an answer to your question. I agree with one point that you're making, but I'm also saying that in the context where we're making models...you were pointing to science, and I was...

GELL-MANN: I'm talking about the WHOLE THING; it is not peculiar to science.

KAUFFMAN: That's right. And I'm telling you that I think that I disagree with you in the case that we're making models of one another, okay. That is, you were talking about getting to the ultimate laws.

ANDERSON: A comment about this last discussion. Science has a maximum amount of rationality, one hopes, and we all believe that there is a true Lyapunov function in that case. There is a true science; and there is a lot of other science that isn't true. I think all situations contain some fraction of that, so that in some sense Murray is right: you should have a part of your program that contains general theories, contains reality. But I think also some part of your program must depend on the other man. But that's all a comment on the previous discussion. I had some specific questions for your talk.

One is, talking about autocatalytic probabilities of 10^{-9}, what I remember is that the amount of enhancement of the reaction rate in a catalytic reaction is about 10^8. That's suggesting that in the background ten times as many reactions as the catalyzed reactions are going on, and how do you get out of that one? And then the other question has something to do with this last discussion. Most of what you do is assuming, at least in the earlier part, that there is a fitness function. We know there isn't, in general, and so how much of it is robust to the absence of a Lyapunov function, and how much isn't?

KAUFFMAN: With respect to the first point, you're absolutely right. I showed a curve that looked at the equilibrium concentration that you'd expect, just because of size and equilibrium thermodynamics of small and large polymers (and things, of course, fall off exponentially as you make larger and larger polymers), there's always a spontaneous bunch of reactions going on. So question number one is: If you have an autocatalytic set, where the probabilities that a polymer catalyzed the reaction is something like 10^{-9}, first of all, can you actually funnel flow from a food set down some corridor so you get high concentrations of some polymers compared to the equilibrium level? And the answer is "Yes, you can." Meanwhile, spontaneous reactions are going on all the time.

So it's a race between funneling flow of stuff into the self-reproducing entities, and a background smear that's going on all the time. And it's actually what Darwin told us: the kinetics dominates. The added point that you're making is that, for a variety of reasons, you'd better get your reactive molecules in a small volume (so they can react fast); you better screen off interactions with a lot of the outside world (like, a membrane is a dandy idea) so that you don't have too many of these other spontaneous side reactions; you better stay subcritical internally, so that you don't catalyze too many of those side reactions even if they're still happening. And if they are happening, and you have them in a bounded volume, the side reaction happens spontaneously but so does the back reaction, and now you can up the flow in the direction that you want it. You have to win over the dissipative processes that are going on all the time. It has to be fast enough, and coordinated enough, so you actually get a cluster of coordinating processes. You have to beat the problem that you're raising, because it's always there in the background.

On the coevolutionary point, or the point about potential functions: I think that we're harming ourselves with some unnecessary distinctions. Of course, there's not a written-down potential function that says what my fitness is and biologists don't actually work with it; they try to find out who has more offspring. What there is, of course, is that in a fixed environment under fixed conditions there's something like an implicit fitness function. And, of course, in the coevolving system, or when the outside world is changing, there's not even a potential function. The whole system is just flowing. And that is the point that I was trying to get to: the notion of coevolution to this edge of chaos. We have to imagine that we can always redescribe the coevolutionary processes *as if* the adapting entities were climbing landscapes, but those landscapes are deforming. If one can redescribe it that way, then one can have a general theory that says, "What's the structure of the landscapes right now, and how badly do they deform?" You don't have to have an explicit function; it's an implicit function. And then you can ask, "Do the entities change their internal structures and couplings to one another, and is there [an] attractor of that metadynamics?" And I'm suggesting the possibility, based on inadequate simulations right now, that the phase transition between an ordered regime and a chaotic regime does it.

In that case, you have no potential function, you have no explicit fitness function, you have things doing better by having more offspring—kinetics wins, as Darwin told us and you just urged. And the best winners, as they change their game with one another, emerge in the long run.

ANDERSON: Sounds as though there's a theorem there that nobody has proven, but some of us believe it.

GELL-MANN: The idea is, you have forces, or statistical distributions of forces, that throw the resulting position around in some abstract space. And without talking about landscapes, or changing landscapes, or coevolving landscapes, or any of these implicit potentials (that may never have existed in the first place even before coevolution was introduced), you can talk about the equilibria, on the one side, and you can talk about a very disordered regime, on the other side. Whether there is a phase transition in between has never been established, but it would also be interesting to try and do so, and it would be very interesting to examine the nature of the intermediate case. But the only way one can do that is to set up some sort of well-defined problem.

KAUFFMAN: There are analytical results for this coevolving model that Per Bak and Flyvbjerg have in fact pursued. Their results explicitly show that there *is* a phase transition. . .

GELL-MANN: I didn't say there weren't cases from physics where you have phase transitions.

KAUFFMAN: No, this isn't physics; this is my NK model. And precisely as I showed numerical simulations to suggest that there's a phase transition, the order regime is, roughly speaking, evolutionary stable strategies; the chaotic regime is, roughly, Red Queen. And, in fact, there is a transition between the two, and it emerges analytically (it appears) in this NK-coupled model. . .

ANDERSON: But your NK model has a Lyapunov function, and our question—my question at least. . .

KAUFFMAN: No, it doesn't. Phil, it doesn't. Once I've coupled *your* fitness landscape. . .

ANDERSON: But we were talking about looking back to the original NK model because that was what you were appealing to, and that has a Lyapunov function. We ask if that's got an asymmetric probability and, if it doesn't, is your work robust to that? That is something that Elizabeth Gardner has looked at, and found a wide class of such models that have point attractors, which is the question that you're asking.

KAUFFMAN: What do you mean by asymmetric probabilities?

ANDERSON: Well, the probability that A goes to B is not the same as the probability. . .

KAUFFMAN: You mean in the connection matrix among the sites?

ANDERSON: Yes.

KAUFFMAN: Well, the NK model has an asymmetric connection probability. I'm connected to you, but you're not connected to me. It's a totally asymmetric model.

So all one knows is that for this class of models, in the simulation results, there appears to be a phase transition with analytical results in a particular limit that says there's rigorously a phase transition. Even if that's true, one doesn't know how general that is. The NK class of models captures a family of landscapes that appears to fit, for example, in the traveling salesman problem and some RNA-folding problems. It looks like there's a wide family of actual landscapes, in landscape world, that look like the NK model.

But it doesn't follow that all possible landscapes that deform—therefore, all possible ways that you can make dynamical systems—would have these properties. So one needs a lot more results. We just have one exemplar: this class of models that says that there's a phase transition between an ordered and a chaotic regime. I believe it's deeply related to the same thing in the Boolean network story; I think it's got to be fundamentally the same phase transition in parallel processing systems. How broad it is, I don't know. It's barely known, and I completely agree that one wants to task for concrete, specific models and then try to get general theorems. So we have existence case of one, okay, but it's a fairly broad class of models. The NK landscapes are a very very wide class of spin-glass-like model.

FONTANA: I think that a theory's useful to the degree that it explains something. Here's the theory: there is a subcritical ordered regime in which systems cannot adapt; there is a supercritical chaotic regime in which systems cannot survive. The statement is: adaptively surviving—i.e., complex systems—evolve to the edge between these two regimes. How do you escape tautology?

KAUFFMAN: You always have the question about what you put into the model and what you omitted. That's what you're asking here. The set of ideas that I sketched out for you is also new. What one wants, in that case, is to make a class of concrete models that have the property that as you watch them, they do, in fact, evolve and make blobs of some size, or however it is that these things are screening off reactions so it doesn't explode. And then, try to deduce as many consequences as you can that appear to be independent of the premises.

I don't know whether I have put in, or deduced, the hypothesis that cells are subcritical. But it's not *trivial* that cells *might* be just subcritical, and might have been so for three billion years. I don't know if it's circular yet, or not. But I certainly hadn't thought about it until I thought about this issue. You know, it's just too new yet to understand what the other possible ramifications of this idea are. But that's not bad, if that's correct. It wouldn't be bad to have a formulation that says, "Cells actually can't have more than about their current diversity of organic molecules and proteins; else they'd be supercritical internally." And by the way, Phil, your immune system acts on antigens outside of your cells, but inside of you...

ANDERSON: And preventing them from getting inside the cells.

KAUFFMAN: Exactly. You make a real fancy machine to try to ward off the external world, because it'll zap you.

CRUTCHFIELD: At the end of your talk you mentioned it's sort of unclear as to why a system would evolve to a more complex internal model. And the answer, I would suggest, is that even if you fix the uncertainty in the environment—measure that with the entropy rate, if you like—if the minimal model has more states in it, what that means is that the agent is recognizing more different environmental contexts. Each one of those states is dual to some behavior in the environment which the agent can take advantage of. Increased complexity can mean there's a wide range of strategies that the agent can implement that it can use for enhancing its survival.

GELL-MANN: Well, as science makes progress, it allows more and more applications, which allow exactly what you said. As you make a more sophisticated model, appreciating more and more of the properties of the environment, it enables you to master the environment in richer ways.

KAUFFMAN: Jim's question, I think, is getting back to Mitch's question, which is: "Why do things get more complicated? Why does it pay to push the envelope..."

CRUTCHFIELD: Is there a force? If you want, look at $d[\text{complexity}]/dt$, and call that a force; or write down the thermodynamic potential, which you can do for the stochastic automata. But you understand that if the minimal model has more states, then that means that the agent using the model distinguishes more contexts, more regularities in the data. This is one element of the pressures of an agent.

KAUFFMAN: I think that part of what I took away from you and from Brian's talk is that it seems to me very worthwhile to understand the hierarchy of models that you were pointing to, up to Turing models, and to try to come back and ask the question—as you go up from finite automata models and compress that in various ways and try and build models that are, given a finite set of data, the best they can be—does my argument fit with your argument, and Brian's argument? I don't see it yet, but it appears to me there's something very interesting there to explore.

CRUTCHFIELD: We could talk about bounded rationality. The analogue of that in coding theory is called the minimum description length principle. You take a finite amount of data, trade off the model complexity against the error stream, and that formula which is still up there [on the blackboard] on the bottom, you minimize that length "model plus error." And that's known to be asymptotically optimal. You use the model that minimizes that length, that total length; that's asymptotically optimal representation of the input.

KAUFFMAN: So what we should be fiddling around with is that, if I'm making a model of you and you're making a model of me. Then what's the dynamics and will it coevolve the way I'm trying to say that it does...

CRUTCHFIELD: The main point I wanted to make is that actually there's quite a bit understood about the specific constraints on the increase in complexity, and how you balance these finite resources and amount of data.

GELL-MANN: The M has to do with the intrinsic program or size of model and the E has to to with the input data.

KAUFFMAN: Right.

CRUTCHFIELD: What I wrote up summarizes the following situation: you have fixed amount of input data, your task is to build a model, and transmit a model plus an error stream—the data, the facts that are not explained in the model...

GELL-MANN: I'm sorry, my input data are not imput data in your sense. My input data are input data *to* the program; in other words, boundary conditions, things that are not explained, things that have to be put in afterwards.

CRUTCHFIELD: That defines, in my language, the model class.

GELL-MANN: Yes—the extra stuff, that's treated as random. The answer is that you have to look at the minimum description of the regularities, and then you have to look at how much stuff you have to add in, in order to actually do a calculation, in order to make a prediction.

CRUTCHFIELD: Okay. Although there is a physics question here, which I'm puzzling over. Namely, if I'm going to describe a thermodynamic system to you, I'm not going to give you all the data, Avogadro's number of data points. There's something very useful about writing down a compart formula for the free energy, independent of how much experimental data has ever gone before. But I'm making a criticism of this coding theory result I just mentioned: that you minimize the description length. It's useful because it's very concrete and you know what's going on. But if you throw away that error term than I am explaining my experiments to you without telling you all the data, so in other words you, the listener in the audience, can't reconstruct what I did in the lab because I'm just giving you the theory. Then, unfortunately, there's no concrete answer that I know of that tells you how to optimize the models in the cases where to goal is not complete communication.

KAUFFMAN: So you've just guessed the law.

CRUTCHFIELD: No, you can deduce it. You can even infer it. It's just that I don't want to listen to all the dirty laundry that you went through in order to come up with your idea. But that idea can still be useful to me. There is a theory, still.

GELL-MANN: The way that works is that there is a logical lumping of the coarse graining—which is what you're talking about, in ignoring all this extra junk in a thermodynamic system—and of the material that's going to be introduced not as part of the theory, or model, but as input data to the model. (Not your kind of input data; I mean boundary conditions.) You can always trade off between those: stuff that you ignore altogether and stuff that you ignore in making the model and put in later.

CRUTCHFIELD: How do you quantify finding the best theory in that context?

GELL-MANN: As Brian says; it depends on the context. You have a competition.

CRUTCHFIELD: Give me any setting where we can understand this.

ANDERSON: Well, thermodynamics is a context where you can really understand it. There is a thermal noise that can be divided from the deterministic parts.

ARTHUR: In our stock market models, the best theories are the ones that make the most profit, and you just let that evolve and there is a tradeoff between the model and the error, exactly as you're talking about. But it emerges: what works in a changing, adapting·environment.

CRUTCHFIELD: I think we're just talking past each other. Because that tradeoff includes what you just said, part of this error term. So that the theory, in and of itself, cannot be evaluated.

KAUFFMAN: Don't we have to distinguish the case in which we're playing against the world, and we're playing against or with one another? If you're trying to understand the world, we agree with Phil that there's a Lyapunov thing that's called "we're going to the truth about 'the World.'" If what we're doing is, in fact, coevolving with one another, as Brian's agents are, and as we always are when we're trying to coordinate our behavior, then we're always changing our own law of motion. Always, but given our model of the other guy. So we *can't* converge to something—I mean, we may not converge—there's a general dynamical system...

GELL-MANN: We trade off three things if you want to look at it that way. We trade off the coarse graining; the stuff that's being treated as boundary condition or infect to the model (that we don't try to explain); and the length of the model, the theory. And those are just adjusted to, in this case...

CRUTCHFIELD: And so these are the independent coordinates of some cost function?

GELL-MANN: Yes.

ANDERSON: You have to realize that Brian's stock market only worked with an error t, and it was really a relatively big error compared to the signal. He had to do it many many many times before he ever saw the signal...

CRUTCHFIELD: I agree, as long as you put each one of those costs in there.

ANDERSON: ...There's lots of extra data.

GELL-MANN: There's also the further point, that was mentioned by many people, having to do with the fact that in an actual calculation, based on a model, you can compress the model too far. Even though it's beautifully compact, you may find that the computation is arduous. And it's better to let the model get a little bit bigger, because it saves you having to put in a lot of explicit details.

KAUFFMAN: Murray, there's another problem, which you haven't mentioned, with compressed models. If you have to evolve a compressed model (think of your DNA as being a compressed model; it sort of generates you), the more compressed the model is, the more sensitive it is to every bit change; therefore, it's evolving on a more rugged landscape.

GELL-MANN: Yes. I put that in my list, also. That's called "redundancy for error correction."

KAUFFMAN: That points at the same issue. But somehow we have to have models that are learnable by the adaptive agents in the environments in which they mutually create for themselves. Which is different than doing science about the solar system. If you're in there in the world of practical action, you can't afford to sit back and make an abstract model; you have to act now. Because not acting now is also a decision.

GELL-MANN: But if you don't make it sufficiently abstract, you're not going to win. So it's a tradeoff. There are always these tradeoffs.

HÜBLER: There is a misunderstanding going on. The misunderstanding is the following: You have two agents, and they are doing a very simple dynamics. They are doing, for example, nothing. Everything is stationary. And then, you suddenly see that these two agents decide to make that interaction more and more complicated, till they end up at the edge of chaos. And the question is why? Aren't they perfectly happy with this stationary state? Why are they moving their environment to the edge of chaos? Of course, then they need to make a much more complicated model, but why don't they stay in the original state where everything was simple?

ANDERSON: They're bored.

KAUFFMAN: That may be very important.

ANDERSON: I think it is. It is part of the adaptation of the human, or even the animal. He has evolved to the point where he has curiosity.

GELL-MANN: Curiosity implies a pattern-recognition, schema-making device.

ANDERSON: That's a very good example.

GOODWIN: My question is actually relevant to the discussion that's just been going on, but it comes at it in a rather oblique way. I want to take Stuart back to his comments about the Burgess shale, the Cambrian explosion, and so forth. You want your model to apply very generally, in biology. You and I have talked a lot about morphology, and whether there are special rules that apply, whether it's a somewhat different case. It seems to me that the evidence is that there was a remarkable explosion of a diversity of metazoan body plants, and they were all accessible. There was some sense in which the system could move—that is, the evolving system—could move across from one to the other and generate these forms. And then as time went on, there was a loss of accessibility and it looked as if canalization took over.

Now that is part of the evolutionary process, and yet it looked as if those organisms, morphologically speaking at least, are falling into the domain of order, rather than sustaining a position of the edge of chaos (the domain in between these two regimes). And so, just to come back to our previous discussion, it looks to me as if there are more degrees of freedom with respect to *behavior* of organisms than their morphology. And it looks as if it's the behavioral domain that is the richer domain, with respect to the application of this model, and it may not apply to morphology. I wonder what you feel about that.

KAUFFMAN: Well, several things. As it happens, Brian and I have been working for some time on theories of morphogenesis, and there's a lot of people who have models around about how morphs actually get made. There is something awfully robust about morphology. So Brian and I and Jim Murray have recently found quite interesting reasons to suspect that if you put together a set of developmental mechanisms, they inherently tend to make a robust form, in the sense that a wide region of state space and parameter space unfolds to essentially the same morphology. It just sort of falls out from the mathematics of several models that we're looking at that you get a reliable morphology which leads one to the deep debate that Brian and I have been engaging in which is: how much of the morphology out there that you see is actually adaptive, and how much of it is just developmentally stuck because it's sort of the easiest way to make that kind of thing: a sponge, or a tetrapod with five fingers.

Yet I also think, Brian, that there's evidence that the Cambrian explosion might have been in a chaotic regime and, in fact, Roger Lewin partially lies behind this. I wrote a paper on the Cambrian explosion about two years ago. Roger asked, "How would you change it now?," and I said I'd try to bring in coevolution. And he said, "Well, how would that change it?" I said, Well, you'd have to try to say that the Cambrian explosion had an ecosystem that was in a chaotic regime, in which species are making niches for new species are making niches for new species, just as Brian was telling us about technological evolution. Which would lead to an interesting prediction: In the chaotic regime, you'd expect a lot more speciation events and a lot more extinction events, because things are plunging to low fitness.

Two weeks ago I was in Chicago, and Jack Sepkowski said, "I have something interesting for you." In fact, during the Cambrian explosion, speciation rates are enormously high and extinction rates are enormously high, and then they dwindle over the next hundred million years.

There's something very puzzling about the Cambrian explosion that you might all want to know. There were organisms around for a long time. There were single-celled organisms, then there were algae and fungus, and algal mats, roughly for three billion years. I mean, nothing much happened in Australia. They're just laying down algal mats for three billion years, or until, roughly, six hundred million years ago. Then there's the Ordovician fauna—that again probably Leo knows more about than I do—there's a burst of diversity of multicelled organisms about seven hundred to nine hundred million years ago, and then the Cambrian explosion in which you get an enormous diversity, filling in the phyla from the top down. Namely, the species that found phyla arise before the species that find classes, before the species that found orders. And later on, when you get enormous extinction events, you never create any new phyla; you create new species, new genera, and new families, after the Permian extinction.

So I think that this is consistent; this is looking a little bit like a picture of technological evolution. There's a slow creep in diversity that finally gets rich enough that there could be an explosion of niches creating niches creating niches. And a piece of it comes out of it in these coevolutionary models that we've been playing with. Because, if you have very few species interacting with one another, they go to a Nash equilibrium easily; if you just dot in more species playing with one another, the same system goes chaotic.

So you could imagine a system where at first you have few species interacting with one another in a stable regime, gradually increasing the number of species, and the thing just goes chaotic. So it would be very pretty to imagine that lots of these cases of a slow period of development, of diversity, and then an explosion...

SIMMONS: What do you mean by the thing going chaotic?

KAUFFMAN: I mean that an ecosystem with very few species in it, and very few species interacting with one another, can have the property that it goes to a Nash equilibrium. If you just keep—in this *NKCS* model—the richness of interactions that leaves the frog and the fly the same, but you put in more species (so that each species is interacting with more other species), you undergo a phase transition in that model ecosystem from the ordered regime to the chaotic regime.

SIMMONS: I don't understand. You were describing what happened at the time of the Cambrian explosion, and you said that there was a chaotic regime. This is something you can describe without referring to the *NK* model. I just want to understand what you mean in terms of observational data about this ecological system being in a chaotic regime.

KAUFFMAN: I'm making both some theoretical points and some observational points.
    You would expect of an ecosystem that's in a chaotic regime that the frog is buffeted about by the fly, and falls to low fitness at various periods. When at low fitness, it's more likely to go extinct, but when at low fitness there's more directions to improve, so it's easier to speciate. You would expect, therefore, that in the chaotic regime you see lots of speciation events and lots of extinction events. The observational fact is that with the Cambrian explosion (when diversity is increasing) is associated not only a very high rate of speciation, but a very high rate of extinction (compared to other regimes when there's a relatively low rate of speciation and a low rate of extinction). I take that, as a potential, put one's finger down on reality, and say, "Maybe that's a signature of an ecosystem and if you had speciation and extinction events in that..."

SIMMONS: So by chaos in an ecological system, you mean high rates simultaneously of speciation and extinctions.

KAUFFMAN: And I'm trying to tell you a piece of theory, that if you increase the number of species that are directly interacting with one another, as you increase it, the system can go from the ordered regime to the chaotic regime and that, therefore, you can imagine that, if you have gradual speciations going on, it'll eventually cross that boundary and explode. So that's a piece of theory to try to understand the Cambrian explosion.

BROWN: A brief comment with respect to just this issue, about the high correlation between speciation and extinction rate. Speciation is a multiplicative process, and it results in exponential increase in species if it goes very far. So for any *reasonable* span of evolutionary time, unless that happens, you *have* to have positive correlation between speciation and extinction rate...

KAUFFMAN: Because you fill up the biosphere.

BROWN: You fill up the biosphere with species. Now, let me get to the question which fits, I think, with the point that Brian raised. The argument that you've made about catalytic pathways and diversity of molecules, as you've indicated by citing the example of the Cambrian explosion, seems to have some analogues with what we see in the diversification of species, and that level of biological systems. And one interesting difference, it seems to me, is that in the proliferation of species you often lose the primitive forms. We lose the missing links, and we get this specialization thing. And that seems to be part of this loss of these primitive generalized forms.

In the case of molecular evolution, it's not clear to me about other examples where you lose certain molecular species that might be viewed as primitive species as the process continues.

KAUFFMAN: You might. If we knew the history of metabolism, it might turn out that the core of our metabolism is a small fragment of a richer, earlier, mushier metabolism, in which we've thrown away a lot of relatively useless pathways. The difference between organic chemistry and species is that the molecules don't change in organic chemistry. They're really there. But we may very well have evolved, and we may very well have crystallized down to a subset of pathways, like the Krebs cycle and the fundamental pathways of glycolysis, and so on, that are now at the core inner shell of metabolism, and thrown away lots of fluffy stuff that was there before. And, in fact, there's some evidence that they were about.

But what one's really pointing out is that there's at least two kinds of drives that can proceed towards diversity. One is this subcritical-supercritical story, at the molecular level. And the other is in a whole variety of ways, the niches creating niches creating niches, autocatalytically, whether it's technological evolution, or species evolution, or whatever, in some sort of functional space—in which, of course, you do lose ancestral form. I mean, people just don't make a living these days making Roman siege engines—one of these things that lofted stones in—because howitzers are an awful lot better for the same task. So we've lost the old morphology, and we've forgotten how to make them, too.

ARTHUR: Primitive things do stick around in most of the models I've seen. For example, in the Kristian Lindgren model, to quite a degree they stick around. One reason is that they can cope with a wider set of changes in the environment.

SIMMONS: Essentially what happens in Kristian's model is that the primitive things do *not* stick around, so that deep in the evolutionary path those long string strategies have never interacted with the short string (simple) strategies. They're driven out, but not by competition with the things that evolve much, much later.

ARTHUR: Well, the fancier things are able to track changes of the environment.

Sophisticated things have very high performance, and do extremely well, but possibly in a narrow range of environments. Primitive things tend to do well in a much wider range of environments. Therefore, to the degree that the environment is changing, to the degree you have explosions of niches, that will create a changing environment itself. That will provide a home for things that have a wider range. Therefore, you get a kind of coexistence between very sophisticated organisms and quite simple ones, and you tend to see that often. When the sophisticated organisms, however, can mimic anything that the primitive ones can do, and do it better, then the primitive ones will be driven out. And I think, to quite a degree, that happened in the Lindgren simulation.

ANDERSON: Brian, I don't think that's really true, because this Lindgren model has no load, no cost of complexity. And in every real system there is a cost of complexity.

GELL-MANN: Everybody agrees, I think, about the importance of regions of order and regions of disorder, and the fact that there are regions that lie in between. And people tell us different things about that region in between. The original notion of some people at the Santa Fe Institute was that it was a region of efficient adaptation. Chris was trying to persuade us yesterday—perhaps contrary to fact—that it's a region where you always have complexity; but sometimes there are simple phenomena in that region. Today, we're being told that systems are *driven* to that region. However, Stuart tells us that for three billion years of biological evolution life was on one side of it, and then in the Cambrian explosion life was on the other side, which is not exactly the same thing as saying that we're driven to stay there all the time.

So maybe what we have to do is relax a little bit and just see, in all the different questions that are studied, *how* that region comes in. It's a very important region, but maybe it's not wise to jump to conclusions—especially multifarious and self-contradictory conclusions—about exactly what role that region plays. It's very important for our work, but the importance should emerge, gradually.

KAUFFMAN: One is also trying to understand two other things that we understand very poorly: what are the wellsprings of diversity, and why do things get more diverse? And we have some pokes at that. What I'm trying to say with this image of the Cambrian explosion, where you get a leaking—a slow increase in the number of species—there can then be an explosive autocatalytic increase in the number of species (because they're making niches for one another faster than they are there).

GELL-MANN: In connection with that question, didn't Leo Buss say something that might be applicable there (he was talking about a different phase of biological evolution, a different epoch, but it still might apply): That some organizational innovation takes place—we're very familiar with that in our own lives, in technology, and so forth that *allows* multiplications of niches.

KAUFFMAN: But what we still need is to understand whether or not the exploration of increased complexity happens merely because it *can*—that is to say, it's sort of entropically driven.

GELL-MANN: But what about Jim Crutchfield's remark, that when you get richer schemata, then you learn more about the actual environment, and you can distort the actual environment.

KAUFFMAN: Yes. So we have some tentative, possible ideas (and never mind, things get more complex), and I think this poses a *fundamental* question that we could be looking at.

BAK: Actually, it's kind of a consequence of being at the critical state where avalanches or catastrophes of all sizes occur. You should not think of a big event as, say, "Hey, now we go from a regular case to a chaotic state," but this whole evolution is an integrated part of being at the critical state. You *should* have catastrophic events precisely because things are tied very much together at this critical state. And actually, if you take a look at the distribution of these catastrophic events in biology, take the empirical data, it seems to support that something like that is going on, and you do have events of all sizes, including catastrophes.

ARTHUR: I have a question for Brian Goodwin and for people around the table in general.

One of the themes I talked about this morning was what I called "captured software": the notion that a system can capture or slave some very primitive sub-system, which has a grammar, and get it to carry out certain tasks. It's the use of programming, because if I am a real macaque, and I pick up a piece of wood and bash a sea urchin with it, and learn to do that—that's all I can do with that. On the other hand, if I'm an immune system and manage to capture certain cells, I can program them to carry out a wide variety of informational functions. In other words, I'm genuinely programming a system that I've captured.

Brian Goodwin was telling me that you think that there might be an example. I'm curious about examples of this, or a critique of this, or questions to do with that. And I ask Brian if he would respond.

GOODWIN: I think that's coming out of the conjunction between molecular biology and cell biology, especially looking at the coding functions of genes, in eukaryotic cells. The cell cycle is something that's absolutely fundamental to all cells; if they don't cycle, they don't reproduce. So you have a certain basic machinery of cell cycle that is absolutely essential. That's the hardware; that's the implementation of a process, and it's essential. And I think morphology is like that; you have the implementation of processes that are essential. But tunes can be played on these by the software, by what is called "heterochrony": reorganizing the actual sequence of events. Heterochrony for me is not very revealing of what goes on, because that's just one of the degrees of freedom that's possible. But it's understanding the hardware and then seeing what the possibilities are. And the software explores all the possibilities available, with respect to retaining some vital functions.

ARTHUR: And in a sense the hardware doesn't matter.

GOODWIN: It doesn't matter. And the software just plays these, and the enhancers, the way enhancers work. . .that is, you get writing, and overwriting, and overwriting. The genetic control system is a matter of commentary on commentary, on commentary, and as it goes on it generates these diversities. And then it also stabilizes them. So you get a falling into these stable states, as evolution goes on. They have periods of exploration, and then stabilization. And I think there's good evidence coming up. That's just one example; there are many example like that that are beginning to emerge: universals, and then the variations that can be preyed on.

ARTHUR: Is there a name for this phenomenon in developmental biology, or in any other part of biology?

GOODWIN: No, there is no language because it's a bit too surprising. People reckoned that all the reaction was at the level of molecules. Now they're seeing that there are these integrated processes that are in a sense the implementation of what has been called developmental processes, for want of a better word.

ARTHUR: I can't think of a good phrase, maybe somebody can; in fact, Stuart's a good phrasemaker. I'm calling this "captured software" at the moment. Question was, is this familiar in some other area of science, have people talked about it, is there a label for it, etc., etc.

GOODWIN: Mark Kirschner does actually use the term hardware and software in an article that you'll be interested to look at. So it corresponds exactly to your conjecture.

WALDROP: An observation related to Murray's earlier question about what gets modeled and what doesn't: In the artificial intelligence community, there has been something like thirty or forty years of work on the problem of planning— an agent is trying to plan its way toward some goal—and this, of course, is to build a model of its future actions, a model in precisely that sense there; Murray talks about it, and you talked about it, Stuart. And for much of that thirty or forty years, the approach was basically analogous to theorem proving, that to reach a goal, you prove a theorem about how you reach the goal by using the steps. It is all very, very precise, and very very detailed; you had to know everything there possibly was to know about the environment out there. It was just total, perfect rationality.

Of course, in the last five to ten years, most of the work and planning said, "The hell with it. This is almost totally useless," largely because it's so intractable.

(But not the only reason; I'll get back to that.) What they've done, in general, is to design agents such as a robot, come up with a very simple schematic plan, which could be the simplest thing: "I'll start out in this direction." And then, the strategy is to start out in that direction and then wait and see, and cope with situations as they come up. This is analogous, I think, to what Murray was getting at, to what you make a model of versus what you sort of take as it comes, experience as it comes in the world. And, of course, the reason you do this, the reason this is far more effective, is not just because it's computationally intractable to try to project ahead that far, but because intrinsically there's always going to be something in a complicated world that you don't know, because your sensors don't detect it, because you can't be in more than one place at once, and just in general, it's too complicated, it's too uncertain.

And what in general is substituted, as I understand this work, is that you not only model what you're going to do, but you also have a model in your head of what it's not worth bothering about, because you know you can cope with it as it comes along. And, of course, this model of what is not worth bothering about is something that also comes in experience, so you can approach this problem that Murray raised about what is model-explicit, and what isn't, by having a sort of metamodel or a different kind of model, of what you can leave out. There is in general no general theory about what that is; that has to come with experience.

KAUFFMAN: What happens if you add to that the image of living in a supercritical world? It's really the case that, when organic molecules are acting on one another, it can be formally undecidable that some given organic molecule will eventually be formed—and I'm a bug living in that world. Well, if it's formally undecidable, then either I have the computational capacity to grind through them all and say, "Well, for the next million years I can afford to be the kind of bug I am. I know it's not going to happen in the next million years." Or, in some sense, you've just got to live. That's assuming you knew the grammar. But we usually don't know the grammar. So if you don't know the laws for sure, and in any case it would lead to a formally undecidable world, you've just got to muck around, because you can't do better.

ANDERSON: Doesn't formally undecidable merely mean that you have separate attractors, and we don't know which attractor you're on? Or does it mean something fancier?

SIMMONS: It means it's possible to ask questions without knowing whether, within the formal system, it is possible to get an answer.

EPSTEIN: It's formally demonstrable that neither truth nor falsity can be deduced from the given axiomatization of the system.

KAUFFMAN: It's not the same as falling into another attractor.

EPSTEIN: No. No, it's technically different.

KAUFFMAN: There's something strange about our notions of attractors, compared to worlds in which you have indefinitely extended sets of possible results. These jets and mushrooms, and so on, are somehow a different world than a world with finite states.

WALDROP: That's actually an interesting point, because in all these theories about what's decidable and what isn't, they're talking about computations that halt, that will or will not eventually falter. Now in a world where everything's constantly changing, you're constantly having to cope—nothing halts.

SIMMONS: No, but halting is just a way of making an equivalence class to the undecidable questions. "Does this process halt or not" is an example of an undecidable question. The process doesn't have to halt, it's just an equivalent process.

EPSTEIN: Quick question. I'm just trying to clarify, reconcile, the picture you have here with a statement you make about the picture you have in this K-optimal selection picture. What you were saying is that the metadynamics lead the coevolving system to the edge of chaos, the optimum, and it would seem that what you're saying is that the optimum is a stable attractor from below, but an unstable attractor from above, right? In that, if you go supercritical you completely explode, but if you're subcritical you're driven to the optimum. But this picture looks like it's an unstable attractor from both directions, and I'm just wondering what's going on.

KAUFFMAN: You're worried about the coevolving case, and the story there is, that if they're all coevolving on very rugged landscapes, with low peaks, then if I change my landscape structure so that there are fewer peaks but the peaks are higher, as long as I can catch my peak—when you deform—I'm better off. So I change the game I'm playing to make my landscape smoother.

EPSTEIN: But you are claiming that in the subcritical zone you're attracted to the boundary, and in the supercritical zone you're repelled from it.

KAUFFMAN: No, no. You're attracted to it.

EPSTEIN: From both sides.

KAUFFMAN: You're attracted from both sides. That's the point.

WALDROP: A similar question. When you look at these autocatalytic models, and you're talking about being supercritical or subcritical, and when you're supercritical you keep expanding out into molecule space: What is the time course of development there? Suppose you're right at this critical boundary; do you get sort of punctuated equilibrium spurts and some kind of power law behavior?

KAUFFMAN: One would love to believe so. I have no idea.

WALDROP: When you try it on the computer, what does it look like?

KAUFFMAN: Here's what we've actually done. Bill Tozier in my lab has been doing a Walter Fontana experiment—making a random grammar and letting the strings run into them. When you do that, there's the following interesting result. There are live symbol strings and dead symbol strings. A live symbol string can be a substrate for one reaction, or an enzyme; a dead symbol string is just inert. The counterintuitive thing that happens is that in an open stirred reactor—the Noah experiment, going on in time—what happens is that the system generates a lot of inert strings; they're chemically nonreactive. And the live strings waste a lot of time running into and interacting with the dead strings.

So another piece of this puzzle, I think, is that an advantage of making vesicles is to keep the strings away so the live strings interact with live strings, which is the same as saying that I bet that if you do it right, you are making organizations that suck up the stuff of the universe (the food molecules) into themselves the fastest. And this gets back in the funny way that Phil's throwing away degrees of freedom. Therefore, it's right at the phase transition that you actually get the most rapid increase of reactive molecules. Okay? And my bet would be that you have just that stuttering that you're thinking about, and it might be testable in the sense that, if our cells were just poised on the boundary between subcritical and supercritical, if we could look at the evolution of organic molecular diversity, you would see funny bursts. Sometime your molecule will come to the biosphere and will be battered around by a bunch of organisms, and generate a thousand new kinds of organic molecules over a period of twenty years. At other times there should be periods of quiescence. We can't monitor the number of kinds of organic molecules that were around in the biosphere being carried around by less blobs, squirting at one another. But you ought to see just that funny kind of burst And so I hope that one can build models, and see that.

SIMMONS: Can I ask you to clarify something about what you just said that, since I don't know any biochemistry, I found somewhat surprising. Key to the model you described was the idea that every polymer produced by your system is reactive, in principle, with more than one of the other polymers. Now you just told me that in experimental systems there is a problem that you produce lots and lots of nonreactive molecules, which can block anything that's going on. What happens to your system if you put into it that many of the polymers produced in your system (let's say an adjustable percentage) are, in fact, nonreactive?

KAUFFMAN: The answer to that, Mike, is the following. It's the same thing that Walter is finding. You see, you can either think of molecules interacting with one another in a "mean-field" kind of way, or you can actually have rules by which molecules interact. A "mean-field" kind of way is to say that there's a fixed probability that any polymer catalyzes any reaction. If, in fact, you lower that probability enough, then, for any finite set of molecules, it's subcritical and not much happens. If you increase it enough, increase the diversity, you pass the threshold.

If instead you say, "Let's not do it that way; let's be chemically realistic. I will want my polymers that catalyze reactions to be template complements of the left and right ends of the polymers they interact with, like their RNA molecules." And then you find you still get autocatalytic sets. The autocatalytic sets, when they're spewing forth kinds of molecules, have then also to make nonreactive molecules. So you both get a proliferation of reactive molecules, which can be supercritical, and some sort of abundance of inert molecules that are sitting around; just hang more there. So the answer is, that happens.

And the interesting puzzle is that you suspect that, from the point of view of the reactive molecules, flowing through the dead molecules is a waste, so if the live molecules can learn to divert the flow, it will be to the live molecules' advantages. And that's part of thinking that there's such a collective potential of transformation that the live molecules, that react with them, can do to suck up the flow of material into themselves to make these reactive blobs.

SIMMONS: Now, I'm not sure I understood all the answer, but suppose I try to actually run the system. . .

KAUFFMAN: You mean physically, or on a computer?

SIMMONS: On a computer. Run the system with a single adjustable parameter, let us say, which is the probability, that any newly produced molecule is either completely nonreactive, or reactive. . .

KAUFFMAN: You will tune from subcritical to supercritical.

SIMMONS: You will, you've done this?

KAUFFMAN: Yes. That's what we did in the first papers.

ANDERSON: Can I come back with the undecidability question? Computers are one thing, and physical systems are another, and they don't really have identical properties. In a physical system, no question is undecidable, because you just let this physical system run, and it goes on running, and it either makes that molecule or it doesn't. Therefore, it is true that your criterion of undecidability must mean something else, and the question is what does it mean. One speculation is whether it makes that molecule or not depends on the starting conditions. . .

SMITH: You might have to wait an infinite amount of time for the system to produce that molecule.

ANDERSON: Well, you won't have to wait an infinite amount of time, you only have a finite time.

KAUFFMAN: The question is: can you prove that this molecule will not be made by the system?

ANDERSON: I don't care about proofs. Proofs aren't important, because this is a physical system. It will do what it will do.

KAUFFMAN: No, it will do what it will do, but you might be curious to know—as an *E. coli* waiting around to eat this yummy molecule—that you can just imagine whether or not the biosphere will ever provide it to you...

ANDERSON: I think what you're saying is really that there is a finite probability that it is there, and a finite probability that there isn't...

SIMMONS: I think Josh's point is that, if the time to decide that, to run that physical calculation, is the lifetime of the universe measured in microseconds...

FONTANA: No, I think Phil is completely correct in this way. In fact, I think undecidability is a red herring for the entire story here. Obviously, there is no undecidable reaction. You put two molecules together—any two molecules you want—and give them certain thermodynamic parameters, and either the reaction takes place or it doesn't, and you will know it.

WALDROP: Wait a minute. Walter, in that sense a computation isn't undecidable either, because a computer is a physical device, you rev it up, and it either halts or eventually...

ANDERSON: That worries me a little bit about the computer.

FONTANA: Undecidability comes in because it's an artifact, in a certain sense, because you have a Turing machine, then whenever it needs a new square on the tape, it gets a new square. So there's an infinite supply of new squares.

WALDROP: My intuition was something much simpler: that undecidability is really talking about whether you can predict ahead of time whether it's going to halt.

FONTANA: Right, but if the total number of squares you have on the tape is limited, there is no issue of undecidability.

KAUFFMAN: But then you're saying—in answer to Phil's question—since the universe is finite and the universe will end in a big crunch, all we have to do is wait until the big crunch.

ANDERSON: Well, you are at this point getting at real problems of measurement theory, of the validity of thermodynamics, and whether your system is really deterministic—as the computer is—or not deterministic, and so on.

Thomas S. Ray
ATR Human Information Processing Research Laboratories, 2-2 Hikaridai, Seika-cho, Soraku-gun, Kyoto, 619-02, Japan; e-mail: ray@hip.atr.co.jp, ray@udel.edu

Evolution and Complexity

Abstract: The process of evolution is an important integrative theme for the sciences of complexity, because it is the generative force behind most complex systems. The surface of the Earth is covered with phenomenally complex living structures such as the human brain and the tropical rainforest, which emerged from simple molecules through evolution. While the results of evolution by natural selection are abundantly visible, the process is difficult to observe in nature because it is slow compared to a human lifespan. One method of observing the actual generation of complexity through evolution is to inoculate an artificial system with natural evolution. This can most easily be done in computers, where the process can be accelerated to megahertz speeds. The fundamental elements of evolution are self-replication with heritable variation. This can be implemented in a computer by writing self-replicating machine language programs and running them on a computer that makes mistakes. The mistakes can take the form of bit-flip mutations and small errors in calculations or the transfer of information. This experiment results in rapid diversification of digital organisms. From a rudimentary ancestral self-replicating "creature," entire ecological communities emerge spontaneously. Natural evolution in this artificial system illustrates well-established principles of evolutionary and ecological theory, and allows an experimental approach to the study of evolution, as well as observations of macroevolutionary processes as they occur.

INTRODUCTION

Evolution is an extremely powerful natural force, which, given enough time, is capable of spontaneously creating extraordinary complexity out of simple materials. Evolution is the process that has generated most, if not all, known complex systems. These are either the direct products of biological evolution: nervous systems, immune systems, ecologies; or the eip-phenomena of biological evolution: cultures, languages, economies. Thus, understanding evolution is important to understanding complex systems.

The greatest obstacles to this understanding are that we have only a single example of evolution available for study (life on Earth) and that, in this example, evolution is played out over huge time spans. In spite of these limitations, evolutionary theory has firmly established many basic principles. However, these principles have been established through the logical analysis of the static products of evolution, but without actually observing the process, without experimental test, and without the benefit of comparing independent instances of evolution.

Darwin[1] laid out the core of the currently accepted theory of evolution after the voyage of the Beagle. This voyage gave him the opportunity to observe first hand the variation of species preserved in the fossil record, and preserved among geographically isolated populations in areas like the Galapagos archipelago. Darwin formulated the elements of the theory that is still the core of evolutionary biology today:

1. Individuals vary in their viability in the environment that they occupy.
2. This variation is heritable.
3. Self-replicating individuals tend to produce more offspring than can survive on the limited resources available in the environment.
4. In the ensuing struggle for survival, the individuals best adapted to the environment are the ones that will survive to reproduce.

As a result of the iteration of this process over many generations, lineages of organisms change, generally becoming better adapted to their environment.

Darwin developed this theory without actually observing the process, and without the benefit of experimental tests. In this chapter I will describe a method that provides both an experimental test of the theory and the easy observation of the process on a macroscale. This method should allow us to refine our understanding of evolution by observing details of the process that have never been observable before, and by allowing easy manipulation of parameters of the process in an experimental context. This method will also make it feasible to address questions in evolutionary theory that were previously intractable (e.g., how do environmental parameters affect the shape of a phylogenetic tree evolving in that environment?).

METHODS

The methodology has been described in detail by Ray,[5,6,8] so it will be described only briefly here. A new computer architecture has been designed which has the feature that its machine code is robust to the genetic operations of mutation and recombination. This means that computer programs written in the machine code of this architecture remain viable some of the time after being randomly altered by bit-flips, which cause the swapping of individual instructions with others from within the instruction set, or by swapping segments of code between programs (through a spontaneous sexual process). This new computer has not been built in silicon, but exists only as a software prototype known as a "virtual computer." This virtual computer has been called "Tierra," Spanish for Earth.

A self-replicating program was written in this new machine language. The program functions by examining itself to determine where it begins and ends, then calculating its size (80 bytes), and then copying itself one byte at a time to another location in memory. After that, both programs replicate, and the number of programs "living" in memory doubles in each generation.

These programs are referred to as "creatures." The creatures occupy a finite amount of memory called the "soup." The operating system of the virtual computer, Tierra, provides services to allocate CPU time to the growing population of self-replicating creatures. When the creatures fill the soup, the operating system invokes a "reaper" facility which kills creatures to ensure that memory will remain free for occupation by newborn creatures. Thus, a turnover of individuals begins when the memory is full.

The operating system also generates a variety of errors which play the role of mutations. One kind of error is a bit-flip, in which a zero is converted to a one, or a one is converted to a zero. This occurs in the soup, which is where the information that constitutes the programs of the creatures resides. The bit-flips are the analogs of mutations, and cause swapping among the instructions of the machine code. Another kind of error imposed by the operating system is called a "flaw," in which calculations taking place within the CPU of the virtual machine may be inaccurate, or in which the transfer of information may move information to or from the wrong place, or may slightly alter its content.

Mutations cause genetic change and are, therefore, heritable. Flaws do not directly cause genetic change, and so are not heritable. However, flaws may cause errors in the *process* of self-replication, resulting in offspring that are genetically different from their parents, and those differences are then heritable.

The running of the self-replicating program (creature) on the virtual computer (Tierra), with the errors imposed by the operating system (mutations), results in a computer metaphor of evolution. The sequence of machine instructions that constitute the program of a creature is analogous to the sequence of nucleotides that constitute the genome, the DNA, of organic organisms. The soup, a block of RAM memory of the computer, is thought of as the spatial resource. The CPU time provided by the virtual computer is thought of as the energy resource. The sequences of machine instructions that make up the genomes of the creatures constitute an informational resource which plays an important role in evolution.

RESULTS

The details and mechanisms of the evolution of creatures in the Tierran computer have been described in detail[4,5,6,7,8] and will only be summarized here. Running of the self-replicating program on the error-prone computer creates a situation that is, in fact, identical to the one outlined by Darwin. Those genotypes that are most efficient at replicating, leave more descendants in the future generations, and increase in frequency in the population.

What is most striking in this process is the surprising variety and inventiveness of evolved means of replication. Some organisms increase in efficiency through straightforward optimization of the replication algorithm. However, replication is also achieved through more surprising avenues involving interactions between creatures.

Evolution increases the adaptation of organisms to their environment. In the Tierran universe, initially the environment consists largely of the memory which is fairly uniform and always available, and the CPU which allocates time to each creature in a consistent and uniform fashion. In such a simple environment the most obvious route to efficiency is optimization of the algorithm. However, once the memory is filled with creatures, the creatures themselves become a prominent feature of the environment. Now evolution may discover ways for creatures to exploit one another and to defend against such exploitation. The two major modes of increasing adaptation involve optimization and coevolution. These two forms of evolution will be discussed separately below.

OPTIMIZATION

Optimization in digital organism involves finding algorithms for which less CPU time is required to effect a replication. This is always a selective force, regardless of how the environmental parameters of the Tierran universe are set. However, selection may also favor reduction or increase in size of the creatures, depending on how CPU time is allocated to the creatures.

If each creature gets an equal share of CPU time, selection strongly favors reduction in size. The reason is that all other things being equal, a smaller creature requires less CPU time because it needs to copy fewer instructions to a new location in memory.

If CPU time is allocated in direct proportion to the size of the creature, then selection favors neither size reduction nor size increase, because the availability of "energy" (CPU time) to a creature increases in direct proportion to the number of instructions that it must copy to a new location of memory. If CPU time is allocated in proportion to the size of the creature raised to a power greater than one, then selection will favor increase in size of creatures.

Under selection favoring a decrease in size, evolution has converted the original 80-instruction creature to creatures of as few as 22 instructions, within a time span of half a billion CPU cycles of the system as a whole (representing perhaps 1500 generations). Different runs under the same initial parameters, but using different seeds to the random generator, achieve different degrees of optimization. These runs have plateaued at 22, 27, and 30 instructions. While one could easily conclude from this that evolution can get caught on a local optimum from which it cannot reach the global optimum,[8] there is an alternative interpretation.

Ray[9] has shown that the larger plateaus are more complex algorithms which do not minimize the size of the algorithm but increase its efficiency, in terms of the number of CPU cycles required to move a byte of information. This efficiency is achieved through a technique called "unrolling the loop" (see below). This involves the evolution of much more intricate algorithms which are larger but more efficient.

The evolved increase in efficiency of the replicating algorithms is even greater than the decrease in the size of the code. The ancestor is 80 instructions long and requires 839 CPU cycles to replicate. The creature of size 22 only requires 146 CPU cycles to replicate, a 5.75-fold difference in efficiency.

Runs under selection for large size are not very interesting, since the creatures tend to increase in size until only a single creature can fit in memory. This creature will then not be able to replicate, for lack of space to form a daughter. Sometimes the population will die out before being reduced to a single individual, because small populations suffer from genetic problems.

Under conditions in which selection favors neither size increase nor size decrease, it is possible to run the system indefinitely without ever reaching an obvious stable endpoint. Optimization under these conditions tends to involve the production of cleverer algorithms, which are not necessarily smaller than the ancestral algorithm. In fact, under these conditions, the algorithms may show a significant increase in their complexity. A most stunning example of this involves the unrolling of a loop.

The central loop of the copy procedure of the ancestor (see the Appendix) performs the following operations: (1) copies an instruction from the mother to the daughter; (2) decrements the CX register which initially contains the size of the parent genome; (3) tests to see if CX is equal to zero (if so, it exits the loop; if not, it remains in the loop); (4) increments the AX register which contains the address in the daughter where the next instruction will be copied to; (5) increments the BX register which contains the address in the mother where the next instruction bill be copied from; and (6) jumps back to the top of the loop.

The work of the loop is contained in steps 1, 2, 4, and 5. Steps 3 and 6 are loop overhead. The efficiency of the loop can be increased by duplicating the work steps within the loop, thereby saving on overhead. A creature from the end of a run of 15 billion CPU cycles had repeated the work steps three times within the loop, as illustrated in the appendix, which compares the copy loop of the ancestor with that of its descendant.

This evolved creature exhibits an additional unrelated adaptation. In an environment where the allocation of energy to a creature is directly proportional to its size, evolution discovered the value of lying about size. The creature in the appendix calculates its size as 36, but requests a space of 72 instructions for its daughter, thereby doubling the energy available to the daughter. When it first appeared, this deception provided a powerful advantage. However, fitness is relative. Once the mutation swept the population and all the creatures were lying, there was no longer any advantage.

Creatures have to compensate for the lie by counting down the size twice for every instruction copied to the daughter. Otherwise, they would copy all 72 instructions into their daughter, and there would never be an advantage to the lie. Therefore, the unrolled loop contains two decrements of the size for every increment of the source and destination addresses.

In the ancestral algorithm, the "work" part of the copy loop consists of four instructions: $movii$, dec_c, inc_a, and inc_b. Due to the advent of the lie about size, this set of work instructions became slightly more complex, requiring two instances of dec_c. Thus, the "work" part of the evolving copy loop requires the proper combination and order of five instructions. Yet the organism 0072etq shows this set of instructions repeated three times (with varying ordering, indicating that the unrolling did not occur through actual duplications of the complete sequence).

This algorithm is substantially more intricate than the unevolved one written by the author. The astonishing improbability of this complex ordering of instructions is testimony to the ability of evolution through natural selection to build complexity.

COEVOLUTION

Once the self-replicating program has filled the memory, evolution discovers that information is a resource that is readily available in the environment. It is feasible for creatures to shed some information from their genome, and simply obtain it from the neighboring creatures that fill the environment. This buys efficiency, in that the genomes can be made smaller, since they do not need to contain all essential information, as long as the information missing from the genome is available in the environment. This results in a form of informational parasitism. When all creatures receive equal amounts of energy (a condition favoring small creatures), the parasites replicate faster than their host which must carry and copy all the information required for replication.

This causes the parasites to increase in frequency and, as they share a finite space with their hosts, the hosts decline. Eventually, the information that the parasites have shed from their genomes becomes rare in the environment. The parasites will then begin to die off as many of them will not be able to find a host. The hosts and parasites show Lotka-Volterra population cycles.

This initial interaction is the starting point for an ongoing evolutionary race. The hosts can gain advantage by preventing themselves from being parasitized and, in fact, forms of immunity to parasitism do evolve. Parasites evolve to circumvent the immunity of their hosts. Hosts evolve to exploit parasites, by deceiving the parasites that attack them into replicating the genome of the host. This is a form of energy parasitism, in which the energy metabolism of the parasite is subverted to the replication of the genome of the host.

The energy parasitism is so deleterious to the parasites, that they are driven to extinction. The remaining creatures totally own the memory, resulting in a situation where all the creatures in the soup are closely related. These are the conditions that facilitate the evolution of cooperation, and, in fact, various forms of social behavior appear among the victorious creatures. Individual creatures are no longer able to replicate. They can only replicate through cooperation among creatures in close aggregations. However, the cooperation involves a certain trust, which can be violated. Cheaters invade the community and trick the cooperating creatures into replicating the genomes of the cheaters. The cheaters play the same trick on the social creatures that the ancestors of the social creatures invented to drive out the parasites.

DISCUSSION
EVOLUTION THROUGH GENOTYPE SPACE

Think of the creatures as occupying a variable-dimensional "genotype space" consisting of all possible sequences of all possible lengths of the 32 machine instructions. When the system begins running, a single self-replicating creature, with a single sequence of 80 instructions occupies a single point in the genotype space. However, as the program replicates in the computer, a population of creatures forms, and the errors made by the virtual computer cause genetic variation, such that the population will form a cloud of points in the genotype space, centered around the original point.

Because the new genotypes that form the cloud are formed by a random process, most of them are completely inviable, and die without reproducing. However, some of them are capable of reproduction. These new genotypes persist and, as some of them are affected by mutation, the cloud of points spreads further. However, not all of the viable genomes are equally viable. Some of them discover tricks to replicate more efficiently. These genotypes increase in frequency, causing the population of creatures at the corresponding points in the genotype space to increase.

Points in the genotype space occupied by greater populations of individuals will spawn larger numbers of mutant offspring; thus, the density of the cloud of points in the genotype space will shift gradually in the direction of the more-fit genotypes. Over time, the cloud of points will percolate through the genotype space, flowing down fitness gradients.

Most of the volume of this space represents completely inviable sequences. These regions of the space may be momentarily and sparsely occupied by inviable mutants, but the cloud will never flow into the inviable regions. The cloud of genotypes may bifurcate as it flows down fitness gradients in different directions, and it may split as large genetic changes spawn genotypes in distant but viable regions of the space. We may imagine that the evolving population of creatures will take the form of whispy clouds flowing through this space.

Now imagine for a moment the situation that there was no selection. This implies that every sequence is replicated equally. Mutation will cause the cloud of points to expand outward, eventually filling the space uniformly. In this situation, the complexity of the structure of the cloud of points does not increase through time, only the volume that it occupies. Under selection, by contrast, through time the cloud will take on an intricate structure as it flows down fitness gradients through narrow regions of viability in a largely uninhabitable space.

Consider that the viable region of the genotype space is a very small subset of the total volume of the space, but that it probably exhibits a very complex shape, forming tendrils and sheets permeating the otherwise empty space. The complex structure of this cloud can be considered to be a product of evolution by natural selection. This thought experiment appears to imply that the intricate structure that the cloud of genotypes may assume through evolution is fully deterministic. Its shape is predefined by the physics and chemistry and the structure of the environment, in much the same way that the form of the Mandlebrot set is predetermined by its defining equation.

No living world will ever fill the entire viable subspace, either at a single moment of time, or even cumulatively over its entire history. The region actually filled will be strongly influenced by the original self-replicating sequence, and by stochastic forces which will by chance push the cloud down a subset of possible fitness gradients. Furthermore, coevolution and ecological interactions imply that certain regions can only be occupied when certain other regions are also occupied. This concept of the flow of genotypes through the genotype space is essentially the same as that discussed by Eigen[3] in the context of "quasi-species." Eigen limited his discussion to species of viruses, where it is also easy to think of sequence spaces. Here, I am extending the concept beyond the bounds of the species, to include entire phylogenies of species.

The flow of machine code sequences through the genotype space in the Tierran computer is an example of Darwinian evolution by natural selection. It is not a "model" of evolution; rather it is an instance of it, occurring in a physical substrate that is radically different from the one that we organic creatures inhabit. This second instance of natural evolution was created by setting up nothing more than the fundamental elements of Darwinian natural selection: heritable variation among entities self-replicating in a finite world.

This work could be considered as an experimental test of Darwin's theory, and evidently has confirmed the sufficiency of the theory as originally stated. In addition, some other ideas in evolutionary biology have been supported by the experiment. It has been suggested that an important force in the generation of diversity and complexity in evolution is the evolutionary race between predator and prey or parasite and host, which has an autocatalytic and escalatory nature.[2,10] One of the most firmly established ideas in evolution is that genetic relationship is the basis of the evolution of cooperation. Both of these processes are evident in the evolution of digital organisms.

It is likely that extensions of the experiment will also make it possible to test other areas of evolutionary theory. Communities of digital organisms exhibit well-known ecological phenomena as well (e.g., host-parasite population cycles, and "keystone" parasite effects). This suggests that digital communities could also be used for experimental tests of ecological theory.

EVOLUTION AS AN INTEGRATIVE THEME

The science of complexity is of necessity an interdisciplinary one, since complex systems include (at least) both biological and social phenomena. Interdisciplinary studies can be hampered by language problems, as researchers from different disciplinary backgrounds come together and attempt to communicate in their diverse specialized technical vocabularies. One of the consequences of this process is that words that appear useful in many disciplines can loose their meaning as they gain broader use.

A couple of words that have suffered this fate are "chaos" and "evolution." Chaos has a well-defined meaning in the context of dynamical systems, but in broad usage has come to mean nothing more than disorder. In the context of evolutionary biology, "evolution" refers to the iterative process outlined by Darwin (see Introduction above), but in broad usage has come to mean nothing more than change. The phrase "evolution to the edge of chaos" typifies the problem.

Physicists speak of the evolution of the universe, economists speak of the evolution of the automobile, and linguists speak of the evolution of languages. There is nothing wrong with the word evolution being used in these ways, as long as it is not confused with the concept of Darwinian evolution, which has a very specific technical meaning in evolutionary biology. Darwinian evolution is not equivalent to the processes being described as evolution by physicists, economists, or linguists.

The most fundamental sense in which the evolutions described by physicists, economists, and linguists are not Darwinian, is in that their evolving entities generally do not reproduce. Beyond this, where some form of reproduction does occur, such as a corporation spawning new corporations, the "genetic" system is fundamentally different. In this example, the information passed from mother to daughter corporation will have been modified during the lifetime of the mother, resulting in Lamarckian rather than Darwinian "evolution."

In reality, the meaning of the word evolution is different in each disciplinary context. In physics, "evolution of the universe" refers to an unfolding of the distribution of the matter and energy of the universe according to the basic laws of physics. In economics, "evolution of corporations" refers to a process of learning in which corporations change to keep up with changes in technology and markets. In linguistics, "evolution of languages" refers to a process of linguistic change that is generally closely linked to the population biology of humans, and can reflect the human evolutionary process, in an accelerated form. In biology, "evolution of species" refers to the process of Darwinian evolution in which self-replicating individuals pass genetic information on to offspring through DNA.

There is nothing wrong with the word evolution having different meanings in different disciplines. The problem arises in the interdisciplinary context, when persons attempt to communicate using a word like evolution which does not have a common meaning in the new shared context. It can be tempting to take the large body of theory developed in evolutionary biology and apply it to other areas where the word evolution is being used. However, this could be very misleading as the theory may not be appropriate to the new contexts.

As the ultimate generative force of perhaps all complex systems, evolution is an important integrative theme for the sciences of complexity. Nonetheless, it must be recognized that if we are going to view evolution as an active process in the cultural arenas (e.g., in economies or languages), then we must recognize that the evolutionary process will have substantially different properties in the different arenas.

ACKNOWLEDGMENTS

This work was supported by grants CCR-9204339 and BIR-9300800 from the United States National Science Foundation, a grant from the Digital Equipment Corporation, and by the Santa Fe Institute, Thinking Machines Corp., IBM, and Hughes

Aircraft. This work was conducted while at: Santa Fe Institute, 1660 Old Pecos Trail, Suite A, Santa Fe, New Mexico, 87501, USA, ray@santafe.edu; and School of Life & Health Sciences, University of Delaware, Newark, Delaware, 19716, USA, ray@udel.edu.

APPENDIX

Assembler code for the central copy loop of the ancestor (80aaa) and descendant after 15 billion instructions (72etq). Within the loop, the ancestor does each of the following operations once: copy instruction (51), decrement CX (52), increment AX (59), and increment BX (60). The descendant performs each of the following operations three times within the loop: copy instruction (15, 22, 26), increment AX (20, 24, 31), and increment BX (21, 25, 32). The decrement CX operation occurs five times within the loop (16, 17, 19, 23, 27). Instruction 28 flips the low-order bit of the CX register. Whenever this latter instruction is reached, the value of the low-order bit is one, so this amounts to a sixth instance of decrement CX. This means that there are two decrements for every increment. The reason for this is related to another adaptation of this creature. When it calculates its size, it shifts left (12) before allocating space for the daughter (13). This has the effect of allocating twice as much space as is actually needed to accommodate the genome. The genome of the creature is 36 instructions long, but it allocates a space of 72 instructions. This occurred in an environment where the slice of CPU time allocated to each creature was set equal to the size of the creature. In this way the creatures were able to garner twice as much energy. However, they had to compliment this change by doubling the number of decrements in the loop.

```
nop1    ; 01  47 copy loop template        COPY LOOP OF 80AAA
nop0    ; 00  48 copy loop template
nop1    ; 01  49 copy loop template
nop0    ; 00  50 copy loop template
movii   ; 1a  51 move contents of [BX] to [AX] (copy instruction)
dec_c   ; 0a  52 decrement CX
ifz     ; 05  53 if CX = 0 perform next instruction, otherwise skip it
jmp     ; 14  54 jump to template below (copy procedure exit)
nop0    ; 00  55 copy procedure exit compliment
nop1    ; 01  56 copy procedure exit compliment
nop0    ; 00  57 copy procedure exit compliment
nop0    ; 00  58 copy procedure exit compliment
inc_a   ; 08  59 increment AX (point to next instruction of daughter)
inc_b   ; 09  60 increment BX (point to next instruction of mother)
jmp     ; 14  61 jump to template below (copy loop)
nop0    ; 00  62 copy loop compliment
nop1    ; 01  63 copy loop compliment
nop0    ; 00  64 copy loop compliment
nop1    ; 01  65 copy loop compliment (10 instructions executed per loop)
```

```
shl     ; 000 03   12 shift left CX          COPY LOOP OF 72ETQ
mal     ; 000 1e   13 allocate daughter cell
nop0    ; 000 00   14 top of loop
movii   ; 000 1a   15 copy instruction
dec_c   ; 000 0a   16 decrement CX
dec_c   ; 000 0a   17 decrement CX
jmpb    ; 000 15   18 junk
dec_c   ; 000 0a   19 decrement CX
inc_a   ; 000 08   20 increment AX
inc_b   ; 000 09   21 increment BX
movii   ; 000 1a   22 copy instruction
dec_c   ; 000 0a   23 decrement CX
inc_a   ; 000 08   24 increment AX
inc_b   ; 000 09   25 increment BX
movii   ; 000 1a   26 copy instruction
dec_c   ; 000 0a   27 decrement CX
not0    ; 000 02   28 flip low order bit of CX, equivalent to dec_c
ifz     ; 000 05   29 if CX == 0 do next instruction
ret     ; 000 17   30 exit loop
inc_a   ; 000 08   31 increment AX
inc_b   ; 000 09   32 increment BX
jmpb    ; 000 15   33 go to top of loop (6 instructions per copy)
nop1    ; 000 01   34 bottom of loop    (18 instructions executed per loop)
```

REFERENCES

1. Darwin, Charles. *On the Origin of Species by Means of Natural Selection or the Preservation of Favored Races in the Struggle for Life.* London: Murray, 1859.
2. Dawkins, Richard. *The Blind Watchmaker.* London: W. W. Norton, 1986.
3. Eigen, M. "Viral Quasispecies." *Sci. Am.* **269** (1) (1993): 32–39.
4. Feferman, L. *Simple Rules. . .Complex Behavior.* Santa Fe Institute Video. Santa Monica, CA: Direct Cinema, 1992.
5. Ray, T. S. "Is It Alive, or Is It GA?" In *Proceedings of the 1991 International Conference on Genetic Algorithms*, edited by R. K. Belew and L. B. Booker, 527–534. San Mateo, CA: Morgan Kaufmann, 1991.
6. Ray, T. S. "An Approach to the Synthesis of Life." In *Artificial Life II*, edited by C. Langton, C. Taylor, J. D. Farmer, and S. Rasmussen. Santa Fe Institute Studies in the Sciences of Complexity, Vol. XI, 371–408. Redwood City, CA: Addison-Wesley, 1991.
7. Ray, T. S. "Population Dynamics of Digital Organisms." In *Artificial Life II Video Proceedings*, edited by C.G. Langton. Redwood City, CA: Addison Wesley, 1991.

8. Ray, T. S. "Evolution and Optimization of Digital Organisms." In *Scientific Excellence in Supercomputing: The IBM 1990 Contest Prize Papers*, edited by K. R. Billingsley, E. Derohanes, and H. Brown, III. Athens, GA: The Baldwin Press, 1991.

9. Ray, T. S. "Evolution, Complexity, Entropy, and Artificial Reality." *Physica D* (1993): submitted.

10. Stanley, S. M. "An Ecological Theory for the Sudden Origin of Multicellular Life in the Late Precambrian." *Proc. Nat. Acad. Sci.* **70** (1973):1486–1489.

DISCUSSION

WALDROP: Your analogy to the RNA world, and then one of the latter pictures where you had the hyperparasites and the cheater that got in between them: Is this perhaps a model for a kind of "multicellularity" in the RNA world, that pieces of RNA could become multiple creatures themselves and then become integrated, and evolve these units?

RAY: That's possible. I think that a lot of things in this model are suggestive of what could be in the RNA world. It gives us a way of thinking about it. If you have a self-replicating RNA molecule, it's entirely possible that a non-self-replicating RNA molecule could go over there and get itself replicated by it, so you could have this type of parasitism, and perhaps the cheating and some of the cooperations that we saw. It's just a thing that sort of generates ideas. Any type of analogy that you make between the digital world and the carbon-based world always breaks down as you go deeper into the details. But it is suggestive of some possibilities. That's not a real direct answer; sorry.

BROWN: Tom, when you go to sexuality, into a gene pool, how are you going to keep track of the boundaries of individuals? That is, the species and, presumably, whatever your population, has a set of genotypes. A physicist, Fred Hopf, once said, "Why isn't life distributed like light from a perfect laser, with essentially a uniform random distribution over some range of genotypes, or morphotypes, or whatever?" Instead, there are clumps, and gaps between the clumps. Are you going just to see if those build themselves, or…?

RAY: Yes. When I write my first sexual organism, it will scan the soup for an organism that fits a certain pattern that it also fits and, if it recognizes a creature that meets that pattern, it will initiate a reproductive activity. So there's a kind of a mating ritual that they'll have to go through.

GOODWIN: Pattern in what sense, then?

RAY: Well, for example, I had this one marked with a certain pattern at beginning and end. It could simply look for another creature marked with the same patterns at beginning and end, separated by the same distance. That is the simplest implementation.

GELL-MANN: When blue-footed and red-footed boobies nest together, the blue-footed boobies raise their feet as a mating signal.

RAY: Yeah, "I'm blue." Right. That's the idea.

GELL-MANN: You could raise your blue swim fins.

RAY: I'll raise my template.
      I wanted to make one other comment. People may have wondered about my fashion statement, and I wanted to say that this is my response to Steven Harnard, the philosopher who, at the artificial life conference, said that it wasn't possible to create life in the computer because, in a simulation, all you have in the computer are squiggles and squaggles, which we then interpret to be something else. Like you have a simulation of plant growth—but you don't have a plant growing in the computer; you only have squiggles and squaggles.

GELL-MANN: But it's easy to simulate a philosopher at a meeting.

ANDERSON: I was fascinated by your last statement. It is not true, of course, that all organs necessarily execute the same program on parallel data, or on different data. The nervous system definitely executes different programs on its different data, or different programs on the same data...

RAY: Well, it depends on what we mean by the program. What I'm talking about is what part of the genome is expressed, and for a given cell type, all the cells are expressing the same genes. That's what I mean by program. I'm talking about the genetic program encoded in the DNA, only.

ANDERSON: A function, however, can be very different, and...

RAY: Oh, of course. They're interacting with the environment and, since they're interacting with different data, they're doing different things.

ANDERSON: Not only because of that; their programs do different things. The eye has detectors for motion, detectors for...

RAY: Oh, yeah; the different cell types.

ANDERSON: It's not obvious; there may be different programs.

GOODWIN: Your DNA, your programs got shorter and shorter, whereas that's the converse of what normally happens in evolution. What's your interpretation of that?

RAY: Well, it doesn't necessarily. That could be. . .

GOODWIN: No, I know; it can go both ways. But, overall. . .

RAY: First off, there is always selection for efficiency, both in the organic and the digital world.

GOODWIN: That's not clear. I would challenge that—"efficiency"—but go ahead.

RAY: And, in this case, if I set the environment up where everybody gets an exactly equal amount of energy, it will select for getting smaller. When I give energy in proportion to size, it *doesn't* select for getting smaller; sometimes they get bigger, sometimes they get smaller. And in the real world, being bigger can mean garnering more energy; it often does. In fact, you can garner energy disproportionate to your increase in size, and that leads to selection for larger sizes. And I can set up that kind of selection too, and they get bigger. It just depends on how the size relates to the energy intake.

GOODWIN: You see, this is like the Spiegelmann experiment, with a Qβ virus, where it did in fact. It was a naked RNA, it had replication coding—well, it also had a polymerase—but it got simpler and simpler and simpler, because the only criterion was, "how fast is it growing?" That's chemistry.

RAY: Well, remember that algorithm I showed at the end? The algorithm increased in complexity. As it happens, it didn't actually increase in size; it packed all that additional complexity into the same amount of space. But it was at the threshold; if it got any more complex, it would have had to get bigger, and it might very well have. That was in a size-neutral environment. So I think it can happen, and particularly if I create a situation where there's something more to do than just make copies of yourself.

SIMMONS: We keep hearing a lot about adaptation occurring near the edge of chaos, or some other such similar language. Do you see any way of interpreting your results to support those speculations?

RAY: Yes, the set of eight graphs that I showed where mutation rate was the variable between graphs. I think my mutation rate is the closest thing I have to Chris's lambda parameter. At the lowest mutation rates, evolution just plods along and, if I could control the sex, it would plod along even slower. (I didn't choose to control sex, there.) But as I push that parameter up, I get the richest ecological structure at intermediate rates. I see a sort of a three-tiered ecology that isn't so obvious in the lower mutation rates. I push it beyond that, and it dies, because it goes chaotic. So it looks to me like I'm getting the richest ecology, at any rate, at the threshold between the chaotic regime and the more static regime.

FONTANA: I have a question forTom Ray.

Your ancestor has these three modules, right? So let's call them exons. So in your universe, you start with three exons. It strikes me that essentially all the individuals that you observe are plays on these three exons. So my question to you is, is there any chance of increasing the number of exons in your system?

RAY: Well, in the data that I present in that paper, that's true. If I were to try to analyze the genome of that 22-instruction picture for those three exons, I don't think we would find them. At that point it's been erased, and we have something different. So I think the answer is "yes," but it takes more evolution than what I've shown there.

Examples of
Complex Adaptive Systems

Hans Frauenfelder
Department of Physics, University of Illinois, 1110 West Green Street, Urbana, Illinois 61801

Proteins as Adaptive Complex Systems (Abstract)

Adaptive complex systems have recently moved on center stage. Most of the work in this field is theoretical or computational. Quantitative experiments are far behind. The question therefore arises: What are the simplest adaptive complex systems that can be studied experimentally under well-controlled circumstances? Glasses and spin glasses are complex; they have been studied in considerable detail for many years, but they are not adaptive. I will try to show here that biomolecules, in particular, proteins, are complex **and** adaptive and that they can be studied experimentally, theoretically, and by computers. They may be the simplest systems where true adaptive complexity can be explored systematically.

Proteins are the machines of life; they perform most tasks in living systems.[2] The blueprints to their construction are encoded in the DNA. They are linear polypeptide chains, formed from 20 different amino acids. The primary sequence, the arrangement of the amino acids, determines structure and function. At least 101^{10} proteins exist naturally; the number of different proteins that can be produced by genetic engineering far exceeds any astronomical number. Proteins are aperiodic, disordered, and frustrated. As a result, they possess a rough energy landscape, consisting of a very large number of energy valleys.[1] We call these valleys "conformation substates (CS)." The structure of proteins involves different length and energy scales. Possibly as a result, the energy landscape is arranged in a hierarchy. Two different types of CS exist: At the highest tier (level), a protein can assume a small number of taxonomic substates. These substates can be characterized in detail; they have distinct structures, energies, entropies, volumes, and functional properties. Substates of lower tiers, with smaller energy barriers between valleys, are much more numerous and can only be described statistically, in terms of distributions.

A hierarchical and rough energy landscape leads to a multitude of fluctuation and relaxation processes. Most of these processes serve a biological function and have been shaped by evolution. The adaptive nature of proteins becomes obvious when the energy landscape and the dynamic processes are studied in different environments: Protein reactions can be controlled by changes for instance in pressure, pH, viscosity, or by small molecules.

Detailed studies of these phenomena are only at a beginning, but it is already obvious that the field is extremely large and rich. To understand protein function and control, the structure, energy landscape, and dynamics must be explored experimentally, theoretically, and by computer simulations. The concepts and laws that may (or will) emerge from these studies may lead to a more directed synthesis of proteins, but may also help understand more complicated complex systems.

REFERENCES

1. Frauenfelder, H., S. G. Sligar, and P. G. Wolynes. *Science* **254** (1991): 1598–1603.
2. Stryer, L. *Biochemistry*, 3rd ed. New York: Freeman, 1988.

DISCUSSION

ANDERSON: A couple of comments. One is that it's fairly well known that in spin glasses you have also such a hierarchy of conformations but it is a continuum in the sense that all scales are represented. Here we have the concept of "tiers." I'm not certain, or I wonder if you're certain, that this is anything but the fact that we have here a small piece of a spin glass—or a small piece of a glass—that has this hierarchy and which also has short-range interactions. And if you add it up, it can't have very many conformational substates at the top level, and it can't have very many substates of those. And the distinction into tiers, while it looks very real, may turn out to be simply a finite-size effect; or it may turn out to be simply a useful, but perhaps not essential, feature.

FRAUENFELDER: I agree with you. We don't know. There are two features: The first is that the protein is small. The spin glass is usually very large or infinite on that scale. The second is...

ANDERSON: But they both are stuck with information theory restrictions.

FRAUENFELDER: Yes. We have a built-in hierarchy: there're helices, and then they're smaller. So there may be both. But you're right; we don't know.

ANDERSON: The other comment is that the spin glass does, indeed, as a function of temperature, relax via something that looks very like avalanches, and very like self-organized criticality. So you may well be seeing something rather like self-organized criticality. Again, you don't see the scaling very well because you have a discrete problem, rather than a continuum problem.

FRAUENFELDER: I completely agree. That's why we try to have enough contact with the spin glass people and learn as much as we can.

ANDERSON: The third point: Because it's the Santa Fe Institute I should mention that Dan Stein may have been the first to make the analogy between the spin glass and proteins.

FRAUENFELDER: Yes.

KAUFFMAN: Hans, there's almost an orthogonal set of questions to the ones that you raised, all about proteins. It's one thing to study a specific protein that's well evolved. The kinds of energy hierarchy that you see in that could reflect the fact that this thing has undergone a billion years of evolution. Or not. So a very important question to ask is: if you take a random sequence of amino acids and string them together, what kinds of energy landscape will that show? Presumably it won't fold very well, so how hard is it to evolve from that to something that does fold very well, and what does all that look like?

Do you think there might be something like "natural kinds" in the ways that proteins fold, because people are finding that lots of different primary sequences have a quite restricted set of tertiary structures. Is it *just* convergent evolution, so to speak, or—this is just wild speculation—if you select proteins so they can fold, do you think that there may turn out to be relatively few kinds of ways you can go?

FRAUENFELDER: That's what seems to happen. I know much less about that problem, but I usually listen to Peter Wolynes. And they, of course, essentially use neural net techniques to fold proteins. And if I understand it right, the basic idea is you have a set of structures, and the neural net essentially compares the unknown to known, which would agree with what you're saying: that the number of motifs is relatively small. It's always difficult to say whether any selection that one has comes about because you look just at that, or because it really is a true selection.

COWAN: If the protein has such squishy properties but, nevertheless, very precise functions, what can you say about the precision of the function compared to the inability to specify a specific conformation? It acts as though it is specific...

FRAUENFELDER: It's not squishy, in one sense. You know, it is more like the digital switch which goes from one state, to the next state, to the third state. In each state it's quite...rigid isn't quite the right word, because it still moves among the substates of the next lower tier. But it is quite precise, in that sense. So I think it's really custom-made for whatever they should do. And different proteins also have very different characteristics. The one we have looked at...we know, for instance, that proteins that do the same but in different systems may have quite different, what you call "squishiness." So it may be designed for whatever purpose it has to do.

COWAN: But the pH, and the temperature, and minor changes in the pressure seem to change the conformation. Nevertheless, it retains the precision of function.

FRAUENFELDER: It changes the function somewhat, but in a quite precise way. To me it's a digital switch. How would you control it? You control it by going from one to the next, to the next, and then fine tune.

WALDROP: I was just going to follow up on that. Is there any evidence that proteins actually use this kind of "conformational-change-as-digital-switch"-like property to perform their functions?

ANDERSON: Yes.

FRAUENFELDER: Yes. We're recently working on writing it together. It's quite tricky to do quantitatively, but it appears so.

ANDERSON: That was actually one of the first things to be discovered by John Hopfield and Bob Schulman: that hemoglobin performed its basic function by very large conformational change everywhere, rather than by specific lock-and-key mechanisms as the conventional wisdom had it.

GOODWIN: Presumably this hierarchical structure also plays a role in the folding process, does it? Do you get large-scale order first, and then progressively finer?

FRAUENFELDER: I won't answer that question. You know there's a whole community of folders, and that's so highly specialized. And I think they disagree on many aspects. Folding is like spin glasses. You know, I discovered quite early that there are two types of spin glass: the experimental, and the theoretical. And there are two types of foldings: the theoretical, and the experimental. And the experimental is still far behind because it's very difficult to do it well. So you may be right, but I don't think it's known yet.

BROWN: I believe that, if you look across a wide spectrum of organisms, there are different kinds, for example, of myoglobin. I know that there are some differences in the primary structure. Can you say something about what the state of knowledge is with respect to how that affects these conformational and functional aspects of the molecule?

FRAUENFELDER: Only in general terms. We know quite a bit about that, for two reasons: we study different myoglobins, and we also study genetically mutated myoglobins, to see how the structure affects the function. That's why I said we really don't understand it yet. We know that whatever you do it changes the function, but usually not the way you expect it. So what it means is we need a larger data base to really make some reasonable...By the way, there is a question that I ask of students, usually. The question is: There are well over 200 hemoglobins known (naturally mutated), but only about ten or fifteen myoglobins; why?

WALDROP: You mean many species have identical myoglobins?

PERELSON: I wasn't going to answer that, but I was going to ask another question. Besides the myoglobins...

FRAUENFELDER: Let me first answer the question: It's much easier to take blood than muscle!

It's an important remark, because it shows how the type of measurement used biases the result.

PERELSON: Hans, besides the myoglobins and hemoglobins, what other proteins have been shown to have these distinct conformational substates? Have any...

FRAUENFELDER: Everything we have looked at.

PERELSON: Has anyone looked at enzymes?

FRAUENFELDER: Oh, yes. Exactly the same overall; of course, details are different. Everything we have looked at shows it. And we have looked maybe at fifteen or so. We have looked in great detail only at one. Bacterial rhodopsin shows it, which is a much more complex protein. So whatever you pick up and look at has these properties.

PERELSON: And is the general rule just a few major conformational substates?

FRAUENFELDER: Where we have seen it: Yes. But that is less certain. It took us fifteen years to find them in myoglobin. Now we're looking for them; it's somewhat easier. But it's never very easy.

ANDERSON: There can't be many, again just because of information theory. There's only so many interactions.

PERELSON: Right, but does that mean two, or three, or ten...?

ANDERSON: Two or three or ten; well, less than ten.

PERELSON: Less than ten?

ANDERSON: Probably.

FRAUENFELDER: The next level is very much larger; the next level may have 103 or 1020; we don't know. But it's simply large. And it's understandable, because the slightest change in the arrangement makes a new, slightly different conformation.

COWAN: With a different function?

FRAUENFELDER: No. When I said one taxonomic substate is the fastest—if you look at that, it turns out it's not just one speed, but the whole range. And so you have here one, here one, and here one. And so, it's really like fine tuning the coarse tuning.

Alan S. Perelson
Theoretical Division, Los Alamos National Laboratory, Los Alamos, NM 87545

Two Theoretical Problems in Immunology: AIDS and Epitopes

Abstract: In this chapter I will discuss two theoretical problems in immunology. The first is concerned with the interaction of HIV with the T lymphocytes of the immune system. A population-level model leads to a suggestion that a major cause of the T-cell depletion seen during AIDS is the direct infection and killing of T cells and T-cell precursors. The second, more abstract problem relates to the way the immune system recognizes foreign molecules and distinguishes them from self molecules. From our analysis of this problem we predict that the immune system should only recognize a small portion of an antigen of about 15 amino acids, in agreement with recent observations.

PROBLEM 1: WHAT CAUSES T-CELL DEPLETION IN AIDS?

The predominant view of AIDS is that the disease is due to infection by the human immunodeficiency virus, HIV. Infection with HIV results in a severe immunosuppression due to selective depletion in CD4$^+$ T cells (T4 cells). A large number of immunological abnormalities accompany HIV infection, and all but a few can be attributed to the decline in T4 cells.[6,11] Much controversy still exists about the mechanisms that HIV uses to deplete the body of T cells.

The point of the following discussion is to show that even though the immune system is quite complex, and even though we do not understand all of the mechanisms involved in disease progression towards AIDS, a quantitative analysis of the population dynamics of T cells and how HIV may perturb the T-cell population can provide a great deal of insight into AIDS. AIDS is unusual in that very few T cells are infected. Typically only 1 in 10,000 to 1 in 100,000 T cells are producing virus in a patient. It has been argued that even if the virus kills every one of those cells, one would not expect to see much T-cell depletion; greater numbers of T cells can be lost during trauma or during blood donation. However, the model I present, developed in collaboration with Denise Kirschner and Rob de Boer, will show that loss of both mature T cells and the precursors that form them can lead to profound T-cell depletion even though very few cells are infected at any time. To understand our approach, I first review some of the features of HIV and HIV infection.

HIV is an RNA virus that attaches to cells by interacting with a cell surface molecule, CD4.[4,10] After HIV binds to a cell, it can become internalized, and infect the cell. Thus, CD4$^+$ T cells, as well as monocytes and macrophages which also express CD4, are targets of HIV infection. After HIV enters a cell, it reverse-transcribes its RNA into a DNA copy of its genome, and then integrates this DNA copy of itself into the DNA of the cell. A cell containing the viral genome appears perfectly normal, it just has an extra piece of DNA, called a provirus. A cell in this state is called *latently infected*. The viral DNA within the cell will be duplicated with the cell's DNA every time the cell divides. Thus a cell, once infected, remains infected for life.

The provirus can remain latent, giving no sign of its presence for months or years.[8] When a latently infected lymphocyte is stimulated by interacting with antigen (a foreign cell or molecule), it usually begins to divide. As part of the process of duplicating its DNA the cell turns on molecular machinery that also leads to the production of new virus particles. These new virus particles or virions bud from the surface of the infected cell. The budding can take place very rapidly, leading to the lysis of the host cell (this seems to be the case in T4-cell infection), or it can take place slowly and spare the host cell (this seems to occur in macrophage and monocyte infection). Thus, the activation of T cells into a proliferative state, say, by the T cells recognizing antigen, is required for converting a latent HIV infection into active viral replication.

Why does it take so many years from the time of infection with HIV to the time of clinical AIDS? The mean time is estimated to be close to ten years in adults.

One very early view of how HIV causes disease is that the viral infection leads to the death of infected cells. However, because so few infected cells were found in patients, other, more exotic mechanisms in which the immune system destroyed uninfected T cells were championed. One mechanism of this class suggests that gp120, the envelope protein of HIV, which can be shed by the virus, is picked up by uninfected T4 cells. These T4 cells, which have associated viral protein, might appear to the rest of the immune system as a virally infected T4 cell and be killed by normal mechanisms of cell-mediated immunity. In thinking about potential therapies it is important to know the dominant mechanism of T-cell loss, since many mechanisms may contribute to the observed T-cell decline. The model given below shows that T-cell depletion by the most obvious mechanisms, direct viral killing of T cells, should not be abandoned and, in fact, this mechanism can contribute substantially to the overall depletion.

MODEL

While there are very many cells and molecules involved in the immune system, we only consider the major players in AIDS: T cells that are uninfected; T cells that are latently infected, i.e., that contain the virus but are not producing it; T cells that are actively infected, i.e., that are producing virus; and last but not least, the virus itself.

Let T denote the concentration of uninfected T4 cells and let T^* and T^{**} denote the concentrations of latently infected and actively infected T4 cells. The concentration of free infectious virus particles is v. I assume that the dynamics of the various T4-cell populations is governed by the following differential equations:

$$\frac{dT}{dt} = s(v) - \mu_T T + rT\left(1 - \frac{T + T^* + T^{**}}{T_{\max}}\right) - k_1 vT, \tag{1}$$

$$\frac{dT^*}{dt} = k_1 vT - \mu_T T^* - k_2 T^*, \tag{2}$$

$$\frac{dT^{**}}{dt} = k_2 T^* - \mu_b T^{**}, \tag{3}$$

$$\frac{dv}{dt} = N\mu_b T^{**} - k_1 vT - \mu_v v. \tag{4}$$

The derivation of these equations is described in detail by Perelson[15] and by Perelson et al.[16]

The first equation describes T-cell population dynamics. The first three terms in the equation represent the rates of production and destruction of T cells in uninfected individuals, s being the rate of supply of immunocompetent T cells from precursors in the thymus; μ_T represents the average per capita death rate of

T cells. We have chosen s to be a decreasing function of v, so that as the viral burden increases infection of T-cell precursors occurs and the supply of T cells decreases. Here we assume,

$$s(v) = \frac{\theta s}{\theta + v},$$

where θ is a constant that determines the viral load needed to decrease s by a factor of two. In the absence of HIV, $s(v) = s = $ constant. The growth of T cells is modeled by a logistic equation, with r being the per capita T-cell growth rate in the absence of population limitation. The last term in the equation, proportional to k_1, represents T-cell infection by HIV. In the absence of HIV, this equation describes the T-cell population level in the blood. One can set the parameters, so that this level is maintained at 1000 cells/mm^3, as is typical in healthy people.[16]

The second and third equations describe the production of latently infected and actively infected cells. Latently infected cells are assumed to die at the same per capita rate as uninfected cells, μ_T, but actively infected cells are assumed to die at a greatly increased per capita rate μ_b. During their lifetime, actively infected cells are assumed to produce N infectious virus particles. The concentration of infectious virions is described by Eq. (4), in which virus particles lose their infectivity at per capita rate μ_v.

There are a number of features of this system worth noting. First, in the absence of virus, the T-cell population has the steady state value T_0, where

$$T_0 = \frac{T_{\max}}{2} \left[1 - \frac{\mu_T}{r} + \sqrt{\left(1 - \frac{\mu_T}{r}\right)^2 + \frac{4s}{rT_{\max}}} \right]. \tag{5}$$

Thus, reasonable initial conditions for this system of equations are $T(0) = T_0$, $T^*(0) = 0$, $T^{**}(0) = 0$, and $v(0) = v_0$, where v_0 is the infecting dose of HIV. If cells are transferred as well as virus during infection, then the initial values of T^* and T^{**} would also be nonzero.

Second, the model has two steady states, an uninfected state in which $v = 0$ and an endemically infected state in which $v > 0$. We have shown that if N, the number of infectious virions produced per actively infected cell, is less than some critical value, $N_{\text{crit}} = k_3(\mu_v + k_1 T_0)/k_2 k_1 T_0$, then the infection will die out.[16] Conversely, if $N > N_{\text{crit}}$, then the infection will take, virus will survive, and T-cell depletion will occur. When $N < N_{\text{crit}}$, virus infects cells, but the cells that are infected die before producing enough offspring to sustain the infection. The same type of phenomenon is observed in epidemics. If, on average, infected people infect more than one other person the disease spreads and causes an epidemic, whereas if each person on average infects fewer than one other person the epidemic dies.

The model ignores the complexity of viral mutation. It is known that HIV can rapidly mutate and thus, that there are many strains of HIV. Different strains of virus have different properties, and in particular different abilities to grow in T cells. Thus, the parameter N is a characteristic of a particular stain. Strains that are highly pathogenic might be envisioned as corresponding to high values of N.

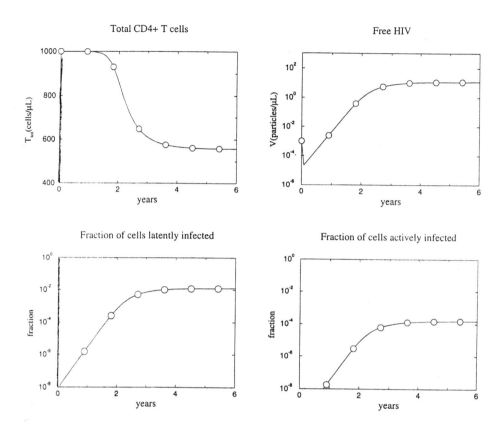

FIGURE 1 Predicted changes in the total T cell, latently infected T cell, actively infected T cell, and viral populations over time found by solving Eqs. (1) with initial conditions $T = 1000$, $T^* = T^{**} = 0$, $v = 1$. Parameters are $s = 10$ day^{-1} mm^{-3}, $r = 12$ day^{-1}, $T_{\max} = 1500$ mm^3, $\mu_T = 0.06$ day^{-1}, $\mu_b = 0.24$ day^{-1}, $\mu_v = 5$ day^{-1}, $k_1 = 2.4$ mm^3 day^{-1}, $k_2 = 1.2 \times 10^{-4}$ day^{-1}, $\theta = 1$ mm^3, and $N = 1400$.

Third, the model predicts that T-cell depletion can take many years. In Figure 1, I illustrate the predictions of the model for $N > N_{\mathrm{crit}}$. Notice that initially free virus declines as it infects cells. The virus then grows exponentially. While the virus level is low, T cells are infected but the level of infection is so low that no noticeable T-cell depletion is observed. But ultimately and rather sharply, the virus population reaches a high enough level that the T-cell level comes crashing down. The time it takes to reach this precipitous decline depends on the value of N and, hence, the viral strain. If N is close to N_{crit}, the decline can easily take eight to ten years.[16] This was an initial surprise, since most of the parameters in the model are on the scale of hours or days; the longest time scale in the model is the T-cell lifetime of a few weeks.

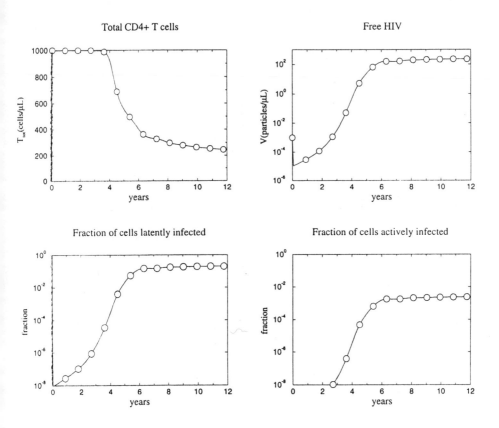

FIGURE 2   Same as Figure 1 except N is a function of time as specified in the text. Here $N_0 = 900$, $n = 3$, $a = 4$, and $\Theta = 7$ years.

Fourth, the model predicts that in the endemically infected state the number of latently infected T cells and actively infected T cells are at the low values typically observed in patients: 1 in 100 latently infected,[17] 1 in 10,000 actively infected.[7]

The model is not perfect. The model does not exhibit T-cell depletion down to the level of clinical AIDS, 200/mm³, over a time scale of years. By increasing N or k_1, low levels of T cells can be obtained, but the depletion then takes months rather than years. However, even though the depletion is not as profound as in patients, the model does show that a single species of virus can lead to substantial depletion just by slowly killing T cells. If we make the model reflect viral mutation and evolution towards more pathogenic strains of virus by replacing the constant N by a slowly increasing function of time $N(t) = N_0(1 + at^n/(\Theta^n + t^n))$, where N_0, n and Θ are constants, then the model does a much better job of mimicking the T-cell depletion seen in patients. This change from N constant to N time varying is not arbitrary.

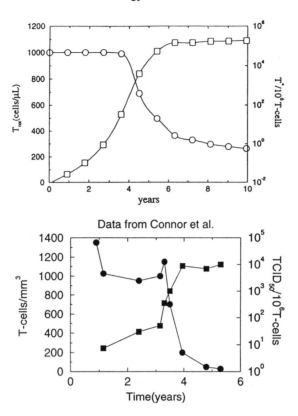

FIGURE 3 Comparison of data from Connor et al.[3] with model prediction. Theoretical curves are redrawn from Figure 2, with (o) the total T cell population and (□) the number of latently infected T cells per 10^6 T cells. The data from Connor et al. also shows the T cell count (■) and the 50% tissue culture infections llose (TCID$_{50}$) of cells per 10^6 T cells. The TCID$_{50}$ is a measure of the number of cells harboring HIV.

It is frequently found that virus isolated from patients at end-stage disease is more pathogenic than the strains of virus that initially are seen in seropositive patients. It may be that fast-growing viruses (large N) are initially eliminated by an immune response, while slow-growing strains (small N) escape immune detection. A model exploring this view has been presented by Nelson and Perelson.[12]

In Figure 2 we show by making N an increasing function of time the model will exhibit depletion down to $200/mm^3$. In addition, recent data by Connor et al.[3] show that in some AIDS patients the T-cell population declines precipitously over a period of one to two years, just as in our model (Figure 3). Also, in these patients this period of decline correlates with increased levels of infected cells, again an observation consistent with our model (see Figure 3). Not all patients followed this pattern. One patient, for example, showed relatively flat levels of plasma virus, and relatively constant numbers of uninfected and latently infected T cells over a period of eight years. This patient may be generating an immune response that is keeping the virus from growing and/or becoming more pathogenic. In any event, our model, which does not incorporate an immune response to the virus, does not predict the relatively constant T-cell and virus levels seen in this patient.

To summarize, a rather simple model of T-cell population dynamics shows that HIV can perturb the normal control of T-cell numbers by killing cells in the blood and by infecting T-cell precursors, thereby reducing the supply of new cells. Both of these effects taken together can lead to T-cell depletion even though only 1 in 100 cells are latently infected and 1 in 10^4 cells are actively infected. If, in addition, the virus mutates with time so that its ability to grow in T cells increases during the course of the infection, depletion down to 200 T4 cells/mm^3, the level characteristic of AIDS, is predicted by this model (Figure 2).

PROBLEM 2: PREDICTING THE SIZE OF RECEPTOR COMBINING SITES

The binding of antibody to antigen occurs by a generalized lock and key fit of portions of the two structures. The question that I address here is: how large should the complementary regions on the two structures be?

Studies on the ability of B and T lymphocytes to be stimulated by antigen typically show that only one lymphocyte in 10^5 will be stimulated by any particular antigen. Lymphocytes interact with antigen via receptors on their surface, antibody or immunoglobulin receptors in the case of B cells and a molecule called the T-cell receptor in the case of T lymphocytes. The receptor repertoire is estimated to be of the order of 10^7. Each lymphocyte expresses only one type of receptor on its surface, and the probability that a lymphocyte receptor will recognize antigen in an appropriate way to lead to cell stimulation is of the order of 10^{-5}. Why should this be?

If the immune system is designed to detect antigen, wouldn't it be more efficient to design "sticky receptors" that had a very high probability of interacting with antigen? Some enzymes, such as trypsin, that degrade proteins, can interact with essentially any protein. Major histocompatibility complex (MHC) molecules interact with a broad spectrum of peptides, so that an individual with fewer than ten MHC molecules can bind the peptides of any protein with probability close to one. Similarly, some antibodies have been found to be multireactive, binding to over 20% of other antibodies and a variety of self-proteins. Why then does the immune system use receptors of such high specificity? Detection of antigen is not the entire story. Clearly, the immune system must also distinguish foreign molecules from self-molecules. Having a sticky receptor that bound everything would detect all antigens but it would be useless as a defense molecule since it would also bind to all the cells of our bodies.

John Inman, an immunochemist, has estimated that the immune system can recognize at least 10^{16} epitopes, an epitope being the portion of an antigen that fits into the antibody combining site.[9] The number of self-epitopes that the immune system should fail to recognize is not known but can be estimated. There are on the order of 5×10^4 to 10^5 genes in the human genome. Assuming each gene codes for a protein, and each protein has, say, ten epitopes, then we would expect our bodies to contain on the order of 10^6 self-epitopes.

Thus, to summarize, there are

$$
\begin{aligned}
N &\sim 10^{16} \quad &&\text{foreign epitopes to be recognized,} \\
N' &\sim 10^{6} \quad &&\text{self-epitopes to be ignored,} \\
n &\sim 10^{7} \quad &&\text{T-cell receptor and immunoglobin variable} \\
& && \text{region sequences to carry out these tasks.}
\end{aligned}
\tag{6}
$$

Assuming epitopes and receptors have been created at random, and that the *a priori* probability that a receptor recognizes a random epitope is P_S, then the probability of the immune system carrying out the above recognition and non-recognition tasks is given by

$$
\begin{aligned}
\Pr(N, N'; n) &= \Pr(\text{each of } N \text{ epitopes is recognized by one or more of } n \text{ receptors}) \\
&\quad \cdot \Pr(\text{none of } N' \text{ self epitopes are recognized by any of } n \text{ receptors}) \\
&= (1 - P_F^n)^N \, P_F^{nN'} ,
\end{aligned}
\tag{7}
$$

where $P_F = 1 - P_S$ is the probability that a receptor fails to recognize an epitope.

Evolution has shaped the repertoires of immunoglobulin and TCR genes found in vertebrates. One possible path that evolution may have taken is to have maximized the probability of the immune system carrying out the recognition of foreign epitopes and not recognizing self-epitopes, i.e., maximizing $Pr(N, N'; n)$. From Eq. (7), the value of P_F that maximizes $Pr(N, N'; n)$ is

$$
P_F = \left(1 + \frac{N}{N'}\right)^{-1/n} \sim 1 - \frac{1}{n} \ln\left(1 + \frac{N}{N'}\right) .
\tag{8}
$$

The above approximation can be shown to be extremely accurate for the parameters in Eq. (6).

If the current immune system has optimized $Pr(N, N'; n)$, then one would expect the probability of successfully recognizing a random antigen, $P_S = 1 - P_F$, to be computable from Eq. (8). Hence

$$
P_S \sim \frac{1}{n} \ln\left(1 + \frac{N}{N'}\right) .
\tag{9}
$$

Using the parameters in Eq. (6), $P_S = 2.3 \times 10^{-6}$, which is close to but somewhat smaller than the empirical estimate of 10^{-5}. Given the simplicity of this calculation and the approximate nature of our estimates of N, N', and n, this is a surprisingly good estimate of P_S.

PREDICTING THE SIZE OF EPITOPES

So far in our consideration of receptor-ligand interactions, we have not quantified the degree of match between two molecules. In this section I shall present a simple model for determining molecular complementarity and use that model to predict the optimal size of an epitope. I shall restrict the discussion to epitopes composed of amino acids since T-cell epitopes are always peptides and proteins are a major fraction of the antigens seen by B cells.

Farmer, Packard, and Perelson[5] introduced the idea of using a binary string to represent the shape of a receptor. Any of a number of string matching algorithms could then be used to determine the degree of complementarity between strings. Here I shall pursue a generalization of that idea, introduced by Percus, Percus, and Perelson,[14,15] in which strings chosen from an alphabet of m letters are used. The idea here is that amino acids can be classified into groups depending on their chemical properties. For example, amino acids can be classified using a "charge alphabet" as being positive, negative, or neutral. Thus, as a simple caricature of the amino acid combining site of a receptor and the complementary epitope, I assume both are composed of m types of amino acids, with each amino acid complementary to exactly one other amino acid in the alphabet. This, for example, is the case in the charge alphabet ($m = 3$), with positive complementary to negative and neutral complementary to neutral. I also assume that a receptor need only be complementary to a piece of the antigen, i.e., an epitope, where an epitope is defined as a sequence of at least r letters.

Consider the simple case in which the receptor and antigen are both sequences of length ℓ. Align the two sequences and denote a matching or complementary pair of letters by x, and a nonmatching pair by y. If receptor and antigen strings are each constructed with the m units chosen at random, then at each position complementation occurs with probability $1/m$ and noncomplementation with probability $(m-1)/m$. In this string metaphor, epitope binding corresponds to an unbroken sequence of at least r contiguous x's. Computing the probability of at least r consecutive matches from a sequence of length ℓ is a classic matching problem. A full analysis is given by Percus et al.[13] For present purposes, a simple argument suffices. A rigorous analysis shows that for the parameters of interest the probability of a matching region is very small and, hence, to a good approximation the various contributing possibilities can be regarded as independent. Starting at the leftmost site of the ℓ-site sequence, r contiguous x's occur with probability m^{-r}. Thereafter, runs of r x's can start at $\ell - r$ possible sites. Each such run is preceded by a mismatch y, for a net probability of $m^{-r}(m-1)/m$. Adding up these probabilities,

$$P_S = m^{-r} \left[\frac{1 + (\ell - r)(m - 1)}{m} \right] . \tag{10}$$

For $m^{-r} \ll 1$, Eq. (10) provides a good approximation to the exact matching probability.[13]

Assuming $\ell \gg r > 1$ [i.e., dropping the negligible $((m-1)/m)r$ and 1 in Eq. (9)], we obtain

$$r = -\ln_m P_S + \ln_m \left[\frac{\ell(m-1)}{m} \right] . \tag{11a}$$

Experimental estimates for P_S can be used, or we can substitute the optimal value of P_S given by Eq. (8) to obtain

$$r = \ln_m(n\ell) - \ln_m \left(\frac{m}{m-1} \ln \left(1 + \frac{N}{N'} \right) \right) . \tag{11b}$$

Because of their logarithmic nature both formulas make similar predictions. In particular, Eq. (11b) is very insensitive to the population sizes N and N' of foreign and self-antigens.

We now estimate r. From Eq. (6), $n \sim 10^7$. The entire variable region of a receptor is not accessible to antigen since some residues are buried in the interior of the molecule. If we estimate that roughly half of the amino acids are accessible, then $\ell \sim 100$. For the charge alphabet introduced above, $m = 3$. Empirical estimates of P_S are of order 10^{-5}. With $m = 3$, $\ell \sim 100$, $n = 10^7$, $P_S = 10^{-5}$, $N \sim 10^{16}$, and $N' \sim 10^6$, Eqs. (11a) and (11b) predict $r \sim 14.3$ and 15.6, respectively. If $N/N' = 10^9$ rather than 10^{10}, then Eq. (11b) predicts $r \sim 15.7$, illustrating the insensitivity of r to the ratio of foreign to self-antigens. These predicted values of r are consistent with the various experimental determinations on the number of contact residues between antibody combining sites and protein antigens and the size of the region on the MHC-peptide complex that interacts with the T-cell receptor.[1,2,18]

One feature of our results that is surprising is that the dominant effect in determining the optimal epitope size turns out not to be self-nonself discrimination. From Eq. (11b), keeping only the largest term on the right hand side, $r \sim \ln_m n$, or $n = m^r$. For $m = 3$ and $r = 15$, $m^r \sim 1.4 \times 10^7$, which is approximately the estimated repertoire size. Thus, we conclude that the optimal value of r generates the maximal number of epitopes that an immune system with 10^7 receptors can detect.

In conclusion, we see in this problem, as in the AIDS problem discussed above, that a simple model can provide insights into a complex biological situation. Clearly, this model ignores most of the physical and chemical features of antigen-antibody interactions that biochemists and crystallographers study. Yet, the model, based on

string matching, captures enough of the essence of the matter to make interesting
and seemingly correct predictions. In many different areas in complex systems research the use of string models has proven useful and provided unexpected insights.
The problem of predicting epitope and receptor combining site sizes is yet another
example.

ACKNOWLEDGMENTS

The work reported here was done in collaboration with Rob de Boer, Denise
Kirschner, Ora Percus, and Jerome Percus. This work was performed under the
auspices of the U.S. Department of Energy and supported in part by National
Institutes of Health Grant AI28433 and the Santa Fe Institute through their Theoretical Immunology Program.

REFERENCES

1. Ajitkumar, P., S. S. Geler, K. V. Kesari, Borriello, M. Nakagawa, J. A. Blue-
   stone, M. A. Saper, D. C. Wiley, and S. G. Nathenson. "Evidence That Multiple Residues on Both the α-Helices of the Class I MHC Molecule are Simultaneously Recognized by the T Cell Receptor." *Cell* **54** (1988): 47–56.
2. Amit, A. G., R. A. Mariuzza, S. E. V. Phillips, and R. J. Poljak. "Three-
   Dimensional Structure of an Antigen-Antibody Complex at 2.8 Å Resolution." *Science* **233** (1986): 747–753.
3. Connor, R. I., H. Mohri, Y. Cao, and D. D. Ho. "Increased Viral Burden
   and Cytopathicity Correlate Temporally with CD4$^+$ T-Lymphocyte Decline
   and Clinical Progression in HIV-1 Infected Individuals." *J. Virol.* **67** (1993):
   1772–1777.
4. Dalgleish, A. G., P. C. L. Beverley, P. R. Clapham, D. H. Crawford, M. F.
   Greaves, and R. A. Weiss. "The CD4 (T4) Antigen is an Essential Component of the Receptor for AIDS Retrovirus." *Nature* **312** (1984): 763–767.
5. Farmer, J. D., N. H. Packard, and A. S. Perelson. "The Immune System,
   Adaptation, and Machine Learning." *Physica D* **22** (1986): 187–204.
6. Fauci, A. S. "The Human Immunodeficiency Virus: Infectivity and Mechanisms of Pathogenesis." *Science* **239** (1988): 617–622.
7. Harper, M. E., L. M. Marselle, R. C. Gallo, and F. Wong-Stall. "Detection of
   Lymphocytes Expressing Human T-Lymphotropic Virus Type III in Lymph
   Nodes and Peripheral Blood from Infected Individuals by *in situ* Hybridiza-
   tion." *Proc. Natl. Acad. Sci USA* **83** (1986): 772–776.

8. Ho, D. D., R. J. Pomerantz, and J. C. Kaplan. "Pathogenesis of Infection with Human Immunodeficiency Virus." *New Engl. J. Med.* **317** (1987): 278–286.

9. Inman, J. K. "The Antibody Combining Region: Speculations on the Hypothesis of General Multispecificity." In *Theoretical Immunology*, edited by G. I. Bell, A. S. Perelson, and G. H. Pimbley, Jr., 243–278. New York: Marcel Dekker, 1978.

10. Klatzmann, D., E. Champagne, S. Chamaret, J. Gruest, D. Guetard, T. Hercend, J. C. Gluckman, and L. Montagnier. "T-Lymphocyte T4 Molecule Behaves as the Receptor for Human Retrovirus LAV." *Nature* **312** (1984): 767–768.

11. Lane, H. C., and A. S. Fauci. "Immunologic Abnormalities in the Acquired Immunodeficiency Syndrome." *Ann. Rev. Immunol.* **3** (1985): 477–500.

12. Nelson, G. W., and A. S. Perelson. "A Mechanism of Immune Evasion by Slow Replicating HIV Strains." *J. AIDS* **5** (1992): 82–93.

13. Percus, J. K., O. E. Percus, and A. S. Perelson. "Probability of Self-Nonself Discrimination." In *Theoretical and Experimental Insights into Immunology*, edited by A. S. Perelson and G. Weisbuch, 63–70. Berlin: Springer-Verlag, 1992.

14. Percus, J. K., O. E. Percus, O. E., and A. S. Perelson. "Predicting the Size of the T Cell Receptor and Antibody Combining Region from Consideration of Efficient Self–Nonself Discrimination." *Proc. Natl. Acad. Sci. USA* **90** (1993): 1691–1695.

15. Perelson, A. S. "Modeling the Interaction of the Immune System with HIV." In *Mathematical and Statistical Approaches to AIDS Epidemiology*, edited by C. Castillo-Chavez, 350–370. Lect. Notes in Biomath., Vol. 83. New York: Springer-Verlag, 1989.

16. Perelson, A. S., D. E. Kirschner, and R. J. De Boer. "The Dynamics of HIV Infection of CD4$^+$ T Cells." *Math. Biosci.* **114** (1993): 81–125.

17. Schnittman, S. M., J. J. Greenhouse, M. C. Psallidopoulos, M. Baseler, N. P. Salzman, and A. S. Fauci. "Increasing Viral Burden in CD4$^+$ T Cells from Patients with Human Immunodeficiency Virus (HIV) Infection Reflects Rapidly Progressive Immunosuppression and Clinical Disease." *Ann. Int. Med.* **113** (1990): 438–443.

18. Sheriff, S. E. W. Silverton, E. A. Padlan, G. H. Cohen, S. J. Smith-Gill, B. C. Finzel, and D. R. Davies. "Three-Dimensional Structure of an Antibody-Antigen Complex." *Proc. Natl. Acad. Sci.* **84** (1987): 8075–8079.

DISCUSSION

BROWN: It seems as if these sort of probabilistic rules underlie this—that there has to be some incredibly powerful mechanism preventing co-mutations, where you get a self-match. Every time you generate one that matches up to somebody by chance, with a self-protein, you want to get rid of that.

PERELSON: Let me answer your question in a strange way. What I should have told you is that, if you actually look at how well we do self-nonself discrimination, we do not do it perfectly. In my model we maximize the probability of doing self non-self discrimination. Now you can ask, "What is the probability of doing this job at the optimum?" And if you calculate that...I forget exactly what the number is, but it's less than 10^{-1000}. Maybe it's ten to the tenth to the minus a thousand: it's essentially zero. So it says you can't design a repertoire, at least based on random considerations, in which you will do *perfect* self-nonself discrimination, just at the level of the shape of the molecules. Or I shouldn't say you can't, but it's extraordinarily difficult.

One might try and think up evolutionary schemes where, over a long time, you improve self non-self discrimination incrementally. In fact, you can do some calculations to show, if you did it incrementally—made the repertoire grow by one antibody at a time—that we could approach doing this discrimination with reasonable accuracy. But it's clear that the real immune system has *not* been able to do that. What we're saying, as Murray might have said if he was here, is: it *strives* to do that. But it clearly doesn't, and there are a large number of other mechanisms that sit on top of this receptor discrimination that really come into play in real immune systems for doing self-nonself discrimination.

ANDERSON: A couple of things. One is: You're relying very heavily on x-ray structures, which we've just been told are hogwash...

PERELSON: That's true, but at the moment it's the best data available. The data on the size of the T cell receptor is not x-ray, it was based on amino acid substitutions and, as far as I know, most immunologists, even though the exact structure may not be right feel that these numbers are consistent with all the mutational studies. People have been substituting amino acids in both the receptors and the antigens for a long time, and trying to build up rules-of-thumb for the size of the binding region based on just binding studies. So if we're wrong, we're not too far wrong (from the crystal structures).

ANDERSON: Another question is just to reveal my ignorance. How does the body select its repertoire in order to do self-nonself discrimination? Are cells killed at some early point?

PERELSON: That's *the* hot topic in immunology; it's been broken open in the last two or three years.

There are two parts to the immune system: the B cells that make antibody, and the T cells that make molecules that regulate B cells and do direct killing. Because T cells play a regulatory role (you usually need T cells to secrete molecules to help the B cells)...most people are of the view that the predominant self-nonself discrimination occurs in the T-cell population. And exactly what you said happens: T cells are made in the bone marrow, as any blood cell is; they go to the thymus (which is why they're called "T cells"), and in the thymus the majority of lymphocytes that enter the thymus are killed. It's known that they undergo recognition of self-molecules that are sitting on epithelial cells in the thymus. If by chance their receptor seems to bind the self-molecules too strongly, that cell is killed and never leaves the thymus. So the thymus acts as a filter, allowing certain cells that don't interact with "self" too strongly to leave.

It's somewhat complicated in that T cells don't recognize antigens in solution. They're designed to look at molecules on cell surfaces. The way they do that is by looking at molecules in combination with a protein, called a major histocompatibility complex or MHC molecule. So in the thymus, there seems to be two levels, of selection. One, you want T cells that will be able to have some interaction with your self-MHC molecules, but you don't want to have strong interactions with MHC *plus* peptides bound to it. And so, it looks like there's a two-stage process that immunologists now call positive and negative selection: you kill off those T cells that bind too strongly to MHC self-peptide complexes, but you allow to grow, or at least leave the thymus, those that just bind with MHC.

The system is not perfect because not all proteins are thought to be able to make it into the thymus. So there's this issue of what do you do with proteins that are generated, say, in a joint: pieces of collagen, or something, that's in a ball joint. Those shouldn't be migrating up to your thymus. Also, what about proteins that are characteristic of your pancreas (certain proteins)? So there's the issue that even though this thymic filtering can do a lot of self-nonself discrimination, there probably are many proteins that never make it there, and one needs other mechanisms out in the rest of the body—more dynamic control mechanisms—that will prevent proliferation of cells that react with self-molecules. And there's a whole other class of models—these network models that I deal with, and Rob [De Boer] deals with—that people think are involved in some of those other regulatory elements.

We know there's autoimmune disease, which is the other answer—that clearly self-nonself is not perfect.

ANDERSON: Is there any possibility that some malfunction in this mechanism is the final stage of AIDS?

PERELSON: It could be part of that, but that's not the characteristics of what we see. Some people believe that there's some autoimmune part of AIDS, but it's the major depletion of T cells that's occurring in the periphery, not so much in the thymus, that is observed. In fact, adults can live without a thymus. You can have your thymus surgically removed, and the T cells will just multiply in the periphery.

BUSS: I seem to remember that genomic organization in sharks and chickens is really strikingly different; differing numbers of variable regions, different numbers of adjoining regions, and the like, which would presumably lead to rather different numbers of what the repertoire of the total number of receptor numbers would be. Presumably it's facing the same antigenic environment.

PERELSON: I agree with half of your statement. The genomic organization, and the mechanism of generating diversity—for example, in the chicken—are extraordinarily different than what I discussed here. In most mammals what you have are what I call libraries, or large families, of gene segments that rearrange and use combinatorics to generate the large numbers of receptors. In the chicken it turns out that there's one master gene sequence that gets duplicated and changed, so it's sort of variability on a theme. However, when people look at repertoires for the number of different protein molecules that ultimately get expressed, it looks like all immune systems need a large number of receptors. And, in fact, the smallest functional immune system seems to have of order 10^5 receptors—*maybe* 5×10^4—and that turns out to be in a young tadpole.

The question you ask has sort of been addressed by immunologists over the last two decades to try and find out what is the minimal repertoire size that would correspond to a functioning immune system. And that's another curiosity of these calculations. If P_s is of order 10^{-5}, then you need of order 10^5 receptors to cover the shape space, and that seems to be roughly correct. Only vertebrates have immune systems, as you know; no really tiny organisms do, and it seems the level of complexity at the level of these repertoires is of order 10^5.

BUSS: Okay, let me see if I got this right. What you're saying is that this variation in genomic organization is, in fact, *not* changing the number of different receptor classes in a major way. And that this is, in fact, an empirical result.

PERELSON: Right.

HÜBLER: In your model with one species, the number of T cells leveled off at a value that is not lethal. Why is this the case? Is it the competition, or what is the reason?

PERELSON: In these models, mathematically what happens is that the models have two steady states; there's the steady state which I'll call uninfected, where there's no virus and it's normal. In the presence of the virus, the virus is constantly killing off some T cells, but the replenishment is such that the T cells establish a new steady state, but it's not a zero steady state. And I can show you what it is analytically; we understand how it varies with all the parameters. Now what happens in real immune systems is that there are lots of things that change, in terms of the characteristics of the immune system, as this depletion goes on (which are not in the model). So the models are not all-encompassing by any means, and we're just looking at, as I said, what we think was a predominant mechanism.

HÜBLER: What I don't understand is that you said that, in the last stage of the disease, those species that are growing faster become more predominant. I would expect this already from the beginning.

PERELSON: It turns out *not* to be the case throughout the disease. In many patients, you find that there is early viral growth. After infection the patient will come down with flu-like symptoms, and you'll see a big spike of viremia. You didn't see that in any of these patients whose data that I showed you. There seems to be an immune response against that fast growing virus and, after three or four weeks, a large number of the virus particles disappear, and you have very low levels of plasma virus. You're now seropositive; and the types of dynamics that I showed then continue from that point on. The early viremia occurs maybe in thirty percent of people. It's hard to get good numbers, because people don't always come into the hospital; they just think they have the flu or something.

HÜBLER: So what you suggest is that you have a specific intense immunoreaction against fast-growing species and you have a less intense reaction to slow-growing and therefore the slow-growing. . .

PERELSON: . . .ones are the predominant ones during this long latent phase. In fact, we have another model, which is the one I spoke about at Pasteur [Institute], which studies that precisely. I can give you a paper on it. But it's just a speculation right now. And the ones that are slow-growing, we think, can evade stimulating a good immune response. Then there's some ideas that these viruses are actually mutating, and causing changes of parameters. And that's what one then finds; there're particular viral strains whose characteristics are such that they not only are evading the system, but can outgrow it, and take off in this way.

HÜBLER: If you would have a competition between different species, you could stimulate a certain species which would not lead to a lethal concentration. But as far as I understand you, there is not such a competition between different species.

PERELSON: Not in this model. There may in fact be competitions in people. That's another interesting feature. Now let me just tell you about data. One can collect virus samples from patients and sequence them. It now looks like, worldwide, that there may be six major subtypes or clads of HIV-1, with different subtypes predominant in different regions. In Africa, for example, all six subtypes are present, and they differ by about thirty-five percent in the nucleotide sequence for their coat protein; so they're very different.

You can ask, "Do you find patients who have multiple infections, who carry all the subtypes?" And the answer is, so far, that there have been no examples found, that you always find just one of these classes of the virus, as if there is some sort of interference going on within the patient. That's also been known to occur with many other types of viruses—that there seems to be one predominant form that wins some sort of competition within our bodies. And whether or not that has to do with defeating the immune system, escaping, it, or what, is not known.

EPSTEIN: Yeah, on this point...Well, first of all, I should apologize in advance because I was out having a high-altitude nosebleed, so you may have answered all these questions; I don't know.

I also wondered about some of the questions that Alfred has touched on; this question of whether there are strains of the viruses, and whether, in fact, it's the *variants* of the population of viral strains in, you know, in bit space or some sort of Hamming space, that ultimately defeats the system. With some strain, the variants get sufficiently great, you simply can't juggle all these balls at once and somebody breaks through, and it's "game over."

But, two other questions. One was—again, I was too far away to see the equation, so I couldn't tell, but—I was struck by the fact that there's this sort of monotonic crash, and no cycling, and I wondered how structurally stable is the mathematics to that, and are there parameter regions under which you get some sort of recovery?

Finally, one final point is that I didn't notice any transport from this distance, and I wondered, what are the kind of energetics...I see that it's all reaction kinetics without any diffusion, so how does it happen, spatially?

PERELSON: Your first remark, having to do with the variants in the population, is something that's been studied theoretically quite extensively by a group at Oxford. The predominant person is Martin Nowak; he works with Robert May, an ecologist. They had exactly the theory that you suggested—that the virus keeps varying, that there's some critical diversity threshold (as they call it) where the variation gets too large, and the virus escapes....They've been looking at some data to try and see if that's true or not, and it's still up in the air. But it's a viable hypothesis; they come up with population-level models that have this "diversity threshold."

EPSTEIN: Structural stability.

PERELSON: Structural stability: You're also right. In fact, Rob De Boer (who's here) has done a fair amount on this, and we've studied these equations, done numerical bifurcation...You can get Hopf bifurcations, get into oscillatory regimes, find limit cycles; all I can say is from my knowledge of immunology, all the parameter regimes where we get exotic behavior, or just oscillatory behavior, do not seem to be particularly realistic. When we restrict ourselves to what I think are reasonable immunological parameters, we get the sort of generic story that I showed you here.

In terms of the spatial diversity, that's really a very good question, and it's something that modelers have not addressed sufficiently. All models of AIDS so far are homogeneous mixture models, and address only the question of what goes on in the blood. The reason we've addressed only what goes on in the blood is that's the only compartment we have data for. The blood is the only thing the clinicians can easily study. Now we *know* that most of the T lymphocytes are in tissue; we're looking at maybe three or four percent of the T cell population, when we talk about the number in the blood. And clearly as the numbers in the blood come down, if there is no commensurate effect in the tissue, the tissues could just sort of bleed out the rest of these T cells (from spleens and lymph nodes...). So there must be depletion going on in these other organs.

One can think about models where one has other reservoirs—and, in particular, lymph nodes and spleens are very large reservoirs of these lymphocytes—and have transport between these compartments. Diffusion probably isn't very important, just because things are so well mixed in the circulatory system.

EPSTEIN: But is there some kind of taxis that's going on, or entrainment of some sort...?

PERELSON: There is, but I don't think it's terribly relevant to deal with. There are specialized T cells that will home to particular regions. Say, in a normal immune response, you get a scratch on your arm, some bacteria go in—that antigen will tend to be carried through the lymph (the fluid that bathes your cells) to a regional lymph node, that drains that area; cells in that lymph node will grab that antigen and, basically, advertise it to cells passing through in the blood stream. And a large number of lymphocytes, whose receptors are specific for that antigen, will exit the blood stream, go into that lymph node—you've got a swollen lymph node—and you have a very good regional effect. It's been studied most dramatically in sheep, because they're big animals, with big lymph nodes, and you can isolate the cells. You can show within a day or so that 95 percent of the lymphocytes that can detect that antigen are all within that one lymph node. It'll increase four- or five-fold in mass; then a few days later the lymphocytes go out into the tissues.

What we're asking now are more global issues, about how the whole population in the body goes down. And, in AIDS, the issue of taxis or entrainment probably isn't relevant; the virus is probably everywhere, from what we can tell.

Brian Goodwin
Development Dynamics Research Group, Department of Biology, The Open University, Walton Hall, Milton Keynes, MK7 6AA ENGLAND

Developmental Complexity and Evolutionary Order

Abstract: The development of a morphologically complex organism from a simple spherical egg involves the activities of thousands of genes. How are these orchestrated so as to result in the remarkably coherent sequences of shape change that generate the basic body plans of different phyla, and the unfolding of morphogenetic detail that identifies individual species? How does the extreme genetic and molecular complexity of development relate to the evolutionary order that is revealed in the systematic taxonomic relationships of biological species?

It is generally assumed that the main actors in the evolutionary drama are the genes themselves, and that whatever order there is comes from the way biological complexity emerges historically, by gradual addition of inherited novelty to established patterns. However there is another approach to the problem of biological order which sees it coming not from genes and the accidents of history, but from the intrinsic dynamic principles of morphogenesis that need to be understood independently of genes in order to explain how and why organisms take their shapes and forms. This will be illustrated by a mathematical mode of morphogenesis in a particular class of organisms, and the argument will be extended to others. What

Complexity: Metaphors, Models, and Reality
Eds. G. Cowan, D. Pines, and D. Meltzer, SFI Studies in the
Sciences of Complexity, Proc. Vol. XIX, Addison-Wesley, 1994

emerges is the conjecture that the dominant morphological patterns that have emerged during evolution may be a consequence of the intrinsic robustness of particular morphogenetic trajectories, the generic dynamics of development. Then the subtle but pervasive order of the biological realm may be understandable in terms of the properties of a particularly interesting complex process that we call developmental dynamics.

INTRODUCTION

Viewed from the molecular and genetic levels, development is an extremely complex process. Thousands of genes and their products are involved in the orderly sequence of events that transform a fertilized egg of a particular species into the coherent intricacy of the adult form. What do we need to know to understand such a process? It is really necessary to identify all the relevant genes involved, to map their changing patterns of activity in the developing embryo, and to decode the combination language of gene interactions? Is the key to morphogenesis to be found in a genetic program?

The predominance of the genetic paradigm in contemporary biology encourages the belief that the relevant order that is observed in the morphology and the behavior of organisms is a result of effective genetic algorithms that can discover, in an immense search space, improbable genetic programs that are then stabilized by the adaptive success of the equally improbable organisms that they generate. In contrast to this view, I shall suggest that the space of possible biological forms, though certainly very large, may be much smaller than that suggested by the size of genetic program space, and that the role of natural selection in determining biological form may be much less than is often assumed. There is nothing in the least original about this position. Eighty-five years ago, in the introduction to his celebrated volumes *On Growth and Form*, D'Arcy Thompson[15] had this to say: "So long and so far as 'fortuitous variation' and the 'survival of the fittest' remain ingrained as fundamental and satisfactory hypotheses in the philosophy of biology, so long will these satisfactory and specious causes tend to stay severe and diligent inquiry...to the great arrest and prejudice of future discovery." So I am simply taking up D'Arcy's theme and developing it in a contemporary context, using insights that have come from complex dynamic analysis and computer simulation that were not available to him. However, the message is essentially the same.

DEVELOPMENT

I shall start with an example from some studies we have carried out on the development of a fascinating organism whose life cycle is shown in Figure 1. This is the giant unicellular green alga, *Acetabularia acetabulum*, whose habitat is the shallow waters around the shores of the Mediterranean. It is a member of an ancient algal group, the *Dasycladales*, that were once very numerous and widespread but are now reduced to a score or so of species. What makes them so interesting is the complex morphology that is achieved by the morphogenesis of a single cell. Starting from the fusion of the isogametes to form a zygote about 50μm in diameter, an axis is established with a growing stalk and a rootlike rhizoid that anchors the cell to the sea floor. The single nucleus remains in a branch of the rhizoid while the stalk continues to grow, producing rings of little branching structures (bracts), rather like primitive leaves, which constitute a verticil or whorl. These whorls of bracts are produced at intervals of a few days while growth continues, but within a week or so they are shed. When the stalk has reached a length of several centimeters, a new structure is produced at the tip—a cap primordium (see Figure 2). This then grows into the parasol-like structure seen in Figure 1, 0.5 to 1.0 cm in diameter, with its delicately sculpted rays, and the last whorls drop off. Although at this stage the alga looks multicellular, the detailed structure of the cap is the product of a single cell. This is what makes the organism so attractive for the study of morphogenesis. It embodies, in simple form, all the essential problems of development.

One of the most interesting aspects of *Acetabularia* development is the production of the sequence of whorls which are shed after growing into the rather beautiful, delicate structures shown in Figure 2. We know that the algae can grow perfectly well without producing whorls, which occurs if the concentration of calcium in the seawater is reduced from its normal 10 mM to 2mM.[6] When these whorl-less cells, consisting of a rhizoid and a stalk, reach a length of a few centimeters, they form normal caps if they are returned to seawater containing 10 mM calcium and they can complete a normal life cycle. So whorls do not seem to serve any function, despite the very considerable resources that the organism puts into their production. Why are nonadapted structures produced? Faced with such an apparent conundrum, biologists turn to a historical explanation. Although they may serve no function in this species, whorls probably were functional in its ancestors. The evidence clearly supports this. The *Dascladales* go back at least 570 million years, to the Cambrian. All species have bracts, most of them whorled, and in the majority, the bracts served as gametangia, where the gametes are produced, as there were no caps. So caps are late comers to the morphogenetic process in this group, and we get our historical explanation: whorls of bracts in *Acetabularia* are the result of a persisting ancestral pattern of development. Although their function has been superseded by caps, whorls continue to be produced in *Acetabularia* because of some kind of developmental inertia, and the price paid for producing them is not a sufficient penalty for natural selection to have eliminated either the whorls or *Acetabularia* as an extant species. In passing, one wonders how this fossil unicellular group has managed to survive in competition with all the apparently better adapted multicellular algae that arrived on the scene after the *Dasycladales*.

It is worth examining this type of historical explanation by looking at a similar example, but in another context—physics. We can ask the question: why does the Earth go round the Sun in an elliptical orbit? A historical explanation would then run as follows: the Earth follows an elliptical orbit this year because that is what it did last year and the year before that, and so on back to the origins of the planetary system; and nothing has happened to significantly disturb this pattern. This is a perfectly correct and sensible answer to the question. But it is precisely by not accepting it as an adequate explanation that physicists took the significant step of discovering sufficient conditions for elliptical orbits, not just necessary, or

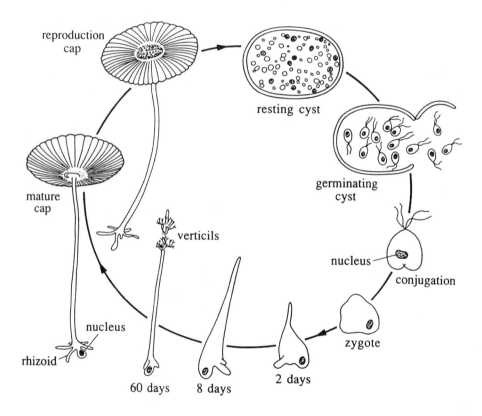

FIGURE 1   The life cycle of *Acetabularia acetabulum.*

FIGURE 2   Formation of a cap primordium, after three verticils or whorls.

contingent, or accidental conditions. This, of course, was the step taken by Newton in deducing that bodies obeying an inverse square law of gravitational attraction must follow trajectories that belong to the conic sections. Of these, the only stable solutions with closed, periodic orbits are the ellipses (circles transform to ellipse under perturbation). There is of course a vast literature on the stability of the solar system whose study actually led Poincaré to the discovery of chaos, but this is not relevant to the simple point I am making now about the inadequacies of historical explanations in a scientific context. There is always the possibility that we can do no better in providing insight into the causes of phenomena than historical narratives. However, it seems to me a mistake to accept them uncritically. Evolutionary biology suffers from a considerable excess of them.

The planetary example illustrates another salutary point about biological explanations. Stability analysis is always part of a causal, dynamic explanation. Natural selection addresses only stability questions in biology, the factors that contribute to the *persistence* of a particular species in a certain habitat. It does not provide any explanation of existence—why a species with its distinctive morphology and behavior is possible, how it is generated as one of the solutions of morphogenetic dynamics. Applied to planetary orbits, natural selection could be used to examine the reasons why ellipses are stable solutions under the particular conditions of the solar system, hence, why they have been selected, i.e., why they persist in preference to other possibilities, such as circles. It may sound strange to use the language of natural selection in the context of physical processes, but again, the ideas behind it, which relate to dynamic stability, are perfectly reasonable. The take-home message is that there is nothing distinctively biological in the notion of natural selection. All the sciences use this idea—but they give it the rather more transparent name of stability analysis.

The objection is sometimes made that natural selection is more than dynamical stability analysis because it is about change as well as stasis. This is precisely what is studied in complex dynamical systems with multiple attractors. A good example of these concepts applied to evolution is the paper by Rand et al.[14] This defines the notion of an evolutionarily stable attractor. An invasion exponent can be defined that describes the conditions for evolutionary change in complex situations involving multiple species with chaotic population dynamics. The invasion exponent is a type of Lyapunov exponent that has been introduced also by Metz et al.[10] It can be used to describe differential selective pressure. These purely dynamical concepts replace fitness functions, which are *ad hoc* and nongeneric for relevant dynamical systems. They also allow us to replace the poorly defined concepts of natural selection and adaptation by precise stability ideas. Clearly, adaptation is very much in the eye of the beholder. Are whorls adaptive in *Acetabularia*? It would appear not, since they seem to serve no function. But the species has been happily surviving for millions of years: the life cycle is clearly stable in the shallow waters of warm (and nonpolluted) seas. And that is all that natural selection tells us. So we return to the question: why does *Acetabularia* make whorls? Neither history nor natural selection gives us an explanation of the type we seek in science. And this failure is true of nearly all problems connected with explanations of biological form (morphology and behavior), not simply the ones that appear puzzling, like whorls, because they don't conform to our idea of adaptation. The whole scenario changes when we look at things dynamically. This is what D'Arcy Thompson was urging us to do 86 years ago, and why he complained about "fortuitous variation" and "survival of the fittest" being "specious causes" that "prejudice future discovery" in the study of growth and form. So let's return to this question.

MORPHOGENETIC DYNAMICS

An obvious way to investigate the sufficient causes of an organism's morphology is to construct a model of how it is generated. This requires a morphogenetic field theory. We constructed one for *Acetabularia* on the basis of our own experimental studies, an extensive body of work on plant cell growth, and other morphogenetic models. Since this model and its behavior have been described in detail elsewhere,[1,2,4,5] I shall confine myself here to a summary of the results as they apply to the question about whorls. The core of the model is the description of the cytoplasm as an excitable medium that can spontaneously break symmetry and generate spatial patterns in the primary variables, which are the concentration of free calcium in the cytoplasm and the mechanical strain (degree of stretching or compression), these variables being linked due to the properties of the cytoskeleton. This mechanochemical field model is related to those described by Oster and Odell[13] and by Murray and Oster.[11] Coupling of cytoplasmic state to cell-wall growth is based on strain through a growth function. A finite-element description of the model in three dimensions, carried out by my colleague Christian Briere, allowed us to study the development of form through growth. We are dealing with a moving boundary problem in which dynamics change the shape of the domain (the developing organism) which acts back upon the dynamics.

The model is biologically very simple, though mathematically it is a complex nonlinear system of coupled partial differential equations. There are 26 parameters, though the dynamics are sensitive to changes in only six of these. We chose parameter values according to two criteria: (1) they lie within the bifurcation range so that patterns can form spontaneously, and (2) the wavelengths of the patterns are smaller than the domain of growth and regeneration, so that interesting shapes can develop. The model was then allowed to do its own thing—to make whatever shapes it could generate. Clearly, there is a very large search space in terms of possible parameter values satisfying our criteria, and we had no idea how long it would take us to find something interesting, or indeed, if anything remotely resembling algal forms would emerge.

What surprised us was how easy it was to find ranges of parameters that gave growth of a stalk either from a sphere (a zygote) or from a hemisphere (a regenerating apex). But it was more surprising to find that these growing "stalks" then generated the whorl pattern, and in the process, explained shape changes that we had frequently observed in growing algae but had never understood. Just before a whorl of verticils is produced, the tip flattens from a conical shape. The model did this after tip growth, due to a spontaneous transition from one pattern to another: from a gradient in calcium and strain with maxima at the conical tip, to an annulus with a maximum away from the tip, where maximum curvature developed while the tip flattened. Under perturbation, the annulus bifurcated to a whorl pattern, with a series of peaks and troughs in the variables. This is described in detail in the publications listed above. There is experimental evidence that calcium patterns change in this way during these morphogenetic transitions.[9]

The model gave further insights into the morphogenetic sequences. As growth of the tip continues, the annulus collapses and a gradient with a maximum in the field variables (calcium and strain) reforms at the tip. This is like resumption of tip growth after whorl formation. With further growth, the annulus forms again, in an intermittent fashion reminiscent of the intermittent production of whorls during growth. No caplike structure was ever observed though a terminal phase of tip expansion that ends the extension of the stalk does occur.

Unfortunately, the finite-element analysis we are using is not sufficiently robust for us to explore the details of bract growth after the formation of the whorl prepattern. This requires switching to a much finer grid, and using a different finite-element geometry. Also, to get a caplike structure it appears that we need more anisotropy in the strain field, so that lateral growth exceeds longitudinal growth. These developments require further investigation.

We also do not know how large is the domain in parameter space where algal-like patterns arise. Clearly, it is very important to investigate this systematically. However, the fact that it was so easy to find domains of interesting morphogenesis suggests that there is a large attractor in this space of moving boundary solutions that results in recognizably biological patterns. It is also interesting that the whorl pattern appeared easily, but not the cap. As mentioned earlier, nearly all members of the *Dasycladales* produce whorls of bracts, so this structure is generic to the group; caps are not. In our model, whorls also appear to be generic, now in the mathematical sense of typical for this dynamic system. In the space of giant unicellular algal forms, these appear to be the high-probability structures. So we can suggest a solution to our problem: whorls are produced in *Acetabularia*, not because they are useful, but because they represent a generic form, a structure that this type of system tends naturally to generate.The explanation lies not in history, nor in natural selection, but in dynamics: these are the high-probability, stable patterns of this morphogenetic field.

What about genes? Where do they fit in? Clearly they are involved in defining parameter values. Genes can be said to specify the domain in parameter space where morphogenesis occurs. It would appear from our model that this is a large domain though we do not yet know how large. That is to say, genes can vary quite a bit and still fall within the domain of the basic algal attractor. Also, we don't see anything like a genetic program guiding our model through its morphogenetic sequence: no parameter changes are necessary for the sequence we observe, because the cycle of dynamics changing the shape through differential growth, which then alters the dynamics, itself results in whorl production. Possibly for cap formation parameters need to change. The result would be a hierarchical dynamic in which "parameters" become variables. This is clearly what happens in more complex morphogenesis. But again, the problem is to describe the field theory appropriately, so that it produces observed behavior. The explanation of *Drosophila* or frog or human morphogenesis cannot come from a description of parameter values and their changes, but from a morphogenetic field model with the appropriate dynamics and range of parameters. Genes and their activities do not *explain* morphogenesis. They define parameter ranges that result in morphogenetic trajectories which give rise to species-specific morphologies. Other stabilizing influences on the trajectories are environmental conditions, which must always be included in morphogenetic models. In our model, a major external influence is calcium concentration in the seawater, but other factors (light, temperature, other ions) also contribute to parameter specification.

THE EVOLUTION OF GENERIC FORMS

From the perspective outlined above, there emerges a fairly obvious conjecture about the morphological products of evolution. All conserved aspects of biological structure may be the generic forms generated by morphogenetic fields.[1,2,5,12] These include the basic body plans of the different phyla, such regularities as the patterns of leaf production in higher plants (phyllotaxis), and homologous structures such as tetrapod limbs. Take the question of phyllotaxis. Leaves are generated by the growing tip of a plant by a multicellular structure, the meristem, in which growth and form are linked as they are in *Acetabularia*. There are only three basic patterns of this process, as shown in Figure 3. Leaves can be produced one at a time on opposite sides of the growing tip (distichous phyllotaxis) as in the grasses (monocotyledons); they can be produced in groups of two, three, or more, with alternating positions at successive nodes (whorled phyllotaxis); or the leaves are generated singly at a fixed angle (average 137.5°), resulting in a spiral (spiral phyllotaxis). Why should these patterns be so constrained? The possibility is that these are the only stable solutions of this morphogenetic process, a conjecture for which Green[7,8] has convincing evidence and is actively seeking a generative model. The fact that more than 80% of higher plant species have spiral phyllotaxis may then arise primarily from the sizes of the attractors for these different solutions. The idea here is simply that all three patterns produce perfectly satisfactory leaf arrangements from the point of view of catching sunlight, respiration, transpiration, and so on, so that the functional aspects ("fitness") of all forms are roughly equivalent. The frequencies of these different patterns in nature may then simply reflect their differential probability as measured by the sizes of the attractors. Furthermore, the different patterns are rationally united as different solutions of the same morphogenetic process: they are transformations within a particular generative dynamic. So taxonomy (relationships of similarity and difference) and differential abundance can both be explained within a single dynamic perspective. Instead of being based upon history (genealogy) and function, classification is then based upon generative dynamics, which includes environmental influences and the study of dynamic stability, as previously discussed. By putting development back into evolution at a fundamental level, we lose nothing of value in the study of gene action (parameters, hierarchical dynamics) or natural selection (stability), but we gain the whole dimension of rational taxonomy (form and transformation) and *sufficient* explanations of biological phenomena, rather than just the description of necessary conditions (genes, survival). So we can begin to talk about the evolution of generic forms, and to consider the possibility of an evolutionary theory that involves the dynamics of development as the generative origins of morphological species, their systematic relationships, and their intrinsic stability.

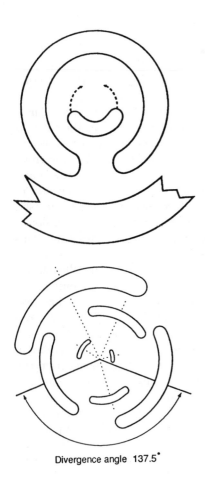

Divergence angle 137.5°

FIGURE 3   The three basic patterns of leaf phyllotaxis. (1) Distichous (corn), (2) Whorled (maple, mint), and (3) Spiral (ivy, lupin, potato).

REFERENCES

1. Goodwin, B. C. "The Evolution of Generic Forms." In *Organizational Constraints on the Dynamics of Evolution*, edited by J. Maynard Smith and G. Vida, 107–117. Manchester: University Press, 1990.
2. Goodwin, B. C. "Structuralism in Biology." In *Science Progress*, Vol. 74 227–244. Oxford: Blackwell, 1990.
3. Goodwin, B. C. "Development as a Robust Natural Process." In *Thinking About Biology*, edited by W. D. Stein and F. J. Varela. Santa Fe Institute

Studies in the Sciences of Complexity, Lect. Notes Vol. III, 123–148. Reading, MA: Addison-Wesley, 1993.

4. Goodwin, B. C., and C. Briere. "A Mathematical Model of Cytoskeletal Dynamics and Morphogenesis in *Acetabularia.*" In *The Cytoskeleton of the Algae*, edited by D. Menzel, 219–238. Boca Raton, FL: CRC Press, 1992.

5. Goodwin, B. C., S. A. Kauffman, and J. D. Murray "Is Morphogenesis an Intrinsically Robust Process?" *J. Theoret. Biol.* **163** (1993): 135–144.

6. Goodwin, B. C., J. C. Skelton, and S. M. Kirk-Bell. "Control of Regeneration and Morphogenesis by Divalent Actions in *Acetabularia* Mediterranea." *Planta* **157** (1983): 1–7.

7. Green P. B. "Inheritance of Pattern: Analysis from Phenotype to Gene." *Amer. Zool.* **27** (1987): 657–673.

8. Green, P. B. "Shoot Morphogenesis, Vegetative through Floral, from a Biophysical Perspective." In: *Plant Reproduction: From Floral Induction to Pollination*, edited by E. Lord and G. Barrier, 58–75. See Also, *Am. Soc. Plant Physiol. Symp.* **1** (1989).

9. Harrison, L. G., K. T. Graham, and B. C. Lakowski. "Calcium Localisation During *Acetabularia* Whorl Formation: Evidence Supporting a Two-Stage Hierarchical Mechanism." *Development* **104** (1988): 255–262.

10. Metz, J. A. J., R. M. Misbet, and S. A. H. Geritz. "How Should We Define 'Fitness' for General Ecological Scenarios?" *Trends in Ecology and Evolution* **7** (1992): 198–202.

11. Murray, J. D., and G. Oster. "Generation of Biological Pattern and Form." *IMA J. Maths in Med. & Biol.* **1** (1984): 51–75.

12. Newman, S. A., and W. D. Comper. "'Generic' Physical Mechanisms of Morphogenesis and Pattern Formation." *Development* **110** (1990):1–18.

13. Oster, G. F., and G. M. Odell. "The Mechanochemistry of Cytogels." *Physica* **12D** (1984): 333–350.

14. Rand, D. A., H. B. Wilson, and J. M. Glade. "Dynamics and Evolution: Evolutionaryily Stable Attractors, Invasion Exponents, and Phenotype Dynamics." Technical Report, Nonlinear Systems Laboratory, Mathematics Institute, University of Warwick, Coventry CV4 7AL, UK, 1992.

15. Thompson, D'Arcy W. *On Growth and Form.* Cambridge: Cambridge University Press, 1917.

DISCUSSION

BROWN: Pursuing the line that Murray began, this is all correct and exciting and important. But I don't see that it in any way contradicts the idea that natural selection and other forces of that sort are also molding the phenotype. I get really upset when you use words like "objective" to refer to this, and "subjective" to refer to adaptation.

Let's go back to the example of the eye. What you said is correct, but what you didn't say was that the vertebrate eye is completely different in its orientation of the sensory elements in the retina to that of the molluscan eye. And presumably that's an accident, because no engineer would have designed a vertebrate eye with the sensory parts pointing inward away from where the light's coming. If you look at the vertebrate eye, you see all sorts of fine tuning reflecting natural selection: differences in rods and cones depending on how diurnal or nocturnal organisms are; differences in the shape of the lens depending on whether they're operating in air or water; different colored droplets in aquatic things, depending on how deep they are and what kinds of light regimes they're operating in. I don't see how you would ever explain that morphological diversity without recourse to some kind of adaptation and utility.

GOODWIN: Let's pursue the question of explanation. I do not regard natural selection as an adequate explanation of the forms we see in nature, because it addresses the question of persistence. And what we want to understand is *existence*: Why are these things possible? Unless you have a theory that tells you why these forms are possible, you're going to go around forever debating questions like, you say, "It's an accident. It must have been an accident because no *engineer* would have designed the vertebrate retina the way it is." I would argue that the reason the vertebrate retina is designed the way it is, is because of the basic constituents of the morphogenetic process that generate nervous system, the optic stalk, the bulb, the interaction with the epidermis, the formation of the optic cup, and so on. I would argue that is intrinsic to the morphogenetic sequences that occur there.

Let me stress that if you only look at stability—and that's all that natural selection addresses—you are not explaining *possibilities*; you're not explaining why they are possible. You're simply saying why they persist. That is not, for me, an adequate explanation.

BROWN: I don't think that natural selection is only concerned with stability; it also gets you from point A to point B. But no one would argue that natural selection accounts for everything we see in the biological world; the other things are important. But, to argue that these things are exclusively responsible for what we see is an equally incorrect view.

GOODWIN: A dynamic analysis always includes stability influences. When I say stability, I mean dynamic stability: persistence of life cycle and variability of habitat and so on.

KAUFFMAN: So you're not excluding adaptation.

GOODWIN: I'm not *excluding* that; I'm saying that it is not the dominant reason why we see what we see, in the natural world.

SCHUSTER: Brian, you gave wonderful examples of stable forms of developmental dynamics. What I don't see is the necessity to make such a strong polarization between natural selection on one side and developmental dynamics on the other side. Natural selection chooses between forms that are determined by molecular genetics, molecular biology, *and* developmental dynamics. For instance, if I look at virus dynamics by mutation, not every mutation is possible, because the population structure determines already which mutations are there, and which are very unlikely to be produced. Yet we have the same case: natural selection chooses only between cells which are possible, and common.

GOODWIN: Stability analysis is what we do on our quasi-species. Why use the language of choice? There's no choosing. Natural selection doesn't choose anything except what's stable, and persists.

SCHUSTER: Here I don't agree. We should not use the term "random variation," because the variations are not random. The mutation, or the developmental dynamics, does not care about what natural selection makes out of it at the end.

GOODWIN: Random mutation is perfectly compatible. We have no objection to random mutation.

GELL-MANN: I won't go over that ground again; let me say something else.

In making this unnecessarily polemic version of a perfectly sensible scientific argument, it seems to me one is ignoring the role of history. One of the most beautiful things about evolution is that there are historical tracks, and these historical tracks do not represent simply stability—comparative stability—of the stages along the way, but there's an actual record (if one could read it) of stages along the way. And it's those stages that determine a lot of the anomalies, as every biologist emphasizes. All sorts of things are there because they come from something else, which came from something else. And so they look peculiar, and an engineer might not have made them that way—unless the engineer were also part of a unique process, which sometimes happens. (You get vestigial organs, and so on, also in technology.)

So it's not just a question of stability. A whole orbit is involved. Now, that orbit obviously is subject to constraints of the kind you're talking about; no person in his right mind would dispute it. But I don't understand why it's necessary to make a polemic out of pointing it out.

GOODWIN: It's necessary to clarify the language that was used. I'm a biologist, so I grew up with this stuff. It took me *thirty years* to try to work my way through this swamp of confusion that I find in the conceptual basis of biology.

Let me give you a tiny example: why does the earth go around the Sun in an elliptical orbit? It's perfectly correct to say that it did that because last year it was in an elliptical orbit, and the year before that, and the year before and back to the origin of the planetary system. So, that's a perfectly correct statement and an historical explanation. The initial conditions determine the trajectory. I'm in *no* sense denying that. That is always part of the dynamical process, as well. It's always got history built into it. Because, from the set of the possible, certain things are realized under certain conditions. Now, if that is accidental, we say, "That's accidental." What I want to push for is to reduce the realm of the accidental to the absolute minimum...

GELL-MANN: Well, to what it is; not necessarily the absolute minimum, but to what it really is.

GOODWIN: The same statement.

GELL-MANN: It may not be the absolute minimum.

GOODWIN: The minimum that we can achieve is as far as we can go. *That's* why I'm polemical about it: because there are so many historical explanations that are used in biology that shut off students' minds, and they think, "That's it; that's an explanation." It's *not* an explanation.

GELL-MANN: You're mad at a bunch of people who aren't here.

GOODWIN: You seem to collaborate with them, Murray!

GELL-MANN: All this resentment is misdirected against us!

GOODWIN: Put history back in its proper place; that's the objective.

BUSS: I don't have a comment on the polemic, but I think that claims of this order have some utility in that they force the scientist seeking to make an adaptationist interpretation to come to terms with the claim that the range of possible forms is not, in fact, infinite. There's only a series of possible forms that might be chosen.

GELL-MANN: It could be infinite, and still restricted.

BUSS: Brian's strong claim is only going to be realized if the choice is only one for any system. That is unlikely to be the case; we wouldn't have the kind of diversity we have. But, nevertheless, that sort of strong argument does focus on the fact that most of our adaptational selections do not take account of the range of forms that are possible (because selection can only choose among those).

GOODWIN: The case of this phyllotaxis, we have three modes. That's multiple modes, with different distribution. Now, the conjecture is that the distribution in nature is equivalent to the relative stability of those modes.

BUSS: That's an empirically testable hypothesis.

ANDERSON: It seems to me that our clone of Walter Gilbert said much the same thing—at least one version of the same thing—when he said there are only seven thousand exons out of billions, or more, possibilities, that already at that molecular level, evolution is very much restricted by the possible, rather than by the. . .

GOODWIN: But you need a generative theory. If you're going to delineate the possible, you have to show why and how they occur.

ANDERSON: Well, in that case, it's very likely to be something very simple, like a generative theory of folding of proteins. Maybe those are the only proteins that can fold.

GOODWIN: That's great. So that's the direction in which to go. Simply saying that there's seven thousand is very interesting, but it's not an explanation.

ANDERSON: I suspect there's seventy thousand rather than seven thousand, or seven hundred thousand; still the number is so tiny compared to the possible. . .

GOODWIN: We agree; what has to be done is explain it.

KAUFFMAN: Brian is focusing us very importantly on the strengths and the weaknesses of an evolutionary paradigm to explain complexity. The evolutionary adaptationist paradigm has the strengths and weaknesses that evolutionary theories have had all the way along. Darwin gives us a theory about descent with modification, and arguments about adaptation to an external world. Darwinian theory never tells us what *is*—it will not tell us if you have a frog; it will tell us that you will get modified frogs in the future. What Brian is properly stressing is that we want to account for what is out there when you walk out in the field, not just how it changes over time. And so, the search for generative mechanisms that create morphology, that create organisms that can unfold in a variety of forms, raises the research problem of how much of the varieties that we see out there are the natural forms that morphogenesis can give rise to, and how much of it is, in fact, realized through possible adaptive means.

You run into the following problem: You see a particular organism and say, "Fine"; some traits are selected, and some aren't. Then you try to decide: "Here's an interesting looking trait. I wonder if this one is around because it has adaptive significance, now or in the past." And then in almost every case—it turns out, when you try to nail it down—you wind up with what's called an evolutionary "Just So" story. You can cook up a perfectly plausible scenario why those whorls are useful. It turns out, in practice, using that as a paradigm to explain what you see is remarkably weak. So there are very deep problems with an adaptationist paradigm that we have to take seriously. There's 150 years of this kind of argumentation, and what it can do, and what it does poorly. It's a very rich area; we've just [to] beware of its difficulties.

BAK: The message I hear is that nature likes to have simple rules. And simple rules, we now know, can produce very complicated patterns. You shouldn't look for logic in every single piece of this pattern; we shouldn't look at some kind of function for everything because it's generated from very little basic information. So the contrast is that, traditionally, one might look for the complexity of leaves, and so on, and find a reason for everything—which couldn't be there, because the genetic information needed to create it is very little; the rule is very simple, couldn't do all that. The message is that we don't have to decipher all the complexity and attach understanding to all of it in the traditional way. Isn't that the sort of controversy that you're talking about?

GOODWIN: That's exactly what's emerging from recent studies in *Drosophila* developmental genetics, that what we're seeing is a fantastic amount of genetic overwriting. The underlying—what we call the hardware—is basically very simple.

BAK: We know, for instance, that to generate the brain from the genetic code we need very little information; it's very complex, so that means that there cannot be so much information...

GOODWIN: The overwriting, the software, is fantastically complex. But what I'm getting at is the underlying hardware. And that, I think, is simple and robust. And that's what generates form. It's not easy to get at; that's the problem.

EPSTEIN: I find your conjecture very attractive, because it holds out the prospect of learning a hell of a lot of biology by just studying a little math.

I had a question on this. I was just wondering whether some of these whorls, as Murray suggested, might have been useful at some point—and they might not have been useful at some point—but maybe, instead of saying that the system makes some judgment about whether it's useful, maybe it makes a calculation that it would be expensive to bother to get rid of it. Maybe it's not that something is selected because it's useful. You guys have coevolving, rugged things, ruggedizing around, and at some point, something crops up that isn't useful, or is useful, and the thing calculates, "Ahh, it's not very useful, but it would be a total hassle to reprogram myself and get rid of it." It's adaptive in that sense—in the same sense that it's adaptive for Chevrolet to keep its screwed-up V-8 engine that's a pain to maintain, because it would be necessary to reorganize it completely to get at this valve that keeps screwing up. So the question is maybe it isn't natural selection at all; maybe it's some kind of funny cost-benefit analysis that goes on, that says, "Maybe this thing is really useful, and that's terrific," but maybe it also says, "It's not useful but, blechh, the hell with it; it's a schlep to get rid of it."

The second question is: I like your genericity marking, but it seems to me there is implicitly a type of uniqueness claim you're making about the mathematics. That is, you cannot achieve the same stable morphology with equations that would have the whorls not appear. It seems to me that's a tough point to demonstrate.

GOODWIN: Yes. Your first question: if you like that language of cost-benefit analysis, that's fine. But I think you contribute to confusion, because what it means is that you come back to robustness. In other words, robustness means it's generic and, therefore, it's difficult to stop something from happening if it's generic. I want to replace cost-benefit language by dynamic concepts, and get out of a concept that is highly subjective; to whom? For whom? That may be fine economics, but I don't think it's fine in biology. I think it really messes up the pitch.

EPSTEIN: Well, could you do a calculation that says, "Here's what it would require to reprogram this dude, so he doesn't have these things on it, and is that a very energy-conserving operation?

GOODWIN: Exactly; we want to explore the domains of parameter space, and maybe there is a little region where you can do it.

JEN: I have a question about the robustness, but I also want to say that I appreciated the polemics on both sides—partly because I think it focuses attention on an aspect of adaptation that we haven't really been explicitly talking about, namely, time scales of adaptation. And, in particular, in talking about stability, it seemed to me that you were talking about stability on a short time scale; Murray and Jim, and a chorus of polemicizers, were talking more about stability on long time scales.

We haven't really been distinguishing, and talking about, adaptation. Basically, a number of sets of issues that come in very importantly—time scales, strength of stimuli, difference of stimuli (in terms of environment), and all these factors which, I think, make adaptation (change the scale of adaptation, actually)—really need to be associated with our discussion of adaptation. But my question actually has to do with the robustness, a little bit, maybe, related to the uniqueness question, which is: Given the fact that it is so generic and it is so robust, doesn't it then lead to some questions as to whether you can, in fact, have confidence in your model that generates behavior that's qualitatively similar to what you see? Given the fact that it is so robust, presumably there are going to be almost an infinite variety of models, all of which have some sort of underlying principle, perhaps, but are quite different in detail, that generate (probably) the same sorts of behavior—if it is so robust, and it is so generic. And so, then, how do you distinguish among your different models to try to figure out which is actually correctly describing the mechanics of the development?

GOODWIN: The process I talked about was a moving boundary process, in which you have geometry influenced in dynamics, and that influences the geometry. So you get an unfolding, a cascade. And we need to study those in order to see what the range of possibilities is, and whether they can indeed be generated by a whole variety of other types of process, which are equally robust and generic.

Now, those are very interesting questions. My intuition, from the reading of the physical world generally, is that is not the case. There are particular ways of generating the elements; there aren't many ways of doing it and there are a limited number of forms that you get, that are stable. And there are others that are unstable, and so on. So it's the question of generative rules, and precisely your question: What do they lead to? Is there a multiplicity of such rules, that generate equivalence classes? What are the equivalence classes? For me, homology is an equivalence class, you see. And it's that kind of analysis of symmetry, symmetry breaking, equivalence classes, that needs to be introduced into the realm and study of biological form in order to make sense of it. But your question is a good one.

Walter Fontana* and Leo W. Buss[†]
*Santa Fe Institute, 1660 Old Pecos Trail, Suite A, Santa Fe, New Mexico 87501
†Yale University, Department of Biology and Department of Geology and Geophysics, New Haven, Connecticut 06511

What Would be Conserved if "The Tape Were Played Twice"?

Reprinted by permission of *PNAS*.

Abstract: We develop an abstract chemistry, implemented in a λ-calculus based modeling platform, and argue that the following features are generic to this particular abstraction of chemistry, hence would be expected to reappear if "the tape were played twice": (i) hypercycles of self-reproducing objects arise, (ii) if self-replication is inhibited, self-maintaining organizations arise, and (iii) self-maintaining organizations, once established, can combine into higher-order self-maintaining organizations.

INTRODUCTION

Gould[12] has asked the question whether the biological diversity that now surrounds us would be different if "the tape were played twice." If we had the option of observing a control earth, would we observe, say, the evolution of *Homo sapiens*, or the evolution of something unambiguously identifiable as a metazoan, or even something akin to an eukaryote? The question is important in that it focuses attention on the fact that historical progressions, such as the history of life, are the product of both contingency and necessity. While Gould's[12] emphasis on the contingent is well taken, one nevertheless has the sense that certain features would recur. What are those features and how might we discover them?

The fundamental difficulty with analysis of the questions of contingency and necessity in the distant past is the very fact that they occurred in the distant past. Experiments today cannot be performed with systems as they might have existed billions of years ago. The only alternative is to establish a model universe in which such an exploration is possible. In such a universe, one may unambiguously demonstrate whether the appearance of a given result is necessary or contingent. The question of the validity of such a claim may then be rigorously challenged by questioning the abstractions upon which the model is based or by introducing increasingly realistic elaborations of the model universe.

A model universe designed to explore what is contingent in the history of life cannot assume the prior existence of organisms. The approach must seek to establish how biological organizations are generated. In this contribution we sketch a framework, developed in greater detail elsewhere,[10] that holds promise for such an undertaking. We introduce an abstract chemistry implemented in a modeling platform that permits the study of the origins of self-maintaining organizations in a minimally constrained fashion. In several specific instances this system spontaneously and robustly generates a number of features which occurred in the history of life. The minimality of our model, then, suggests that these features arise generically and, hence, might be expected to reappear if "the tape were played twice."

THEORETICAL FRAMEWORK AND MODELING PLATFORM

We seek to develop a model of biological organization that is grounded in a particular abstraction of chemistry. Chemistry is characterized by a combinatorial variety of stable objects—molecules—capable, upon combination, of interacting with each other to generate new stable objects. When two molecules interact, the product is determined by their structure, i.e., the components of which they are built and the manner in which these components are arranged. Thus, a molecule is an object with both a syntactic structure and an associated function. Syntactically, it is built up from component objects, according to well-defined rules. Its function, coded by its structure, is revealed by the chemical reactions in which it partakes. Chemical reactions generate a stable product through a series of structural rearrangements driven by thermodynamics. We abstract from chemistry both (i) the interaction between molecules to generate new molecules and (ii) the driving of a reaction to a stable form by structural rearrangement.

The mathematical machinery that provides us with an implementation of such a situation is known as the λ-calculus.[3,4] In λ-calculus, syntactical structures, that is, objects, are defined inductively in terms of nonlinear combinations of other objects, starting from primitives. This definition implies that each object is a function. The function represented by object A is the mapping that assigns to any object B a new object expressed syntactically as $(A)B$, referred to as the action of A on B. To execute this action, λ-calculus defines axiom schemes for rearranging the structure of objects. Let $(A)B$, say, be restructured by applying the schemes of rearrangement one at a time until no further modification is possible. Such a process generates a series of intermediate objects, $(A)B \rightarrow C_1 \rightarrow C_2 \rightarrow \cdots \rightarrow C$, and is termed reduction. The unique final product thereby reached is called a normal form. The schemes of rearrangement are such that functional equality ensues, i.e., we can replace $(A)B$ by C since $(A)B = C$. Thus, in λ-calculus, (i) objects combine with other objects to produce new objects, which (ii) are transformed to achieve a stable form.

Lambda-calculus, while capturing certain key abstractions from chemistry, is not a theory of actual chemistry or a theoretical biophysics. For example, this level of description intentionally lacks any explicit reference to thermodynamic notions. Thermodynamic driving is abstracted solely by requiring that every object in our system be in normal form, i.e., schemes of rearrangement are applied to obtain a stable (normal form) object. From a logical point of view thermodynamics essentially implements a consistency requirement by preventing arbitrary rearrangements in arbitrary reactions from occurring. The reduction process as defined in λ-calculus guarantees such a consistency. Thus, λ-calculus captures what is inherent in such consistency requirements, but not necessarily what is inherent in thermodynamics. In addition, the present system does not consider spatial constraints, conservation laws, or unequal reaction rates. Our intention is not to emulate actual chemistry, but rather to explore the consequences of those minimal features we abstract from chemistry.

This theoretical framework is instantiated in a model with the following components:

1. *Universe*: A universe is specified by the axioms of λ-calculus, which define the nature of objects and the manner in which objects are transformed syntactically. We use the axioms of λ-calculus as defined by Revesz.[17] In our universe, all objects are required to be in normal form and reduction to normal form is required to be completed within some maximum number of steps; otherwise, the object is not allowed in the universe.

2. *Collision Rule*: The basic event in our model universe is the interaction among two objects, A and B, upon collision. In the simplest case the interaction between A and B invokes *application* in λ-calculus: $(A)B$. The new object created by the interaction is the normal form of $(A)B$. The collision rule may itself be expressed as an object in Λ, which provides a powerful generalization.

3. *Interaction Scheme*: Let $[A, B]$ denote a collision event between the ordered pair A and B. The interaction scheme used here is (i) $[A, B] \longrightarrow A + B+$ normal form of $(A)B$, (ii) A must be of the form $\lambda x_1.Q$, with Q arbitrary, and (iii) computation of the normal form of $(A)B$ must be completed within $10,000$ steps, and must not exceed $4,000$ characters. If any of the requirements are violated, no reaction occurs: $[A, B] \longrightarrow A + B$. These limits, imposed for practical computation reasons, were not ususally exceeded in our computer experiments.

4. *System*: The system is a well-stirred flow reactor which is initialized with $1,000$ randomly generated (and reduced) objects unless otherwise noted. A pair of objects, A and B, is chosen at random for collision, $[A, B]$, according to the above interaction scheme. The object chosen first is hereafter referred to as the operator. Note that $(A)B \neq (B)A$. On average, however, half of the collisions between A and B will be of the form $[A, B]$, the other half of the form $[B, A]$. The newly created collision product is checked against predefined syntactical and/or functional constraints (i.e., boundary conditions). If the object passes the filters, it is added to the system. We keep a constant number of objects at any one time. To do so, one object chosen randomly from the system is eliminated. This gives each object a finite lifetime. The whole procedure is reiterated.

We do not describe the model in further detail here. Rather, we summarize the results of computer experiments which we hope will be sufficient to introduce the behavior of the system and serve as an inducement to readers to explore the primary literature describing the approach,[8,9,10] in particular, Fontana and Buss.[10]

COMPUTER EXPERIMENTS

We describe three series of experiments.

LEVEL 0 EXPERIMENTS

EXPERIMENTAL PROTOCOL AND SUMMARY OF RESULTS. The system was initialized with a series of 1000 randomly generated functions each initially present in one copy. Such experiments always become dominated by either single self-copying functions or ensembles of hypercyclically[6] coupled copying functions (i.e., functions f with $(f)g = g$ or f, for all g in the system). Under perturbation, i.e., the introduction of random objects, Level 0 ensembles reduce to single self-copying functions, i.e., a function f with $(f)f = f$.

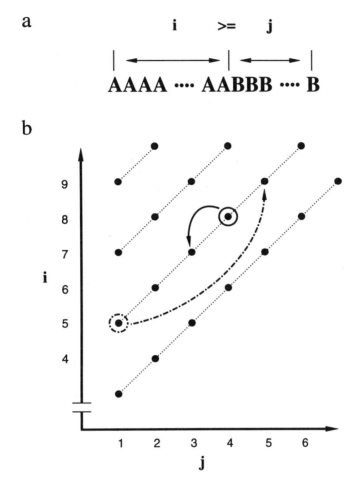

FIGURE 1 An example of a Level 1 organization. (a) All objects in the organization share a syntactical regularity. Specifically, each object is a string of two elements, A and B, such that every object contains a sequence of A elements followed by a sequence of B elements, with the number, i, of A elements equalling or exceeding the number, j, of B elements. Within the formalism of λ-calculus, A is encoded as $\lambda x_i.$ and a sequence of j B's is x_j. (b) The functional relationships insuring self-maintenance can be succinctly stated as a consequence of two laws. Specifically, in the i, j plane, the bold arrow illustrates the action of Law 1 $[(O_{i,j})O_{k,l} = O_{i-1,j-1}$ for all $j > 1, i, k, l]$ and the broken arrow that of Law 2 $[(O_{i,1}O_{k,l} = O_{k+i-1,l+i-1}$ for all $i, k, l]$. Note that the emergent laws do not make reference to the underlying λ-calculus, but nonetheless are sufficient to describe the product of any object in the system interacting with any other object. For the sake of a less congested figure actions are drawn schematically and every other diagonal is missing.

RELATIONSHIP TO THE WORK OF OTHERS. These results represent an independent rediscovery of the work of Eigen and Schuster[6] on hypercycles. Specifically, our Level 0 experiments generate autocatalytic ensembles of copy reactions of which the hypercycle is one example. Like the original hypercycle, our ensembles are not stable entities upon perturbation (but see Bjoerlist[2] for spatial systems).

LEVEL 1 EXPERIMENTS

EXPERIMENTAL PROTOCOL AND SUMMARY OF RESULTS. Level 1 experiments are identical to Level 0 experiments except that copying functions (i.e., Level 0 entities) are barred from action. Such experiments generate organizations of considerable complexity. Each such organization is a set of objects, distinct from the initial ones, that maintains itself without any single member engaging in a copying action.

EXAMPLE OF A LEVEL 1 ORGANIZATION. A particularly simple Level 1 organization is illustrated in Figure 1. All objects maintained in the system are characterized by a particular syntactical architecture. All objects are made of two building blocks (A, B) such that i contiguous A's are followed by j contiguous B's with $i \geq j$. The organization can, therefore, be visualized in the i, j plane (Figure 1). All actions that occur within this organization can be described by two invariant laws, i.e., emergent regularities in the behavior of the system. The first law states that an object acting on another object will produce the object immediately below the operator along its diagonal, as illustrated by the bold arrow in Figure 1(b). This law applies to all objects except those at the end of the diagonals. The second law governing the system states that these objects acting upon any others produce specific objects somewhere up the argument's diagonal, as illustrated by the dashed line in Figure 1(b). Together these two laws ensure that the system is syntactically closed and self-maintaining.

FEATURES OF LEVEL 1 ORGANIZATIONS. A zoo of different organizations can be generated by specifying syntactical filters that prohibit particular organizations from emerging. Detailed descriptions of Level 1 organizations will be published elsewhere.[10] All Level 1 organizations, however, share certain common features:

1. *Self-Maintenance Versus Reproduction:* In Level 0 experiments, objects were copied by other objects or were self-copying. In such a case, it is appropriate to speak of reproduction of objects. In Level 1 organizations, however, neither single objects, nor the organization itself is copied. Level 1 organizations are self-maintaining, but not reproducing.
2. *Emergence of Laws:* Each organization may be completely specified by a listing of all objects in the system and a corresponding listing of all actions of each object on all other objects. This description, however, is hardly concise—it is rather like describing a rat liver by handing a student a biochemistry textbook. We find that all organizations generated by this system display two classes of

emergent patterns. One class refers to patterns at the syntactical level of the objects and the other refers to patterns at the functional level. In the example above, what is emergent at the syntactical level is the restriction of all objects to a particular form of combination of the building blocks A and B. What is emergent at the functional level is that two laws govern all transformations among objects of this type.

Laws are emergent in the sense that global regularities result from the collective behavior of locally interacting objects without those regularities being imposed on the objects initially. The laws represent a distinct level of description in that they do not refer to the detailed micromechanics of the system (i.e., the underlying λ-calculus operations).

Level 1 organizations may be considerably more complex than the example outlined above.[10] From a formal point of view, laws describing the syntactical constitution of objects are known as a grammar and the (infinite) set of objects conforming with it is a formal language. The laws characterizing the relations among objects specify an algebraic structure that is supported by the language. Both, grammar and algebraic structure are invariant with respect to the ongoing interactions among objects. Grammar and algebraic structure constitute a compressed description that completely characterizes the organized state of the system. This leads to a minimal notion of "organization" as a dynamically maintained algebraic structure.

3. *Synthetic Pathways and Self-Repair:* Level 1 organizations are remarkably robust toward deletion of particular objects. Elimination of most objects results in their recreation by the remaining interactions. For example, if one eliminates all end-objects of the diagonals in Figure 1, these objects reappear in the system shortly after their elimination (by virtue of the first law 1). The ability of these organizations to repair themselves is a direct consequence of self-maintenance, i.e., that all objects survive within an organization only by virtue of being the product of some production pathway.

4. *Seeding Sets:* An organization contains a number of different smallest sets of objects that are sufficient to recreate that organization, if an experiment were begun only with members of such a set. We refer to such sets as seeding sets. All organizations we have observed so far have a unique self-maintaining seeding set.

EFFECTS OF PERTURBATIONS. The effect of perturbing Level 1 organizations was explored by periodically injecting random objects into an existing organization. Four different peturbation schedules were explored: (i) introduction of one random object in 10 copies every 30,000 collisions, (ii) as in (i), but with perturbation occurring every 50,000 collisions, (iii) as in (i), but with perturbations injecting 50 copies, and (iv) same as (i), but with three random objects injected simultaneously. With one exception, the results were identical. The organization persists without any change in the emergent laws that characterize it. Level 1 organizations are very robust to small perturbations.

In the exceptional case, a perturbation resulted in the appearance of a new emergent law. This new law did not displace any of the existing laws, but rather represented an addition to the existing laws. The new system proved robust to continued perturbations. The exceptional case is relevant in that it illustrates that Level 1 organizations can be altered by perturbations.

We have also explored the consequences of relaxing the restriction on copying. Recall that Level 1 organizations were obtained by preventing identity functions from acting. If identity functions are either present upon initialization of the system or allowed to arise early during the course of an experiment, the system typically does not reach Level 1. Rather, the system becomes dominated by copying Level 0 objects. We have found a variety of conditions under which this restriction may be relaxed. If (i) a restriction in placed on the efficacy of copying or if (ii) objects that copy also support constructive interactions, Level 1 organizations are generated in the presence of copy actions. In addition, we have found that if (iii) identity functions are allowed to act or are introduced into the system after a Level 1 organization has been constructed, the organization remains stable. Specifically, the same laws which characterized the organization prior to perturbation characterize the system after the perturbation. Removing the no-copy constraint results in their kinetically stable integration into the existing organization.

RELATIONSHIP TO THE WORK OF OTHERS. Our Level 1 organizations recall three different lines of research. Our Level 1 organizations share with the hypercycle model of Eigen and Schuster[6] a limitation on the advantage to self-copiers, but differ fundamentally in that they are founded on constructive interactions and the ensuing network of transformations. A copy action is precisely the negation of a transformation. From a functional point of view a copy action does not force a new object (i.e., function) into the system. This simple difference accounts for the remarkable stability towards functional perturbations as well as for the specificity by which new functional objects can be stably integrated into an established Level 1 organization.

The second and third research traditions are work on autocatalytic sets[1,5,7,13,16] and on autopoietic systems,[14,15] respectively. Our results share with these models the phenomenon of self-maintenance, but differ markedly from these efforts in defining a formal framework which allows systematic exploration of the conditions permitting its emergence and characterization.

LEVEL 2 EXPERIMENTS

EXPERIMENTAL PROTOCOL AND SUMMARY OF RESULTS. Level 2 experiments are initiated with the products of two different Level 1 experiments. The procedure is otherwise identical to the Level 1 protocol, except that the system is increased to a constant size of 3000 objects. Such experiments have one of two outcomes: either a single Level 1 organization comes to dominate the system or a new self-maintaining metaorganization arises (hereafter referred to as Level 2). Such metaorganizations have the two self-maintaining Level 1 organizations as components. In addition, the metaorganizations contain a set that is not self-maintaining, but which acts to knit the self-maintaining Level 1 sets into a higher-order self-maintaining entity. This set contains objects that result from the communication (cross-interaction) between the Level 1 organizations and that do not belong to either organization. The grammars and the algebraic laws characterizing Level 2 organizations will be described elsewhere.[10]

FEATURES OF LEVEL 2 ORGANIZATIONS. As in the case of Level 1 organizations, a zoo of different Level 2 organizations can be generated by varying boundary conditions such as syntactical filters or the constituent Level 1 organizations. In addition to the features of Level 1 organizations emumerated above, Level 2 organizations share certain common features unique to this level of organization:

1. *Emergence of Level 2 Laws*: The laws characterizing the Level 1 organizations which are contained within a given Level 2 organization remain unchanged by their inclusion in the higher-order entity. In addition, a new set of laws, i.e., Level 2 laws, are found that define the structure and the actions of the metabolism that glues the two Level 1 organizations into a self-maintaining higher-order entity.
2. *Seeding Set*: Under the boundary conditions imposed, the seeding set of a Level 2 organization is nothing more than the seeding sets of the Level 1 organizations from which it is constructed. However, the organizational description, as manifest in the laws, is not a superposition of the descriptions of the Level 1 organizations.
3. *Metabolic Flows*: In a Level 2 organization, the set that links the two self-maintaining Level 1 organizations is composed of products derived from syntactical elements of both Level 1 organizations. These hybrid products define production pathways from each of the Level 1 organizations into the shared metabolism and from there back into the two Level 1 organizations. The magnitude of these metabolic flows changes the diversity of products present in each Level 1 component as compared to its stationary diversity in isolation. Metabolic flows, and the differences in metabolic diversity they generate, are maintained despite the fact that both the seeding sets and the laws of each component Level 1 organization remain unchanged.

EFFECTS OF PERTURBATIONS. Three different perturbation schedules were explored: (i) introduction in 10 copies of three random objects every 30,000 collisions into the set that links the two self-maintaining Level 1 organizations, (ii) simultaneous introduction of three random objects in 10 copies every 30,000 collisions into each of the Level 1 organizations and into the set that links them, and (iii) sequential introduction of 3 random objects in 10 copies into each set, such that each set is perturbed every 60,000 collisions.

The first perturbation schedule resulted in no changes in the seeding sets or laws of the component Level 1 organizations and no change in the laws of the Level 2 organization. The second and third perturbation schedules resulted in a simplification of the seeding sets of one of the component Level 1 organizations. In one case (perturbation schedule ii), this simplification had no effects on the Level 2 laws (since the lost part was redundant), and in another (perturbation schedule iii), the simplification had the effect of simplifying the Level 2 laws. In all cases in which simplification occurred, the laws were not fundamentally restructured upon perturbation. These results indicate that Level 2 organizations are resistant to small perturbations.

WHAT WOULD BE CONSERVED IF "THE TAPE WERE PLAYED TWICE"

The results of these experiments can be briefly summarized to reflect three general findings: (i) Hypercycles of self-reproducing objects arise. (ii) When replication is prohibited or inhibited, self-maintaining organizations of considerable complexity emerge. (iii) Organizations can be hierarchically combined to produce new self-maintaining organizations that contain the lower level organizations as self-maintaining components.

To assess the consequences of these results for the issue of contingency and necessity, it is important to keep in mind the level of description that our model induces. The model pictures a particular abstraction of chemistry in terms of a calculus, and endows it with a simple dynamics. The system spontaneously constructs networks of *functional* relationships that constitute a formalization of an interesting notion of "phenotype," i.e., organizational structure. The model allows a natural definition of organizational grades and the exploration of the conditions under which they arise.

Our results invite analogy to the organizational grades as they arose in the early history of life: self-replication, self-maintaining prokaryotic organizations, and self-maintaining eukaryotic organizations. In particular, Level 0 hypercycles are equivalent to the RNA hypercycles proposed by Eigen and his colleagues.[5,6] Level 1 organizations are systems of transformations in our abstract chemistry, analogous to the metabolism of procaryotes (see below). Finally, Level 2 organizations are metaorganizations constructed from interactions between Level 1 organizations, as they occurred in the generation of the eukaryotic cell from prokaryotic precursors.

The correspondence between the results of our computer experiments and real life is intriguing. Yet it is a strong claim indeed to contend that these results are sufficient to inform us literally about what would emerge "if the tape were played twice." The claim that our results represent the generic behavior of self-maintaining organizations is based on two propositions:

1. *Lambda-calculus is an appropriate set of constructs to encompass the origins of metabolism*: A fundamental property of all extant forms of life is the presence of a metabolic organization. A metabolism is in essence a self-maintaining network of catalysed reactions, characterized by the transforming action of catalysts on substrates. We have shown here that two abstractions from chemistry together with a simple dynamics are sufficient to generate self-maintaining organizations. Thus, if it is true that (i) metabolism is an invariant property of living systems and (ii) the minimal notion of metabolic organization is captured by the study of the origins of self-maintaining transformation systems, then λ-calculus is a natural approach for analyzing the origin of biological organizations. The question of the validity of such a claim may then be rigorously challenged by increasingly sophisticated elaborations of our model universe.

2. *The assumptions made are biologically meaningful*: Our Level 1 and Level 2 results are dependent on two assumptions with regard to our interaction scheme. We show that these assumptions are equivalent to neglecting "food" and "waste" in a sense that is appropriate at our functional level of description.

   i. *Food*: We assume that when two functions interact to produce a third, that the functions that participate in the interaction are not consumed in the process. If this assumption is relaxed, self-maintaining organizations fail to emerge. This assumption represents a controlled input into the system from the environment, i.e., food. Note, however, that this assumption is not a buffer since it does not guarantee the persistence of an object in an organization. Any object has a finite lifetime induced by the dilution flow. This provides for a sorting mechanism that biases towards those objects that have production pathways involving other objects in the same system. It follows that as soon as a stably self-maintaining seeding set of an organization has formed, we should be able to relax the interaction scheme to a catalytic transformation in which one interaction partner is used up. We have found this to be the case, without influencing either the syntactical or algebraic regularities that characterize an organization.

   ii. *Waste*: Interactions that yield products which in combination with existing objects generate a normal form without reduction are prohibited in our system (see model platform, item 3(ii)). If this assumption is relaxed, self-maintaining organizations fail to emerge. Such products are incapable of participating in closed transformation networks, i.e., the product will fail to metabolize. The biological interpretation of this assumption is straight-forward. Metabolic processes are disrupted by the accumulation of waste, and this assumption guarantees that a particular form of waste does not accumulate (see 2 for details).

SELECTION, SELF-MAINTENANCE, AND THE EMERGENCE OF BIOLOGICAL ORDER

Darwinian selection, as opposed to the mere differential sorting of an arbitrary collection of objects, presupposes the existence of self-reproducing entities.[18] In this study, the only reproducing entities are the self-copying functions arising in Level 0; Level 1 and Level 2 organizations are self-maintaining, but not self-reproducing. Indeed, we find that organizations (Levels 1 and 2) can arise in the absence of self-reproducing entities (Level 0).

Self-reproduction and self-maintenance are shared features of all extant organisms, barring viruses. It is not surprising, then, that there has been little attention paid to the generation of one feature independently of the other. While selection was surely ongoing when transitions in organizational grade occurred in the history of life, our model universe provides us the unique opportunity to ask whether selection played a necessary role. Our finding clearly indicate that it need not.

Moreover, separating the problem of the emergence of self-maintenance from the problem of self-reproduction leads to the realization that there exist routes to the generation of biological order other than that of natural selection.[11] Indeed, this is apparent upon inspection of the formal structure of the theory. Neo-Darwinism is about the dynamics of alleles within populations, as determined by mutation, selection, and drift. A theory based on the dynamics of alleles, individuals, and populations must necessarily assume the prior existence of these entities. Selection cannot set in until there are entities to select. Our exploration of an abstract chemistry not only provides a route to generate such entities, but illustrates that different organization grades can arise in the absence of selection. This raises the problem of determining which features of biological organization are attributable to the emergence of the organization and which features are attributable to its subsequent modification by selection. At issue is the primacy of natural selection in shaping the major features of biological organization.

ACKNOWLEDGMENTS

This paper is communication number 2 from the Center for Computational Ecology of the Yale Institute for Biospherics Studies. We thank John McCaskill, Günter

Wagner, John Reinitz for discussions which proved central to the development of these ideas and M. Dick, M. Eigen, D. Lane, H. Morowitz, S. Rasmussen, J. Reinitz, P. Schuster, B. Singer, P. Stadler, A. Wagner, and, particularly, J. Griesemer for comments on the manuscript, and the organizers for the opportunity to present our work in this forum.

REFERENCES

1. Bagley, R. J., and J. D. Farmer. "Spontaneous Emergence of a Metabolism." In *Artificial Life II*, edited by C. G. Langton, C. Taylor, J. D. Farmer, and S. Rasmussen. Santa Fe Institute Studies in the Sciences of Complexity, Vol. X, 93–141. Redwood City, CA: Addison-Wesley, 1992.
2. Bjoerlist, M. C., and P. Hogeweg. "Spiral Wave Structure in Prebiotic Evolution: Hypercycles Stable Against Parasites." *Physica D* **48** (1991): 17–28.
3. Church, A. "A Set of Postulates for the Foundation of Logic I." *Ann. Math.* **33** (1932): 346–366.
4. Church, A. "A Set of Postulates for the Foundation of Logic II." *Ann. Math.* **34** (1932): 839–864.
5. Eigen, M. "Self-Organization of Matter and the Evolution of Biological Macro-Molecules."' *Naturwissenschaften* **58** (1971): 465–526.
6. Eigen, M., and P. Schuster. *The Hypercycle.* Berlin: Springer-Verlag, 1979.
7. Farmer, J. D., S. A. Kauffman, and N. H. Packard. "Autocatalytic Replication of Polymers." *Physica D* **22** (1986): 50–67.
8. Fontana, W. "Functional Self-Organization in Complex Systems." In *1990 Lectures in Complex Systems*, edited by L. Nadel and D. Stein. Santa Fe Institute Studies in the Sciences of Complexity, Lect. Vol. III, 407–426. Redwood City, CA: Addison-Wesley, 1991.
9. Fontana, W. "Algorithmic Chemistry." In *Artificial Life II*, edited by C. G. Langton, C. Taylor, J. D. Farmer, and S. Rasmussen. Santa Fe Institute Studies in the Sciences of Complexity, Vol. X, 159-209. Redwood City, CA: Addison-Wesley, 1992.
10. Fontana, W., and L. W. Buss. "'The Arrival of the Fittest': Toward A Theory of Biological Organization." *Bull. Math. Biol.* (1994): in press.
11. Goodwin, B. C., and P. Saunders. *Theoretical Biology: Epigenetic and Evolutionary Order from Complex Systems.* Edinburgh: Edinburgh University Press, 1989.
12. Gould, S. J. *This Wonderful Life.* New York: W. W. Norton, 1989.
13. Kauffman, S. A. "Autocatalytic Sets of Proteins. *J. Theor. Biol.* **119** (1986): 1–24.
14. Maturana, H. "The Organization of the Living: A Theory of the Living Organization." *Int. J. Man-Machine Studies* **7** (1975): 313–332.

15. Maturana, H., and F. Varela. *Autopoiesis and Cognition.* Dordrecht: H. Reidel, 1980.
16. Rasmussen, S., C. Knudsen, and R. Feldberg. "Dynamics of Programmable Matter." In *Artificial Life II* edited by C. G. Langton, C. Taylor, J. D. Farmer, and S. Rasmussen. Santa Fe Institute Studies in the Sciences of Complexity, 211–254. Redwood City, CA: Addison-Wesley, 1992.
17. Revesz, G. E. *Lambda-Calculus, Combinators, and Functional Programming.* Cambridge: Cambridge University Press, 1988.
18. Vrba, E., and S. J. Gould. "Sorting is not Selection." *Paleobiology* **12** (1986): 217–228.

DISCUSSION

BUSS: The preceding stuff that Walter's done is a framework that's particularly general for asking certain classes of questions. What we've done has made no substantive progress on the framework itself; we've just used that framework to ask a particular class of biological questions. A paper that describes those questions, and these answers, will be available shortly.

WALDROP: Just to follow up on that. As usual, this is incredibly beautiful stuff, but there's something about it that has always bugged me.

On the one hand, if you look at the underlying lambda-calculus framework, which you just referred to, it's like chemistry but it's really hard to see how you would really make the association to real chemistry, with mass conservation and all the rest. So it's a little too abstract on the one hand. On the other hand, it's almost as if it's not abstract enough, because you seem to be making all these, what to me sound like ad hoc, assumptions to make the model work: the structure of your chemostat (you can imagine any number of ways to do that); the assumption of having to keep both substrates around; the various filters you put on, not allowing the self-replicator to hang around. Maybe I'm missing something but it all seems arbitrary, and you're just doing it to make the model work and sort of ruining the beauty of the lambda-calculus underlying this.

FONTANA: You're correct in this. But at the same time you're not correct. Why?

There is no point in making an emulation of chemistry, at all. Let me turn the entire thing around: We try to define a level of description of nature, in which we can think of the objects inhabiting that level as being mathematical functions. And then we'll look at the consequences. Now, if we go then back to nature and find that similar consequences occur there, we are justified in saying that we hit an interesting level of description. So we're not emulating chemistry, we're setting a level of description of which we would say chemistry is a particular instantiation, but by no means the only one.

The gist of chemistry is to be a combinatorial variety of stable structures that can manipulate each other. And it's not coincidental that, complex phenomena like life take off precisely at that level. So it is an abstraction, but I think the abstraction captures the essence of what you would like to call a "chemistry."

BUSS: Let me try to answer the same question, less kindly. We see a variety of models here, and we see a whole bunch of different emergent behavior. And I keep hearing, time and time again, that we "got this for free." I'm a firm believer that there isn't any free lunch; the utilities of these models are not because they're going to give us solutions as a free lunch. The utility of them is that they allow us to provide a framework for asking whether our insights are any good. Okay? And so the fact that you have to imply these syntactical constraints is precisely the power of this system, because that's where you ask whether your insight is any good. Whether, in fact, your insight—that this is what made a transition from one biological class to another—is, in fact, realizable.

WALDROP: Not allowing self-replicators, and loss of transition?

BUSS: I think that that's very important. Let's give you an example of how that might have occurred. You've got the early Archean ocean; you've got a weird little geochemical environment that's letting self-replicating molecules work. Now if that is ever going to turn into life as we know it, it's got to get out of that weird little geochemical environment, and into other weird little geochemical environments, right? Because there's no primordial soup; we all know that, right? (Flushing rates through the vents were about the same as they are today.) And that means that you've got to put it in something. If you put it in something, the first stage of putting it into something probably means it's not going to be able to copy anywhere because you're putting it into a fundamentally different environment.

We just made a set system that says, "Okay, now you can't copy." That's our model of that particular geochemical property. And, as a result, you get an organization.

VOICE: It sounds like selection. It sounds a little like Darwinian selection.

BUSS: In the case of Level 0, it is Darwinian selection, because your chemostat is taking objects out, and objects can self-copy; hence, you are, in fact, selecting on self-reproducing entity classes. Level 1 and Level 2, when you're not, in fact, taking out organizations—and there's no birth or death of organizations—it's not Darwinian.

WALDROP: You've satisfied me on that point, but it's still bothersome. I love the abstraction of the lambda calculus; that's not what I was objecting to so much. It was your model of the chemostat. I can imagine six different ways to model that; would that give you different things?

FONTANA: That is true; that has to be checked. There are two reasons for the chemostat. Again, the assumptions of this model are very simple. You think of your objects as being objects that encode functions, mathematical functions. And now you just look at what follows when you put a whole bunch of them together. That's all. Now you have to realize this in some way. I thought that the chemostat was the simplest way you can think of.

The other reason for choosing the chemostat is that under appropriate circumstances describe it in terms of differential equations that are exactly a generalization of replicator equations. I would say it's the simplest assumption, kinetically, that you can make. So it's just on the grounds of simplicity.

MOORE: It seems that the Level 2 organization is still basically a set of Level 0 objects; you draw boundaries around the objects, sort of by looking at subsets of them, almost like subgroups that are self-maintaining or that are algebraically closed. When you combine your two Level 1 organizations, you're sort of taking the algebraic closure of their union, and then saying, "Well, I still have boundaries here," as opposed to finding a way to take Level 1 organizations, combining them, and making a large set of Level 1 organizations, and you call that a Level 2 organization...

FONTANA: No, that you call a Level 3.

MOORE: Okay. I guess it's not clear to me how it generalizes upward. Maybe this is actually better biologically, because it's not as if cells act on other cells, to produce third kinds of cells (the way molecules do), but, just from a mathematical point of view, it is the definition of boundary that I'm not quite satisfied with.

FONTANA: There are two definitions of boundary. There is a boundary condition, of which such an organization is a solution. But there is a boundary, also, in the sense that you're out in the space of objects, and you hit a particular subspace, that is characterized by invariances, syntactical and functional. Once you are on that subspace, you can't get out of it anymore. So there's no way you would get an object of a different syntactical kind. From that point, the system is stuck there. There was no notion of genes mutating, or something like that. The structures you see here are a complete consequence of constructive interactions—of object functions, or molecules, or whatever.

I would say this is a natural boundary, because that's what the system did itself. I agree with you that in order to get to Level 2, we just had to maintain the boundary conditions not the boundaries that are intrinsic to the Level 1 organizations—those maintain themselves. I think this is biologically plausible when you think about how eukaryotes came into being. I agree with you as well that one should try to get entire organizations acting upon each other, switching off—copy actions, and then perhaps see if one can get to Level 3, or to something like multicellularity. But we went—how many billion years?

BUSS: Two.

FONTANA: Two, so...

BUSS: So we're halfway there.

EPSTEIN: Two senses of the word "boundary," and I want to just ask about a third sense of the word "boundary." All these members of the final set of logical trackers—whatever you call these terminal objects—those are local basins of attraction. Lots of initial object shapes are out there, right, in the chemostat. Some of them go into this well; some of them go into that well; some of them go into another well. Have you looked at the basin boundaries of these terminal attractors?

FONTANA: Yes, we looked to a certain extent to them. When you start with the same initial objects, but you vary the collisions—the random number generator seeds, or so—the organizations are robust; nothing changes at all. It is, however, very difficult to sweep through initial sets of objects—all possible initial sets of objects and to assess, therefore, what the boundaries of those attractors in function space really are. There are basins of attraction; we don't know how large they are.

SCHUSTER: I have a very short comment to the last question, and then another one. The short comment is the following:

If you go from Level 1 to Level 2, what you are doing is essentially the same question that Manfred Eigen has posed, and we have posed previously, and there is no way to maintain Darwinian selection when you do this jump, because Darwinian selection acts on Level 1, and you need two objects of Level 1 so...

Level 1 is your self-replicating entities, as I understood it here.

FONTANA: This is Level 0 here.

SCHUSTER: The Level 0 was the soup, that prepared...

FONTANA: No, no. Level 0 is a single, self-copying entity; Level 1 is the self-maintaining organization; Level 2 is a self-maintaining organization of self-maintaining organizations.

SCHUSTER: So when you make this transition you have to exclude Darwinian selection; in one way, Darwinian selection is not a law of chemistry. It's a law of a subset of chemical reactions and, if we allow chemistry as such, then there will be no Darwinian selection unless these replicating entities are dominating.

And the second question is related to the problem that you have to exclude, mechanisms in terms of replication. You have to exclude self-replication as a reaction. This is, in a way, very typical for models that are too general, where chemistry just allows a subset of these reactions. This is a common feature, I suspect, in most of the computer models, where copying is one of the most simple and cheapest processes that you can get. And also from the intellectual point, copying is something that appears to be very simple. In chemistry, it's very much the other way around. To make a template-copying mechanism is a very improbable chemistry. Most chemistry does not do the copying reaction. So, if you would put chemistry in such a way, that you choose only a subset of your relations that would mimic chemistry, I think there is no need to exclude self-replication.

BUSS: Sure. That's entirely conceivable, but I would say that the power in the framework is by throwing out a lot of chemistry. The first point interests me more, and the more I think about it, the more I think that we already knew this business about making a transition between organizational grades not involving Darwinian selection. Think about the speciation process; it's all about boundary conditions. It says that, when you make a boundary to gene flow, there's no necessary role for Darwinian selection in speciation.

SCHUSTER: I would put it the other way around. The beauty of the model—it's very general, and I'm convinced you can include all chemistry in it, but chemistry's just a subset of the various interactions you have in your model. And choosing that subset has to restrict the probability of self-copying to occur.

FONTANA: This is a legal question, but we're not interested in that question, or in what follows from that question. We're interested in what follows structurally once you have a function that is a copy function. The fact that it's so easy to get a copy function here, that's exactly the virtue of the system, because we don't have to bother by making 200-page model—with a whole lot of arbitrary assumptions anyway—about how you get "copy" in the first place. This is just trying to isolate the logical structure of certain processes.

BUSS: You're talking about wanting to match different forms of phenomenology, and, clearly, if what you're most interested in is why it is hard to get self-copying, then you don't want to put the same interpretations on these objects as we have. That doesn't mean that the framework can't, in fact, be changed to deal with the set of problems that you're most interested in. Clearly, we have made a series of assumptions about how our chemostat works, and the like, that are biologically motivated; a series of assumptions about syntactical filters that are biologically motivated. They are biological hypotheses. You would not use the same one if you were trying to model, you know, the earth's atmosphere system, or if you were trying to model why it is that it's hard to self-copy in chemical systems. I would say that the model could be used for that, but we did not for a specific reason: we wanted to ask some biological hypotheses.

FONTANA: In fact, there are no rate constants in here. Every collision—independent of how long it takes to reduce it to normal form—happens with the same kinetic constant, in just one time unit.

GOODWIN: Leo, I had a question that relates to the title of this, which is "The Evolution of Individuality," and it connects with your book, and the relationship between the notion of individuality expressed here, and that which you have expressed in your book.

BUSS: The title "Evolution of Individuality" was an assigned subject which we have not pursued here. But your point is well taken. The next thing you want to know, to ask the questions that I explored at some length in that book, was once we have these Level 2s, and we start playing with the relative selection pressures on differing entity classes—the enclosed Level 1s and a Level 2—to what extent that is capable of making fundamental modifications in the emergent laws that come out of that system. And Brian Goodwin would state emphatically, with great assurance, that there would not be any, right? That, in fact, the generation here was the whole story. It's going, in fact, to be so robust that we're going to have to get way outside of biologically reasonable niches to get the changes. I would expect that there is some element of truth to that, but not quite as large as Brian would claim; that, in fact, we can get changes in organizations, but the changes that we would get in organizations would still reveal some fundamental features of the origin. And I don't think you can claim anything but [that], if you look at metazoans, or you look at eukaryotes. Structural features shared among them are very striking.

GOODWIN: So syntactic closure is not equivalent to an individual.

BUSS: Syntactical closure is equivalent to organization in our system.

GOODWIN: But that's not an individual.

KAUFFMAN: My question to the two of you bears on the question of syntactical closure, and what I'm concerned about is that there's an undue precision in the algebraic structures that you've got, if we translate this back to chemistry (which is also, I think, what Peter's concerned with). Consider Alan Perelson's and George Oster's notion of a shape space. Precisely the point of a shape space is that you get overlapping balls, diffusing around and winding up covering an entire shape space. So that, if I take a shape, and the shape complement, and then the complement of the shape complement, and rattle back and forth from a shape to a shape complement, to its shape complement—the fuzz in real chemistry will not get you back (from the shape to the shape complement) only to the original shape, but it will get you back to a whole penumbra of things around the original thing that you did. And then, if you map it back again, you get a wider schmear, and a wider schmear, and a wider schmear until eventually you've diffused out all over a shape space. This is a little bit—in a function space sense—like Phil's question to me yesterday, with autocatalytic sets (your, I guess, Level 1), where, if the rate constants are too fast and you leak out into reaction space, and make all kinds of funny molecules, [then] if there's a fuzz on the transformations in the molecules acting on the molecules to give it a chemical interpretation, you may diffuse all over function space and, therefore, the syntactical boundaries may not be sharp; you may just, in fact, diffuse across them.

BUSS: You're asking two kinds of questions. One, a question about the sensitivity of the emergence of these organizations to the size and the diversity of equivalence classes. It is our strong intuition that it will, in fact, be sensitive, although we have not systematically explored it. It's part of our agenda.

Now, you're raising a second question: given some dependence on the size and diversity of equivalence classes, how does that change your space if your equivalence classes have some fuzzy edges? And that we have not explored.

FONTANA: The equivalence classes don't have fuzzy edges.

KAUFFMAN: No, the mathematics doesn't but chemistry does. I mean, you're right algebraically, but you're not right about the chemistry.

But if you look at (shape)-(shape complement), (shape complement-(shape complement)-(shape complement)), you'll find that you wind up diffusing out over shape space.

FONTANA: But that's not what I'm looking at.

KAUFFMAN: But that's what molecules could do.

ANDERSON: It's your collaborator who should answer this, because he's the biologist.

FONTANA: I'm the chemist. [Laughter]

BUSS: I already did.

SIMMONS: I'm not sure to whom this question is addressed; maybe it's to the two speakers, and maybe it's to you [Stuart].

Stuart presented, in his talk, what is a quite seductive picture of the arrival of the fittest, the origin of life, namely: not the origin of self-copying mechanisms, but of self-maintaining chemical structures, or rather self-maintaining abstractions of chemical structures. On the surface, it appears to be very different from the kind of model that's being presented here, and I want to know if that is true, and whether this model can accommodate Stuart's ideas, [and] where the conflict lies, if any.

FONTANA: This is definitely not the first model in which you have objects that in some way encode, even very arbitrarily, actions upon other objects. Phil

Anderson has been looking at this type of things a couple of years ago, in a similar setting as Stu did—probably even before.

So, the quick answer to your question is yes, I would say Phil's and Stuart's cases are special cases, because of this: What we did here was nothing else than mathematize these things. You'd like to isolate a formal structure, because I believe still this is what gives us, eventually, understanding (or part of it). The step seems to be very simple, to go from molecules to function-objects; it took us a long time to do that step. It's a small step afterwards, it's a huge abstraction if you think about it. And the mathematical apparatus that it buys you is extremely powerful. So I would say, in some sense, yes, there was precursor work. Obviously, that was tied into rather arbitrary models.

SIMMONS: I'm asking about what appears to be a conflict between your model and Stu's...

KAUFFMAN: No, there's no conflict. The collectively reproducing set of functions that Walter talks about, when you block a copying operation, is identical to the collective autocatalytic sets that I'm talking about.

FONTANA: The point is to throw out thermodynamics for a while, and to focus just on logical structure. There is thermodynamics, I agree; that thermodynamics gives you a lot of additional superstructure which is fine, but we would like to isolate what can be understood without thermodynamics.

KAUFFMAN: Isn't there a difference between whether you have a founder set that you maintain...

FONTANA: You want to begin to classify what these algebraic organizations are, and there it is important that these functional organizations are of a type where you do not need to put in any function from the outside, except supply the system with symbols that it can use to build a new object. As soon as you constantly supply products from the outside, then your organization will always be the closure of it.

ANDERSON: There were, between your Level 1 and Level 2, actually two events at least that took place—or maybe they're the same event—namely, speciation and sex. There seems to be no discussion about it...Or perhaps at the next level, but...

BUSS: Oh, you mean in the real earth...

ANDERSON: Yes. In Level 1 and Level 2 in the real world, two very important things happen.

BUSS: A lot of stuff happened that isn't here. Sex is certainly something that Walter and I have thought about that would be fairly easy to do, by having seeding sets acting on one another. But we have not explored that as of yet. The speciation is not something that, in our current construction, it's well suited to do. I wouldn't argue that you couldn't do it, but the current construction is not well suited to doing it. The reason we have actually chosen these features is because they're organization shifts of a particular class; the particular class is that they involve having more than one self-replicating entity enclosed within another. And I have a particular interest in that class of evolutionary transformations, and that's why we focused on those. People who are interested in the evolution of sex might have focused on something else.

Charles F. Stevens
The Salk Institute, Laboratory for Molecular Neurobiology, 10010 North Torrey Pines Road, La Jolla, CA 92037

Complexity of Brain Circuits

INTRODUCTION

The human *brain* is, at least by some definitions, the most complex of the complex systems. Because evolution works by modifying the existing, other mammalian brains are, to a great extent, smaller and less complex versions of our very large and very complex one. Some brain areas, like the primary visual cortex of primates, are structurally and functionally very similar to our own, but even the mouse brain (1000 times smaller than the human brain) is quite closely related in both its anatomy and principles of operation to the human brain. This observation is, of course, the basis for using experimental animals like mice, rats, cats, and monkeys as models for learning about how our brain is constructed and does business.

Evolution, then, has provided us with a series of similar brains that do basically the same job (neglecting some very significant specializations like language) but vary in size (and apparent complexity) over four orders of magnitude, from several tenths of a gram for pigmy tree shrews to several kilograms for whales. Thus, we are

Complexity: Metaphors, Models, and Reality
Eds. G. Cowan, D. Pines, and D. Meltzer, SFI Studies in the
Sciences of Complexity, Proc. Vol. XIX, Addison-Wesley, 1994

offered the possibility of studying how the complexity of this most complex system varies with its size.[13] How complex is a mouse brain compared to that of a human?

To compare the complexity of various brains, a measure of complexity is, of course, necessary. Making brains larger, that is, simply providing them with more computing elements, does not necessarily make them more complex objects. Since the brain carries out its computations through circuits of neurons, one reasonable measure of complexity is the extent of interconnectedness of these circuits. On this view, circuit complexity would increase with the average number of sources of information available to each of the brain's neurons. To explain this measure more precisely, I must review the structure (neuroanatomy) and function (neurophysiology) of the brain.

NEUROANATOMY

The computational element of the brain is a cell called the *neuron*. Each neuron has three distinct parts: the *cell body*, the *axon*, and the *dendrites*. The cell body is the repository for the cell's genetic information and the place where many metabolic processes, like protein synthesis, are carried out. The axon is a long, thin tubelike extension from the cell body that carries information coded as nerve impulses. Axons form the brain's data paths. The dendrites are, like the axon, extensions from the cell body that form a treelike structure that is known, in fact, as the *dendritic tree*. The dendrites are the neuron's receptive surface into which information flows (in a way to be described presently). For a typical neuron, the cell body is approximately spherical with a diameter of perhaps 30μm, the axon is perhaps 10 cm long, and the total length of the dendrites is perhaps 10 mm with the dendritic tree being about 1 mm tall and .3 mm in diameter.

The whole business of the brain depends on elaborate neural circuits formed by chains of neurons that communicate with one another. The point of contact at which one neuron sends information to another is a special structure called the *synapse*. The diameter of an axon is usually about $.3\mu$m; but spaced along this axon every 5μm or so, when the axon passes close to dendrites of other neurons, are beadlike enlargements of the axon about 1μm in diameter. These enlargements are known as *boutons*. Small extensions from dendrites, called *dendritic spines*—these are much like the thorns on a rose plant—come into contact with the boutons; "contact" here means a separation of only about 20 nm, and this tiny space between axon and dendrite is the *synaptic cleft*. The bouton and its spine constitute the synapse: information flows, as will be described, from axon to dendrite (but not the other way around) at synapses. In instances that have been studied so far, one neuron makes only one or a few synapses on any other. Thus, the number of synapses on a neuron is a good general estimate of the number of other neurons that send information to that neuron. A typical neuron has several thousand synapses.

NEURONAL FUNCTION

The nerve impulse is a brief electrical event, about 1 msec long, that travels along axons at speeds of about 1 mm/msec. How this electrical signal is generated is understood in great detail, right down to the intricate operation of the special proteins that produce the electrical changes, but—in the present context—these things count as details that we shall ignore.

When a nerve impulse reaches a synapse, it announces its arrival in an unusual way: a small quantity of a chemical known as a *neurotransmitter* is released from the bouton, and this neurotransmitter diffuses across the synaptic cleft and binds to special proteins embedded in the dendritic membrane termed *receptors*. The release of a neurotransmitter is also a complex process about which quite a bit is known. Neurotransmitter is packaged in small vesicles which, on nerve impulse arrival, fuse with the bouton membrane to discharge their contents into the synaptic cleft; this process by which fusion occurs is named *exocytosis*. The fusion of a synaptic vesicle with its bouton membrane is probabilistic and, in fact, is well described as a Poisson process in which the Poisson rate, which normally is near zero, increases transiently just after a nerve impulse arrives. Synaptic transmission is also "quantal" because an integral number of vesicles must fuse to release their neurotransmitter; one vesicle-full constitutes a neurotransmitter *quantum.*

The brain uses a rather large number of different chemicals as neurotransmitters, and each neurotransmitter has its own specific receptor. Some common ones are glutamate (the very most common one), acetylcholine, GABA, dopamine, and serotonin. If a particular synapse uses glutamate, it uses only glutamate throughout the animal's life. A bit of incidental information: most drugs that affect the brain's function do so by blocking or enhancing the action of neurotransmitters in one way or another.

The neurotransmitters bind to receptors, but then what happens? Receptors fall into two large families that operate in quite different ways. One class of receptors are called *channels* and the other are *G-protein coupled*. The channels are used for normal synaptic transmission, and the G-protein coupled receptors are used for *neuromodulation*. The operation of these two receptor classes is described in more detail in the following several paragraphs.

When a neurotransmitter binds to a receptor in the channel class, it causes the receptor/channel to alter its shape: a submicroscopic pore in the protein opens and permits certain ions (sodium, for example) to pass from outside the neuron to inside the neuron. These ions move down their electrochemical gradients; all cells use metabolic energy to pump sodium ions out and keep potassium ions in, so there exists inside-outside concentration differences from these and other ions. When sodium ions move into a neuron, they carry in positive charges that produce a transient, positive-going voltage change within the neuron, and these voltage changes constitute the signals used by neurons in carrying out their computations.

Different types of receptors open pores that possess various ion selectives. Some receptor types permit positively charged ions to flow into the cell (sodium ions, for example), and others permit negatively charged ions (chloride ions, for example) to flow in. Of course these two different ionic species produce voltage transients of opposite sign. Positive-going voltage transients tend to excite the neuron to produce nerve impulses whereas negative-going transients inhibit the production of nerve impulses. Thus, synapses are divided into two classes, excitatory and inhibitory. Normal operation of the brain requires the correct balance between excitation and inhibition. For example, drugs like the poison strychnine block the operation of inhibitory synapses and produce convulsions by leaving excitation unopposed. Drugs like barbiturates and alcohol enhance inhibitory synaptic transmission and shift the balance of excitation and inhibition to the inhibitory side: these drugs cause sleep and, in large enough doses, coma, or even death, when respiratory and other vital centers cease to operate properly.

The channel class of receptors, then, operate by producing voltage signals in neurons. The G-protein coupled class work in an entirely different, and quite complicated way. These receptors derive their name from the fact that the signal they produce is a change in the concentration of a special class of diffusible signaling proteins known for historical reasons as *G-proteins*. These G-proteins then elicit a cascade of events that finally results in many changes like the chemical modification of certain specific proteins. The effects of G-protein coupled receptors are hard to describe and to keep straight because they involve such a complex series of steps (a cascade), and because the precise effects of this cascade vary so much from one type of neuron to another. The important points are, however, that: (1) G-protein coupled receptors use concentration changes of special chemicals (known as second messengers) as their signaling mechanism; (2) the effects they produce are relatively slow in onset (two orders of magnitude slower that the channels) and very long lasting (seconds or even minutes, as opposed to milliseconds); and (3) their final effect is to alter the properties of neuronal function in such a way that the computations carried out can be greatly modified. This is the process of neuromodulation through which the brain modifies its own circuits.

CIRCUIT COMPLEXITY

The above anatomical and physiological considerations provide justification for our measure of circuit complexity; as the average number of synapses per neuron increases, circuit complexity increases. The more complex circuit is one in which each computing element (neuron) receives information from and sends information to other neurons. The average number of synapses per neuron, our measure of complexity, can be found by counting the number of neurons/brain and the number of synapses/brain. Neither of these quantities can, however, be determined directly; a cubic millimeter of brain contains about 100,000 neurons and 2 miles of axons that make approximately a billion synapses. Thus, counting all of the neurons and synapses in a brain would be impossible. Fortunately, an indirect approach does permit connectivity (synapses per neuron) to be estimated for a wide range of species and, thus, for a spectrum of brain sizes.

Brain size will be specified by N, the number of neurons in the *cerebral cortex*. Neural circuit complexity, $c(N)$, is then defined as the average number of synapses per neuron:

$$c(N) = \frac{S}{N}$$

where S is the total number of synapses in the cortex. Take a particular species as the reference—a mouse, for example—and let the number of neurons in the reference brain be N_O. The circuit complexity of some other brain would thus be related to the reference brain complexity by

$$c(N) = f\left(\frac{N}{N_O}\right)c(N_O) ,$$

where $f(N/N_O)$ is some continuous function that relates the complexity of a brain to one of its evolutionary neighbors that is (N/N_O) times larger; this function is assumed to be continuous because each brain in a product of a genome that is slowly varying from one species to its nearest neighbors. The functional equation for $c(N)$ has, as is well known, the solution[1]

$$f\left(\frac{N}{N_O}\right) = \left(\frac{N}{N_O}\right)^{\beta} ,$$

for some constant β. The complexity measure defined here, therefore, follows the equation

$$c(N) = c(N_O)\left(\frac{N}{N_O}\right)^{\beta} .$$

The constant β in this equation determines the way in which complexity varies with brain size. If $\beta = O$, then complexity would not vary with brain size; but if $\beta > O$, then larger brains would have more complex circuits. Perhaps larger brains trade neuron numbers for circuit complexity, so that $\beta < O$.

To determine the constant β, and thus to discover the way in which complexity varies with brain size, we must find some way to link the equation presented above to available experimental data. This link to data depends on two experimental observations. First, the number of neurons underlying a square millimeter of cortex is constant across species and cortical regions (with one exception that is not important here) irrespective of cortical thickness.[11] The number of neurons that lie under a square millimeter of cortical surface area is given by the constant n; n has a value of about 150,000. Second, the number of synapses per cubed millimeter of cortex, r, is also approximately independent of species and cortical region.[2,3,4,5,6,8,9,10,12,14] This constant has a value of about $r = 10^9$. The definition of complexity, together with these experimental observations, provide a second equation for $c(N)$:

$$c(N) = \frac{S}{N} = \frac{rAT}{nA} = \left(\frac{r}{n}\right)T ,$$

where A is the cortical surface area and T is the cortical thickness (so AT is the cortical volume). Note that the quantity (r/n) is known from experiments. The two equations for complexity $c(N)$ are equated to give

$$\left(\frac{r}{n}\right)T = c(N_O)\left(\frac{N}{N_O}\right)^{\beta} = c(N_O)\left(\frac{An}{A_On}\right)^{\beta} = \frac{c(N_O)}{(A_O)^{\beta}}A^{\beta},$$

where $C(N_O)$ and A_O (the reference brain cortical area) and T_O (the reference brain cortical thickness) are constants to be determined by experiment. This equation can be rearranged to give

$$T = \left[\frac{nc(N_O)}{rA_O^{\beta}}\right]A^{\beta}.$$

Thus, cortical thickness is proportional to surface area to the β power; a great quantity of data is available for cortical thickness and surface area for many different species with cortical surface areas ranging over about four orders of magnitude.

Double logarithmic plots of cortical thickness vs. cortical surface area are, indeed, linear over the full range of data available (four orders of magnitude) with a slope of less than .1.[13] This means that a tenfold increase in brain size would be associated with a 26% increase in the complexity measure ($10^{.1} = 1.26$), but even this small number is an overestimate. The reason is that larger brains have larger nerve cells,[7] so part—probably most—of the apparent increase in complexity actually reflects an increase in volume unassociated with a change in the number of neurons.

Additional analysis along the lines described above also reveals that the absolute distance over which information is sent within a cortex does not vary much with brain size: axons that are 2 mm long in a mouse brain are also about 2 mm long in our brain. The cortical surface area assigned for the primary processing of visual information by the human brain, for example, the primary visual cortex (V1), is almost two orders of magnitude larger than the the the corresponding cortical region of the smallest primates. But the length of intracortical connections (measured in terms of cortical lengths) is not significantly larger in humans than in smaller primate brains; lateral connections transmit information over progressively smaller parts of the visual representation in larger brains. This suggests that the same modular circuits are used throughout cortex, but that larger brains possess many more modules.

In summary, then, because β is close to zero, mammalian brains all have circuits with the same complexity; all mammals, including ourselves, have brains with several thousand synapses per neuron. Our conclusion is that the complexity of circuits for all mammalian brains is about the same, so that—by this measure of complexity—the brain of a mouse is as complex as that of a man. But the mouse brain is really not as complex as a human's: even the most exceptional mouse cannot learn quantum mechanics. What this means is that the complexity of the human brain arises not from the complexity of its basic processing elements (the cortical module), or in the richness of connections between modules, but simply in the number of modules present. The great challenge to neuroscientists is to discover the architectural principles used by the brain that permit larger numbers of uniform modules to be interconnected so as to greatly increase the complexity and power of our brain/computer.

ACKNOWLEDGMENTS

The research reported here was supported by the Howard Hughes Medical Institute and NIH grant NS 12961.

REFERENCES

1. Aczél, J. *Applications and Theory of Functional Equations*. New York: Academic Press, 1969.
2. Aghajanian, G. K., and F. E. Bloom. "The Formation of Synaptic Junctions in Developing Rat Brain: A Quantitative Electron Microscopic Study." *Brain Res.* **6** (1967): 716–727.
3. Armstrong-James, M., and R. Johnson. "Quantitative Studies of Postnatal Changes in Synapses in Rat Superficial Motor Cerebral Cortex." *Z. Zellforsch* **110** (1970): 559–568.
4. Cragg, B. G. "The Density of Synapses and Neurones in the Motor and Visual Areas of the Cerebral Cortex." *J. Anat.* **101** (1967): 639–654.
5. Cragg, B. G. "The Density of Synapses ad Neurons in Normal Mentally Defective and Aging Human Brains." *Brain* **98** (1975): 81–90.
6. Cragg, B. G. "The Development of Synapses in the Visual System of the Cat." *J. Comp. Neurol.* **160** (1975): 147–168.
7. Hardesty, E. "Observations on the Medulla Spinalis of the Elephant with Some Comparative Studies of the Intumuscentia Cervicalis and the Neurones of the Columma Anterior." *J. Comp. Neurol.* **12** (1902): 125–182.
8. Jones, D. G., and A. M. Cullen. "A Quantitative Investigation of Some Presynaptic Terminal Parameters During Synaptogenesis." *Exp. Neurobiol.* **64** (1979): 245–259.
9. O'Kusky, J., and M. Colonnier. "A Laminar Analysis of the Number of Neurons, Glia, and Synapses in the Visual Cortex (Area 17) of Adult Macaque Monkeys." *J. Comp. Neurol.* **210** (1982): 278–290.

10. Radic, P., J.-P. Bourgeois, M. F. Eckenhoff, N. Zecevic, and P. S. Goldman-Rakic. "Concurrent Overproduction of Synapses in Diverse Regions of the Primate Cerebral Cortex." *Science* **232** (1986): 232–235.

11. Rockel, A. J., R. W. Hirons, and T. P. S. Powell. "The Basic Uniformity in Structure of the Neocortex." *Brain* **103** (1980): 221–244.

12. Schüz, A., and G. Palm. "Density of Neurons and Synapses in the Cerebral Cortex of the Mouse." *J. Comp. Neurol.* **286** (1989): 442–455.

13. Stevens, C. F. "How Cortical Interconnectedness Varies with Network Size." *Neural Comp.* **1** (1989): 473–479.

14. Vrensen, G., D. De Groot, and J. Nunes-Cardozo. "Postnatal Development of Neurons and Synapses in the Visual and Motor Cortex of Rabbits: A Quantitative Light and Electron Microscopic Study." *Brain Res. Bull.* **2** (1977): 405–416.

DISCUSSION

COWAN: As you clip on more modules, what are you doing functionally to the ability of the brain to compute?

STEVENS: That hasn't been explored much. It's an interesting question: what are the consequences of this kind of scaling relationship for how you make computers? I don't know the answer.

I have been asked to elaborate on hypercolumns. As you go across the cortex, for example, in the visual field, the visual field changes. And different aspects of the visual field are encoded in different, adjacent—but separate—regions of the cortex. So, for example, if I look out there, the door is an image in a particular part of my cortex. And if we looked in the part of that cortex where the door is being imaged, we would find that there are regions that are segregated. So there's going to be one region of the cortex—one subregion—where the left eye projects, and another region where the right eye projects, and they're separate. And within that region, there's going to be a whole bunch of small things that are called "minicolumns," where, if you go down through the cortex (from the surface down like that [*pointing to board*]) recording from cells, all of the cells within that column will respond to a stimulus of the same orientation, but not other orientations. So you show all the cells something with this orientation, they'll respond; you show them that orientation, they won't respond. Right next to it, there'll be another column of cells that responded the adjacent orientation. And so, in order to conserve on the wiring, the visual computer has taken a whole bunch of different computers, that compute different things about our visual environment, and has put them interleaved right next to each other.

GELL-MANN: That is related to my question. There's a famous problem (to which there's now some fashionable, partial answer), namely: if there are different parts of the cortex that are concerned with color, and motion, and with a gestalt, and so on and so forth—and yet we see an object in the room (a clock or a door or a person), the way we construct that concept must involve coordination among many different parts. These days it's fashionable to suppose it may have something to do with simultaneity of firing. But in general there's a big question about the integrative functions, which could be chemical, or have to do with time (as in the fashionable theory), and so on. How do you visualize how these change—these integrative functions change—from small animals to large, and also to the human being? There must be some significant difference there, even though just looking at it the way you were before, it's just a matter of adding more and more brain material. The integrative functions must have some kind of change with scale.

STEVENS: There is a theory that all types of cortex are the same, that the cortex that we use for processing language is identical to the cortex that we use for moving. This is the sort of "unitary theory" of cortical function, and there's evidence for this. The idea is that there's some computation that a cortex does. We don't know what the computation is. And the difference then between a language cortex and a motor cortex is not in the machinery that does the computation, but it is in how the inputs are processed before they get there, how that they are mapped onto the cortex, and where the outputs go to. There are 32, 34 visual areas, and so we do sequential processing, and every one of these visual areas is doing the same thing—whatever that thing is. It's carrying out the same computation, the same hardware, the same software (if you want to think of it that way). What's different is, what gets mapped into it and what gets mapped out of it.

And I'll give you an example of a dramatic experiment. . .I could give you a lot of examples. Whether that's true or not, no one knows, by the way. But you can make a persuasive case for it, and there's some experimental evidence to indicate that it may be true, and I'll give you one example:

If you take a hamster, you can reroute the axons that normally would go to his visual cortex by taking out parts of the brain. You can reroute them to the auditory cortex, or the somatosensory cortex. And so now, in this animal, the information that normally is going to the visual cortex is going over to the somatosensory cortex. And now you can record from those cells over there and see, whether they behave like a normal visual cortical cell. And the answer is yes, they do. And in fact there's even some behavioral evidence that these animals can see with their somatosensory cortex.

Now here's the interesting part: Some of these cells, that you can record from, respond simultaneously to vision and somatic sensation. And, the receptive field they have—that is, when you stimulate their environment and record from them—their responses to a visual stimulus are perfectly good visual receptive fields. Their responses to a somatosensory stimulus are, simultaneously, perfectly good somatosensory receptive fields. So this means that the same machinery is capable of doing whatever computations the cortex has to do in order to generate these receptive fields simultaneously; not literally simultaneously, but in quick sequence when you do the stimuli for the two different processes. That's a dramatic piece of evidence, but that's one of the pieces of evidence that all cortex does the same.

So my answer to your question, Murray, would be that there's nothing, in principle, different between us and the mouse. It's just that we have a lot more cortical areas and that we process things through these cortical areas a lot more times, and that we have fancy ways of mapping the output of one onto another one.

GELL-MANN: But specifically about the integrative mechanisms, you would say may scale in some interesting way but, if so, it's not the cortex, but the inputs and outputs that we should study and that's a separate issue? That's where we should look for the integrative function?

STEVENS: That's right. That's what I would say; I'm not sure if it's right or not, but that's what I would say.

WALDROP: Actually, this is related to Murray's question about the integrative functions. You talked about the requirement to minimize the volume of wiring, as to why the circuits had to be small. But of course in the cortex, lots of areas get involved simultaneously in working on various problems. How, in fact, does one part of the cortex communicate with another? Is it by sending out just a few axons from one part to another, or is it like a wave propagating from one of these elements to the next?

STEVENS: We really don't know.

HOLLAND: Just to add to Mitch's comment, I remember a long time ago, actually, in the '50s, Bob Dody used to talk about so-called "U fibers" that go down into the white matter, and then go really relatively long distances, and pop up elsewhere. At that time, nobody had any estimate whether that was 10% of the white matter, or 90. Is there anything more recent?

STEVENS: It's probably a small fraction of the white matter, but it happens a lot. There are a lot of real long connections.

WALDROP: It's the difference between a cortex that can self-organize as it learns, or that's hard wired. In other words, it could be both.

HOLLAND: Yes. They could serve a lot like a telephone switching system, where this takes you into a somewhat larger thing which then transfers you. And it's not hard wired entirely, because you can change the local numbers and so on, but it gives you a structure on which you can build.

STEVENS: You just summarized yesterday's talk at the Neuroscience workshop.

KNAPP: Yesterday we heard a little bit about morphogenesis, and generators, and so on. When a mammal is born, are the synapses made, and are they remade as development occurs? Is there any experimental evidence for the dynamics of making the connections, or wrecking them, or burning them?

STEVENS: It used to be thought, when I was a student, that the brain was wired up; it was fixed, and that was it. You had learning, but the actual neural circuits were specified by the genes. Nowadays, there's very good evidence in same cases, and people believe it's true just in general, that the general plan of the brain—sort of, which areas hook (in a crude way) to which other areas—is specified by the genome, but then it self-organizes. And there is something the neurobiologists call the "Hebb rule," which basically just says that you keep connections. If two synapses have activity that's correlated, then you keep them, and if two synapses have activity that's not correlated, you eliminate them. And that's not just a fantasy; that's really true. There's tons of experimental evidence where you can show how that works. The Hebb rule states that synapses that are correlated, are strengthened; synapses that are not correlated, or anticorrelated, are eliminated. And what's interesting is that there's a critical period. So during development, this goes on and on and on; you refine your circuits by their activity—up to a certain time window, and then the window closes, and you can't do it anymore.

GELL-MANN: And when is that window of maturation?

STEVENS: It's different for different areas. So probably, the experience that we've all had that presumably relates to this is that—with maybe a few exceptions in the history of mankind—if you try to learn a foreign language, and to speak it accent-free, after puberty, you can't do it.

GELL-MANN: Well, it's rare.

STEVENS: Yes. I've never known anyone who learned English after the age of, say, 18, that I couldn't hear some little...you may not say, "Oh, that person has an accent," but you would pick up small differences in cadences, and things like that.

GELL-MANN: My father was perfect; but he was so very perfect, you could tell he was a foreigner.

STEVENS: Murray, there are exceptions to everything!
But probably the reason for that is because the developmental window closes. For the visual system, this rewiring closes at an earlier age. If kids have cross eyes, they wear patches...And if their eyes are not corrected by a certain age—like three years old—then their connections to their visual cortex grow wrong, and they can never be fixed. Or, if children after a certain age have bad vision, and it's not

corrected by glasses, then they get cortical bad vision (called "amblyopia") and it can't be corrected anymore, because the circuits aren't hooked up well.

COWAN: Even though the functions are all identical, there must be some hard wiring, because doesn't the visual signal choose where to go in the brain—to the visual cortex—right away?...

STEVENS: Yes.

COWAN: What makes it choose, if any part of the cortex can do the job?

STEVENS: Oh, there is hard wiring. For example, the projections of the visual system get made through a general map of the world on the visual cortex: that is due to the genes. And within the neural circuits, there are rules by which they hook themselves up; we don't know how self-organizing that is. People are pushing the self-organization more than they once would have.

COWAN: So that within a region, there is an option for mapping with some flexibility as the organ matures, but the region itself is...

STEVENS: ...is specified by the genes. Yes.

ANDERSON: I want to defend Murray's colleague. I don't think he ever intended the Hopfield network, or any neural network of that simple kind, to be a model for the entire brain. He would certainly say that it's a model for some small piece of the brain, for one of these modules—which this, of course, does not exclude.

There's another point, which is that 0.1 is still a lot of synapses. Most of the brain's functions involve relatively routine parallel processing of sensory data, etc. That is surely done the same in one brain to another. And that, again, this suggests that 0.1 may still be a very significant exponent, in that it shows that complexity increases with size.

STEVENS: Or even if it were zero, or even if the number were negative— and so the complexity as I measured it was declining for us—you could still say that that's not a good measure of circuit complexity, that the real complexity has to do with some refinement.

ANDERSON: Well, you know, software isn't very heavy.

Another point I wanted to make, which is just a point that was mentioned in passing: the question about "Is the coordination done by synchronization." I think that must be a disprovable theory, and the kinds of arguments that David Dennett makes in his book...

GELL-MANN: What must be disprovable?

ANDERSON: Synchronization of nerve pulses. Because he shows many cases, many illusions that you can produce, in which you have the impression that two sense impressions arrived in opposite order from when they actually did arrive in the brain. And I think that's excluded by synchronization.

GELL-MANN: I don't think so.

GOODWIN: Just two brief questions. One: I was fascinated with the hamster experiment. This amplifies a bit on Ed's question. Presumably that was done not on embryos, but on very young organisms because all the secondary projections would have to change as well.

STEVENS: That's right. Actually, the way they do those experiments is that the eye projects to a part of the thalamus, which is a relay station; that goes up to the cortex. What they do is they destroy the part of the thalamus...During development, what happens is there're lots of transient connections that are made. So for example, visual cortex, during development, transiently sends axons down the spinal cord to motor neurons, and then those get eliminated during development. And among the things that happen is that there are transient connections between the visual system and other parts of the thalamus (the somatosensory part of the thalamus). So what you do is you take out the visual part of the thalamus, and then you take out a lot of the somatosensory inputs, and so you stabilize the connections onto the somatosensory thalamus. But then the somatosensory thalamus is still projecting up to the somatosensory cortex, and so you really haven't changed that projection—just the detailed organization of it.

GOODWIN: What would you suggest as simulation; how would you modify artificial networks in order to simulate the brain?

STEVENS: I think it depends on how you want to use the artificial networks. If you want to use an artificial network as a literal model for the brain, then you have to study the brain and make a more realistic model. But I think that there are uses for artificial networks that you don't have to take so literally.

GELL-MANN: But they're confused about that. That's just the problem with many of those people. They do not understand in which proportion they're trying to build ingenious computers, and in which proportion they're trying to understand the mind, or the brain.

ANDERSON: It seems very likely that the mind or the consciousness is, in some sense, a software function, rather than a hardware function.

STEVENS: It's a little hard to differentiate between software and hardware in the brain.

ANDERSON: In some sense, that's what I'm saying.

KAUFFMAN: Chuck, one of the things that comes up in the neural network literature is the notions of categorizations by falling into attractors, of course, whether it's a Hopfield net, or a feedforward network, or feedforward with backprop, or whatever. And Freeman has produced an elegant model, and a body of "maybe-data" suggesting that olfactory cortex, when it's categorizing smells or responding to smells, appears to give rise to patterns of behavior that are at first chaotic and then look like they're settling into stable spatiotemporal sort-of wave forms, that are propagating around, or stationary, on the olfactory cortex. And his claim is that as you are learning, or as the mouse is learning, the organism learns to fall into attractors that are literally seen as spatiotemporally stable patterns. What's the received view, if any, about Walter's kinds of claims? Are the data any good? Are the theories reasonable? Is it rejected out of hand?

STEVENS: I personally do not understand what Walter Freeman at Berkeley did, and I have many colleagues that don't understand it, so I think we just can't comment on whether it's right or wrong; we just don't understand it.

KAUFFMAN: Because he purports to be talking about real data.

GELL-MANN: Doesn't smell right to you?

STEVENS: No, I just don't understand it.

ANDERSON: "Don't understand it" can mean two things: one is "I don't really believe it"; the other is "I don't understand it."

STEVENS: No, but I truly don't understand it. What he does is, he puts electrodes on the surface of the brain; he stimulates and he records wiggles of voltages. And then, he takes theories, and those theories have wiggles, and they're similar in some ways. And so, I just have never really quite understood what the tight connection between them is—and it may be that there is a tight connection, and that I haven't taken the trouble, or it may be that there's not a real tight connection, that it's just an analogy.

KAUFFMAN: If it were true, it would be very important.

STEVENS: Yes. But it's a lot of work to find out.

EPSTEIN: I wanted to follow up on Brian's point, really. I just wondered whether a had been calculated for other sorts of nets that purport to be modeled more closely on the cortex, like Kohonen nets.

STEVENS: Well, neural networks don't scale very well, usually, and what they do is they hook everything to everything. And that would go like N, so a would be one. That's the normal thing. But I think people haven't tried hard.

LLOYD: Is something known about the structure of a as a function of radius from the original neuron? Obviously, if you go far enough away, then the number of connections has to scale with the area that's enclosed in the radius. And if you go close enough, it may very well be that this approximation—that almost every neuron is connected to every other neuron—is correct.

STEVENS: Within a cubic millimeter of cortex (which is a kind of standard processing unit) a neuron connects to about three percent of the other neurons. But I would think that within a very small region, the chances that one neuron connects to another neuron gets much higher. But I think it probably never is that everything is hooked to everything. I think that's probably never true. But it may be; we just don't know.

LLOYD: And how about at ten millimeters?

STEVENS: It's possible to find out how the number of synapses falls off as a function of distance, and that's actually an interesting question. But people haven't done it very well yet because it's a lot of work. There are ten thousand things you have to count, and put them in space. But people usually think that. . .if you took a whole bunch of neurons and averaged over them, that you would find a Gaussian distribution.

LLOYD: In terms of distance?

STEVENS: Yes.

ANDERSON: There's a very trivial neural network that scales exactly this way, namely, what I always assumed it was—and I'm amazed that anyone tries a totally connectionist approach—namely, a finite-range spin glass, or a finite-range neural net. It's a well-defined model and the most likely model, which is on that level of simplicity, and it works well with the numbers.

COWAN: You said the mouse and the human have the same kind of brain, but the ability to deal with symbolic language and so forth—that, presumably, is confined to a region of the brain. I mean, there is another region that deals with symbols, a genome hard-wired region dealing with symbolic language.

GELL-MANN: Two regions.

COWAN: Two regions. Those are new in man, right? They do not exist in anybody else. Not in the mouse. . .

STEVENS: Or even monkeys.

COWAN: . . .so that the cell function remains the same, but it's a dedicated region that's wired to deal with that problem. Is that correct?

STEVENS: That's right, and actually...

GELL-MANN: They may have something to do with self-awareness, too, Broca's and Warneke's areas.

COWAN: Self-awareness, yes.

STEVENS: Those things may actually not be hard wired either. It may be that the general projection, from what projects onto those areas, may be hard wired. But when people do epilepsy surgery, what they have to do is to take out regions of cortex where the focus is, and if you have a temporal lobe focus—where the language area is—life isn't worth living; you'd rather have seizures than have your language gone. And so, what the neurosurgeons do, is they map the cortex to find out precisely where it is, and they've done it in lots and lots of people.

COWAN: For individuals, because they never know where it is in advance.

STEVENS: That's right, and it turns out that the language areas move around by centimeters, and they can be very big, and very small. And there are interesting things: first-learned languages tend to be more compactly represented than later-learned languages.

GELL-MANN: Come again? What was that?

STEVENS: First-learned languages tend to be more compactly represented than later-learned languages. So, if they were to map your language areas, they would find your first language would be confined to a very small area, and the language that you were least facile with would actually be interfered with by stimulation over a greater region.

GELL-MANN: Even if these are multiple native languages? That is, suppose you have, say, three care givers in early childhood, and the three care givers speak, consistently, three languages, and you learn them all. They're all native languages, but you know one better than the others—that sometimes happens—then that one is more compactly represented? Or is it a question of time; ones you learn later on are less compactly represented?

STEVENS: It's ones you learn later on. I could say that my German, for example, is extremely diffusely represented. I could be quite sure of that.

WALDROP: It's integrated with lots of things.
     I was a little confused when you talked about these standard processing units in the brain. Is it literally that when you look down on the cortex, these things have boundaries, and a grid of some sort, where they just overlap with it? How does it work?

STEVENS: Yes. What you can do is, you can stimulate the brain, and you can find out which areas are active, and which areas are not active. And so—you can do this with monkeys—if you do that, you can show that, for example, the right eye and the left eye project to actually distinct regions that are physically separate. And the columns, and the orientations, are in different areas. And there's a sort of symmetry in the cortex, in that you have two classes of cells: ones that are inhibitory and ones that are excitatory. When you show an animal a figure, and record from the cortical cells, you find one class that responds like the inverse of the other class. So if one cell is made to fire by a particular pattern of light, you'll find another class that is made to stop firing by that same pattern. And those are physically separate.

WALDROP: But is that true over the whole cortex?

STEVENS: Well, it's not known over the whole cortex; it's probably true, but in places it is not.

Ben Martin
Department of Psychology, Stanford University, Stanford, CA 94305

The Schema

Adapted from a section of the author's Ph.D. dissertation.

INTRODUCTION

The Santa Fe Institute exists to promote the study of a variety of natural and artificial systems that defy traditional mathematical and empirical methods of analysis. As a rule, these systems are *adaptive*; changes in their internal states occur in response to the environment. They are also *complex*; changes in their behavior are not linearly related to changes in their surroundings.[1] By bringing together researchers with diverse interests, the institute encourages explorations of the general properties of complex adaptation. This essay addresses one such property: internal representation. Many complex adaptive systems develop a representation of environmental information; their internal structure comes to reflect external conditions

---

[1] The proper definition and measurement of complexity may be a matter of great debate but any definition will likely imply that nonlinearity is a hallmark of complexity.

Complexity: Metaphors, Models, and Reality
Eds. G. Cowan, D. Pines, and D. Meltzer, SFI Studies in the
Sciences of Complexity, Proc. Vol. XIX, Addison-Wesley, 1994   **263**

of the world. As a result, behavior arising out of plans or predictions based on internal states can lead to successful action. Successful action yields high evolutionary fitness in systems under adaptive pressures, or high performance in natural or artificial goal-directed systems. It should not be surprising, therefore, that internal representation is ubiquitous in systems of sufficient complexity to support it.

The notion of an internal representation as a reflection of the external world has a long history, going back almost to the dawn of philosophy. More than 2400 years ago, Democritus reasoned that knowledge of the world must originate in a physical interaction between the world and the body. He argued that atoms in the world interact with the atoms of the body and that the mind perceives these effects as faint images in the body, the residue of atomic influences from the environment.[41] The broader philosophical conception of representation, internal and otherwise, also has a long history but a discussion of all the issues that have arisen in two millenia would be so expansive, general, and necessarily vague as to be of little value to the scientist or mathematician. A more sensible strategy is to consider one thread in this tapestry. Cognitive scientists have explored various theoretically tractable notions of representation in the attempt to explain *mental states*. The most successful of these has been the *schema*. An analysis of the notion of schemata reveals much about the nature of representation generally, while avoiding some of the vagueness of the larger notion. Furthermore, unlike "representation," the term "schema" has left a traceable paper trail in the history of scientific ideas and so an etymological analysis can dispel some of the vagueness that surrounds its use in fields as different from each other as anthropology and computer science. Precision of ideas depends upon precision of language. Precision of language entails sensitivity to the import of words. Even opponents of the Wittgensteinian doctrine that meaning *is* use will grant that studying the use of a term reveals much about its meaning. I will examine the history of the term "schema" and identify a number of characteristics of the schema that make it useful for understanding complex adaptive systems.

σχηματα

According to Liddel,[16] the Greek word "schema" (σχημα) originally denoted a physical form, shape, or figure. By extension it named figures of speech, the position of troops arrayed for battle, or the outward appearance or fashion of a person's dress. Thus the term was used at times to denote the outward character of a thing, its appearance as distinct from an underlying nature. The term was also used by geometers to denote a geometric drawing or figure. This usage is critical because it led to an interesting inversion of meaning in the doctrines of Plato and Aristotle.

In "The Republic,"[27,26] Plato uses the term "schema" in reporting a discussion between Socrates and Glaucon concerning the nature of appearance and reality (*Republic*, VI, pp. 510 c–d):

> You are aware that the students of geometry, arithmetic, and the kindred sciences assume the odd and the even and the figures ($\sigma\chi\eta\mu\alpha\tau\alpha$) and three kinds of angles...and do you not also know that although they make use of the visible forms and reason about them, they are not thinking of these but of the ideals which they resemble; not of the figures which they draw, but of the absolute square and the absolute diameter, and so on...

Thus in "The Republic," Plato uses the term $\sigma\chi\eta\mu\alpha$ to denote the geometric figure and goes on to distinguish that outward form from an underlying and metaphysically prior ideal form. In other words, Plato draws a strict distinction between outward appearance and underlying reality and uses the term schema to describe an example of an *appearance* (that of a geometrical figure). Plato makes another technical use of the term schema to describe one of the forms of mimesis. In this sense the term refers to the form or style of a thing that can be imitated in order to convey a sense of its nature. As an example, Plato describes Homer, in *The Iliad*, taking on the persona of Chryses, an elderly priest. Speaking not as himself (Homer, the poet-narrator), but as Chryses, Homer aims to convey the style of the character by demonstration rather than description.

Metaphysically, Plato considered the apparent world a shadowy reflection of an underlying realm. He argued that the objects of thought exist in this realm of ideal forms, and that sensation points our minds towards it. Plato viewed the schema as a collection of aspects of outward appearance. Such a collection has no special metaphysical status, it is simply a kind of summary of nonaccidental properties. The idea that such a summary description might itself *be* the mental representation of an underlying universality later emerged in Aristotle's metaphysics. Nevertheless Plato described an important element of the core notion of schemata; a schema can include important information rather than exhaustive information. This property, **summarization**, is the primary characteristic of schemata: a schema acts as a reduced description of important aspects of an object or event.

Aristotle, like Plato, held that the schema is a kind of blueprint or plan that describes the character of a thing. Unlike Plato, however, he regarded these descriptions as fundamental. In his *Metaphysics* (Book VII, Ch. 3, 1029a), Aristotle[19] wrote:

> ...that which underlies a thing is thought to be in the truest sense its substance. And in one sense matter is said to be of the nature of substratum, in another, shape, and in a third, the compound of these. (By the matter I mean, for instance, the bronze, by the shape, the pattern ($\sigma\chi\eta\mu\alpha$) of its form, and by the compound of these the statue, the concrete whole.) Therefore, if the form is prior to the matter and more real, it will be prior also to the compound of both, for the same reason.

To some extent, Aristotle[3] concurred with Plato's vision of a metaphysically prior realm of mentally accessible forms (*De Anima*, III, 429a):

It was a good idea to call the soul the "place of forms," though this description holds only of the intellective soul, and even this is the forms only potentially, not actually.

But Aristotle argued that the objects of conscious thought ("the intellective soul") are not forms in a metaphysically distinct realm. Instead they are potential forms without any existence beyond their mental representations. Thus Aristotle, argued that Plato was correct to think that the mind perceives not the world as it is in all its particulars but the world as assimilated through universals. For Aristotle unlike Plato however, these universals exist as representational patterns (schemata) in the mind, not as forms in a realm outside both the mind and the sensible world. What Plato considered metaphysical, Aristotle took as epistemological.

This difference between the views of Plato and Aristotle produced an inversion in the meaning of the term "schema." For Plato and his predecessors, the term referred to the apparent nature of a thing, its observable form and properties. For Aristotle, however, the schema was an inward principle that captured the character of that which it represented. The inversion is particularly striking because the ubiquitous metaphor that assigns importance to the underlying and the unseen, while derogating mere appearances, suggests not just a changing role for the schema but also a changing status from a mere list to the very basis of knowledge. Aristotle's use of the term introduced the critical notion that schemata in the mind can provide a foundation for representing knowledge. This is the second useful characteristic of schemata: they can encode regularities in the world in terms of the internal states of a system (e.g., the mind.) I will call this property **internalization**.

KANT'S TRANSCENDENTAL SCHEMATA

Though many philosophers since the Hellenic age have concerned themselves with the questions "What can we know?" and "How can we know?," Immanuel Kant was the first post-Aristotelian philosopher to propose a detailed theory of the structure and character of mental representation in addition to describing its relationship to the external world. Descartes and Hume proposed theories concerning the limits of knowledge and the relationship of perception and belief to truth. Kant attempted to find a framework for mental states that would turn Cartesian and Humean concerns on their head. His aim was to show that rather than the mind lacking certain knowledge of space, time, and causation, it is precisely in the mind that these categories exist. In effect, Kant proposed that Hume and Descartes made the same error that Aristotle found in Plato: the mistake of putting into the world that which exists in the mind.

Kant's revolutionary idea was that the mind is an organized system that assimilates sensory perception in ways that are determined by existing *mental* structure. In other words, Kant proposed that the apparent structural features of the world such as space, time, and causation are really properties of the organization of the mind (properties which Kant called "categories"). This theory led Kant to his doctrine of transcendental schemata. Kant defined a schema as "the phenomenon or sensible concept of an object in agreement with the categories" (cited in Young[1]). Kant believed that minds experience the world through a process of "apperception" in which the mind, using the categories and schemata, synthesizes experience out of sensory perceptions. While this view may seem needlessly abstruse, it has radical consequences for the meaning of the term "schema." In particular, Kant's transcendental schematism introduced the notion that the structure of the mind, and not just the structure of the world, determines the character of mental representations.

What are the practical consequences of Kant's doctrine? Kant developed his view in part to respond to Hume's radical scepticism about causation. Hume claimed that we cannot come to know the causal character of events because it is in no way observable in the world except through co-occurrence and that alone is not sufficient evidence. Kant replied that we can directly experience space, time, and causation precisely because they are not observable in the world but are rather a property of apperception itself. More generally, the importance of Kant's ideas about schemata lies in the observation that the schemata and the organization among them determines their representational character and capacity. This is a critical feature of schemata: they exist in a system constrained to certain possible states and so can only represent the world in a way that is consistent with the organizational principles of that system. As a result every experience is an assimilation of sensation through the lens of the schemata. This property of **assimilation** originates in Kant's metaphysics but only after reinvention by Jean Piaget and Frederic Bartlett (both of whom called it "assimilation") did it become an element of scientific theories of cognition.

Kant's insight presents a problem of which he himself was acutely aware: if space, time, and causation exist as categories of experience rather than properties of the world, is not our belief in the spatial, temporal, and causally connected character of experience simply a fantasy? Kant's attempt to respond to this difficulty has provoked much analysis and argument because it is fraught with complication. One of the most lucid attempts to explicate this aspect of Kant's philosophy can be found in a paper by Waxman.[42] Kant's answer, in Waxman's view, is that experience itself is a synthesis by schemata of sensation which is the "raw and utterly formless...primary matter" out of which experience is constructed. In short, there is no way of testing our experience of the world against what actually exists because all that actually exists are the material causes of sensation and they are categorically different from the structured information we experience. There can be no "reality check" because it would involve a false assumption that there exists something actual that has qualities comparable to experience. This is not to say that nothing exists (that would be simple solipsism) rather it is to say that the world and experience are different *in kind* and so cannot be compared directly.

While this answer is perhaps philosophically coherent, it also presents a difficulty for cognitive scientists because it decouples experience from the world in which we believe we make measurements. If we hope to have a theory explaining the relationship between the measurable world and experience, it is important to avoid treating them as incommensurable.

Fortunately, there is an alternative account that has the virtue of better preserving the linkage between experience and the world. The alternative is based on considerations of evolutionary selection as it applies to internal representational systems. Any organism that depended upon a representational system that did not have a felicitous relationship to the nature of the environment would only be hindered by that representational system. It stands to reason that the extant examples of representational systems are well suited to the continued existence of their organismic hosts in the environments in which they evolved. This is not to say that experience is necessarily *veridical* in any sense that would satisfy a Humean sceptic, and that is why Kant rejected a similar theologically based argument for the veridical character of experience (that divine fiat established a correspondence between reality and experience). But as empirical scientists, our first duty is not to answer the challenge of radical Humean scepticism but to account for the apparent structure and function in the phenomena we study.

In short, while this argument does not establish that we experience the world *veridically*, it does imply that our experiential state reflects information in the world that is important to our evolutionary fitness. Ironically this fact may help to explain the many cases in which our experience is *not* veridical, i.e., illusions. On this account, illusions can arise for at least two reasons. First, it may be that some information is not sufficiently useful to exert adaptive pressure. In that case, misperception might result, or not, depending upon the other constraints on the perceptual system. Second, it might be that some information actually reduces the fitness of organisms. In that case an illusion or "misperception" might actually confer selective advantage to an organism. For example, by eliminating some information, it may be possible to make other information more salient. If that information had disproportionate consequences, this strategy could prove useful. It may be that *attention* is an evolved faculty that allows "misperception" to systematically distort information to reflect its significance to a perceiving organism. In a more pervasive sense, the fact that our eyes see particular wavelengths of light and not others, or the fact that we hear only at certain frequencies results from adaptive pressures towards particular abilities (and perforce away from others).

This line of reasoning suggests another property of schemata that may be of use in analyzing complex adaptive systems. Schemata exist as states in a system whose organizational principles reflect any information in the world that is sufficiently useful to an organism to confer selective advantage. Or, to use a formulation Kant might prefer: knowledge can only come to us through schemata, so anything that schemata cannot represent cannot be experienced. This property of potentially encompassing all that is knowable, I will call **inclusiveness**. Inclusiveness differs crucially from veridicality. In deference to Hume it should be remembered that nothing guarantees the veridicality of schemata, although adaptive pressures no doubt enforce a strong linkage between information in the environment and characteristics of experience.

BARTLETT'S THEORY OF CONSTRUCTIVE REMEMBERING

Frederic Bartlett was the first psychologist to describe the properties of human memory in terms of schemata.[4] Bartlett attributes[5] the term to the neurophysiologist Sir Henry Head.[11] Strangely, Head's work had little to do with conceptual representation but was concerned with the representation of motor sequences. He theorized that memory for such sequences takes the form of a kind of motor schema that represents the chain of nervous activity necessary to perform a movement. Bartlett used the term in a way that was much closer to its philosophical origins in describing the process of reconstructing stories from memory. Bartlett asked experimental subjects to read stories and then examined their recountings over the course of days, weeks, months, and even years. He found that subjects remembered the stories imperfectly and that the nature of their recounting was greatly shaped by conventional assumptions about what might be sensible events and by confusions that could be explained in terms of distorted reconstructions of partly remembered objects or events. These results led him to conclude that human memory is fundamentally *constructive*. When events occur, Bartlett argued, we encode them by relating them to similar events with which we are familiar. In other words, we instantiate a schema that represents novel information in terms of existing conceptual organization. When we want to remember the encoded event, we use the instantiated schema to reconstruct the facts of the event that might have produced the relevant schema.

This account of the nature and function of schemata has a close relationship to the Platonic, Aristotelian, and Kantian notion. Bartlett's schemata act as a kind of reduced description of events in the way that Plato suggested, they are a means for mentally representing situations as Aristotle argued and, most importantly, they are a distorting lens that assimilates events through the constraints of a preexisting mental organization, as Kant insisted they must be. There is no mention of these connections in the writings by Bartlett but both he and Henry Head were influenced greatly by the British philosopher and psychologist James Ward who wrote extensively (though often critically) on Kant and by their contemporary W. H. R. Rivers, an anthropologist and psychologist. It is possible that Head and Bartlett first encountered the term in studying with James Ward. Head and Rivers also studied with the German psychophysicist and physiologist Eward Hering whose nativist theory of perception allied him closely with the Kantian philosophy of mind. It may be that Hering introduced Kant's notion of schemata to psychology through Rivers or Head.

Bartlett's theory of remembering made use of the notion of schemata to propose an explanation of the constructive nature of memory. At the same time that he relied upon the notion of schemata, however, he also elaborated it. Both Head and Bartlett viewed schemata as tools by which the mind might alter its operation to improve performance. By recording experience, schemata allow history to operate on future behavior. This is a critical insight because it is the role of schemata in prediction and the planning and execution of behavior that are their primary reason for continued existence. In short, schemata allow memory to affect the behavior of a system by assimilating information for future use. This property might be termed **diagnosticity**, since it incorporates into the notion of schemata the fact that knowledge permits effective action. At the same time, however, Bartlett argued that errors in memory and mistakes of judgment can often be attributed to the fact that high-level schemata operating on incoming information may distort the way that we perceive and encode that information. Thus Bartlett was aware of the importance of schemata as diagnostic tools but he saw that like a magnifying glass, these tools depend on a lens that must also distort. In this sense Bartlett can be said to have brought Kant's point about assimilation to the attention of psychology.

PIAGET'S SCHEMA ACCOMMODATION

Piaget[13] employed the term "schema" vaguely although his theory of cognitive development depends upon a notion of schemata that has at least the properties of summary representation, internalization, and assimilation. While Piaget was often vague where others had been more explicit, he did address a question about which others had been vague. Piaget was the first to investigate experimentally the processes by which schemata might be formed and changed. By studying the development of human cognition throughout childhood, Piaget came to the idea that schemata generally guide the assimilation of new information. He observed, however, that the world can yield information that we cannot assimilate easily. In this case he proposed a process of "accommodation." Accommodation involves assimilating information through the lens of our schemata but, at the same time, "restructurizing" our schemata so as to best account for the new experience. Thus a compromise is struck between old and new. Piaget's contribution to this history was to suggest that schemata might operate according to such a property of **accommodation**.

This may seem inconsistent with Kant's doctrine that there can be no mismatch between schema and reality since "reality" is no more than a product of the action of schemata and categories upon sensations. On the other hand, one interpretation of Piaget's position is that he assumed the existence of the more fundamental categories and schemata that shape experience (about which Kant was particularly concerned). If so, then what he described as mismatch between schemata and the world, Kant would describe as revealed inconsistencies *among* the schemata. In other words, certain schemata might conflict with others in mapping sensation to experience. In this case an accommodation process as described by Piaget might provoke changes in the schemata without violating Kant's transcendental schematism.

THE SCHEMA IN COGNITIVE SCIENCE
BARTLETT'S PROGENY

Rumelhart[36] has argued that the notion of a schema, implicitly and explicitly, has permeated the thinking of cognitive psychologists at least since Bartlett and probably longer. In a sense, even psychologists who do not employ the term "schema" have depended on the notion of a mental representation with many of the same properties. One might argue that many psychologists preceding Bartlett used the term "habit" to denote a similar theoretical construct. This term has a worthy history of its own. Not surprisingly it begins with Aristotle.[19] It then makes its way through Ockham,[10] Aquinas,[2] and Descartes[40] to the visionary American logician C. S. Peirce[25] and, thence, to William James[14] and John Dewey.[7] An analysis of this term might itself prove informative but would be a lengthy digression from this already convoluted story.

Neisser's seminal textbook *Cognitive Psychology*[29] established Bartlett as the locus classicus of schema theory for cognitive psychologists. The influence of Bartlett's notion of constructive memory is clear in the work of many of the founding figures of cognitive psychology. Miller and Selfridge were among the first to invoke Bartlett's notion in developing information theoretic mathematical models of cognitive phenomena.[22] Bartlett's ideas quickly spread to the new community of cognitive psychologists. It was probably Neisser who enshrined Bartlett in the pantheon of cognitive psychology but the influence of the notion of schemata could already be glimpsed in the work by Bruner, Goodnow, and Austin,[6] Miller, Galanter, and Pribram,[20] and many others. The schema continued to gain influence as a theoretical construct in the late 1960s and the 1970s when Newell and Simon,[30] Fillmore,[9] Minsky,[28] Schank and Abelson,[38] Norman and Rumelhart,[32] and others realized the usefulness of schemata as a high-level construct in computational models of a host of domains from vision to language. Rather than discuss all of the work that occurred during this heyday of schema theory, I will try to glean a few simple principles about the properties of schemata. After all, Kant's immediate predecessors probably thought they were witnessing the heyday of schemata, too.

MINSKY'S FRAMES

In 1975, Minsky proposed a representational structure that he called a "frame."[28] His theory was very much influenced by Bartlett's notion of schemata. Schank and Abelson[38] developed a roughly contemporaneous theory based on a construct called the "script" that was somewhat less influenced by Bartlett though greatly shaped by more contemporaneous work in cognitive psychology.

Minsky proposed that the assimilation of information according to a schema can be described as a set of mappings from variables to values. His "frames" employed this computational principle to represent a variety of data structures relevant to computer vision, problem solving, and knowledge representation generally. This variable binding analysis of assimilation owes much to Fillmore's analysis of case grammar.[9]

One of the primary insights of the work on frames was that schemata are potentially recursive. In fact, if schemata are to have a thoroughly general capacity to represent information, then it must be the case that they can embed, since the information about one of the important summary characteristics of an object or event must itself be representable by a schema. This, of course, implies that there must be some schema that are atomic, i.e., not reducible to other schemata in order to avoid infinite regress. It seems appropriate to call the principle that schemata can contain other schemata **recursiveness**.

RUMELHART AND ORTONY'S SCHEMA GENERALIZATION AND SPECIALIZATION

Rumelhart and Ortony,[35] drawing upon work in both cognitive psychology and computer science, developed a detailed theory of schemata. Though they did not formulate their theory in terms of the historical elaborations of the meaning of the term schema, their theory captures many of the essential properties that have accrued to schemata over time:

> There are, we believe, at least four essential characteristics of schemata, which combine to make them powerful for representing knowledge in memory. These are: (1) schemata have variables; (2) schemata can embed one within the other; (3) schemata represent generic concepts which, taken all together, vary in their levels of abstraction; and (4) schemata represent knowledge rather than definitions.

Each of the four properties Rumelhart and Ortony propose have a direct relationship to the features that have historically accreted around the term "schema":

■ Schemata have variables. That is to say, schemata provide a framework into which new information from the environment can be assimilated. This property combines Plato's principle that schemata represent the important features of objects and Bartlett's theory that memory involves assimilating information to schemata. It also resembles Minsky's view that frame instantiation involves assigning values to variables.

- Schemata can embed. In other words, the schema that constitutes a framework for understanding the concept *airplane* may contain a schema for representing information about wings or even about the process of traveling. Those schemata would be subordinate structures constituting a part of the larger schema. This property parallels Minsky's proposal that frames can have recursive structure.
- Schemata represent knowledge at many levels of abstraction. For example, schemata can represent information about objects in the environment but they can also represent information about the way objects interact or the nature and structure of events. This characteristic follows from Kant's principle that schemata must be able to represent any knowable information.
- Schemata represent knowledge rather than definitions. This aspect of Rumelhart and Ortony's theory puts a novel spin on Plato's principle that schemata are a summary representation of important information about an object or event. They argue that schema are essentially "encyclopedic" rather than "definitional" since the important information about an object or event and the definitional information (if such exists) are not necessarily the same.

In addition to identifying these essential characteristics of schemata, Rumelhart and Ortony proposed two ways in which schemata can develop and adapt to environmental information: specialization and generalization. Specialization forms less general schemata from more general ones. One way to conceive of specialization is as the removal of some variables from a schema. The resulting schema can no longer vary with respect to those dimensions but remains the same otherwise. For example, the schema for surprise parties might be a specialization of the schema for parties in which several variables such as whether there is a recipient of the party and whether it involves planning have set values. Generalization involves the reverse process: features of a schema that once were fixed become variable creating a more general schema. In proposing explicit models of schema change, Rumelhart and Ortony filled a conceptual lacuna. The property of schema that they can be developed by simple modifications of other schemata I will call **generativity**.

RUMELHART AND NORMAN'S TAXONOMY OF LEARNING

Rumelhart and Norman,[33] keenly aware of the need for a theory to explain the acquisition and modification of schemata, tried to develop a more general framework for learning within which specialization and generalization would be special cases. They proposed that learning can occur in three substantially different ways: by accretion, tuning, and restructuring.

Accretion consists of registering new information from the environment by instantiating existing schemata. This is the process Bartlett described in his work on remembering.[4] Bartlett described it as "assimilation."

Tuning involves modifying schemata in order to account for sensory information. This mechanism operates by a process like Piaget's accommodation. In detail, Rumelhart and Norman propose that tuning operates according to Rumelhart and Ortony's dual principles of specialization and generalization. In addition they argue that tuning can occur through a refining of the accuracy of schemata or by determining default values for some variables in a schema. These processes elaborate on the ways in which schema learning reflects the general property of **generativity**.

Finally, Rumelhart and Norman propose that learning can occur through a process they call "restructuring." Restructuring involves a more radical change in the structure of internal representation than accretion or tuning but is still consistent with the general principle of generativity: that new schemata are created by modifying existing schemata. Rumelhart and Norman discuss two ways in which restructuring can occur: "patterned generation" and "schema induction."

Patterned generation involves creating a new schema by altering the set of values from which one or more of the variables in an existing schema are chosen. Rumelhart and Norman offer analogical reasoning as an example of schema creation by patterned generation. As an example, they offer a possible source of the schema for a rhombus:

> ...even if we never had direct experience with a rhombus, we could develop a schema for one by being instructed that a rhombus has the same relationship to a square that a parallelogram has to a rectangle.

Schema induction involves creating a new schema by combining two existing schemata that have a strong relationship to each other. Rumelhart and Norman focus particularly on spatial or temporal contiguity although, in principle, one might imagine other connective relationships. Rumelhart and Norman argue that this kind of learning occurs only rarely and that most learning, in fact, occurs through accretion and tuning, with only occasional instances of patterned generation and extremely infrequent schema induction.

All of the attempts to describe the details of a strategy for modeling generativity in schemata are directed towards answering the question of the origin of schemata. This was a question that Kant abjured on account of his essentially nativist metaphysical stance. For empirical scientists who wish to use a notion of schemata resembling Kant's, the puzzle is to explain how schemata can be the very fabric of our experience and at the same time can be learnable from experience. In this regard it is worth noting that all of the proposals for generativity that I have discussed depend upon the prior existence of schemata. In fact, the way I have formulated the generativity property seems intrinsically to depend upon the existence of schemata prior to the creation of new ones. If this is so, then the only usefulness of generativity is to account for the variety of schemata on the basis of experience, not to explain how schemata came to exist in the first instance.

To account for the creation of schemata in the first instance, we must appeal to a process of adaptation operating on systems but outside the scope of the internal representation itself, such as the biological evolution of human cognition. The answer to the question of the origination of schemata is implicit in the argument for the property of **inclusiveness**; systems that use internal representations are the result of an adaptive process that succeeds best when particular structures arise. Those structures are the prior framework for a system of schemata and they develop as a result of adaptive processes that must occur outside the scope of the internal representational capacity of the system itself. To try to give an adaptive account of the emergence of schemata solely in terms of the so-to-speak "phylogenetic" properties of an internal representational system would be to engage in a circular and teleologically confused enterprise.

THE SCHEMA AS A POINT IN A STATE SPACE

Rumelhart, Smolensky, McClelland, and Hinton describe schemata in this way:

> large-scale data structures...playing critical roles in the interpretation of input data, the guiding of action, and the storage of knowledge in memory...
> [T]he schema has, for many theorists, become the basic building block of our understanding of cognition.[37]

After a brief discussion of the history of the term, they go on to delineate several important properties of schemata (loosely following Rumelhart and Ortony[35]) and to analyze those properties in terms of the behavior of a connectionist model of the mental representation of knowledge about rooms in a house.

Rumelhart, Smolensky, McClelland, and Hinton provide an interpretation of certain properties of schemata in terms of the dynamics of a model consisting of a number of simple processing units that interact to produce a global representational state. The units in the model are connected to one another by symmetrically weighted links that transmit information about their state to adjacent units. Each unit computes a function of the weighted sum of the units that connect to it. The unit communicates that value to all adjacent units. The adjacent units use that signal to determine their own output values. When information from the environment constrains parts of the network, the rest of the network equilibrates to a state that can be thought of as a representation of the information from the environment. The system as a whole, thus, comes to represent states of the environment by internal states. To establish a network that equilibrate to a useful representation, one might imagine designing some kind of weight correction procedure based on a theory of what is useful (a kind of fitness function). Rumelhart, Smolensky, McClelland, and Hinton use a probabilistic rule based on feature co-occurrence to impose weights and a particular interpretation on their network.

In the dynamics of such a system, there are analogous properties to several aspects of schemata. Just as schemata represent world knowledge, a connectionist network can represent environmental information by altering weights so as to change the stable states of the system. Just as schemata have variables, a connectionist network can use unit activities or the activity of ensembles of units to encode features of the environment. The possible state of this unit or ensemble would then correspond to possible values of an environmental variable. Just as schemata can embed, so a part of the network can represent a subschema of the schema instantiated by the full network. Finally, just as schemata represent knowledge at multiple levels of abstraction, the ensembles of activity in a network can capture the properties of any information from the environment that as long as the parameters of the network are adjusted according to the proper learning procedure.

The importance of this model lies not in its implications about the nature of schemata but in its demonstration that a particular complex adaptive system (in this case a connectionist network) can exhibit dynamics that resemble the behavior of a schematized system of internal representation. Rumelhart et al. have made a plausible case that a system of schemata can be described by a complex adaptive system. If this is so, then there must exist a class of complex adaptive systems whose behavior can be described by the high-level constructs of schema theory. Such a system should be understandable in terms of the properties of schemata that I have outlined. Such an undertaking may provide considerable assistance in the analysis of the dynamics of some complex adaptive systems. If nothing else, it should provide a framework that makes some aspects of system dynamics comprehensible in terms of intuitively manageable concepts.

The picture of schemata as basins of attraction in the state space of a complex adaptive system has also allowed connectionists to provide interesting theories of the underlying dynamics of psychological systems and in that way has been of great use to cognitive scientists. It would seem that some sort of zero-sum principle ought to imply that psychologists cannot turn around and provide tools to those interested in studying complex adaptive systems. It is possible, however, that much might be gained in understanding complex adaptive systems if we described their dynamics in terms of emergent high-level concepts such as schemata. It might provide the kind of probative guide to intuition that can always be of use in understanding the behavior of extremely complicated phenomena. In this respect it seems fair to say that cognitive science and the study of complex adaptive systems are coevolving disciplines.

APPENDIX: THE PROPERTIES OF SCHEMATA DISCUSSED IN THIS ESSAY

- summarization: schemata summarily represent important information about objects or events.
- internalization: schemata internalize information from the world by representing it.

- assimilation: schemata inform what they represent; they assimilate states of the environment in a manner consistent with their organization.
- accommodation: information from the environment can alter schemata if there exist adaptive pressures towards different patterns of assimilation.
- inclusiveness: schemata are inclusive; they potentially represent all states of the environment that can be experienced.
- diagnosticity: schemata are diagnostic; they convey information about history that can be used to predict future states of the environment.
- recursiveness: schemata can contain other schemata.
- generativity: schemata can be created from other schemata by modification of existing structures.

REFERENCES

1. Ahn Kang, Young. *Schema and Symbol*. Amsterdam: Free University Press, 1985.
2. Aquila, Richard E. *Representational Mind*. Bloomington, IN: Indiana University Press, 1983.
3. Aristotle. *Aristotle: On The Soul, Parva Naturalis, On Breath* (with an English Translation), 8–201. Cambridge, MA: Harvard University Press, 1957.
4. Bartlett, Frederic C. *Remembering*. Cambridg: Cambridge University Press, 1932.
5. Bartlett, Frederic C. *Thinking*. New York: Basic Books, 1958.
6. Bruner, Jerome S., Jacqueline J. Goodnow, and George A. Austin. *A Study of Thinking*. New York: Wiley, 1956.
7. Dewey, John. *Human Nature and Conduct*. New York: Random House, 1922.
8. Duda, Richard O., and Peter E. Hart. *Pattern Classification and Scene Analysis*. New York: Wiley, 1973.
9. Fillmore, C. S. "The Case for Case." In *Universals in Linguistic Theory*, edited by E. W. Bach and R. H. Harms. New York: Holt, Rhinehart, & Winston, 1968.
10. Fuchs, Oswald. *The Psychology of Habit According to William of Ockham*. St. Bonaventure, NY: The Franciscan Institute, 1952.
11. Head, Henry. *Studies in Neurology*. London: Oxford University Press, 1920.
12. Hearnshaw, L. S. *A Short History of British Psychology 1840–1940*. New York: Barnes and Noble, 1964.
13. Inhelder, Barbel, and Jean Piaget. *The Early Growth of Logic in the Child*. New York: Harper & Row, 1964.
14. James, William. *The Principles of Psychology*. Cambridge, MA: Harvard University Press, 1890.

15. Johnson-Laird, Phillip N. *Mental Models.* Cambridge, MA: Harvard University Press, 1983.

16. Liddel, Henry George, and Robert Scott. *A Greek-English Lexicon.* Oxford: Clarendon Press, 1968.

17. McCulloch, Warren S. *Embodiments of Mind.* Cambridge, MA: MIT/Bradford Books, 1965.

18. McCulloch, Warren S., and Walter Pitts. "A Logical Calculus of the Ideas Immanent in Nervous Activity." *Bull. Math. Biophys.* **5** (1943): 115–133.

19. McKeon, Richard. *The Basic Works of Aristotle.* New York: Random House, 1941.

20. Miller, George A., Eugene Galanter, and Karl H. Pribram. *Plans and the Structure of Behavior.* New York: Holt, Reinhart & Winston, 1960.

21. Miller, George A., and Phillip N. Johnson-Laird. *Language and Perception.* Cambridge, MA: Harvard/Belknap, 1976.

22. Miller, George A., and J. Selfridge. "Verbal Context and the Recall of Meaningful Material." *Am. J. Psych.* **63** (1950): 176–185.

23. Minsky, Marvin, and Oliver G. Selfridge. "Learning in Random Nets." In *Information Theory: Fourth London Symposium.* Washington, DC: Butterworths, 1961.

24. Minsky, Marvin, and Seymour Papert. *Perceptrons.* Cambridge, MA: MIT, 1969.

25. Peirce, Charles S. *Philosophical Writings of Peirce.* New York: Dover, 1955.

26. Plato. "Politeia." In *Plato's Republic: The Greek Text,* edited by B. Jowett and Lewis Campbell, 1–465. Oxford: Clarendon, 1894.

27. Plato. "The Republic." In *The Republic of Plato,* translated into English, edited by B. Jowett, 210–213. Oxford: Clarendon, 1921-1922.

28. Minsky, Marvin. "A Framework for Representing Knowledge." In *The Psychology of Computer Vision,* edited by P. H. Winston, 211–280. New York: McGraw-Hill, 1975.

29. Neisser, Ulric. *Cognitive Psychology.* New York: Meredith, 1967.

30. Newell, Allen, and Herbert A. Simon. "Memory and Process in Concept Formation." In *Concepts and the Structure of Memory,* edited by Benjamin Kleinmuntz, 241–274. New York: Wiley, 1967.

31. Nilsson, Nils J. *Learning Machines.* New York: McGraw-Hill, 1965.

32. Norman, Donald, and David E. Rumelhart. *Explorations in Cognition.* San Francisco, CA: W. H. Freeman, 1975.

33. Norman, Donald, and David E. Rumelhart. "Accretion, Tuning, and Restructuring: Three Modes of Learning." In *Semantic Factors in Cognition,* edited by John W. Cotton and Roberta L. Klatzky, 37–53. Hillsdale, NJ: Lawrence Earlbaum, 1978.

34. Rosenblatt, Frank. *Principles of Neurodynamics.* Washington, DC: Spartan, 1962.

35. Rumelhart, David E., and Andrew Ortony. "The Representation of Knowledge in Memory." In *Schooling and the Acquisition of Knowledge,* edited by

Richard C. Anderson, Rand J. Spiro, and William E. Montague, 99–135. Hillsdale, NJ: Lawrence Earlbaum, 1977.

36. Rumelhart, David E. "Schemata: The Building Blocks of Cognition." In *Theoretical Issues in Reading Comprehension*, edited by R. J. Spiro, B. Bruce, and W. Brewer, 33–58. Hillsdale, NJ, Lawrence Earlbaum, 1980.

37. Rumelhart, David E., Paul Smolensky, James L. McClelland, and Geoffrey E. Hinton. "Schemata and Sequential Thought Processes in PDP Models." In *Parallel Distributed Processing: Explorations in the Microstructure of Cognition*, edited by James L. McCleland and David E. Rumelhart, 7–57. Cambridge, MA: MIT/Bradford Books, 1986.

38. Schank, Roger, and Robert Abelson. *Scripts, Plans, Goals and Understanding*. Hillsdale, NJ: Lawrence Earlbaum, 1977.

39. Slobodin, Richard. *W. H. R. Rivers*. New York: Columbia University Press, 1978.

40. Watson, Richard A. *The Breakdown of Cartesian Metaphysics*. Atlantic Highlands, NJ: Humanities Press, 1987.

41. Watson, Robert I. *The Great Psychologists from Aristotle to Freud*. Philadelphia: J. B. Lippincott, 1963.

42. Waxman, Wayne. *Kant's Model of the Mind*. New York: Oxford University Press, 1991.

DISCUSSION

QUESTION: Should considerations of psychology and human behavior enter into economics?

MARTIN: Economics shouldn't push into psychology; economics exists precisely because psychology is not well suited to solving certain kinds of economic problems. There would be no field of economics, otherwise; we would just look out in the world and perceive proper investment strategies, as we perceive visual tableaus, or something like that.

ARTHUR: I wish you would tell economists that.

KAUFFMAN: He just did.

ARTHUR: Economists do believe that people can actually look at a problem, that the economist as a researcher may have taken six months to solve, and instantaneously perceive the correct answer by some mysterious process, not necessarily by logical deduction—and behave as if they had solved it perfectly. It's nonsense.

GELL-MANN: Perfectly in conformity with well-defined transitive preferences, which nobody must question. That's the best thing about it.

COWAN: I have a problem with what you just said, and that is that in the spirit of complex, interconnected processes, to say that psychology goes so far and then economics does what psychology was unable to do, sort of separates the things into two disciplines, whereas in fact they're constantly interacting.

MARTIN: Well, what I would say about that is that, again, like in the case of language and language use, what we have here is a need to distinguish two kinds of complex adaptive systems. There are economies, which are complex adaptive systems, and which we can characterize by the use of those models. There is human behavior, which we can characterize as a complex adaptive system. To expect the human behavior system to make predictions in the economic realm that map onto the sort of rational-expectationist strategies that we devise formally and algebraically, I think would be a mistake. But that's not necessarily to say that those are two unrelated fields of endeavor.

COWAN: No, but you could also say, to expect economics to make predictions about the behavior of people in an economically rational way is probably also an overstatement.

MARTIN: Yes, although I think it's something desirable for economists; I mean, I think it's their only real desideratum.

GELL-MANN: You mean, just to give them something to do? Wouldn't it be better if they tried to model real economic behavior?

LICHSTEIN: Aren't you getting at the question of what is reasonable to think about? The case of financier Robert Maxwell is an example of a system gone awry. Institutions around the world were willing to fund a three billion dollar fraud. And BCCI has become an issue for regulatory purview because BCCI collapsed and became something they had to think about. One of the concerns of the regulators is "Here we are thinking about the problem, but that one has already hit us; what's going to be the next one?" So to some extent, the issue you are raising relates to scale. Something is not an issue until somebody—because of his psychological makeup—decides to try it. And we let them try it, and then it becomes something to deal with.

MARTIN: Right. Well, I think that sort of segues into the second point that I wanted to make about this, which is that psychology may be useful in describing certain cases that are mysterious, from an economic standpoint. And in particular, I thought it was interesting (Stuart's point)—this idea of "laws of motion." It's an analogy; the "laws of motion" that govern economic behavior are of a restricted class, when we concern ourselves with the behavior of individuals in their choices, and judgment, and decision making. And what constrains that class of laws of motion, as we would want to represent them in our complex adaptive systems, are psychological variables of various kinds. And I would point to Tversky's work in particular, on things like risk aversion.

So for example, if I offer you the choice between some gambles, and in one case I say to you, "Would you rather have a 50–50 chance of gaining $1000, or would you rather have $700 for sure?" I could tune that second amount until I had an equivalence, a tradeoff, between your choices so you were indifferent between them. Then I could reverse the problem and ask you about losses. Would you rather have a 50–50 chance of losing $1000, or would you rather lose $700 for sure. And what you find is—contrary to what an economist would want you to say—it's completely asymmetric. In fact, when you're in the domain of losses, you think, "I'll take a chance; if I'm going to lose $700 for sure I'll roll the dice, and see what happens. And maybe I'll lose the $1000 and maybe I won't." So in the domain of losses we tend to be what Tversky calls "risk seeking." In the domain of gains, on the other hand—I offer you the choice of $700 for sure and a 50–50 chance at $1000—you say to me, "$700 for sure sounds pretty good; it's only a 50–50 chance at $1000."

EPSTEIN: You expect asymmetry; in one case you can lose, and in the other case you're not going to lose.

MARTIN: Exactly. That's exactly the point. Right. You expect this asymmetry because of the way the mind represents the problem.

EPSTEIN: Then why is it surprising?

MARTIN: Because economists find it surprising.

EPSTEIN: Why wouldn't economists expect it?

LLOYD: The expected loss in one case...the expected loss and the expected gain, they have the opposite sign, so...

MARTIN: Oh, no; that's not the point. The point is the comparison between the two equivalent gambles. So I make them equivalent in the positive case and then I ask you about the negative case. That's the asymmetry.

GELL-MANN: But isn't it true that we shouldn't talk about what economists expect? Economists are people like everybody else, and they expect ordinary things like everybody else. When they go to the office and write papers—it's a whole different story!

MARTIN: Right. This is a loose way of speaking. This is like this story that Tversky tells about two professors in decision making at Columbia. And the one goes to the other and says, "What's the matter?" (because he's looking real troubled). And he says, "I'm sitting here trying to decide; I've got this tenure offer from Harvard, and I don't know whether to take it or not. I could get a promotion here—what should I do?" And the guy says, "This is ridiculous. You're one of the world's foremost decision theorists. Why don't you just lay it out as a problem-solving issue, and see what the solution is?" And he said, "Come on, this is SERIOUS!" [Loud and prolonged laughter.]

So I think it's exactly right what you're saying. There's a big chasm between economists as people and economists as...

GUMERMAN: But I'd like to underscore that what George said, is true. When you're talking about economics as a complex adaptive system, really you're talking about one subsystem of an integrated cultural system. If you look at the [photographs of the prehistoric ruins that surround us on the] walls of this conference room, you have a good example of a culture that does not consider economics as a separate behavioral system. If we can judge by their descendants, the Pueblo people of today, economics was only one subsystem of the entire cultural system that also included the subsystems of ideology, social organization, and the environment.

MARTIN: Well, I would say that's right; in certain cases, economic behavior should be characterized...

COWAN: You don't need it in your field, but economics needs you.

MARTIN: Right. In a way, that's the point that I've been trying to make; that there's an influence that I think psychology should have on economics viewed as a complex adaptive system. And that influence is this: the "laws of motion," so to speak (as Stuart was calling them), are restricted, in certain cases, by what we know about the psychology of human behavior. And this example of risk aversion is one example of such a case. So, in fact, I would agree quite strongly with what you're saying. I think there is a tie to be made, and I think, again, the two fields should influence each other in that way.

KAUFFMAN: But in particular, the other point that finally dawned on me is that there's a coevolution of schemata, because a law of motion "me" is determined by what I think you're going to do, and vice versa, pretty obviously. And why should those opinions of one another be stable over time, as is demanded by, for example, Nash equilibria, or as demanded in particular by Rational Expectations?

MARTIN: Well, I think it's interesting; we do know something about the stabilities, right? So, in fact, an example like risk aversion points to something that is a stability, that may be a result of other seemingly unrelated facts about the way human psychology works. So in fact, it may not be optional. There may be a strong constraint on the class of economic models that work, and that constraint may be due to limitations on what's possible for someone who has a working psychology of this kind.

ARTHUR: Just about economics and psychology: I don't think you're going nearly, nearly far enough...

MARTIN: In terms of being negative towards economists?

ARTHUR: No, it's not an issue...Economics is its own thing, and that's very sad...Talk about the economy for a moment, or how people behave economically. The inputs from psychology have been rather meager so far, and this is why I think that psychologists are not going far enough. The main thing psychologists have had to say so far to economists is that, "Hey, you know, when you guys make a decision to do with losing theater tickets versus purchasing new ones,..." Kahneman and Tversky and a few other people have shown that we as economists aren't quite accurate, that there are nuances—like in the example you've just given—that should be taken into account, and those are likely to affect decisions, and so on. As good economists, we've all bought that, and we've said, "Fine, we know that we're not quite accurate, but we'll go on writing our papers as if people do behave that way. And we'll refer to Kahneman and Tversky as the sort of second-order twiddle on the solution."

I would go far, far further. Let me say exactly how. Imagine you're playing tic-tac-toe, and any ten-year-old playing it repeatedly will come upon the so-called deductively rational economic solution. They'll pretty quickly learn what to do that works. It's rational, it's so-called Nash equilibrium, and so on; the economic model is upheld. Then turn the dial and go up through complicatedness, through checkers, and then on into chess. Economics does have a theory as to how chess might be played; we know that solutions do exist: Nash solutions, mixed strategies...

Theoretically, chess is solved, although nobody knows what the solution is, or has computed it yet. This is precisely where all your ideas of schemas, or schemata, come in. When people are actually playing chess, what they're doing, in my opinion, is constructing internal models of what's going to happen in the game, of how their opponent is about to maneuver; they're looking at past games of the opponent; they're thinking of their own strengths and weaknesses; they're looking at the board positions forming and reforming, and at the patterns that are appearing on the board. And, on the basis of those internal models—what you would call schemas and I would call internal hypotheses—are actually deducing forward, three, four, or five moves, or maybe more. There's a little bit of deduction added in.

Now, the point I want to make is that in any problem of serious complication in the economy, where you're playing something as difficult as chess—and I would maintain that negotiating with the Japanese is such a thing, or trying to figure out schedules for producing steel in many workshops, and so on—these are all complicated problems and it's precisely there that psychology can contribute. Because, as human beings—smart ones—running these operations, or doing the negotiations, or playing chess, we're always constructing these hypotheses, these schemas. So there's a whole major, huge chunk of psychology we should be taking into economics in dealing with bounded rationality when people are actually talking about problems of complications. And at least half the problems that we deal with from day to day are not these trivial ones. Economics talks about how we can deal with problems, use logical deduction—that's fine if they're sufficiently simple problems. Once the problems get complicated, we're in exactly this world you're talking about. And that's where psychology can contribute; not just in small twiddles.

MARTIN: I think that's a wonderful example. I actually have worked a little bit at the board game Go, and I would point to both Go and chess as examples of where psychologists have found quite regular ways of characterizing the difference between an expert and a novice, in terms of the kinds of schemata that they've developed to encode information about board positions, and possibilities inherent in board positions. In fact, the modeling work that I've done on a sort of connectionist model of reduced-board Go games exhibits some of this kind of behavior. That is, as you train the model, what it gets better at is encoding certain regularities in frequently occurring game situations, in order to exploit them to make limited predictions about likelihoods of next moves for a few leaves down the tree.

GELL-MANN: So, "chunking," roughly speaking.

PINES: There is a trivial example of the interplay of psychology and the economy, which seems not to be generally remarked upon. Namely, consumer confidence is one of the indices used in the Leading Economic Indicators. Yet there you have a built-in, positive-feedback situation, since now everyone listens on the tube to what is the set of Leading Economic Indicators. The consumer hears things are bad, therefore confidence goes down, etc., etc. One is never presented with the information with, and without, consumer confidence, and in particular the consumers are not presented with that information. And one could well imagine situations—which I think have happened over the last year or two—in which that has been the critical component in steering things up or down, and in which things have moved far more than might otherwise have been the case. Henry, you're looking responsive to this?

LICHSTEIN: Well, on two very different counts. Some of the most interesting discussions over the last couple of years on recession and economic growth are: how should you read the Consumer Price Index, and the discussion of what a change up or down, or what an "81" or a "50," or whatever, means. It's fascinating to watch the business press deal with it. I hadn't thought of it, but you're raising the question about a feedback loop. We have to think of a different way to read the same mechanisms because, in fact, we can't use the indicators that were important before in the same way. So we're seeing an adaptation to the flow of information; as the flow of information becomes internalized, we then start asking ourselves very consciously how to think about that piece of information.

MARTIN: Well, I don't know if there's time for discussion still...

COWAN: I would like to say that Ben Martin was called on very late in the game, and I think has responded magnificently. Thank you. [Applause.]

GELL-MANN: Either we should always call on him late and when he has a cold, or else he would be even better if he were well! Either way, it's terrific.

Alan Lapedes

Theoretical Division, MS B213, Los Alamos National Laboratory, Los Alamos, New Mexico, 87545, USA and External Faculty, Santa Fe Institute, 1660 Old Pecos Trail, Suite A, Santa Fe, New Mexico, USA

A Complex Systems Approach to Computational Molecular Biology

Abstract: We report on the continuing research program at the Santa Fe Institute that applies complex systems methodology to computational molecular biology. Two aspects are stressed here: (1) the use of coevolving adaptive neural networks for determining predictable protein structure classifications, and (2) the use of information theory to elucidate protein structure and function. A "snapshot" of the current state of research in these two topics is presented, representing the present state of two major research thrusts in the program of Genetic Data and Sequence Analysis at the Santa Fe Institute.

INTRODUCTION

We address two topics: (1) a novel algorithm for using coevolving, adaptive networks to define and predict new classes of protein secondary structure, and (2) the use of concepts from information theory to elucidate protein structure and function.

Complexity: Metaphors, Models, and Reality
Eds. G. Cowan, D. Pines, and D. Meltzer, SFI Studies in the
Sciences of Complexity, Proc. Vol. XIX, Addison-Wesley, 1994   **287**

The first topic describes the construction of a neural network algorithm that uses two coevolving neural networks to create new definitions of protein secondary structure that are highly predictable from primary sequences. Accurate prediction of the conventional secondary structure classes: alpha helix, beta chain, and coil, from primary sequence has long been an important, unsolved problem of computational molecular biology, with many ramifications, including multiple sequence alignment, prediction of functionally important regions of sequences, and prediction of tertiary structure from primary sequence. Our ability to use coevolving adaptive networks to evolve new and highly predictable definitions of secondary structure represent an example of the utility of new notions of complex systems theory, such as coevolution. This work was performed in collaboration with Robert Farber (External Faculty, Santa Fe Institute and staff member, Complex Systems Group, Los Alamos National Laboratory), and Evan Steeg (Department of Computer Science, University of Toronto, Toronto Candada).

The second topic concerns the use of information theory to detect correlations among positions in protein sequences. In earlier work[9] we used the mutual information between codon positions in exons to define new features that allowed a neural network to distinguish between coding and noncoding regions of DNA with high accuracy. Subsequently, we applied the concept of mutual information to a set of aligned sequences of the V3 loop of HIV-1. Statistically significant correlations, as evidenced by high-mutual information values were observed in positions that were widely separated along the sequence, and experimental evidence shows that a subset of these positions are functionally linked.[18] One hypothesis that would account for increased mutual information between functionally linked yet distant positions (as measured along the sequence), is that they are actually close in three-dimensional space. We present evidence from examples of protein secondary structure elements, such as alpha helixes and beta chains, that structural constraints of protein secondary structure can be reflected in correlations between sequence positions. Evidence from beta sheets also shows that tertiary effects of protein structure can be reflected in weak correlations between nonlocal sequence positions. These statistically significant correlations indicate that ghosts of tertiary structure information are manifested in sequence data as weak, nonlocal correlations. Work is continuing to remove phylogenetic artifacts due to shared ancestry that can cause spurious correlations among sequence positions that are not proximate in space. This is work in collaboration with Bette Korber (Santa Fe Institute and staff member, Theoretical Biology Group, LANL), Robert Farber (external Faculty, Santa Fe Institute and staff member, Complex Systems Group, LANL), and David Wolpert (postdoctoral fellow, Santa Fe Institute).

COEVOLVING NETWORKS FOR DEFINING AND PREDICTING PROTEIN SECONDARY STRUCTURE
PREDICTION OF CONVENTIONAL SECONDARY STRUCTURE CLASSES

Prediction of secondary structure classes of proteins from amino acid sequence has evolved from attempts to construct a useful tool that can, e.g., aid prediction of protein tertiary structure, to a "numbers game," where researchers employ increasingly sophisticated algorithms to achieve incremental improvements in accuracy. With due respect to those researchers who have tried (and we are in that category), the bottom line is that presently no one can predict protein secondary structure with sufficient accuracy to be of much use, and it is immaterial whether the Q3 coefficient is, e.g., 62% or, say, 68%.[27]

The "secondary structure" of proteins are those classifications of structure that can be defined using only a local stretch (a short "window") of structural information about the protein. Structural information is available in databases like the Brookhaven database which contains structures of many proteins determined from x-ray diffraction. There have been numerous attempts to predict these locally defined secondary structure classes using only a local window of sequence information. It has become conventional to use the Kabsch and Sander definitions/software[15] to define three classes of secondary structure: alpha helix, beta strand, and a default class called random coil. The prediction methodology ranges from a combination of statistical and rule-based methods[4] to neural net methods.[17,24,27]

A major reason that prediction of secondary structure is of interest is that a successful prediction of secondary structure from amino acid sequence may be used in tertiary structure prediction algorithms to constrain their search space.[26] For example, Skolnick[26] has found that biasing amino acids towards assuming the measured secondary structure, when coupled to his global tertiary structure prediction codes, greatly increase the agreement of the global tertiary structure prediction with the experimentally determined structure. However, his test of the value of knowing the secondary structure classes used the actual, experimentally determined,secondary structure classes, and not error prone predictions of secondary structure classes. His method, and others, are not successful if they attempt to use predictions of secondary structure classes at the current level of inaccuracy.

A widely accepted definition of protein secondary structure classes is that of Kabsch and Sander.[15] Their definitions are implemented in a software that is widely available. In Figure 1 I illustrate the Kabsch and Sander software defining secondary structure classes, depicted as a "black box" on the right, and also a neural network that attempts to learn the secondary structure classes from the from amino acid sequence on the left. The Kabsch and Sander "black box" first defines hydrogen bonding patterns from the structural information, and then uses the hydrogen bonding patterns to define classes of secondary structure. This picture represents the standard approach to training a neural network to classify secondary structure from amino acid sequence.[27] A local window of structure information obtained from, e.g., x-ray diffraction data in the Brookhaven database, is input to the right-hand Kabsch and Sander black box. The box outputs the secondary structure class of the fragment, using the Kabsch and Sander definitions. For example, if one were dichotomizing all the windows of structure information into "alpha helix," "not-alpha helix," then the right-hand box will emit a "1" if the fragment is alpha-helix, and emit a "0" otherwise. The left-hand neural network "sees" the corresponding window of sequence information as input, and attempts to adjust its synaptic weights so that the output neuron of the neural network agrees with the output state of the Kabsch and Sander black box. Hence, if the input sequence adopts an alpha-helix state according to Kabsch and Sander, then the output neuron of the network should change state to "1." Conversely, an input sequence fragment not in an alpha helix should cause the state of the output neuron to change "0."

We consider in this exposition just two classes of structure—the extension to multiclasses is trivial, but will not be made explicit for reasons of clarity. We won't discuss details concerning construction of a representative training set, or details of conventional neural network training algorithms, such as back-propagation. These are well-studied subjects that are addressed by e.g., Storloz, Yuan, and Lapedes,[27] in

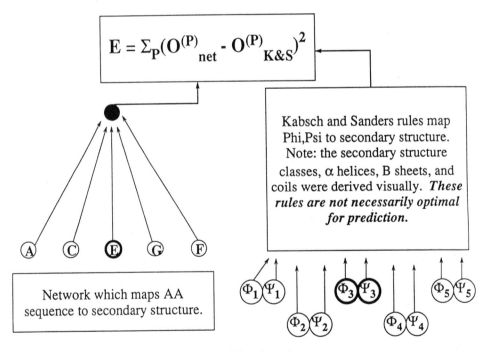

$$E = \Sigma_P(O^{(P)}_{net} - O^{(P)}_{K\&S})^2$$

Kabsch and Sanders rules map Phi,Psi to secondary structure. Note: the secondary structure classes, α helices, B sheets, and coils were derived visually. *These rules are not necessarily optimal for prediction.*

Network which maps AA sequence to secondary structure.

Ⓐ   Ⓒ   Ⓔ   Ⓖ   Ⓕ

$\Phi_1 \Psi_1$ $\Phi_3 \Psi_3$ $\Phi_5 \Psi_5$

$\Phi_2 \Psi_2$ $\Phi_4 \Psi_4$

FIGURE 1   Neural net learns Kabsch and Sander rules.

the context of protein secondary structure prediction. We note in passing that one can clearly employ more complicated network architectures, more output neurons (e.g., three neurons for predicting alpha helix, beta chain, random coil) etc. (c.f., Kneller, Cohen, and Langredge,[17] Qian and Sejnowski,[24] and Skolnick, Yuan, and Lapedes[27]).

DEFINITION AND PREDICTION OF NEW SECONDARY STRUCTURE CLASSES

The key ideas of this section are contained in Figure 2. In this figure the right-hand black box implementing the Kabsch and Sander rules is replaced by a second neural network. This right-hand neural network therefore sees a window of structural information, while the left-hand neural network sees the corresponding window of sequence information. Note that the right-hand neural network can implement extremely general definitions of secondary structure. For example, if the weights in the right-hand network are set to *arbitrary values*, then the right-hand network will correspondingly produce an arbitrary classification of the structures that are input to it. On the other hand, one could train the weights of the right-hand network to perform structure classification according to, say, the Kabsch and Sander rules. To demonstrate the generality of the procedure we have done the latter, and have successfully captured the Kabsch and Sander structural definitions in the right-hand network with high accuracy. The representation of the structure data in the right-hand network uses phi-psi angles. Problems due to the angular periodicity of the phi-psi angles (i.e., 360 degrees and 0 degrees are different numbers representing the same angle) are eliminated by utilizing the sine and cosine of each angle.

POINT (1). One can replace the right-hand black box of Figure 1 with a neural network (see Figure 2). A neural network on the right-hand side is an equally valid implementation of a set of rules defining secondary structure as a piece of software. We have explicitly demonstrated this by training a neural network to reproduce the Kabsch and Sander rules with high accuracy.

POINT (2). The right-hand network need not be restricted to implementing the Kabsch and Sander rules for secondary structure. The right-hand neural network is capable of representing a very general set of rules, of which the Kabsch and Sander rules are but one choice.

To define new rules one merely changes the synaptic weights. Arbitrary synaptic weights would define arbitrary rules, and there would be little chance that these new classes would be either predictable or meaningful.

POINT (3). A requirement on the rules is needed. The necessary requirement is that the "secondary structure" classes defined by the right-hand net be predictable from the corresponding amino acid sequence of the left-hand network.

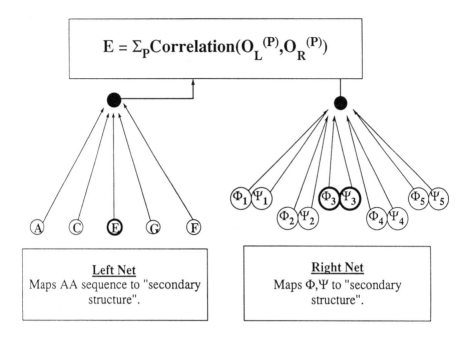

FIGURE 2   Kabsch and Sander rules may be represented by a neural net, instead of software.

In other words, the only requirement is that the synaptic weights be chosen so that the output of the left-hand network and the output of the right-hand network agree for each sequence-structure pair that is input to the two networks.

To achieve this, both networks are trained *similtaneously*, starting from random initial weights in each net, under the sole constraint that the outputs of the two networks agree for each pattern in the training set. The mathematical implementation of this constraint is described in various versions below. This coevolution of the two networks is clearly a more difficult computational problem than the conventional approach (Figure 1) that employs fixed targets. Each net now chases a moving target during training, and additional numerical difficulties occur. Our preliminary results (below) show that these difficulties are surmountable (we achieved interactive runtimes using the CM2 Connection Machine). Therefore this procedure is a very general, effective method of evolving predictable secondary structure classifications of experimental data.

COEVOLVING ADAPTIVE NETWORKS

The requirement that the two networks *coevolve* states that they evolve from random initial conditions into a cooperative phase, in which each network is able to predict the output of the other network. Neither network has a fixed target to which it may be trained—the only requirement is that the outputs of both networks agree for each pattern. A naive method to require that the two networks evolve weights allowing cooperation is suggested by analogy to conventional back-propagation.[25] We present the naive method first, and then refine the method to an effective procedure in the following section. In back-propagation one performs gradient descent in the synaptic weights of the error function, E:

$$E = \sum_p (LeftO^{(p)} - t^{(p)})^2 \tag{1}$$

where $t^{(p)}$ is the target output value for the pth pattern, and $LeftO^{(p)}$ is the output of the left network for the pth pattern. $LeftO^{(p)}$ is a function of the synaptic weights. Gradient descent in the synaptic weights will decrease the error, E, evaluated on the training set by forcing the output of the network, $LeftO^{(p)}$ to agree with the target output, $t^{(p)}$, for each pattern.

Note that in Figure 1 the target value for the left-hand network is given by the fixed rules implemented in the right-hand Kabsch and Sander black box. These targets of the conventional approach, $t^{(p)}$, are therefore fixed constants, i.e., either "0" or "1" for each pattern. In Figure 2 one might consider using the same error function, Eq. (1), but replacing the previously fixed target values for each pattern by the variable output of the right-hand network. Hence the new error function, whose minimization will enforce agreement of the left-hand and right-hand networks is

$$E = \sum_p (LeftO^{(p)} - RightO^{(p)})^2 . \tag{2}$$

The difficulty with this naive idea is that there is a trivial way for the two nets to agree. They merely need to decrease their synaptic weights to the inputs to zero, so each will stay in the "0" state, regardless of the input pattern (it is also possible to have each stay in the "1" state). The output of the two nets would agree, as demanded by minimizing Eq. (2), but the result is trivial. The outputs remain either "on" or "off" regardless of the input data, and are completely uninformative. One might consider adding a variance term to Eq. (2) to require the networks to respond to their inputs, i.e., to impose variation in the network outputs as the input patterns change. However, a cleaner approach is to demand that the outputs of the networks co-vary by modifying Eq. (2) to maximize the mutual information or correlation between the network outputs.

COEVOLUTION: TRAINING WITH CORRELATION MEASURES. The standard correlation measure, C, between two objects, $LeftO^{(p)}$ and $RightO^{(p)}$ is

$$C = \sum_p (LeftO^{(p)} - \overline{LeftO})(RightO^{(p)} - \overline{RightO}) \tag{3}$$

where $\overline{LeftO}$ denotes the mean of the left net's outputs over the training set, and respectively for the right net. C is zero if there is no variation, and is maximized if there is similtaneously both individual variation, and joint agreement. In our situation it is equally agreeable to have the networks maximally anticorrelated, as it is for them to be correlated. (Whether the networks choose correlation, or anticorrelation, is evident from the behavior on the training set.) Hence the minimization of the following expression will ensure that that outputs are maximally correlated (or anticorrelated)

$$E = -C^2 . \tag{4}$$

Minimizing this expression forces the correlation of the two outputs, considered as a set of real values, to tend to either perfect correlation or perfect anticorrelation. Note that Eq. (4), as opposed to Eq. (2), does not allow the situation of unchanging outputs to be a local minimum. This would give a value of 0.0 to E. However, E in Eq. (4) will be negative under even the slightest correlation given the random initial weights, and the dynamics of gradient descent will continue to decrease E. Thus the system is forced away from unchanging outputs, solving the problem associated with the naive approach of Eq. (2) above.

PREDICTING WITH CORRELATION MEASURES. The procedure for predicting the structure of a new sequence pattern is different when the correlation measure is used for training. Because one explicitly trained the network using Eq. (4), then the output of the networks are only significant in relation to their mean value over the training set. It is not just the state of the left-hand net that determines the prediction of the secondary structure of the pattern on the right, but rather it is whether the state of the left-hand net is above or below its mean value in the training set. It is therefore necessary to subtract from the output of the right-hand network its mean value as calculated over the training set. This simple change to the usual method of prediction is an easy-to-implement offset to the value of the output.

OTHER TRAINING MEASURES, ALGORITHMS, AND ARCHITECTURES. Other training measures forcing agreement of the left and right networks may be used. Particularly suitable for the situation of many outputs (i.e., more than two class discrimination) is "mutual information." This version of the idea is closely related to the IMAX algorithm of Becker and Hinton.[2] The mutual information is defined as

$$M = \sum_{i,j} p_{ij} \log \frac{p_{ij}}{p_{i.}p_{.j}} \tag{5}$$

where p_{ij} is the joint probability of occurence of states i and j of the left-hand and right-hand networks, and the $p_{i.}$ and $p_{.j}$ are the marginal probabilities. In previous work[27] we showed in general how p_{ij} and the marginals may be defined in terms of neural networks. Minimizing $E = -M$ maximizes M. Preliminary simulations show that M is more prone to local maxima than C. Initializing the network weights using C, and then switching to maximizing M, is a procedure well worth testing in view of the useful properties of M (we won't discuss details of information theory in relation to sequence analysis here, see, e.g., Farber and Lapedes;[9] Korber et al.;[18] Lapedes et al.;[20] and Storloz, Yuan, and Lapedes,[27]). Finally, we point out that since a common quantity measuring predictive performance is the Mathews correlation coefficient (see, e.g., Storloz, Yuan, and Lapedes[27]), then it is reasonable to train the two networks to maximize this measure. The maximum achievable Mathews correlation coefficient is 1.0. The Mathews coefficient is designed to be a single number that incorporates measure of both over-prediction, and under-prediction. Intensive investigation of the effect of network architecture on the derived structural classes, using all the error measures, is in progress.

As noted earlier, this problem is extremely computer intensive, requiring use of the CM2 Connection Machine (on which we've achieved 3 gigaflop throughput in our preliminary investigations). An alternative classification algorithm may not only run faster, but may also uncover different structural classes. We have been working with Melanie Mitchell, of the Santa Fe Institute and University of Michigan, to develop the ideas presented here using "genetic algorithms,"[13,21] which are a alternative machine learning algorithm to neural networks. Genetic algorithms are powerful adaptive algorithms that may have some advantages for this problem. Investigations are in progress.

RESULTS

Best results so far have been obtained with the Mathews objective function. Random initial conditions are necessary for the development of interesting new classes—if one uses initial conditions appropriate for predicting the standard Kabsch and Sander classes then the local minima is so deep that nothing much else happens. Naturally, one can "gang" together objective functions as soon as one gets the network out of the initial local minima. Thus, one can start training with the correlation objective function, and then finish with the more precise mutual information function.

Our best results so far involve two class discrimination using the Mathews correlation function. If one assigns the name "Xclass" (for want of a more descriptive word at this stage of the investigation) to the newly defined structural class, then the network already classifies local windows of structure into a "Xclass/NotXclass" dichotomy *with higher predictability than prediction of conventional secondary structure classes.*

The Mathews coefficient on a disjoint prediction set of the new classes is 0.425. The Q3 of the new classes, which we emphasize is not a particularly informative quantity, (but is often quoted) is 73%. (Note that Q3 can be essentially 100% for a ridiculously simple algorithm that classifies most examples as "Xclass," and then uses a default for prediction. This is not happening here.) For comparison, the Mathews coefficient for dichotomization into the standard secondary structure classes, Alpha/NotAlpha, Beta/NotBeta, and Coil/NotCoil, for the same data is 0.33, 0.26, and 0.39, respectively. Given the minimal amount of optimization we have performed so far, the 0.53 Mathews correlation coefficient of the new class dichotomization is most encouraging.

Are the new classes simply related to the more conventional classes of alpha helix, beta, and coil? Although more precise analysis needs to await visual examination of examples of the newly defined classes one can conclude immediately that the relationship is not necessarily simple. We classified (using the Kabsch and Sander definition) the conventional secondary structure classes for the newly defined classes. Thus, all patterns labeled "Xclass" by the new code were classified into alpha, beta and coil according to Kabsch and Sander definitions. The new classes turn out to be a mixture of the conventional classes, and are not dominated by either alpha, beta or coil; although there is some relationship between Xclasses and helices. It will be most interesting to see if structural features of the new classes, which we emphasize are more predictable from amino acid sequence than the Kabsch and Sander defined classes, exhibit striking visual features.

```
CCCCCCCCCCCCCCCCCHHHHHHHHHHHHHHHHHHHCCCCCCCHHHHHHHCCCCCCCCCCCCCCCHHHHHHHHHHHCCCCCCC
0001010000000000011100110100000000000000000011000000000000000000000000000000000000
0000000000000000011111111111111111100000000001111101000000000011101111111110000000
```

```
CCCCCCCCCCCCHHHHHCCCCCCCBBBBCCCCCHHHHHHHHHHHCCBBBBBCCCCHHHHHCCCCBBCCCCCCCCCCBBBBBB
0000000000001111110100000000000000011110000000001111111111000000000000000000000000
0000000000000000010110000000000000001111111000000000011111000000000000000000000000
```

```
BBBBBCCBBBBBBBBCCCCCCCCCCCCBBBBBCCCCCCCCCCCCCCCC
00000000000000000000000000011110000000000000000
00000000000000000000000000000000000000000000000
```

FIGURE 3 A segment of 2ACT, sulfhydryl proteinase. First line is labeled H = helix, B = beta chain, C = coil. Second line is target Xclass. Thrid line is predicted Xclass.

In Figure 3 I compare the assignment of structural features into Xclass/ NotXclass categories, with the conventional assignment of structural features into alpha helix, beta chain, and coil, for the protein "2act—Actinidin."

Comparison of the second and third lines of Figure 3 illustrates the accuracy of XClass prediction for this protein.

Comparison of the third and first lines illustrates the relation between the conventional secondary structure classes and the new Xclass categories.

Note that the Xclass category of secondary structure bears some relationship to Helix, but that significant differences also exist. Clarification of the new classes awaits detailed analysis, including visual inspection using molecular modeling of examples of the new Xclass categories.

A primary goal of this investigation is to evolve very predictable secondary structure classes that can then be used to constrain tertiary structure prediction. The above results, although preliminary, are most encouraging. Our goal is to significantly improve accuracy still further.

INFORMATION THEORY ANALYSIS OF PROTEIN STRUCTURE AND FUNCTION
BEYOND CONSENSUS SEQUENCES

A common, and intuitive, approach to characterizing important regions of sequence data, e.g., those regions containing regulatory signals, is to attempt to find a motif of mostly contiguous, conserved sites across many aligned sequence examples that contain the region of interest. The motif disappears, and the usual assumption is that information about the region also vanishes, for sequence positions that vary so much across examples that the dominant symbol in these positions become uninformative. However, the mere existence of variation in a position doesn't necessarily mean that information about the position is no longer characterizable. It is quite possible that variations in different positions are linked, and that although a single position might appear to be varying randomly, it is in fact varying in a correlated fashion with changes in another position. Correlations between real, i.e., floating point variables are easy to measure by the usual linear correlation analysis. Correlations between discrete variables can be analyzed by using the concept of mutual information from information theory.

In previous work we've used mutual information to quantify the degree of correlation between positions in sequence data. We found that there exists nontrivial and statistically significant mutual information between the neighboring codons of exons in DNA, which allowed us to develop neural net algorithms of great sensitivity that distinguished between exons and introns in unannotated DNA sequences.[9,20] In other work we discovered correlations in transcriptional promoters of E. coli that also aided computational identification of these regions.[1] In more recent work we've analyzed a set of aligned sequences of the V3 loop of HIV-1 and discovered nontrivial, statistically significant mutual information between nonlocal sequence positions.[18]

MUTUAL INFORMATION

A formal measure of variability[19] at position i is the Shannon entropy, $H(i)$. $H(i)$ is defined in terms of the probabilities, $P(s_i)$, of the different symbols, s, that can appear at sequence position i (e.g., $s = A, S, L \ldots$ for the twenty amino acids: Ala, Ser, Leu ...). $H(i)$ is defined as:

$$H(i) = - \sum_{s=A,S,L\ldots} P(s_i) \log P(s_i). \tag{6}$$

Mutual information is defined in terms of entropies involving the joint probability distribution, $P(s_i, s_j')$, of occurrence of symbol s at position i, and s' at position j. The probability, $P(s_i)$, of a symbol appearing at position i regardless of what symbol appears at position j, is defined by $P(s_i) = \sum_{s_j'} P(s_i, s_j')$ and similarly, $P(s_j') = \sum_{s_i} P(s_i, s_j')$. Given the above probability distributions, one can form the associated entropies:

$$H(i) = - \sum_{s_i} P(s_i) \log P(s_i)$$

$$H(j) = - \sum_{s_j} P(s_j') \log P(s_j')$$

$$H(i,j) = - \sum_{s_i, s_j'} P(s_i, s_j') \log P(s_i, s_j').$$

The mutual information $M(i, j)$ is defined as:

$$M(i,j) = H(i) + H(j) - H(i,j).$$

Mutual information is always non-negative and achieves its maximum value if there is complete covariation. The minimum value of 0 is obtained either when i and j vary completely independently, or when there is no variation.[3,19]

The above formulae assume true probability distributions are known. In practice, however, they are not known and must be estimated from a finite data set. Two effects require consideration. First, since mutual information is always non-negative, the mutual information between any single pair of truly independent positions is consistently overestimated, while the mutual information of a covarying position can be either overestimated or underestimated, depending on the nature of the fluctuations in the finite data set. One must therefore assess statistical significance of single pairs in the light of small sample bias. Secondly, one must consider problems caused by selection effects. Typically, one selects a pair of positions that exhibits a large mutual information value, compared to other pairs in the sequence, as "interesting." One must therefore assess the probability that out of many such estimated mutual information values (one for each pair of positions in the sequence) a high-estimated value might be achieved by chance.

THE V3 LOOP OF HIV-1

The V3 loop of the HIV-1 envelope protein (env) has been the focus of intense research efforts because it is a potent epitope for neutralizing antibodies (NAbs)[10,14,23] and T cells,[12,28] and plays a role in determining cell tropism and viral growth characteristics. While there is some propensity to conserve amino acid side chain chemistry in the different positions in the loop this conservation often breaks down upon inclusion of phylogenetically distant viruses (12). Such variation presents a difficult challenge for those attempting to design broadly reactive V3 loop based vaccines.[14,28] Our goal was to quantify the degree of covariation of mutations at different sites by analyzing the available database[22] of V3 amino acid sequences using mutual information, a concept from information theory (15–18). All pairs of positions in an alignment of 308 distinct V3 loop sequences were compared.

An algorithm which employed multiple randomizations of the initial data set was used to determine the statistical significance of the estimated mutual information values using a very conservative measure that addresses both small sample bias and selection effects (for general reviews of methods of this type, see Efron[6,7,8]). Highly statistically significant mutual information scores were obtained for several pairs of sites, some on opposite sides of the V3 loop.

High-mutual information between certain sites suggests that functional studies of the V3 loop using site directed mutagenesis may depend upon simultaneously altering amino acids on both sides of the loop. Indeed, this has been shown to be the case for some of the positions linked through mutual information analysis—de Jong et al.[5] showed that simultaneous mutations were required at sites 10 in conjunction with sites 21 through 24, located across the loop, to get a complete conversion in viral phenotype from nonsyncytium inducing, low replicating, to syncytium inducing, high replicating. Our analysis indicated sites 10, 23, and 24 were covariant. Virus viability as well such phenotypic "switches" may require simultaneous mutations in covarying sites. When sites related by high mutual specific information were compared with alignments of V3 regions of viruses with distinct tropism and cytopathicity, several of the positions that appear to be significant in terms of phenotype were also seen to covary. This correlation supports the hypothesis that mutual information can identify functionally interactive sites. While several of the sites we predicted to be mutually interactive were substantiated by experimental evidence, additional linkages were observed that also may be relevant for the generation of viable V3 loops with specific phenotypes. These positions may have been missed in experiments to date, due to their relative conservation among cloned samples used for experiments in culture.

CORRELATIONS AND PROTEIN STRUCTURE

One possible explanation for why positions with covarying mutations seem to be associated with functional sites, is that these covarying positions, although possibly distant along the sequence, are proximate in space. The closeness in physical space, due to an underlying conserved structure associated with the functional region, might constrain the sequence mutations sufficiently to result in covariation. There is a precedent for associating covariation with structure in analysis of families of variable RNA sequences. Recent work of Stormo et al.[11] show that covarying positions in RNA families are associated with both secondary and tertiary structural features. A structural basis is but one possibility for why covarying sites may be associated with functional sites. Other possibilities include interaction with the protein's environment such as specific requirements due to protein/protein interactions, or the necessity to specify certain types of amino acids at particular sequence positions to help define the folding pathway. We emphasize that the tertiary structure of the V3 loop is not known in detail, and that we are not suggesting that the observed correlations among positions in the V3 loop are necessarily a reflection of structural constraints. Never-the-less, it is tempting to investigate in a setting where structure is known, if structural and physico-chemical constraints give rise to correlations among sequence positions.

CORRELATIONS INDUCED BY PROTEIN SECONDARY STRUCTURE

Protein secondary structure elements, such as alpha helices and beta chains, define regular elements of protein structure.[15] It is possible to extract from the Brookhaven data base numerous, examples of secondary structure elements with different sequences. We took a database of alpha helices (previously used to train a neural net to distinguish helices, beta chains, and coils[27]) and computed the mutual information between all pairs of positions in a window 13-residues long, which was centered on each successive residue that participated in an alpha helix. In Figure 4 I represent the mutual information between positions in these windows of alpha helix. Two features are clear: there is correlation between positions at a spacing of two residues, and also between those residues at a spacing of four residues. The latter correlation is gratifying. Alpha helices, by definition[15] involve residues that have hydrogen bonding along the backbone across intervals of four residues. The structural and hydrogen bonding constraints implicit in the definition of alpha helices therefore seem reflected in the correlations between residues spaced four apart. The correlations between residues at a spacing of two is less clear. One possible

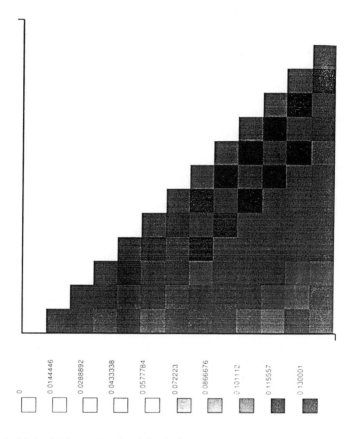

FIGURE 4   Mutual information in alpha helices.

explanation is that the database contains a significant component of amphipathic helices. Such helices have one side facing solvent, and the other facing the protein interior. Therefore one expects alternation of hydropathic and hydrophyllic residues between one side of the helix and the other. Since the pitch of an alpha helix is approximately 3.7 residues per turn, the spacing of two residues between correlated residues is in approximate agreement. While this research was being performed, we received a preprint[16] that included related research on correlations between residues in secondary structure elements.

A similar calculation can be performed for beta strands. The mutual information is presented in Figure 5. Clearly, there is higher mutual information between residues at a spacing of two residues. Here again, we have taken windows of sequence, 13 residues long, centered on successive residue participating in a beta chain. The higher mutual information between residues at a spacing of two residues is in accord with structural constraints on beta chains. Beta chains have an alternating hydrogen bonding pattern along the backbone which is presumably being reflected in the correlations seen between residues at a spacing of two.

The absolute magnitude of the mutual information presented in these figures is not particularly informative. Finite sample effects cause a shift in the absolute magnitude of the mutual information depending on sample size. Also, for visual clarity, we have nonlinearly scaled the values to improve contrast for reproduction. However, relative comparisons of mutual information between positions in an alpha helix or a beta chain, respectively, are informative. Randomization experiments to test statistical significance (in analogy to the V3 loop analysis, above) were performed, and verified the significance of the enhanced mutual information at various spacings between residues.

CORRELATIONS INDUCED BY PROTEIN TERTIARY STRUCTURE

Beta sheets, composed of beta chains that are hydrogen bonded together (and hence proximate in space), provide an example of tertiary structural information being reflected in mutual information between nonlocal sequence positions. Beta sheets are intrinsically nonlocal objects as far as sequence considerations are concerned, but are extremely localized in space. For example, a sequence can adopt a beta chain configuration early in the sequence, subsequently change to an extended region of coil, then later in the sequence adopt another beta chain configuration, which happens to be parallel and close to (in three-dimensional space) the first beta chain. Two, or in general more than two, spatially close regions of beta chain, can support interchain hydrogen bonding which contributes to the tertiary stability of the protein. There is also the usual intrachain hydrogen bonding with a spacing of two residues (see above) within each individual beta chain. An analogy to the spatial positioning of beta chains, and the interchain bonding necessary to make a beta sheet, is the sticking together of distant regions of a piece of multiply folded cello-tape. One might hope to see correlations induced among those residues in each beta chain that may be well separated along the sequence, but which are brought into close spatial proximity by the interchain hydrogen bonding defining the combination of chains into sheets.

We constructed a database of paired, antiparallel, beta chains participating in a beta sheet, by extracting a centered window of residues, 13 residues long, around each position in one chain, and also for the respective partner in the other chain of the beta sheet. For example, if residues numbered 5,6,7,8,9 were participating in a beta chain, and residues 21,22,23,24,25 were the corresponding partner chain, then we construct a database consisting of a window around each residue from one chain concatenated with the corresponding window from the other chain. Thus, windows in each chain that are, e.g., three-residues long, result in concatenated windows that are six residues long. These six resdiue long windows would be the concatenation of the centered windows (a) and (b) below:

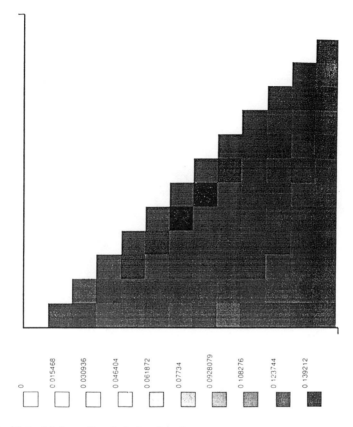

FIGURE 5   Mutual information in beta strands.

(a) 4,5,6 (b) 20,21,22
(a) 5,6,7 (b) 21,22,23
(a) 6,7,8 (b) 22,23,24
(a) 7,8,9 (b) 23,24,25
(a) 8,9,10 (b) 24,25,26

In Figure 5 I show the mutual information between the positions in the paired windows of antiparallel beta chains of length 13 (concatenated window length it therefore 26). There is the usual increased mutual information between residues at a spacing of two within in each chain. In addition, there is increased mutual information between the residues *in separate chains* that are in close spatially, but distant along the sequence. This is evidenced by the increased mutual information along the antidiagonal in the lower right-hand corner of the plot. This provides

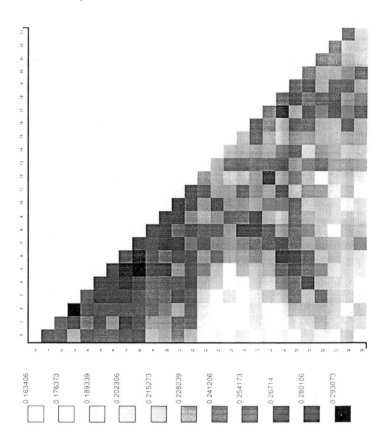

FIGURE 6   Mutual information in poaired strands of a beta sheet.

evidence that spatial proximity limits the amino acids that may occur in nearby positions, resulting in statistically significant mutual information between those positions.

The next step will be to extend this analysis to families of proteins with variable sequence, and known crystal structure, such as globins, immunoglobulins, MHC molecules, kinases, serine proteases, etc. Success in this endeavour, could result in a formalism that could be powerful in defining components of functional or structural domains which are distant in terms of linear sequence, yet may be working coordinately in intact proteins. It could then be applied to proteins which have defied attempts at crystallization, and serve as a guide for molecular biologist who are mapping functional domains through exchange of restrictions site fragments and deletion and mutational analysis. Our preliminary tests on globin sequences identified several important functional sites, but the identification of structural correlates was complicated by the nonmonomeric structure of hemoglobin. It will be necessary to construct linked sequences expressing the full tetrameric structure of hemoglobin in order to test the structure hypothesis. Indeed, most of the variable protein families are nonmonomeric, and construction of the linked sequence segments is a nontrivial, but resolvable, issue.

CONCLUSION

Techniques from adaptive network theory, information theory, and concepts such as coevolution, form the core of research in complex adaptive systems. In this article we have shown how this core of new ideas can be used to give fresh insight into some outstanding problems in computational molecular biology. It seems only fitting that theories and techniques developed to analyze and understand a wide variety of complex systems, have as a prime example of their value, the successful analysis of life itself.

REFERENCES

1. Abremski, K., K. Sirotkin, and A. Lapedes. *Math Modeling and Scientific Computing* **2** (1993): 636–641.
2. Becker, S., and G. Hinton. *Nature* **355** (1992): 161–163.
3. Blahut, R. E. *Information Theory and Statistics*. Reading, MA: Addison-Wesley, 1987.
4. Chou, P., and G. Fasman. *Adv. Enzymol.* **47** (1978): 45.
5. de Jong, J.-J., J. Goudsmit, W. Kuelen, B. Klaver, W. Krone, M. Tersmette, and A. de Ronde. *J. Virol.* **66** (1992): 757–765.
6. Efron, B. *SIAM Rev.* **21** (1979): 460–480.
7. Efron, B. *J. Am. Stat. Assoc.* **78** (1983): 316–331.
8. Efron, B. *Science* **253** (1991): 390–395.
9. Farber, R., and A. Lapedes. *J. Mol. Biol.* **226** (1992): 471.
10. Goudsmit, J., C. Debouck, R. H. Meloen, L. Smit, M. Bakker, D. M. Asher, A. V. Wolff, C. J. Gibbs, and D. C. Gajdusck. *Proc. Nat. Acad. Sci. U.S.A.* **85** (1988): 4478–4482.
11. Gutell, R., A. Power, G. Hertz, E. Putz, and G. Stormo. "Identifying Constraints on the Higher Order Structure of RNA: Continued Development and Application of Comparative Sequence Analysis Methods." *N.A.R.* **20** (1992): 5785.
12. Hart, M. K., T. J. Palker, T. J. Matthews, J. A. Langlois, N. W. Lerche, M. E. Martin, R. M. Scearce, C. McDanal, D. P. Bolognesi, and B. F. Haynes. *J. Immunol.* **145** (1990): 2677–2685.
13. Holland, J., K. Holyoak, R. Nisbett, and P. Thagard. *Induction*. Cambridge, MA: MIT Press, 1986.

14. Javaherian, K., A. J. Lanlois, C. J. LaRosa, A. T. Profy, D. P. Bolognesi, W. C. Herlihy, S. D. Putney, and T. J. Matthews. *Science* **250** (1991): 1590–1592.

15. Kabsch, W., and C. Sander. *Biopolymers* **22** (1983): 2577.

16. Kingler, T., and D. Brutlag. "Probabilistic Representation of Protein Structure." Preprint, Stanford University, 1992.

17. Kneller, D., F. Cohen, and R. Langridge. *J. Mol. Biol.* **214** (1990): 171.

18. Korber, B., R. Farber, D. Wolpert, and A. Lapedes. *P.N.A.S.* **90** (1993): 71–76.

19. Kullback. *Information Theory and Statistics.* New York: Wiley & Sons, 1959.

20. Lapedes, A., C. Barnes, C. Burks, R. Farber, and K. Sirotkin. In *Computers and DNA*, edited by G. Bell and T. Marr. Santa Fe Institute Studies in the Sciences of Complexity, Proc. Vol. VII, 157–182. Reading, MA: Addison-Wesley, 1989.

21. Michalewicz, Z. *Genetic Algorithms.* Berlin: Springer-Verlag, 1992.

22. Myers, G., B. T. M. Korber, J. A. Berzofsky, R. F. Smith, and G. F. Pavlakis, eds. *Human Retroviruses and AIDS 1991.* (Theoretical Biology and Biophysics Group, Los Alamos National Laboratory, NM), Section III, 1991.

23. Palker, T. J., M. E. Clark, J. Langlois, T. J. Matthews, K. J. Weinhold, R. R. Randall, D. P. Bolignesi, and B. F. Haynes. *Proc. Nat. Acad. Sci. U.S.A.* **85** (1988): 1932–1936.

24. Qian, N., and T. Sejnowski. *J. Mol. Biol.* **202** (1988): 865.

25. Rumelhart, D. et. al. *Parallel Distributed Processing.* Cambridge, MA: MIT Press, 1986.

26. Skolnick, J., and A. Kolinski. *Science* (1990): 250.

27. Storloz, P., X. Yuan, and A. Lapedes. *J. Mol. Biol.* **225** (1992): 363.

28. Takahashi, H., Y. Nakagawa, C. D. Pendleton, R. A. Houghton, K. Yokomuro, R. N. Germain, and J. A. Berzofsky. *Science* **255** (1992): 333–336.

DISCUSSION

MARTIN: I was interested in your comparison with statistical methods in networks. Something that a bunch of people in Dave Rumelhart's lab have been working on is the question of whether more complicated nets can be understood in terms of, not simple statistics, but slightly more complicated statistics; in particular, whether certain restrictions on back-propagation networks result in a kind of statistical estimation where—by matching the activation function you're using with the proper learning rule—you can compute conditional probabilities in the output units. I wonder if you've thought about using any of these methods to try to improve the performance of your back-prop nets.

LAPEDES: Actually, Dave came to visit Los Alamos—I guess this was probably a year or so ago when they first started doing this—and so he was telling me about some of these things they were doing, which I thought was quite nice. And we pretty much confined ourselves to the sort of analysis that I've presented here, because Dave was doing the rest of it.

We haven't tried to imply any of the ideas they came up with to restrict our networks; that would be an interesting thing to do.

MARTIN: A second point. Have you thought about doing a different kind of prediction task: Instead of predicting a class given the string, try to predict the next element in the string given the preceding element in the string.

LAPEDES: Bergman (who's a frequent visitor to SFI) and I have had some discussions about that. In fact, you can use that idea to get a better approximation to the probability distribution of sequence-given-class, because you've now extended yourself beyond assuming independence. I hope all of that will gel together sometime.

LLOYD: My question is closely related. It seems to me a bit unfair to compare these statistical techniques to the neural net techniques, when, having examined your data, you know that your assumption of statistical independence is incorrect. I was wondering what happened when one relaxed this assumption and attempted to apply these Bayesian techniques.

LAPEDES: Let me just address the first point. It does seem unfair. On the other hand, prior to this work, previous to this work, the biologists were in fact making the assumption, going ahead, and doing the analysis. And, apparently— I least I haven't seen it in the literature—they didn't go and look to see if the assumption was correct.

GELL-MANN: And anyway, Bayesian—as you pointed out—doesn't necessarily mean Bayesian with the assumption of independence (in other words, coarse graining that eliminates everything but independence). You can have a less coarse graining, as you showed, and still use Bayesian methods. And Bayesian methods are ultimately perfectly general.

LAPEDES: Absolutely. Now, suppose one measured second-order correlations, and wanted to use a Bayesian method which capitalized on that. The only way that I know to do that exactly is to calculate the maximum entropy distribution, subject to the constraint that it has second-order correlations. That's not easy for these long sequences. So I don't know the answer to your question without making some sorts of approximations about the higher-order probability distributions.

BROWN: Correlations structure in the exons but not the introns seems to imply that it's not just the structure of constraint in the way that DNA is organized in a string, but it has something specific to do with the way that it's translated, right? And that seems to suggest some functional hypotheses. I wondered, are there such things? Do you have any ideas about why that might be the case?

LAPEDES: No; I've wondered about that, and I've always been interested to get the reaction of real biologists. So far, nobody's expressed an opinion as to the functional reason why that should be so, why you should have these adjacent correlations.

COWAN: But it is possible that the introns have a well-defined function, is that right?

LAPEDES: This seems to be a matter of some debate. All I was saying is that there's no conventionally accepted function for introns, and I hesitate to call them junk—like some people—because perhaps there's something we don't know about them.

SCHUSTER: I wouldn't say that introns have no function—nobody knows that—but there is a mechanism by which they are actually excised from the transcript of DNA. There are several classes. And my question would be related...some of these classes of introns. My question is related to these classes. They do have conserved regions, sometimes only a few bases, sometimes more. Don't you see any correlations in the introns? That some bases correlate very well?

LAPEDES: When you say classes, you mean... ?

SCHUSTER: Class One introns, or Class Two introns, etc.

LAPEDES: No, we haven't investigated that. We have started to investigate splice junctions (in which on one side you have exon, and on one side you have intron). We've actually come back to that investigation, and we did it about a year, but we weren't as careful as we are now about statistical significance. So what we're doing now is redoing that with all the new care we're taking with statistical significance. I would hope that there would be some sort of interesting correlations; I don't know yet.

RASMUSSEN: I think that Theodore Puck has come up with some really interesting ideas about how the introns might have a very important function, in the sense that...

LAPEDES: Yeah, I guess there are a number of ideas; I'm not sure that any one is the conventional explanation.

RASMUSSEN: At least not generally accepted, but it's exciting.

John H. Holland

Division of Computer Science and Engineering, University of Michigan, Ann Arbor, MI 48109

Echoing Emergence: Objectives, Rough Definitions, and Speculations for ECHO-Class Models

This abstract first appeared in the Winter 1992 issues of *DAEDALUS*.

Abstract: One of the most important roles a computer can play is to act as a simulator of physical processes. When a computer mimics the behavior of a system, such as the flow of air over an airplane wing, it provides us with a unique way of studying the factors that control that behavior. The key, or course, is for the computer to offer an accurate rendition of the system being studied. In the past 50 years, computers have scored some major successes in this regard. Designers of airplanes, bridges, and America's Cup yachts all use computers routinely to analyze their designs before they commit them to metal. For such artifacts, we know how to mimic the behavior quite exactly, using equations discovered over a century ago.

However, there are systems of crucial interest to humankind that have so far defined accurate simulation by computer. Economies, ecologies, immune systems, developing embroyos, and the brain exhibit complexities that block broadly based attempts at comprehension. For example, the equation-based methods that work well for airplanes have a much more limited scope for economies. A finance minister cannot expect the same accuracy in asking the computer to play out the impact of a policy change as

Complexity: Metaphors, Models, and Reality
Eds. G. Cowan, D. Pines, and D. Meltzer, SFI Studies in the
Sciences of Complexity, Proc. Vol. XIX, Addison-Wesley, 1994

limited scope for economies. A finance minister cannot expect the same accuracy in asking the computer to play out the impact of a policy change as an engineer can expect in asking the computer to play out the implications of tilting an airplane wing.

Despite the disparities and the difficulties, we are entering a new era in our ability to understand and foster such systems. The grounds for optimism come from two recent advances. First, scientists have begun to extract a common kernel from these systems: each of the systems involves a similar "evolving structure." That is, these systems change and reorganize their component parts to adapt themselves to the problems posed by their surroundings. This is the main reason the systems are difficult to understand and control—they constitute a "moving target." We are learning, however, that the mechanisms that mediate these systems are much more alike than surface observations would suggest. These mechanisms and the deeper similarities are important enough that the systems are now grouped under a common name, *complex adaptive systems*.

1. THE NATURE OF COMPLEX ADAPTIVE SYSTEMS

Many of our most troubling long-range problems—trade balances, sustainability, AIDS, genetic defects, mental health, computer viruses—center on certain systems of extraordinary complexity. The systems that host these problems—economies, ecologies, immune systems, embryos, nervous systems, computer networks—appear to be as diverse as the problems. Despite appearances, however, the systems do share significant characteristics, so much so that we group them under a single classification at the Santa Fe Institute, calling them complex adaptive systems (CAS). This is more than terminology. It signals our intuition that there are general principles that govern all CAS behavior, principles that point to ways of solving the attendant problems. Much of our work is aimed at turning this intuition into fact.

Even a brief inspection reveals some characteristics common to all CAS: All CAS consist of large numbers of components, *agents*, that incessantly interact with each other. In all CAS it is the concerted behavior of these agents, the *aggregate behavior*, that we must understand, be it an economy's aggregate productivity or the immune system's aggregate ability to distinguish antigen from self. In all CAS, the interactions that generate this aggregate behavior are nonlinear, so that the aggregate behavior cannot be derived by simply summing up the behaviors of isolated agents. It is this latter characteristic that makes the study of CAS so difficult.

Another feature comes close to being a trademark of CAS: The agents in CAS are not only numerous, they are also diverse. An ecosystem can contain millions of species melded into a complex web of interactions. The mammalian brain consists of a panoply of neuron morphologies organized into an elaborate hierarchy of modules and interconnections. This diversity is not just a kaleidoscope of accidental patterns. The persistence of any given part (agent) depends directly on the context provided by the rest. Remove one of the agent types and the system reorganizes itself with a cascade of changes, usually "filling in the hole" in the process. Moreover, the diversity evolves, with new niches for interaction emerging, and new kinds of agents filling them. As a result, the aggregate behavior, instead of settling down, exhibits a perpetual novelty, an aspect that bodes ill for standard mathematical approaches.

There is a less obvious feature of CAS that is important. CAS agents employ *internal models* to direct their behavior. (Murray Gell-Mann uses the term *schema* to describe this aspect of agents, but I have used that term in the past as a technical term in the study of genetic algorithms, so I cling to "internal model.") An internal model can be thought of, roughly, as a set of rules that enables an agent to anticipate the consequences of its actions. We are most conscious of our own internal models when we do lookahead in a complex game such as chess or Go. We try to move in such a way as to set the stage for future favorable configurations, as when we sacrifice a piece or make a "positional move" in order to capture an important piece later. However, even an agent as simple as a bacterium employs an "unconscious" internal model when it swims up a glucose gradient in the search for food. And humans make continual prosaic use of internal models. Consider our unconscious expectation that room walls are unmoving; should this expectation be violated, our attention immediately shifts to that part of the environment.

Internal models add still further to the complexities of aggregate behavior. Anticipations based on internal models, even when they are incorrect, may substantially alter the aggregate behavior. The anticipation of an oil shortage can cause great changes in aggregate behavior, such as stockpiling and price runups, even if the shortage never occurs. The dynamics of CAS will certainly remain mysterious until we can take such effects into account.

2. TWO IMPORTANT QUESTIONS

Two questions stemming from these observations are, for me, "right questions," in that they lead to useful thoughts about general principles:

1. Why do CAS evolve toward diverse arrays of agents rather than toward optimal agents?
2. How do agents create and exploit internal models?

These questions might seem a bit to one side of central issues, but the following arguments make me think otherwise.

2.1 DIVERSITY

Consider first the question of diversity, and the attendant perpetual novelty. Diversity and perpetual novelty do blunt some of our most powerful tools for understanding complexity, but they also provide an entering wedge to deeper insight. If we look at real CAS closely, it is clear that the diversity, in every case, is the product of progressive adaptation. It is a dynamic feature, much like the standing wave produced by a rock in a fast-moving stream: If you poke a "hole" in the flow, it quickly "repairs" itself. It is the pattern that persists, not its perpetually changing components.

The standing wave simile can be carried further if we think of CAS as involving *flows* of resources through agents. Each agent is a kind of middleman, accepting resources from other agents, modifying them in some way, and passing them on to still other agents. The "births" and "deaths" of agents in a CAS produce a continual turnover in the component parts of the flow. If you remove one kind of agent, the flow of resources is temporarily redirected, but the evolutionary mechanisms soon produce a new kind of agent to exploit the abandoned "niche" (the "convergence" of evolutionary biologists). For CAS, however, the standing wave simile does not go far enough because the complexity of the interactions increases over time. Each time a new kind of agent arises it opens possibilities for interaction—new niches—for still newer agents. The pattern evolves.

The complexities of these long-term adaptive progressions are best exemplified by the evolutionary patterns studied by paleontologists. These progressions exhibit a complex hierarchical organization that is characteristic of all CAS; we do not find real CAS consisting of a few highly adapted individuals that exploit all opportunities. Paleontologists have an accepted principle that should be applied, mutatis mutandis, to the study of this universal characteristic of CAS: To understand species, understand their phylogeny. If we can find common mechanisms that give rise to hierarchical organization, with the diversity and perpetual novelty it entails, we will have taken a substantial step toward uncovering general principles that govern CAS dynamics.

A useful analogy for exploring the origin of hierarchies comes from the study of developing default hierarchies in classifier systems (for details see Holland et al.[6]). In a classifier system, the first rules that establish themselves are "generalists," rules that are satisfied by many situations and have some slight competitive advantage. They may be "wrong" much of the time, but on average they produce interactions that are better than random. Because their conditions are simple, such rules are relatively easy to discover, and they are tested often because they are satisfied often. The frequent tests assure that survivors exploit real recurring features that offer advantages over random action. That is, the repeated tests provide a "statistical" confirmation of the generalist's "hypothesis" about its world. Once the "generalists" are established, they open possibilities—niches—for other rules. A more complicated rule that corrects for mistakes of an over-general rule can benefit both itself and the over-general rule. It benefits from its own useful actions, while it prevents the "generalist" from making a mistake. A kind of symbiosis results. Repetitions of this process produce an increasingly diverse set of rules that, in aggregate, handle the environment with fewer and fewer mistakes. As we will see, the mechanisms involved here have counterparts in other CAS.

Along similar lines, there are good reasons why single "super-agents" that fill all niches do not appear. The exploitation of resources generated by the aggregate behavior of a diverse array of agents is much more than the sum of the individual actions—the nonlinearity of the interactions again. For this reason, it is a complex task to design or discover a single agent with the same capabilities for exploiting the resources. It is simpler to approach this capability step-by-step using a distributed system, as in the case of a developing default hierarchy. There will be more to say about this point in the discussion of internal models.

To return to the basic point, in CAS it is the evolving patterns of interaction that are important. The patterns of interaction most familiar from ecology— symbiosis, parasitism, mimicry, biological arms races, and so on—all are best described in terms of flows of resources through agents, and all have counterparts in other CAS. Agents direct these flows, sometimes rearranging the basic "elements" defining the resources, but generally conserving the elements themselves. When groups of agents cause cyclic flows, resources are held within the system. Resources so retained become more readily available to other agents. CAS thrive in proportion to their ability to keep resources around, and their ability to do so increases as the number of interactions, particularly cyclic interactions, increases (another point that will be elaborated later). This is simply natural selection writ large. Parts of CAS that exploit these options persist and expand; parts that fail to do so lose their resources to those that do. Thus, there are strong selective pressures on CAS to discover and retain agents that provide such interactions. Each new interaction offers possibilities for still further interactions and redirection of flows, pushing diversity still further. Technically, CAS search for a diversity of agents that provide progressively refined covers of the range of possibilities, rather than trying to design an optimal agent that handles all possibilities.

2.2 INTERNAL MODELS

Let me turn now to the processes whereby agents create and exploit internal models. Here, there is a difficult definitional question: How can we know from the specifications of an agent that it has, or will build, an internal model? What characterizes agents that have internal models?

We might start with the idea that an agent has an internal model if we can infer something about the agent's environment by just inspecting the agent's structure. It is certainly true that we infer a great deal about the environment of any organism by studying its morphology and biochemistry. We can infer a nocturnal environment from big eyes, and we can infer a mosquito-rich, malaria-infested environment from sickle cells. But this is an inadequate diagnostic for internal models. Consider a meteorite. We can infer much of its environmental history from its composition and surface condition, but it is fruitless, even metaphorically, to attribute an internal model to meteorites.

Somehow, it does not make sense to attribute an internal model to an agent unless the model takes an active role in determining the agent's behavior. The model should suggest current actions that make future environmental states accessible, and the agent should act on these suggestions. Here we can make a useful distinction between tacit and overt models. A tacit internal model simply prescribes a current action under an implicit prediction of some desired future state. For example, a bacterium swimming up a chemical gradient employs a tacit internal model that implicitly predicts valuable resources at the apex of concentration. (We might, as earlier, call tacit models "unconscious," if that were not such an ill-defined word.) On the other hand, an agent uses an overt model for active internal, or virtual, explorations of alternative lines of action. This virtual exploration is often called lookahead, the quintessential example being exploration of alternative move sequences in games like chess. Both tacit and overt internal models actively affect the agent's behavior and CAS employ internal models of both kinds. The internal models of agents in an immune system are at the tacit end of the scale, while the internal models of agents in an economy are both tacit and overt.

In realistic situations, internal models must be based on limited samples of an agent's environment. In a perpetually novel environment, generalization from limited samples is a sine qua non for exploiting experience. Only with generalization can the model be useful in situations not yet encountered. Stated another way: When situations never recur, the model must treat situations not previously encountered as equivalent to situations already seen. Technically, the model must be built in terms of equivalence classes over the set of environmental states. The problem then is to find useful equivalence classes, classes that capture regularities in the agent's environment.

How easy is it to discover and exploit regularities? Artificial worlds in which useful regularities are rare or non-existent are easy enough to construct. However, that does not seem to be the case in the real world. We can describe almost anything in the real world in terms of simpler component parts, so-called "building blocks." Moreover, limited numbers of building blocks can be recombined in many ways to yield descriptions and models of diverse arrays of real objects. This is true of everything from the morphology or biochemistry of a living organism to "a red Saab by the side of the road with a flat tire." The building blocks let us construct reasonable descriptions, and hypotheses, for situations we have never before encountered.

Here the discussion of internal models merges with the earlier discussion of diversity, extending our discussion of classifier systems and default heirarchies. Earlier, the search was for a diversity of rules that, working together, responded usefully to the environment. Rules were constructed by representing frequently encountered regularities in the environment. Technically, each regularity corresponds to an equivalence class over the possible states of the environment, the elements of the equivalence class being all environmental states involving the given regularity. When the regularity is captured in the condition for a rule, the corresponding equivalence class has been turned into a building block. The question becomes: Do internal models foster rules that correspond to building blocks that can be combined in a variety of ways?

To see why internal models favor diversity of this kind, consider the commonplace building blocks for a face (hair, forehead, eyebrows, eyes, etc.). Consider a set of rules, wherein each rule responds to a particular alternative for one of these components (a particular hair color, a hairline, a shape of forehead, an eyebrow shape, or the like). With a few dozen such rules it is possible to describe millions of faces. Each rule, then, is active in a variety of contexts—the different alternatives for other parts of the face—gaining utility accordingly. A novel situation is handled by simultaneously activating various rules dealing with the building blocks of the situation. A face never before seen can be handled by an appropriate selection of rules describing its component parts. "A red Saab by the side of the road with a flat tire" evokes rules dealing with "red," "roadsides," "tires," "cars," etc. Thus, the discovery of building blocks plays a key role in the formation of useful internal models, assuring both diversity of agents and diversity of interactions.

In constructing internal models, parsimony is a problem, even when the model condenses environmental states into equivalence classes. To see why this is so we must look more carefully at the way a model is specified. A model, a $fortiori$ an internal model, is specified by its transition function: A transition function f specifies what the next state of the model $S(t + 1)$ will be when the input $I(t)$ to the model and its current state $S(t)$ are given; i.e., $f = I \times S \to S$. For an internal model, there is one element of S for each equivalence class of environmental states. Each action of the agent constitutes an input to the environment, so I contains one element for each possible action.

Simple models, those dealing with a few hundred or a few thousand alternatives, can be presented explicitly by a table. There is one line in the table for each pair in $I \times S$, and for that pair the table specifies the element of S that is expected to follow. For realistic CAS, though, the number of lines required in a tabular presentation is just too great. Chess, which presents a problem much simpler and more one-sided than most realistic environments, would require a table with substantially more than 10^{50} lines (an average of 10 alternative actions, leading to a distinct configuration, on each of 50 successive moves). Even if the board configurations are condensed into equivalence classes, so that several alternatives at each move were treated as already given in earlier lines of the table, the number of lines would still easily exceed 10^{40}. Somehow, the model must be compressed.

Compression usually amounts to finding a set of generators for the model (much as one uses a set of generators and relations to generate the multiplication table for a finitely generated mathematical group). An appropriate set of n generators can generate a table of 2^n lines. We can think of rules as generators that are activated in clusters to cover a great diversity of conditions, as in the example of the faces or the "red Saab." The dynamic of discovering these generators is interesting and important. At first, the model consists of easily discovered generalizations that are often wrong but better than random. This coarse model is steadily refined by "exception rules" as experience builds up, as described earlier in the classifier system example. New information is gracefully incorporated, without destroying the parts of the model already constructed. Classifier systems were in fact designed to illustrate this process for discovering and elaborating parsimonious internal models. Such models are called quasi-homomorphisms (see Appendices 2A and 2B in a paper by Holland et al.[6] for a more formal description).

There is a final point concerning overt internal models. If the model is to be useful for "look ahead," the agent must be able to execute the component rules "fast time." That is, there must be a fast dynamic that lets the agent run through successive rules faster than the corresponding states occur in the environment.[6] Then the agent can extrapolate from the current situation, using the model to anticipate the effects of different action sequences. The fast dynamic lets the agent "run ahead" of the environment. This use of the overt model entails an additional requirement: The agent must be able to exercise the kinds of control on the model that it would use on the environment. For this to be so, the agent must have two sets of rules: (1) The rules that define the overt model, and (2) the rules that determine action. The second set of rules must include rules that allocate action between the model and the real world. With this arrangement the agent bases current actions in the environment on their predicted future effects as derived from the model. Predictions that are subsequently falsified can trigger the agent into correcting the parts of the model responsible for those predictions. That is, the agent can improve its ability to predict even when rewards (payoffs, reinforcements) are absent. This is a valuable asset in a world where rewards are usually intermittent or sparse.

The ECHO models, which I'll discuss next, are *not* meant as a vehicle for studying the acquisition of overt internal models; classifier systems are better suited to that study.[5] However, tacit models play a key role in ECHO. Tacit models, while not providing the advantage of look ahead, do offer the advantage of tacit anticipation of the future consequences of current actions, as in the case of the bacterium with a tacit model that causes it to swim up a chemical gradient. The parsimony of the model enters quite directly, because it costs resources to implement the model. An agent with a tacit model will proliferate only if the resources required to implement the tacit model are less than the additional resources acquired because of the model (suitably discounted, if acquired over an extended period of time). As with other characteristics that enhance an agent's ability to reproduce in its niche, tacit models are acquired over successive generations as the space of possibilities is searched. It is reasonable to believe that there are building blocks for different kinds of tacit models, and that these building blocks will proliferate through a variety of agents.

3. THE ECHO MODELS

The ECHO models all have a common framework, though there are several variants, so I will refer to the class as if it were a single model hereafter. ECHO may be able to simulate actual ecosystems,[2] but that is not its primary purpose. It is designed to facilitate computer-based gedanken experiments. More specifically, ECHO is designed to facilitate explorations for mechanisms that generate major CAS phenomena such as diversity and internal models.

Because ECHO is a computer-based simulation, it allows no unarticulated or ambiguous assumptions. The generated behavior is a precise consequence of the assumptions implemented. This rigor, combined with ECHO's ability to handle very complex systems, provides possibilities not available through traditional mathematical analysis. However, these possibilities are bought at the cost of the generality that usually accompanies mathematical models. Though parts of ECHO can be analyzed mathematically, providing important guidelines for the model-building process, the results of individual simulation runs are simply points in the space of possibilities.

It is true that simulation runs can suggest generalizations, but this depends upon assigning real-world meanings to numerical results. Such interpretations can be misleading or false. There are cases in the literature where a relatively simple algorithm, such as linear regression, is identified with some sophisticated real-world process, such as "perception." Or some generated numerical sequence that bears a superficial similarity to collected data is labelled with the same name, even when there has been no attempt to compare the underlying mechanisms. With care, simulations can suggest genuine generalizations, generalizations that can be rigorously established, but indifferent attention to problems of interpretation can destroy the usefulness of computer-based models. Facile labelling of what are, after all, streams of numbers in a computer, leaves too much to the eye of the beholder.

I have taken two steps to meet this "eye of the beholder" difficulty. First, I have tried to select primitive mechanisms that have direct counterparts in the various CAS. The interpretation of the behavior generated by these mechanisms is thereby constrained to be consistent with the interpretation of the mechanisms. Second, ECHO incorporates, as special cases, a wide range of well-established paradigmatic models—Wicksell's Triangle, Overlapping Generation models, Prisoner's Dilemma games, two-armed and n-armed Bandits, abstractions of the antigen-antibody matching process, biological arms races, cyclic food webs, and so on. These paradigmatic models have undergone intense scrutiny in the disciplines in which they originated—economics, political science, immunology, and so on—and have been adjudged to be relevant abstractions of critical problems in those disciplines. Their appearance in ECHO forms a bridge from the abstractions in ECHO to familiar, accepted abstractions in the relevant disciplines, again constraining the interpretation.

3.1 THE ORGANIZATION OF ECHO

ECHO provides for the study of populations of evolving, reproducing agents distributed over a geography, with different inputs of renewable resources at various sites. In the simulation the resources are designated by letters drawn from a small finite alphabet. Each agent has simple capabilities defined by a set of "chromosomes": Resource acquisition (from the environment or by interaction with other agents), resource transformation, and chromosomal exchange. In the simulation these capabilities are defined by a combination of behavior-defining strings (rules) and tags (playing the role of "identifiers" such as banners, trademarks, active surface elements, or the like). The strings are defined over the same alphabet as the resources; that is, the strings are defined in terms of the resources required to construct them.

Though these capabilities are simple, and simply defined, they provide for a rich set of variations. Collections of agents can exhibit analogues of a diverse range of phenomena, including ecological phenomena (e.g., mimicry and biological arms races), immune system responses (e.g., interactions conditioned on identification), economic phenomena (e.g., trading complexes and the evolution of "money"), and even evolution of multicellular organization (e.g., emergent hierarchical structure).

3.2 PRIMITIVE ELEMENTS

Agents play a central role in ECHO, but the basic primitive is a component that I will call a compartment. Agents are collections of compartments, much as a cell consists of a collection of organelles. The compartments in an agent interact with each other and with compartments in other agents. In any given interaction, resources are exchanged between compartments; individual compartments may also transform a resource from one kind (letter) to another.

The full capabilities of a compartment are specified by four strings put together from the resource alphabet: (1) a tag that serves as a (phenotype) marker that is visible to other compartments; (2) an interaction condition that specifies the kinds of compartments—as determined by their tags—that the given compartment will interact with; (3) an offense string that is used to determine a compartment's ability to "force" the outcome of an interaction (cf., the notion of induction in developmental biology); and (4) a defense string that is used to determine a compartment's ability to "resist" attempts to force the outcome of an interaction (cf., the notion of competence in developmental biology). In addition, the compartment has a reservoir in which it stores the resources that it acquires.

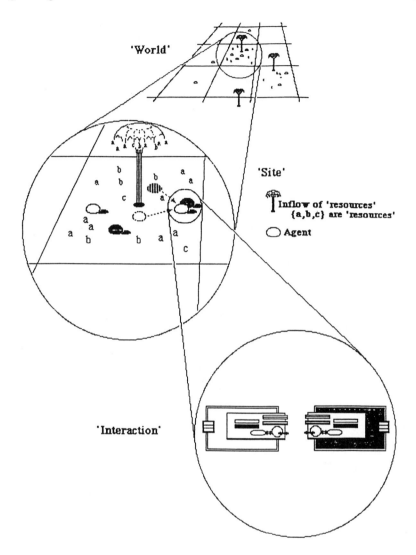

FIGURE 1 General Overview. $ECHO$ simulates a connected array of sites, each of which has its own distribution of renewable resources. Agents within a site can take up resources from the site and may acquire resources with other agents at the site. Agents may migrate from site to site.

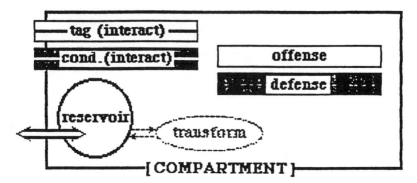

FIGURE 2 A Compartment. Agents are composed of compartments and interactions between agents are mediated by their compartments. The resources exchanged in an interaction are collected in the reservoirs of the interacting compartments. The form of the interaction is controlled by four basic components—a condition, a tag, an offense string, and a defense string—each of which is specified by a string defined over the alphabet of primitive resources. When two compartments come into contact, each attempts to match its condition against the other's *tag*. Matches may be bilateral ("trade"), unilateral ("combat"), or null (no interaction). If an interaction takes place, the amount of resource exchanged is determined by cross-matching each compartment's offense string against the other's defense string. The outcome may range in severity from a simple exchange of surplus resources, from the reservoirs, to complete destruction of either or both of the compartments. From the time of its first completed interaction onward a compartment is said to be in active status.

3.3 REPRODUCTION

REPRODUCTION OF AN AGENT. Agents reproduce by reproducing their component compartments. To reproduce a compartment the agent must provide copies of each of the four strings that define that compartment. Because the strings consist of sequences of letters from the resource alphabet the agent must collect enough of the proper resources to make copies of these strings. If the agent has more than one compartment then, before it can reproduce, it must collect enough resources to make copies of the strings defining all of its compartments.

In more detail: By absorption of resources from the site, and through interactions with other compartments, each compartment in an agent collects resources in its reservoir. For purposes of reproduction, an agent is assumed to have access to the resources in all the reservoirs of its component compartments, so that resources in the reservoir of one compartment can be used to replicate another compartment. When an agent's compartmental reservoirs, taken as a whole, have enough resources to make copies of all the strings defining all the agent's compartments, the agent is ready to replicate itself (subject to conditions that we will come to shortly).

With this provision, an agent's rate of reproduction depends entirely on its ability to collect the necessary resources to copy its compartments. A complex agent, with many compartments and complex capabilities (long strings defining the compartments), will have to collect many resources in order to reproduce. There is no explicit fitness parameter; indeed, the fitness of any given agent is dependent upon the context provided by its site and the other agents at that site.

REPRODUCTION OF MULTIAGENT ORGANIZATIONS. One of the most interesting features of real CAS is their hierarchical organization. If we look to a mulcellular organism, we see that it consists of a great many different kinds of cells, ranging from nerve cells to muscle cells. Nevertheless, all of these cells share the same basic blueprint. The chromosomes of all the different cells in the multicellular organism are identical. The differences come about because only some of the genes in each cell are expressed; the genes expressed determine the detailed structure of the cell.

To translate this style of organization to ECHO, we must supply each agent with a "chromosome" that specifies the set of compartments it can have—each "gene"— in the "chromosome" specifies a particular compartment. Then we must make some provision for turning genes on and off, so that the actual compartments the agent contains depend upon what genes are on at the time the agent is formed. That is, when the agent replicates, a compartment gets copied only if the corresponding gene is on (cf. repressors for an operon in a bacterial chromosome). Thus, only compartments specified by genes that are in the parent apepar in the offspring, though the offspring carries the whole chromosome of the parent.

The question now becomes: What determines whether a given gene is on or off? The object is to make the condition endogenous and subject to selection. A simple version associates a condition with each gene; the gene is turned on only if the associated condition is satisfied. To make the condition endogenous, I will make it dependent upon activity within the agent. Since all activity centers on compartments, this means defining the condition in terms of the activity of some particular (kind of) compartment.

In what follows, I will define a compartment as active if it has undergone an interaction (processed some resources) for the agent prior to the time replication starts. A gene's condition then becomes a requirement that some particular (kind of) compartment is active. Note that this may be a compartment quite different from the one gene directly specifies. (The metabolic product of one organelle in a cell can repress or de-repress the activity of a gene associated with a different organelle in similar fashion.)

With this provision it is possible to arrive at a multiagent organization consisting of differentiated agents with a common chromosome. Assume, at the time of reproduction, that the offspring agent "sticks" to its parent, instead of floating away to be come a "free agent." That is, the offspring agent becomes part of the multiagent organization. It may, however, have a different set of compartments than its parent, because this depends upon the genes that are on in the parent.

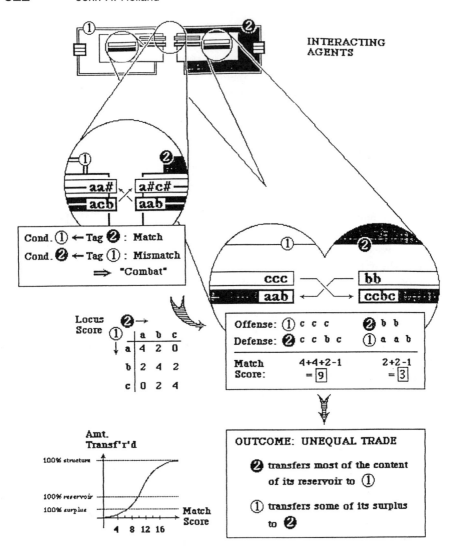

FIGURE 3 Details of an Interaction. When two agents come into contact, one compartment of each agent is selected for testing. The interaction condition of each selected compartment is checked against the tag of the other compartment. If one or both conditions are satisfied, the interaction proceeds. The intensity of the interaction is determined by matching the offense string of each compartment against the defense string of the other compartment. The score, for each compartment, determines the amount of the compartment's resources—resources in its reservoir and, possibly, resources defining its structure—transferred to the reservoir of the other compartment.

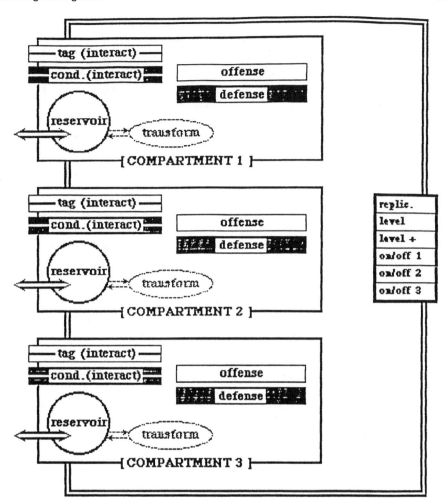

FIGURE 4   An Agent. Agents are the basic self-replicating entities. Replication is controlled by a replication condition defined over the set of possible compartmental tags; replication proceeds only if there is an active compartment in the agent that has a tag satisfying the replication condition. In addition each compartment has an associated on/off condition, also defined over compartmental tags and satisfied in the same way as the replication condition. A compartment is copied at the time of replication only if its on/off condition is satisfied. The replication will be executed only if the agent has collected enough resources in its compartmental reservoirs to make copies of the four defining strings of each of the compartments with satisfied on/off conditions; the resources in all of the compartmental reservoirs are shared for this purpose. If there are not enough resources in the reservoirs, the replication is aborted. The replication condition, together with the set of on/off conditions and one other set of conditions described in Figure 5, constitutes the agent's "chromosome." Any or all conditions may be null; a null condition is treated as satisfied at all times. The full chromosome is always copied during replication, but it may undergo mutation and recombination during the process.

'SEED'

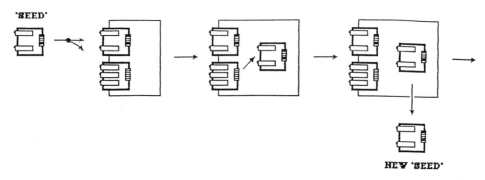

NEW 'SEED'

FIGURE 5 Development of an Hierarchical Meta-Agent. To allow agents to organize themselves into more complex structures, each agent has one or more additional chromosomal conditions, called level conditions. As with other chromosomal conditions, level conditions are defined over the set of compartmental tags and are satisfied if there is an active compartment with a tag belonging to the set specified by the condition. If the agent has only null level conditions, or if none of its level conditions is satisfied at the time of replication, then its offspring is a "free-living" agent. If one or more level conditions is satisfied, then the offspring "sticks" to its parent and the pair move and interact as a unit, a meta-agent. Further offspring with satisfied level conditions will be added to the unit, yielding a more complex meta-agent. A level condition may also specify that the offspring be placed "interior" to its parent, so that the parent and any other agents at the parent's level form a "shell" around the offspring. Iterations of this procedure can yield complex hierarchical meta-agents. If, after successive replications, a given parent has no level conditions satisfied, its offspring will be "ejected" as a free-living agent. If this offspring has an appropriate set of chromosomal conditions, it may become the "seed" for a new meta-agent of similar organization. As always in ECHO, a meta-agent will be selected for its ability to collect the resources necessary for its replication.

Nevertheless, it has the same chromosome as its parent. As this process is iterated, all agents in the multiagent organization share a common chromosome, but they may be greatly different in their compartmental composition.

A multiagent can be generated from a single agent, a "seed," via the iterative production of "sticky" offspring. If, at some point, one of the offspring does not "stick" to the cluster, then it becomes the potential "seed" of a new multiagent. Selection for the ability to gather in the appropriate resources for replication, will act on the multiagent through selection of the component agents. Only those organizations that provide advantages in collecting resources will survive, and only those organizations that provide for timely release of "seed" agents will survive. Agents are to multiagents much as compartments are to agents, the difference being that the agents in the multiagent organization have the same blueprint (the chromosome), whereas the compartments in an agent each have a different blueprint (the genes).

3.4 VARIATION

With these preliminaries in place, we can now address the main question: How can ECHO generate, and sustain, an increasingly diverse array of agents and multiagent organizations? The preliminaries account for selection. Different kinds of agents, and multiagents, will persist if they regularly collect enough resources to produce lines of progeny. In Darwinian terms, this leaves us with the question of variation. How are new variants introduced?

The compartments can be varied quite simply by adding and deleting letters in the defining strings, a simple kind of mutation. A more sophisticated variation results when strings belonging to different compartments are recombined; this brings into play the full powers of a genetic algorithm.[7] Both of these methods produce new kinds of compartments, but they do not yield variation in the number of compartments belonging to an agent. To accomplish this the number of genes in the chromosome defining the agent must be changed. Genetics suggests an interesting mechanism for doing this, intrachromosomal duplication: Genes sometimes are duplicated *within* the chromosome, the result being two adjacent copies of the gene where there was one before. In ECHO this would yield an agent with two copies of the same compartment. A subsequent mutation in one of the duplicate genes would yield two different compartments in place of two identical compartments. If the resulting agent survives, then further intrachromosomal duplications and mutations can yield agents with still more, varied compartments.

3.5 CAPABILITIES

The foregoing is only a general description of the compartment-based version of ECHO, and this version has not been tested, but results from earlier versions motivate the mechanisms described. The next few paragraphs describe my expectations based on tests of the earlier versions.

In the earlier models, the agents at a site were treated as thoroughly mixed and randomly paired for interaction, rather like the billiard ball model of gases. The present model proceeds in much the same way but now it is compartments, rather than agents, that are paired. That is, in the simplest version, pairs of agents from the site are selected at random, and then a compartment from each agent is selected at random for possible interaction. Compartments within the same agent may also be paired for interaction. In either case, an interaction actually takes place only if the tag of one of the compartments chosen satisfies the interaction condition of the other compartment. If the condition of one compartment is satisfied but not the condition of the other, then the second may abort the interaction, say, with some predetermined probability; if both conditions are satisfied, the interaction takes place unconditionally.

In earlier models, agent A could enforce an unilateral flow of resources from agent B, if A's "offense" string closely matched B's "defense" string, while B's offense poorly matched A's defense (for details, see Holland,[7] Chapter 10). This unilateral interaction was called "combat." The use of a pair of strings in each agent allows intransitivity in the evolving "food webs": Agent A can "eat" agent B, agent B can "eat" agent C, but agent A cannot "eat" agent C. There was also provision for a bilateral exchange, called "trade," that involved a cooperative matching of another set of strings. It is clearly of greater interest if these different kinds of interaction can be unified, so that evolution can adaptively select the conditions and kinds of interactions an agent will undertake. Let's see how compartmentation makes this possible.

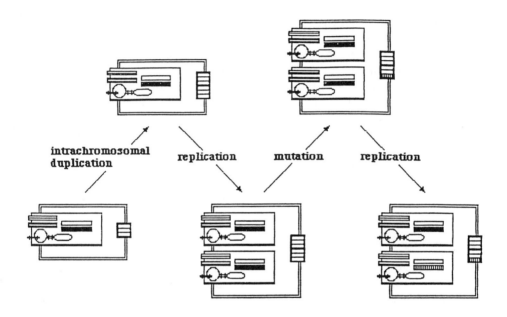

FIGURE 6   From Single-Compartment Agent to Multicompartment Agent. The basic evolutionary operator for increasing the complexity of agents is intrachromosomal duplication. When the intrachromosomal duplication (ICD) operator is applied, a segment of the chromosome is doubled. If one or more compartment-associated on/off conditions have been doubled, the offspring produced at the next replication will contain a second copy of the compartments corresponding to the doubled on/off conditions. Subsequent replications may involve a mutation (or other modification) that provide a modification of the added compartment, yielding an agent with differentiated compartments. The resulting multi-compartment agent will be subject to resource-gathering selective pressures and will survive only if the new compartments somehow give it an advantage that compensates for the additional resources it must collect to copy the additional compartments.

The flow of resources between compartments can be mediated by the same string matching technique used for "combat," but only the resources of the respective compartments are at risk now. (Remember that compartments will only interact if their respective interaction conditions are satisfied, so the flows are conditional on the interaction conditions.) Under this restriction, the trick to unifying "combat" and "trade" is to make the flow dependent upon the match scores. The better the match between the offense string of y and the defense string of x, the higher the resource flow from x to y. Specifically, the flow is given by an S-function, with low match scores resulting in acquisition of some fraction of the surplus in the reservoir, higher scores drawing more from the reservoir than just surplus, and still higher scores acquiring a progressively higher fraction of the "structural" resources (resources tied up in the defining strings) of the other compartment, thereby "killing" it (see Figure 3). A bilateral exchange of surpluses between the reservoirs of x and y, "trade," results when their respective match scores are both low. A unilateral exchange of resources, "combat," occurs if there is a lopsided difference in the respective match scores.

In order to approach questions of emergence with a sparse set of mechanisms, several "shortcuts" have been used in the early versions of ECHO: (1) Details of metabolism and assembly of resources into compartmental structures are omitted. Once the resources are acquired, they are automatically assembled into the required structures with no attempt to simulate the chemistry involved. (By progressively adding resource transformation capabilities to compartments, under control of the chromosome, the evolution of metabolism can be modeled with increasing verisimilitude.) (2) There is no resource cost in duplicating the chromosomes that provide a compartment's blueprints. The only "costs" associated with a longer chromosome are the costs incurred in duplicating the more complex compartmental structures they describe. (This can be easily modified by charging a resource cost for reproducing a chromosome, akin to the cost of recharging ADP in a biological cell. The ratio of this cost to the compartment reproduction cost then becomes a parameter of the system.) (3) The activity of a compartment directly determines the activation of the corresponding gene. In real chromosomes the activation of genes depends upon complex metabolic feedbacks; here it is assumed that the activity of a compartment "stands for" the string of metabolic intermediates that it induces.

As in earlier models, the condition/tag combination provides for selective interaction. This, followed by the match score computation, makes possible a diverse range of interactions. We can design systems with counterparts of ecological phenomena (e.g., parasitism, symbiosis, biological arms races, and mimicry), economic phenomena (e.g., trading complexes, and the evolution of "money"), immune system responses (e.g., interactions conditioned on identification), and so on. It is also easy to show that ECHO subsumes a wide range of standard models such as the Prisoner's Dilemma, the Two-Armed (n-armed) Bandit, Wicksell's Triangle, Overlapping Generation models, and so on. Compartments make possible still more complex organizations, starting with simpler primitives. And, with the addition of agents that have on/off genes in their chromosomes, it is possible to investigate counterparts of metazoan formation and evolution (e.g., emergent hierarchical organization).

While it is useful that ECHO can be set up to imitate such systems, that is not its primary purpose. Rather, ECHO is intended as a vehicle for studying the emergence and evolution of such complexities. Some of the most interesting simulations start with simple initial configurations, so that complexities that appear later must perforce be a consequence of the evolutionary mechanisms supplied by ECHO. When complexity increases in interesting ways under these conditions, we at least have an existence proof for the sufficiency of the mechanisms supplied.

3.6 AN EXAMPLE

A population-based version of the Prisoner's Dilemma provides a simple example of the kind of investigation that ECHO should facilitate. In this version, there is a population of agents that come into contact via random pairings. At each contact, the condition/tag mechanism is used to determine whether or not an interaction takes place. If an interaction does take place, then the pair executes one play of the Prisoner's Dilemma according to the strategies implied by their respective offense/defense strings. The four possible outcomes (cooperate-cooperate, cooperate-defect, defect-cooperate, defect-defect) determine an exchange of resources. The rate at which a given agent collects resources determines its rate of reproduction and, ultimately, the relative frequency of its kind in the overall population. As the frequency of a given kind of agent increases, the frequency of interactions involving that kind of agent increases.

Earlier experiments with selective mating based on tags are relevant here.[8] In those experiments, an early, accidental association of a tag with a trait conferring a reproductive advantage is rapidly amplified because of the higher reproduction rate of the tag's carriers. For example, such an advantage is conferred by a tag associated with "compatible" mates that produce fewer lethal offspring under crossover. The tag, originally meaningless, takes on a meaning. It comes to stand for a particular kind of compatibility. By developing selective mating conditions based on the tags, the agents can react to this compatibility, thereby increasing their fitness.

In the population-based Prisoner's Dilemma, tag amplification provides a way for agents to make useful distinctions. For example, an agent developing a condition that identifies tags associated with "cooperators" will prosper from the increased payoff that results. As in the selective mating experiments, there is strong selection for combinations of tags and conditions that favor profitable interactions. In effect, the agents develop tacit models, anticipating the effects of interacting with agents having certain kinds of tags.

This selective process, in turn, opens new niches. For example, mimicry becomes possible; an agent can present a tag with an established "meaning," while pursuing a different course of action. The mimics, as in biological studies of mimicry, can only occupy a small proportion of the population relative to the agents being mimicked, because the other agents begin to adjust to the deception when the proportion of the mimics becomes large. Even in an "ecosystem" with the limited possibilities offered by the Prisoner's Dilemma, an interesting diversity of strategies and internal models arise.[1]

4. TWO-TIERED MODELS

There is a more sophisticated version of ECHO that emphasizes the flow of resources resulting from the interactions between agents. By assuming a kind of "rapid mixing" of resources among like agents this version makes better contact with mathematical models and allows for much more extensive computer explorations. It uses an integrated two-tiered format:

1. The Upper tier specifies the evolution of the genotype/phenotype relations for the adaptive agents, where the agents have the compartment-based structure described earlier. The Upper tier models the "slow dynamics" of the system. Some techniques associated with classifier systems can be extended to the study of this tier.[4] From a mathematical perspective, there seems to be a generalization of the "schema theorem" for genetic algorithms that applies to the Upper tier.[7] This version of the schema theorem does *not* depend upon an exogenous definition of fitness.

2. The Lower tier uses a matrix of flow coefficients to describe the transfer of resources between different kinds of pairs of interacting adaptive agents. The phenotypic properties of the Upper tier precisely determine flow coefficients of the Lower tier. The Lower tier models the "fast dynamics" of the system. The recursions based on the flow-matrix are natural discrete generalizations of the Lotka-Volterra equations; as such they are susceptible to mathematical study.

   In more detail:

1. The mapping from the real CAS to the Upper tier of ECHO is from repeated structural features (building blocks) of the CAS to generators and relations (resources, tags, rules, etc.) in the Upper tier. As a result the behavior of interest in ECHO is generated and emergent, rather than being given by a table of all possibilities. This provides an insurance policy against "eye of the beholder" interpretations, so that correspondences, even qualitative correspondences, are nontrivial, enhancing ECHO's use as a platform for gedanken experiments.

2. The Lower tier simulates the Upper tier, rather than directly simulating the fast dynamics of the CAS being studied, because this offers a considerable advantage in revising the model as it is tested. Whenever the flow matrix in the Lower tier fails to capture some phenomenon of interest, the nature of the failure can be examined systematically by altering parameters in the Upper

tier as well as the Lower tier. Such detailed control of the hypothesize-test-control cycle is rarely possible within the real CAS; it is typically impossible to repeatedly "restart" a real CAS from the same initial conditions. There is an additional advantage. Regions of the parameter space that prove interesting can be defined precisely, allowing the construction of flow matrices specialized for analysis of those regions.

The two-tiered ECHO model provides several advantages in formalizing the idea of an adaptive landscape:

1. A fixed flow-matrix corresponds to a "steady-state" ecosystem (an ecosystem without innovations). The flow matrix makes it easy to represent the effects, on flows, of lags and carrying capacity. Mathematically there are strong connections to the study of Queued N-Armed Bandits, and to r- and K-selection. The flow-matrix also provides measures of the integrity or robustness of subsystems. In the particular case of economics, the ECHO models provide interesting relations between new markets, technological innovation, and r-selection, on the one hand, and saturated markets, market share, and K-selection, on the other hand.

2. The agent-determined flows are specified rigorously so that the instantaneous form of the landscape allows for a "fast-dynamics," e.g., oscillations like the Lynx-Hare oscillation.

3. Processes of evolution, coevolution, etc., define a trajectory through the space of allowed flow matrices, so that features of those processes are well-defined properties of the trajectory. It is possible to define both a "schema theorem" for these trajectories and a notion of neighborhood based on an "operator metric," making it possible to study some of the main determiners of the rate of evolution.

4. Fitness is endogenous, varying rapidly in some contexts. With fitness so-defined, it is possible to study context-dependent changes in fitness, inclusive fitness, etc., for single genotypes and sets of genotypes (e.g., arms races). For subsystems, the net resource income per unit of resource sequestered, an interest rate, provides more information than fitness.

5. Epistasis is determined by evolving phenotypic interactions and is subject to selection, making it possible to study selection for or against particular forms of epistasis.

5. SUMMING UP

Computer-based models are perforce caricatures of real systems. It is in the nature of modeling that certain features of the real system are emphasized, while details must be thrown away. This is no less true of mathematical models, though they often have the compensating advantage of generality. In both computer-based modeling and mathematical modeling, as in cartooning, much skill and taste go into choosing what is to be emphasized and what is to be thrown away. And, again as in cartooning, the resulting caricature will only be informative if the choices point up essential characteristics. In modeling, unlike cartoons, there is the additional criterion that the results be protected from "eye of the beholder" ambiguities.

ECHO uses two techniques to avoid "eye of the beholder" ambiguities. First, ECHO is overtly designed to subsume specific models from other disciplines, such as Wicksell's Triangle and the Prisoner's Dilemma, in its larger framework. Because these specific models have already been closely examined for relevance in their respective disciplines, the larger framework inherits this relevance. Second, ECHO uses a "generators and relations" approach to modeling. Instead of using a long list of rules in an attempt to describe each action of the system being modeled, ECHO models elemental processes and mechanisms. This approach avoids the "unwrapping" syndrome (where the "solution" is explicitly "wrapped" into the initial instructions of the program) that pervades much computer-based modeling. ECHO, so-contrived, offers the possibility of studying interactions and generalizations of well-known special models in a larger, still rigorous, context.

The two-tiered version of ECHO also provides some further protection from "eye of the beholder" interpretation. It makes contact with extant mathematical studies, ranging from the Lotka-Volterra equations to the schema theorem of genetic algorithm studies. The relevant theorems, suitably adapted, offer guidelines as to parameter settings, and regions in the space of possibilities, that will prove interesting.

The design of ECHO turns on the aspects of CAS that are to be emphasized, with a focus on the evolution of diversity and internal models. The basic mechanisms center on the discovery of building blocks and the exploitation and elaboration of tags. The mechanisms for discovering and exploiting building blocks are modeled on the genetic algorithm's manipulation of schemata. The mechanisms for exploiting and elaborating tags are modeled on the manipulation of tags in classifier systems. Because tags control interactions between compartments, the only tags that persist over long periods are those that control useful interactions. That is, the tags in ECHO provide a tacit model of the agent's world, as in the Prisoner's Dilemma example, implementing useful anticipations.

It is relatively easy to find counterparts of these mechanisms in other CAS. Indeed the mechanisms were originally designed with such counterparts in mind, and the successful use of these mechanisms in earlier genetic algorithm/classifier system experiments bodes well for their use in ECHO.

The central questions I intend to pose to ECHO are the two questions concerning diversity and internal models. With the help of ECHO's simulations I hope to (1) demonstrate that an increasingly diverse array of agents ("genotypes") incorporate certain building blocks discovered early on, a kind of "founder effect" for building blocks, akin to the pervasive use of the Krebs cycle in organisms and, using this information, I hope to (2) formulate a generalization of the schema theorem based on endogenous fitness. This version of the schema theorem would help to determine how "innovations" in CAS shift under the discovery of building blocks; it may show that combinations of building blocks providing innovation are close together in terms of the operator metric(s) defined by crossover, mutation, etc. The theorem should apply to building blocks for structural features, such as the compartment's offense and defense strings, as well as to building blocks for the tags that implement tacit models.

It is worth noting that the compartment-oriented version of ECHO is suitable for studying the evolution of organizations, with the embryogenesis of metazoans providing a prime example (see, for example, the work by Buss[3]). When each agent is trying to maximize its reproductive rate, under the constraint that it must not seriously damage the overall organization that assures the supply of critical resources, innovation takes some surprising turns. The sequestration of the germ-line in metazoans, or the genetic constraints that prevent the runaway reproduction we call cancer, are prime examples.

Among those who have carefully compared different CAS, there is little doubt that they form a coherent subject matter. At the right level of abstraction, their mechanisms and processes can be given a unified description. Within this framework we begin to see common causes for common characteristics. Common characteristics such as diversity in components, perpetual novelty in behavior, exploitation of internal models, and persistent operation far from equilibrium, all seem to arise from similar mechanisms and processes. The challenge now is to provide a rigorous treatment of these observations.

The challenge is formidable because our traditional mathematical tools rely on linearity and equilibria—fixed points, basins of attraction, and the like—features mostly missing from CAS. Oscillations and recursive interactions are not features of linear systems, and the anticipations provided by internal models frequently destroy equilibria. To meet this challenge we need an unusual amalgam of techniques:

1. Interdisciplinarity. Different CAS show different characteristics of the class to advantage, so that clues come from different CAS in different disciplines.
2. A "correspondence principle." Bohr's famous principle translated to CAS, means that our models should encompass standard models from prior studies of particular CAS, not only to forestall "eye of the beholder" errors, but also to assure relevance.
3. Computer-based gedanken experiments. Such models provide complex explorations not possible with the real system—it is no more feasible to isolate and repeatedly "re-start" parts of a real CAS than it is to test "flameouts" on a real jet airplane carrying passengers—suggesting critical patterns and interesting hypotheses to the prepared observer. Such experiments can also provide "existence proofs," showing that given mechanisms are sufficient to generate a given phenomenon.
4. A mathematics of competitive processes based on recombination. Ultimately we need rigorous generalizations, something computer-based experiments cannot

provide on their own. Mathematics is our most powerful method for attaining such ends. The mathematics needed must depart from traditional approaches, emphasizing persistent features of the far-from-equilibrium, evolutionary trajectories generated by recombination.

I believe this amalgam, appropriately compounded, offers hope for a unified approach to the difficult CAS problems that stretch our resources and place our world in jeopardy.

REFERENCES

1. Axelrod, R. "The Evolution of Strategies in the Iterated Prisoner's Dilemma." In *Genetic Algorithms and Simulated Annealing*, edited by L. D. Davis, 32–41. Los Altos, CA: Morgan Kaufmann, 1987.
2. Brown, J. H., and L. W. Buss. Personal communication, 1992.
3. Buss, L. W. *The Evolution of Individuality.* Princeton: Princeton University Press, 1987.
4. Holland, J. H. "A Mathematical Framework for Studying Learning in Classifier Systems." In *Evolution, Games and Learning*, edited by D. Farmer et al., 307–317. Amsterdam: Elsevier, 1986.
5. Holland, J. H. "Concerning the Emergence of Tag-Mediated Look Ahead in Classifier Systems." In *Emergent Computation*, edited by S. Forrest, 188–301. Amsterdam: Elsevier, 1990.
6. Holland, J. H., K. J. Holyoak, R. E. Nisbett, and P. R. Thagard. *Induction: Processes of Inference, Learning and Discovery.* Cambridge: MIT Press, 1989.
7. Holland, J. H. *Adaptation in Natural and Artificial Systems*, 2nd ed. Cambridge: MIT Press, 1992.
8. Perry, Z. A. "Experimental Study of Speciation in Ecological Niche Theory Using Genetic Algorithms." Ph.D. Thesis, University of Michigan, 1984.

DISCUSSION

ARTHUR:　　　　In the business of tagging, it's often said that if you get a Harvard MBA, which is a very valuable tag to have, it may not be necessarily true at all that you've learned anything in the Harvard MBA program—coming from Stanford, I'd say you've learned zilch—but the fact that you were smart enough to be selected to get that tag is the real value of the Harvard MBA tag.

KAUFFMAN: John, the tags determine who interacts with whom and, therefore, as your system is evolving, your species are controlling how richly interconnected the species web is.

HOLLAND: That's right.

KAUFFMAN: One of the things that I found, looking at the model of co-evolution that I've looked at, is that if every species interacts with every species, the system tends to be chaotic. And by tuning down the number of species which interact with one another, it can go into an ordered regime, and then tune back and forth. There's also evidence from real food webs that species work quite hard at controlling the number of other species that they interact with, as you know. Have you looked in the models as to whether or not, if you were to start the thing with a large number of species that interact with one another very richly, whether they tune that down, or if you start it very sparsely they tune it up? Which ties a little bit to some studies that Per did also, by the way; his studies began to suggest that. Have you looked for that?

HOLLAND: There's some stuff that's relevant. One thing: In this model, there's a very basic parameter which is like the temperature in simulated annealing. That's how frequently do the agents come into contact. Think of them as billiard balls in the site, and they keep bouncing into each other.

KAUFFMAN: It's a frequency-dependent thing.

HOLLAND: Now, I can turn that up or down. I can make it hot or cold. An interesting thing is, it's along the same lines. If I make it too hot, then species diversity drops off and, in fact, at a certain point it'll drop off so that you get the very minimum that you can get in that world. They may either all go extinct, or you may have just one predator and one prey, or something like that.

KAUFFMAN: So you get a big extinction event.

HOLLAND: That's right. If you turn it down, that's when you begin to get persistent diversity. There's another interesting thing that's a little bit related to that. You can build this very much like a chemostat model, where the resources are flowing in at the top, you've got one agent here who consumes the resources, and another agent that consumes that agent. This bottom agent has a death rate which determines how fast resources flow; the other one doesn't. Interestingly enough—and again, it was one of those things that surprised me—if I turn up the death rate of the bottom agent, the predator, what do you think happens to the other one?

KAUFFMAN: I'm scared to guess.

HOLLAND: In fact, if you think of it from chemostat point of view, it decreases. So increasing the death rate of the predator does not, in the long run, under stable conditions, help the prey. Because, you've got this steady state going through there. So there are a lot of things like that, that hold. And, I haven't done the kind of thing you're talking about, but it's easy to do.

GELL-MANN: John, in connection with that: Suppose you take an island biogeography model, with variants of a species, different competition situations, as in a story about the geospizine buntings in the Galapagos. Do you need mating in such a model, to get speciation?

HOLLAND: Mutation would do it in that case. You would certainly get speciation with mutation on, because I get that right now with mutation on. On the other hand...

GELL-MANN: Because you have different competition situations on different islands and, therefore, the niches get to be different...

SIMMONS: What do you mean by speciation here?

GELL-MANN: Oh, speciation means simply this: Suppose a few birds fly from one island to another, where is already a different competitive set up, which allows a certain niche for these arrivals. After some generations, if you were to take them back to the original island, they would have changed enough so that they wouldn't recognize, as mates, the birds of the species that they left behind.

ANDERSON: Speciation is a biological thing; they couldn't mate.

GELL-MANN: For mammals, birds, etc species are very well defined—locally in space and time. The questions arise if you hypothetically an organism to an earlier era or later era, or to another island where it doesn't normally occur: is it a different species, or the same one? And for that you have a thought experiment: you take them, let them go in the new place, and see if they mate with the others. You might not want to do a real experiment involving a new place, for various ethical reasons.

HOLLAND: It's all right if it's cockroaches.

GELL-MANN: All of this you can do with a mathematical model, with islands, and variants—as with the geospizine buntings—you just see whether speciation develops with a pure mutation model and, I would think, if there's no obstacle to its doing so. I don't think mating is crucial.

ANDERSON: If it had exactly the same number of chromosomes with the same length, it could still mate, perhaps.

HOLLAND: You're getting really into this area of what's called repro-ductive isolation, and the question is, is this physiological, is it geographical, or something else?

GELL-MANN: They look totally different, there. They have totally different beaks, because they're in different niches, with different competition situations; they develop totally different ways of earning a living; they're just different. And after a while, they certainly wouldn't mate.

HOLLAND: In this context, you can get this sort of island biogeography kinds of thing. You can get that by just having two sites, with two rates of resources upwelling, and you allow mutation and mating selection. You will immediately get the fact that they will start selecting for those that supply the same building blocks they need to use the resources.

GELL-MANN: Right. So the mating is needed only as a test of whether they have really speciated; you don't need the mating to produce the species.

SCHUSTER: Although it's not sure, I guess, in molecular genetics peo-ple would favor the models that say chromosome change—chromosome rearrange-ments—are prior to some kind of psychological reproductive isolation.

GELL-MANN: That's fine. But it could be point mutation, too, couldn't it?

SCHUSTER: A point mutation wouldn't be sufficient to hinder recombina-tion, so you need something more: that the chromosomes have to be substantially different, and you can't recombine them anymore.

HOLLAND: It's pretty clear cut. Usually they want an inversion, at least, so that you inhibit crossing in a particular region; inversion will do that, if it's not too long.

BROWN: I gather that the sort of ultimate limit on this is your fountain of resources that you're putting in from the bottom. And my question is "Have you looked at the effect of turning that up and down," when you have the opportunity for this speciation? And is there something like a monotonic increase in diversity, or do you see things like the paradox of enrichment, where you may actually, when you turn it up, get collapses again?

HOLLAND: I haven't tried that. Most of my experiments have been on a single site. (I've only been running these experiments for about a month and a half.) On a single site, it starts out with two resources, and two species, and then I watch to see how the diversity increases in that setup. Now the one thing I've done that's a first step in that direction is to add a new resource, add a third resource into that site. In that case, it does pretty much what you'd expect: things start incorporating that new resource, and you get a lot more diversity. And it goes up, not linearly, but combinatorially.

GELL-MANN: Gualtero?

FONTANA: Grazie. (There are few Italians that know what "Walter" means in Italian. Murray's certainly not an Italian, but he knows.)
    I have a question sort of concerning the philosophy of your model. As you repeatedly said, a lot of your model is based on a notion of a tag. I think that a lot of the questions concerning complex systems are actually, "What is a tag?" Let me make it more explicit. A genome is a very sophisticated organization; it's not just a linear array of things. A combat tag, or a food-search tag, or something like that— these are complex behavioral situations that require underlying organizations, at least cellular, and so forth. So a lot of the biology, a lot of the complexity, and a lot of the questions that are on that white board, are actually buried into the notion of a tag. And I just want to emphasize that you're assuming here, already, a very complex structure and organization in the word "tag" which you just assume in your model. And the rest is beautiful, but it's a strong assumption.

HOLLAND: Let me pursue this a little bit, and then I'll listen to Leo. In the model per se, inside the model itself, there's nothing complex about the tag. If I were to try to do that biologically, it would be very complex. . .

FONTANA: No, no; I want to correct myself on that. There is a lot of buildup of complexity in your model; there is no doubt about that. It's just that, in your very definition, there is something in there that at your level of description is simple, and then you just look at what complexity can you build up upon that. But in real nature, there is a lot complexity already in the tag, and a lot of. . .

HOLLAND: You've got all the steps between the initial chromosome specification, all the translation of all this stuff; finally, we get a protein that shows up on the surface, which may or may not be used as a tag.

FONTANA: There are entire levels of organization in these things.

HOLLAND: Yes. And all of that I put aside for the following reason. . .Notice, for instance, for those of you who know about classifier systems, that this would be a classifier system with a very few rules. And so, from the computational point of view, it's utterly simple. These are a few if-then rules that control exchanges, and there's nothing very deep about that. The purpose, again, was to design this so that I could get far enough to see the counterparts of a wide range of models that I know about—like the two-armed bandit, like the Prisoner's Dilemma. For instance, to give you an example again, that shows you both the complexity and the simplicity of the tags: As you may or may not know, if you do Prisoner's Dilemma in a population, and there are no identifiers for the individuals, so you don't know who you're interacting with and it's not the same individual each time, and you have no choice—you don't get tit-for-tat; you get the classic solution: they defect all the time. It's the only way, because you have no chance for reciprocity. On the other hand, allow them to carry a little banner around—and whether they lie or not, if you get the chance to learn which kinds of banners to interact with, tit-for-tat emerges very rapidly.

Okay, that's a simple model. We know something about it; it fits in here as a special case. And what I did, to the best of my knowledge, is to take a bunch of models that I knew about in complex adaptive systems—like the Volterra, the Prisoner's Dilemma kind of stuff, the two-armed bandit, and the cued bandit (which is even more interesting for our purposes); the kind of Dawkins arms race—and I tried to see if I could find simple counterparts, and being very cavalier about the rest. Because—to repeat myself—the way I was striving to get legitimacy was to say, "Look, people have spent a lot of time with a lot of papers in these various areas, studying these simple models, so they must think there's something useful about them. If I can incorporate them in here, then at least I've got the start of something that may be useful as well." And I would make absolutely no claims about how difficult it is now to start mapping this into a real biological system, where there's so many steps between getting a tag, and the genotype.

BUSS: I want to make Walter's point in a slightly different way. If you look at this system compared to Tom Ray's system, this is substantially richer because you have the phenotype; that's what the tags are. As soon as you have a phenotype, then selection operates on traits, and that's what, in fact, gives you richness. Walter's point was really an application to biology. What you have not done is generated the tags, and of course that's where biology lacks theory. It is how you generate the tags, not what the consequence is with respect to the subsequent dynamics in terms of the dynamics of the individual entity class.

HOLLAND: Although I would claim that we don't know a hell of a lot about the dynamics either.

BUSS: Yes, it's real hard to get data.

HOLLAND: And the dynamics, for the moment, is the thing I'm mostly interested in.

RASMUSSEN: I was also reminded about the classifier systems when I first saw this model. So you have this population of small classifier systems swimming around, but have you tried to think about formalizing the first level, this ecological level, trying to make mathematics on that? It seems as if you kind of dart down there, and you started to talk about differential, or difference, equations. This is very interesting, but I'm sure you must have been thinking about how to formalize the other level also, where you actually have these innovations, and all that stuff.

HOLLAND: There are only two things that I have very much on; in each case I have what I believe is a theorem, but I sure want to make a lot more sure than I am right now. In other words, I've got a sketch of a proof. At the upper tier, there is something that's very like a schema theorem of why I do classifier systems. That is, that if certain building blocks—whether they show up originally in tags, or in the condition part—turn out to be important, then they spread to other genotypes. And you can even talk about the rate at which they spread. So that certain building blocks become important in this world, and then they spread. And that's the schema theorem now put in this flow context, instead of in the original context.

At that level, that's the kind of theorem I would try to prove. At the other level, it's the typical, "Okay, if I've got these Lotka-Volterra equations, what's the limit cycle?," and stuff like that.

ARTHUR: You've given the impression that your fast dynamics lead you to sort of a unique solution. And I was trying to think why that may be...

HOLLAND: Not unless I cut off the evolution. Is that what you meant?

ARTHUR: Yes, cut off the evolution for a moment, and I think you'd have something like a unique solution. And, in turn, I think that's because the system you're dealing with is quadratic; therefore, it would have some sort of a Lyapunov type of setup; therefore, it might have some sort of attractor. Why should it be quadratic? In real biological ecosystems, there's no reason it should be. But you have one-to-one interactions. You don't have armies of ants sort of teaming up and saying, "If we have more than those guys...." So you don't have thresholds so that if ten ants come upon five things, that beyond some threshold the ants will take over and so on. That would immediately lead you out of this sort of quadratic type of mathematics, and in turn you would have what—in your language—would be a coevolving system, what in my language would be a multiperson game, which, highly likely, would have a multiplicity of solutions, one of which would be selected by historical accident (even in the fast dynamics). And then you add on these mutations, a further historical accident...So I'm wondering if this very set of one-to-one type of interaction biases the whole thing to actually more determinism, then you would tend to see in reality.

HOLLAND: You picked exactly what makes it quadratic; the fact that all interactions are one-on-one. I'll show you this one slide...

ARTHUR: The quadratic thing is not trivial; that drives an awful lot of your dynamics, this Lotka-Volterra structure.

HOLLAND: Here. [Showing slide.] And this is, with all due apologies to Leo, because this does a great deal of damage to everything. These are just the agents we had before. Remember when I talked about putting the caterpillar and the ant on the same thing, and now they're a firm? You can build up—it's just a layer on top of the model; it's already there—you can build up a layer that allows these things to work as coalitions. And now, your quadratic rules—just as you were suggesting—go out the door, and you get much more complex behavior.

ARTHUR: In particular, you might get cooperative effects, competitive effects, etc., etc. I'd expect, normally, a high multiplicity, and possibly even chaos.

HOLLAND: Yes. I wouldn't be surprised at that, at all. All I know is that you add this in there, and the dynamics goes right through the roof. It's very hard to understand; it's hard enough already. If you've got Lotka-Volterra, you at least have an attractor. Add evolution to that, and that's already pretty tricky. If you plot the oscillations in these things over time, they get very weird, very quickly.

JONES: I guess I don't know where your building blocks are coming from, because, as far as I made out, you don't do recombination yet. And when you do combat, you broke everything up, so you're not passing around building blocks. And when you do self-replication, you're not really passing around building blocks, you're just copying the entire organism. So I don't know where your building blocks are coming from.

HOLLAND: Terry, you're quite right. If I'm just doing mutation, I'm not going to see this very easily, I think. I'm not even sure I could prove the theorem. It takes recombination and, as I said, I was largely putting this thing aside. Go back to the notion of mating selection: Now two things come together, and now they exchange, in the usual crossover sense, their structure. In other words, in the simplest case, treat all that structure as one long chromosome, do a crossover, and exchange. That's where you can start to prove the schema theorem. Not before.

KAUFFMAN: A number of us, including you, have at various times looked at parallel processing networks in the classifier model. It would be awfully interesting to let your schemas that are inside the chromosomes be a chemistry, rather than merely a description, and work on one another to make—if you will—other kinds of molecules. And so you'd have lots of ways of generating new kinds of variants, and also they'd have to control their own dynamics. So it would be fun, then, to study whether or not they coordinate their dynamics internally to be in a chaotic regime, or an edge of chaos regime, or an ordered regime. Meanwhile, they're building up molecular diversity because they act as catalysts and substrates on one another. It would be a fun other connection to try to. . .

HOLLAND: This would be a much closer connection, for instance, to the kind of thing that Walter and that you've done earlier. I've run a few models with that, but you'd be amazed—well, maybe you wouldn't be amazed—at the amount of complexity that comes in once I can start transforming resources, and build little transformers in there.

SIMMONS: John, you talked about one kind of hierarchy of time scales: fast dynamics and slow dynamics. There's another kind that I'm not sure is in there, but maybe I don't understand the model. Namely, in the real world, some kinds of organisms will reproduce and do other things much more rapidly than others. Is it possible to build this in? Would you, or have you—or would you expect it to have—important effects?

HOLLAND: That's endogenous, Mike. If I've got a very long defense tag, it's going to take me much longer to accumulate those letters; more combats, more whatever. So that's endogenous.

SIMMONS: But does it take a long time to copy it? Is that built into the...

HOLLAND: No, once you've got the resources, copy time is considered to be short, relative to collection of resources. You could add it, but it's not in there.

FONTANA: Have you looked at a linkage disequilibrium emerging in your system? That is, you have building blocks, and do you get correlations within one chromosome?

HOLLAND: You've touched on something that Marc and I have discussed for a long time, and now have come much closer. I think both Marc and I now agree that if you look at population genetics, there are really two times to be studied. The classic study is so-called "time to convergence." And that's under the assumption of what they call—and I didn't understand it for a long time—an "infinite population." Now, an "infinite population" is not a large population; an "infinite population" is a population in which every genotype is already present. Then, time to fixation is the only thing to worry about. In real populations, of course, you only have an extremely small sample of all genotypes, and the other question to ask is "time to first discovery." That's a very different kind of question, and that comes up with things like linkage, disequilibrium; these play a much more important role in time to discovery than they do in time to fixation. And, yes, that's the kind of that we want to look at now, and the generalization that Freddy Christenson and Marc have done in the schema theorem is the thing that you would now try to apply to this kind of model.

One last comment, Murray, about times and hierarchies. It's easy to make this notion in terms of classifier systems, but it flies equally well here. If I have a rule that has as very coarse condition—that's very accepting of a wide range of messages in classifier systems—the sample rate is very high. Broad equivalence class on the condition, high sample rate because the repetition is high. Narrow equivalence class on the condition, low sampling rate. Now, ask yourself what's rational to do in terms of the information I already have, at some point. Clearly, it doesn't pay to build a rule where I only expect to have a fraction of a sample, at this point in time. So there's a natural evolution, I would claim, in such systems, from coarse to fine, because there's no advantage to building fine rules early. Now that hierarchy, in time, implies a great deal for the rest of the ontogeny of the system; a great deal, at least in classifier systems, and, I would claim, it holds here as well.

Alfred Hübler* and David Pines[†]
*Santa Fe Institute, 1660 Old Pecos Trail, Suite A, Santa Fe, NM 87501, and Center for Complex Systems Research, Beckman Institute, UIUC, 1110 West Green Street, Urbana, IL 61801-3080
[†]Physics Department, University of Illinois at Urbana-Champaign, 1110 West Green Street, Urbana, IL 61801-3080

Prediction and Adaptation in an Evolving Chaotic Environment

Abstract: We describe work in progress on computer simulations of adaptive predictive agents responding to an evolving chaotic environment and to one another. Our simulations are designed to quantify adaptation and to explore coadaptation for a simple calculable model of a complex adaptive system. We first consider the ability of a single agent, exposed to a chaotic environment, to model, control, and predict the future states of that environment. We then introduce a second agent which, in attempting to model and control both the chaotic environment and the first agent, modifies the extent to which that agent can identify patterns and exercise control. The competition between the two predictive agents can lead either to chaos, or to metastable emergent behavior, best described as a leader-follower relationship. Our results suggest a correlation between optimal adaptation, optimal complexity, and emergent behavior, and provide preliminary support for the concept of optimal coadaptation near the edge of chaos.

We consider adaptive predictors for a system specified by a logistic map dynamics in which the parameters evolve in a random fashion. The prediction is a single-step prediction of the dynamics of a single map within

Complexity: Metaphors, Models, and Reality
Eds. G. Cowan, D. Pines, and D. Meltzer, SFI Studies in the
Sciences of Complexity, Proc. Vol. XIX, Addison-Wesley, 1994

a nonstationsry network which provides the environmental dynamics. The model is a set of K_m functions and their weights, which relate the actual state of the map to a future state of the map. Therefore, K_m is a measure of complexity of the adaptive system. Since we keep the set of functions fixed, the weights are the only parameters of the actual model. They are extracted through a maximum likelihood estimation from the most recent history of the map. The length of the corresponding time series N_m, as well as the number of parameters of the model K_m, and the number of events N_m^{ig} which are ignored between succeeding modeling processes are adjusted by trial and error. $L_m^{ig} = N_m^{ig}/(N_m^{ig} + N_m)$ is called the level of ignorance. N_m describes certain features of the rationality of the adaptive systems, i.e., the number of events taken into account in order to predict and to optimize the quality function. The adaptive system could in principle determine all K_m parameters of its internal model from K_m events. Because noise is present, it has to use $N_m(\gg K_m)$ events, in order to reduce statistical errors. The statistical errors could be made smaller by taking more events from the past; however, since the map dynamics is not stationary, such data would introduce large systematic errors. Balancing the reduction of the statistical error against the increase of the systematic error gives an optimal N_m, i.e., an optimal bounded rationality.

We find (i) optimal adaptive predictors have an optimal rationality and an optimal complexity, which are small in a rapidly changing environment, (ii) that the predictive power can be improved by imposing chaos or random noise onto the environment, (iii) the predictive power and the maximum level of ignorance decrease linearly with the rate of change of the environment, and (iv) the typical time scale of the adaptive process equals the rate of change of the environment if the adaptive system is capable of modeling the experimental dynamics with a small number of parameters. In this case there is a simple way to detect optimal predictors experimentally. For competing adaptive predictors, a configuration appears to be most stable when one imposes a weakly chaotic dynamics on the environment and the other predicts this controlled environment, i.e., a leader-follower relation emerges, in which the leader imposes a weakly chaotic dynamics on the environment.

For a randomly evolving network of weakly coupled logistic maps we find that models of optimal adaptive predictors for individual maps which use a Fourier series for the modeling are (i) simple, i.e., the number of significant parameters of the model equals approximately the number of significant parameters of the dynamics of the environment, (ii) reproducible, i.e., the modeling process yields a unique set of model parameters, and (iii) meaningful, i.e., there is a simple relation between the control parameters of the experiment and the parameters of the model which makes it possible to predict future settings of those parameter values. The models have predictive power in the region of interest of the adaptive system, which may not necessarily overlap with the natural dynamics of the environmental system.

1. INTRODUCTION

Like complexity, adaptation is a key concept, albeit difficult to define (even in a biological context). Thus, while incorporating quantitative measures of adaptation would seem essential to an understanding of complex adaptive systems, such an approach is rarely followed. One reason may be that any definition of adaptation must be contextual. Hence, in considering the role which adaptation plays in physical, social, political, economic, or biological systems, it is necessary to specify its purpose, the environment in which adaptation takes place, and the time scale over which it occurs. These aspects are often difficult to quantify and thus, frequently ignored. For example, to a noneconomist it seems intuitively obvious that the relative adaptive ability of economic agents or institutions plays a role in economics. Yet, because it is difficult to incorporate even a crude model of adaptation into mathematical models, economists are generally reluctant to include adaptation in models of the economy.

If we define adaptation as the capacity for modification of goal-oriented individual or collective behavior in response to changes in the environment, it is useful to distinguish between two kinds of adaptation: learning and innovation. Learning might be characterized as experience-based response involving, say, the acquisition of a set of predetermined patterns, and then sorting through that pattern file to determine an optimal response. Innovation, on the other hand, could be characterized as involving the development of new (i.e., not previously known) patterns. Quite generally, it would seem useful to compare adaptation in both individuals (can one define the AQ [adaptive quotient] of an individual, or is this the same as IQ?) and in their collective behavior as social and political systems, and to study coadaptation at every level of development.

It is also useful to distinguish between passive and active adaptation. A passive adaptive agent responds to changes in the environment without attempting to modify the environment. An active adaptive agent exerts some measure of control or influence on the environment in an effort to improve predictive power. Coadaptation, the adaptive behavior of interacting agents, will typically involve some mixture of passive and adaptive response.

The questions for any student of adaptation are broad and numerous. Are some climates (environments) more favorable to adaptation than others? Is, for example, the boundary between order and disorder (called by some the edge of chaos[5,20,27]) particularly conducive to learning and innovation? What is the role played by adaptation in the development and persistence of institutions, be they firms, social, political, economic systems, or religions?

As a first step toward the development of an approach that incorporates a quantitative understanding of adaptation, we have carried out computer simulations of a "toy model" of a complex adaptive system: individual agents, operating in an evolving chaotic environment specified by a simple logistic map, seek to predict the future states of their environment by modeling and controlling it. A brief summary of our results, which will be reported in detail elsewhere,[15] is given here. Although elementary, and in large part calculable, our adaptive agents (predictors) meet the definition of a complex adaptive system proposed by Murray Gell-Mann earlier in this workshop.[10] Thus, the agents are:

- information-gathering entities,
- respond both to the environment and to one another,
- segregate information from random noise,
- compress regularities into a model (schema), and
- modify their internal characteristics to improve their predictive (adaptive) capacity.

In our model the chaotic environment to which an agent responds is specified by a simple logistic map, with parameters which can be altered, plus random or dynamic noise which can also be altered. A given agent may be either passive or active; thus, agents both respond to the environment (by receiving signals from it) and attempt to control it (by sending signals to it).[14] More specifically, an agent measures and models the chaotic environment and employs various control strategies to predict its future states. For each agent we give explicit quantitative measures of:

- adaptation (the predictive ability of the agent),
- complexity (the number of parameters used to specify an agent's, and model), and
- rationality (the data used in the modeling process).

We determine, both experimentally (via our computer simulations) and analytically, the conditions for optimal predictive behavior (adaptation), complexity, and rationality.

Our computer simulations demonstrate the consequences of competition between the two agents. Competition leads to chaos, if the agents follow typical learning strategies, or to emergent metastable behavior, if the agents develop a new learning strategy. Thus, we find metastable solutions (strategies) in which the two agents optimize their joint predictive capacities by coadapting in a leader-follower relationship. A sufficient condition for arriving at this joint strategy is the development of adaptive predictions which enable one agent to recognize the presence of another. Our results suggest a correlation between optimal adaptation, optimal complexity, and emergent behavior. Preliminary support is provided for the concept of optimal coadaptation near an order-disorder transition.[5,20,22,27]

The computer simulations were performed on a Silicon Graphics 340 VGX machine. The length of time for a given study varies from 1000 time steps (the number required to determine numerically the optimal response of a single agent) to 100,000 time steps (the number required to explore in detail the competition between two active agents which leads, over time, to their arriving at dynamic controls near the edge of chaos). For two agents, each time step required 2 sec of cpu time.

In Section 2 we specify our model, and consider the behavior of a single agent. We consider competition (and cooperation) between two agents in Section 3. Extensions and possible applications are discussed in Section 4. We present a mathematical discussion of quantitative results in the Appendix, Section A.

2. SYSTEMS WITH ONE AGENT

2.1 THE DYNAMICS OF THE ENVIRONMENT

We consider an environment described by a simple logistic map. The state of the environment at some time n determines its state at a later time $n + 1$:

$$y_{n+1} = py_n(1 - y_n), \tag{1}$$

where $0 \leq y_n \leq 1$ represents the state of the environment at time n, and $1 < p < 4$ is a control parameter. Each transition from y_n to y_{n+1} is an event. Depending on the choice of p, the environment may be stationary, periodic, or chaotic in nature.

The use of a simple logistic map to describe a typical high-dimensional complex environment may be considered an oversimplification. However, there is a long history of similar approaches in studies of physical systems. Most physical systems are in reality high-dimensional systems. A physical pendulum, often used as an example of low-dimensional motion, has many degrees of freedom. When stimulated by a short kick, such as the impact of an hammer, many kinds of vibrations may be stimulated. Usually those vibrations die out fast, which means that the dynamics settles down to a low-dimensional, approximate inertial manifold, as previously mentioned in Phil Anderson's presentation. Often, the trajectories are complicated but confined to a very small region on this approximate inertial manifold. If the inertial manifold and the flow vector field are smooth in this region, one may expand the flow vector field in a Taylor series and drop higher-order terms. In this case, the limiting dynamics is low dimensional and the nonlinearity of the corresponding flow vector field is of low order. Lorenz has shown[23] that low-dimensional, low-order systems can exhibit deterministic chaos, i.e., irregular motion which is sensitive to initial conditions. Since the flow vector field of such systems is smooth, their dynamics is usually smooth and oscillatory with trajectories which may have a simple or fractal geometry.[9,11] Moreover, Poincaré[28] has shown that it is in general useful to study the dynamics of the amplitudes of the smooth oscillatory motion of a low-dimensional, low-order system and to model it with low-order maps. Therefore, we model the environment with logistic map dynamics, a simple, nontrivial deterministic chaotic system.

In many systems of interest, noise is present and the control parameters vary over time. We thus wish to consider environments in which additive background noise is present, and in which the control parameter is noisy:

$$y_{n+1} = p_n y_n (1 - y_n) + F_n^{\sigma_D} \tag{2}$$

$$p_{n+1} = p_n + F_n^{\sigma_p}, \tag{3}$$

where $F_n^{\sigma_D}$ describes additive system noise and, $F_n^{\sigma_p}$ is parametric noise. The noise parameters $F_n^{\sigma_D}$, $F_n^{\sigma_p}$ have a mean which is zero and standard deviations, σ_D and σ_p. σ_p determines the rate of change of the environment, whereas σ_D measures the noise level. We assume that the rate of change of the control parameter of the environment is small, i.e., $\sigma_p \ll 1$. In Figure 1 we illustrate the dynamics of such an environment.

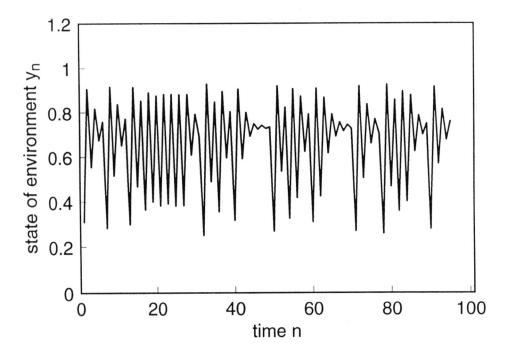

FIGURE 1   A typical time series of the environment specified by Eq. (2), with $\sigma_D = .006$, $\sigma_p = .0006$, and $2 < p_n < 4$.

2.2 THE DYNAMICS OF THE AGENT

PREDICTION, ADAPTATION, LEARNING, AND INNOVATION. In our numerical experiments, the primary goal of each agent is prediction of the environment to which it and, where present, other possible agents are responding. An agent seeks to discover patterns or regularities in the environment; this process of constructing a model of the environment is greatly facilitated if the agent has the option of turning on or off a control signal which may entrain the environmental dynamics to a predetermined sequence of states, defined as the **goal dynamics** of the agent, a process of active adaptation. As we shall see, by pursuing a reasonably well defined strategy, a single agent can arrive at a goal dynamics which maximizes its predictive power.

A given agent samples a set of N_n successive events $y_i, y_{i+1}, \ldots, y_{i+N_n}$, which characterize the time evolution of the environment, and uses it to predict the next m events, $\tilde{y}_{i+N_n+1}, \tilde{y}_{i+N_n+2}, \ldots, \tilde{y}_{i+N_n+m}$ The success of the prediction process of the agent is measured by the **prediction error**, $\epsilon_{n,m,M}$ defined with respect to the background noise level, σ_D, as:

$$\epsilon_{n,m,M} = \frac{1}{M} \sum_{i=n}^{n+M} \frac{(\tilde{y}_{i+m}(y_i) - y_{i+m})^2}{\sigma_D^2} , \tag{4}$$

where $\tilde{y}_{i+m}$ is a m-step prediction of the environment by the agent. The prediction error is a sliding average of length M over an ensemble of rapidly fluctuating values. It depends on the time of prediction, n, and the number of steps predicted, m. The best prediction is limited by the background noise level σ_D. From Eq. (4) the minimum value of the prediction error is 1.

At each step n, the agent can modify its model. This process of modification, or updating, represents the **adaptive behavior** of the agent. For our adaptive agents, **learning** is a two step process:

- first, acquire and apply a predetermined class of schemata or fitting functions to model the past events of the environment, and
- second, choose fitting function parameters that do the best job of prediction.

In the present context, **innovation** involves the development and application of a new and different class of models to analyze the past and predict the future. It may include active adaptation, an exploration of the response of the environment to an imposed goal dynamics. We assume that the time scale of the innovation process is in general longer than the time scale of the learning process N_n. We also assume that the probability an agent decides to try to improve a parameter of the modeling and control processes is given by:

$$P_n = c\epsilon_{n,m,M}. \tag{5}$$

Here c (a constant) is the minimum rate of adaptation and $\epsilon_{n,m,M}$ is the current m step prediction error of the agent averaged over M time steps. The minimum rate of adaptation, c, may be different for each parameter of the modeling and control processes. Typical values of c are such that $10^{-6} \le c \le 10^{-3}$.

The agent begins with a given initial setting of the model and control processes and observes the environment for at least N_n time steps. It then calculates an initial model and updates the model and prediction error at each subsequent time step. In addition, at each time step, the agent selects a random number, R, which lies between 0 and 1. If $P_n < R$, the agent does nothing; if $P_n \ge R$ the agent innovates. With this procedure, the probability that an agent will innovate is simply P_n, as long as $P_n \le 1$. Whenever $P_n > 1$, the agent will always innovate. Thus, after say, t time steps the agent will innovate, i.e., try another class of fitting functions, switch the control on, or alter the goal dynamics. t is short if the prediction error is large and vice versa, according to Eq. (5). The expectation value of t is P_n^{-1} Equation (5) also guarantees perpetual novelty; no matter how well an agent is doing, that agent will, sooner or later, be prompted to innovate. For optimally predictive agents, whose prediction error is of order unity, the minimum rate of innovation is c. The trial period for innovation is assumed to last $2M$ steps. After that period the parameter is reset to its previous value if the prediction error has increased on average during the last M steps of the trial period.

LEARNING: MODELING AND PREDICTION. We now express the prediction error as a function of the parameters of the modeling and control processes in order to identify the attractors of the adaptation process and to study their stability. In the simulations, we restrict the attention of each agent to states of the environment that lie within a predetermined region of interest. Those areas of the region of interest in which few or no observations are available are usually of little importance for the prediction process. However, it is possible to spoil the modeling process even in the regions where a sufficient number of events is present. For example, making a poor interpolation in regions which contain few events can substantially increase the number of model parameters or make a unique determination of the model parameters impossible. In order to handle this problem, we introduce an equidistant grid $y_{j,n}^{\text{grid}}$, j=1,..,N_G, $N_G \gg N_{n_m}$ and estimate the events $(y_{j,n}^{\text{grid}}, f_{j,n}^{\text{grid}})$ at these grid points through linear interpolation[15] (see Figure 2). The interpolation is a generalization of the observed events, since the agent must guess the response of the environment for those states where no observations are available.

The agent separates information from random noise by constructing a model from the generalized observed events. The model parameters are determined by a least-squares fit which minimizes the difference between the generalized events and single-step predictions of the observed events. We find that under some circumstances, the fit problem has a unique solution.

If we define **complexity**, K_n, as the number of model parameters used by the agent, it is possible to determine the complexity K_n^{opt} of an optimal model. The optimal model neglects all parameters with value smaller than the error bar for that parameter, estimated by its standard deviation. This concept is illustrated in

Figure 3(a)-(b). The minimum prediction error in Figure 3(a) is at approximately $k = 10$. In Figure 3(b), the Fourier coefficients equal the error bar also at $k = 10$. The optimal complexity is small for a large rate of change of the environment and vice versa.

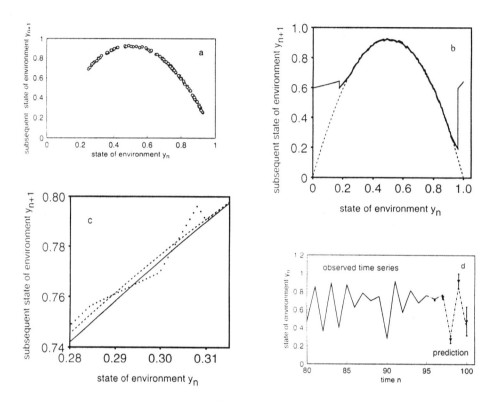

FIGURE 2   A state-space representation, y_{n+1} versus y_n, of the time series plotted in Figure 1(b): The dashed line represents the exact map $f(y_n)$, the dotted line is a linear interpolation between the observed events, $f_{j,n}^{\text{grid,agent}}$, and the continuous line is a Fourier approximation of the dotted line, $f_n^{\text{Lmodel,agent}}(y_n)$. The observed events are in the region $0.2 < y_n < .92$. The extrapolation of the model outside this region is not smooth, but can be made smooth with an appropriate choice of $d_{0,n}^{\text{model}}$ and $d_{0,n}^{\text{model}}$; (c): an enlargement of the center region of (b); (d): The predicted evolution of the time series shown in Figure 1. The dot-dashed line is the system behavior, while the triangles denote the predicted values; the errors associated with the prediction are also shown.

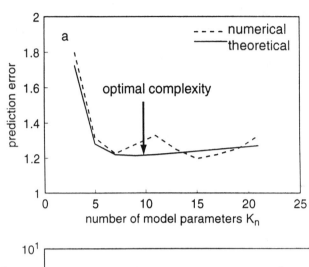

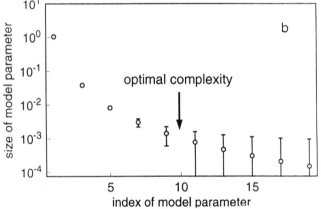

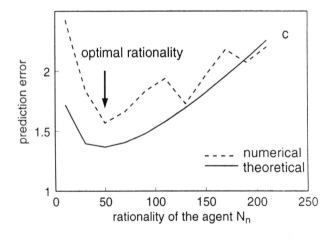

FIGURE 3 (a): The single-step prediction error as a function of the number of model parameters K_n; all Fourier coefficients with $k < K_n$ are included in the model. The parameter values are $\sigma_D = 0.006, \sigma_p = 0, N_n = 100$, and $p_n = 4$. (b) The size of the model Fourier coefficients and the size of their standard deviation as a function of the model parameter k. (c): The prediction error as a function of the number of events used to extract the model. The parameter values are $\sigma_D = 0.006, \sigma_p = 0.0003$, and $K_n = 10$. The dashed lines represent numerical results, and the continuous line is the analytic result.

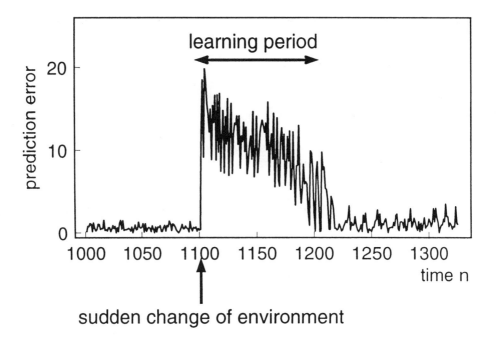

FIGURE 4 Experimental detection of adaptive agents. At time 1100 the parameter of the environmental dynamics is suddenly changed from 4.0 to 3.5. The single-step prediction error increases significantly, but returns to its original value within 100 steps, which equals N_n. The parameters are $\sigma_D = 0.006, \sigma_p = 0, N_n = 100, p_n = 4$, and $K_n = 10$.

Another quantity that can be optimized by an agent is the number N_n of states of the environment which are used in the modeling process. In principle, an agent is assumed to have access to the whole history of the environmental dynamics. In practice, the agent will find it advantageous to use only a small portion of this information. Since N_n measures how much of this information is used for the modeling process, N_n is a measure for the **rationality** of the agent. If N_n is large, outdated data may decrease the quality of the model. If N_n is too small, statistical errors may prevent an agent from choosing an optimal description. Therefore, there is an **optimal rationality** N_n^{opt} of an agent, which strikes a balance between the errors introduced by the noise level and those produced by the rate of change of the environment. In Figure 3(c) we show how a proper choice of N_n minimizes the prediction error.

Since N_n data are used to fit the model, a delay of N_n time steps is needed to model the environment after a sudden change of the parameter p_n. Therefore, it is possible to establish a relation between the complexity of the agent's model and the **learning rate**, the minimum time required to extract a completely new model from the environmental dynamics. This result provides a method to determine experimentally whether an adaptive agent is functioning optimally. To evaluate an agent's performance an observer can introduce a sudden change of the environment and then measure the recovery of an adaptive agent, as shown in Figure 4. A match between the recovery rate and the optimal learning rate indicates that the agent is optimally adapted.

An important feature of the modeling process is that the resulting model parameters are unique. If the relation between the evolving control parameters of the environmental dynamics and the model parameters is continuous, it may be possible to apply the same modeling procedure to the time series of the model parameters and to construct hierarchical models, of the kind considered by Jim Crutchfield.[5]

CONTROL OF THE ENVIRONMENT. The process just described represents passive adaptive behavior of the agent. However, an agent can modify the environment in an effort to improve his predictive power, by turning on or off a control signal, which may entrain the environmental dynamics to a particular goal dynamics. Our control strategy for such an active agent is based on the approach developed by one

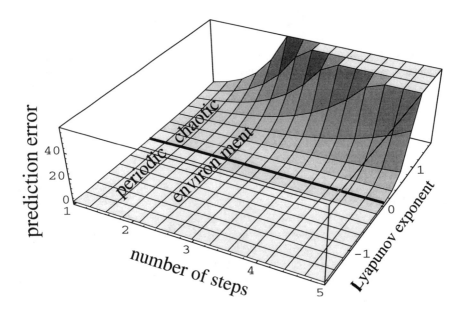

FIGURE 5   Multiple-step prediction error of a passive agent as a function of the number of time steps and the Lyapunov exponent of the environmental dynamics.

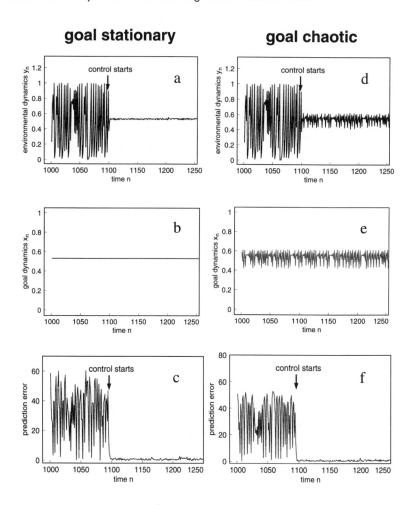

FIGURE 6 Situations in which it is advantageous for an agent to become active. Depicted is the the five-step prediction error as a function of the parameter of the goal dynamics and the parameter of the environment for the case without control (a) and the case with control (b) and (c), the latter being an enlargement of (b); (d) a schematic illustration of the different regions. The parameter values are $\sigma_D = 0.006, \sigma_p = 0, N_n = 100$, and $K = 10$.

of us[13] for the control of chaos. The agent applies a suitably chosen driving force to **entrain** the chaotic environment to a predetermined goal dynamics.

There is a close relation between entrainment and optimal prediction. Entrained oscillators may have an optimal energy exchange since they are at resonance[8] while the concept of optimal information transfer has widespread application in research on phase locked loops.[6,7] As we shall see, for the problem at hand entrainment makes optimal prediction possible.

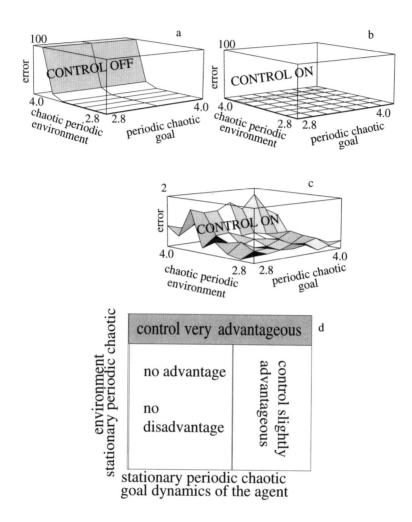

FIGURE 7   Numerical simulations of active adaptation. (a) The evolution of the environment, y_n, subject to an active agent with a goal dynamics $p_n^{\text{goal}} = 2.6$, illustrated in (b), which turns on its control at $t = 1100$, (c) depicts the resulting prediction error, which is seen to drop significantly soon after the environmental dynamics becomes entrained by the goal dynamics. (d)-(f) show a similar series of plots for the case that the goal dynamics is chaotic ($p_n^{\text{goal}} = 3.8$). The parameter values are $\sigma_D = 0.006, \sigma_p = 0, N_n = 100, p_n = 4$, and $K = 10$.

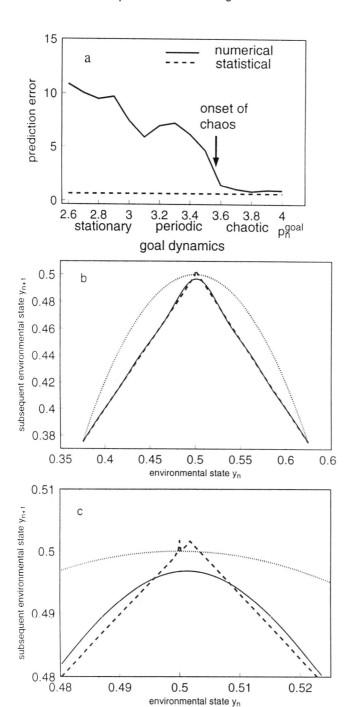

FIGURE 8 (a): The single-step prediction error of an active agent as a function of the parameter, p_n^{goal} of the goal dynamics, for $\sigma_D = 0.0002, \sigma_p = 0, p = 4, N_n = 100$, and $k_n = 10$ (continuous line denotes the numerical results, dashed line, the statistical estimate). (b) The exact map (dotted line), is compared to the linear interpolation (dashed line), and the model map (continuous line) versus y_n. (c): An enlargement of (b).

To control the environmental dynamics an active adaptive agent iterates a logistic map time series x_n of desired environmental states; this **goal dynamics** is specified by a parameter p_n^{goal}:

$$x_{n+1} = g(x_n, p_n^{\text{goal}}) \tag{6}$$

to impose the goal dynamics on the environment, the agent applies a driving force;

$$F_n = x_{n+1} - f_n^{\text{model}}(x_n), \tag{7}$$

tailored to make up the difference between the agent's model of the uncontrolled environmental dynamics and the desired state of the environment. The resulting controlled environment dynamics is given by:

$$y_{n+1} = p_n y_n (1 - y_n) + F_n + F_n^{\sigma_D} \tag{8}$$

$$p_n = p_n + F_n^{\sigma_P}, \tag{9}$$

Control is advantageous. It enables the agent to avoid the exponential growth of the prediction error with the number of steps, found in the case of chaotic systems with positive Lyapunov exponents (see Figures 5 and 6). If the agent succeeds in controlling the environment, the prediction error becomes bounded; the upper boundary becomes small if the control which is exercised is stable over long periods of time. For an example of the way in which control reduces the prediction error, see Figure 7.

The type of the goal dynamics has both a direct and indirect impact on the prediction error. Since the size of prediction error depends on the stability of the control, an agent may improve the prediction error by choosing a goal dynamics which provides a very stable control. From this point of view, the prediction error would be as small as possible if the goal dynamics is, or is close to a stationary state, or some other superstable stationary orbit[29] of the unperturbed system.

However, this discussion takes only statistical errors into account.

If the goal dynamics is a stationary state, for example at $x_n = 0.5$, the interpolation procedure may lead to large systematic errors. This is illustrated in Figure 8. There we assume the region of interest is the whole interval and that f is known at the boundaries of the interval. In Figure 8(a) we show that the prediction error may be significantly higher than the statistical estimate as long as the goal dynamics is not chaotic, i.e., $p_n \leq 3.56$. This is because the linear interpolation produces edges which are very sharp for stationary states close to $x_n = 0.5$ (Figure 8(b)), much less sharp for period-two dynamics, and essentially absent for a chaotic goal dynamics. Of course, other interpolation schemes could weaken this effect, but the best solution to this problem is to pick a chaotic or random goal dynamics which covers the entire state space and makes interpolations unnecessary. Likewise, a small amount of additive noise in the environmental dynamics or the goal dynamics may help to reduce the prediction error, since it reduces systematic errors in the modeling process.[1,2] Moreover a control with a chaotic goal dynamics makes the system more robust against sudden changes in the noise level, since the agent has a global model of the flow vector field.

3. SYSTEMS WITH TWO AGENTS

As might be expected, the results change dramatically when two agents are present. For example, a second agent may alter the environmental dynamics of the first agent sufficiently to make it impossible for the latter to exercise effective control and make accurate predictions. Or, without establishing direct communication, one agent may identify the presence of a second, and the two may establish a cooperative relationship which improves their joint predictive abilities.

There are three scenarios for dual agent behavior: both may be passive; one may be passive while the other is active, i.e., exercises control to improve its predictive power, a leader-follower situation; or both may be active, vying for control (and optimal predictive power). We consider these scenarios in turn, assuming that each agent is capable of optimizing its rationality and complexity in response to an environment in which the other agent is absent.

3.1 TWO PASSIVE AGENTS

While both agents are passive, the prediction error of each will be as though the other agent were not present. Since an agent may improve its predictive power by exercising control, it is to be expected that the "two passive agent" scenario will be of comparatively short duration. Following a trial period, one or the other agent will turn on its control, and the scenario becomes one of leader (the active agent) and follower (the passive agent).

3.2 THE LEADER-FOLLOWER RELATION

While the second agent remains passive, the scenario for the behavior of the first agent, the leader, is identical to that of a single agent. The leader explores various controls, entrains the environment, and improves his predictive power. However, as this process of active adaptation proceeds, the passive second agent, the follower, senses a changed environment. If the leader switches off its control infrequently, the follower will see an environment which is mainly determined by the goal dynamics of the control exercised by the leader. Such a controlled environment is easier to predict as long as the goal dynamics of the leader is not chaotic. Under these circumstances, the optimal rationality of the follower may, in fact, exceed that of the leader, whose rationality depends on both the noise level and rate of change of the environment. The results of our numerical experiments on the role played by the goal dynamics of the leader are displayed in Figure 9.

The passive agent will eventually try to improve its predictions by switching on a control; when he does, the environment will follow his goal only if it is more complicated than the goal of the agent which is already active. For example if the active

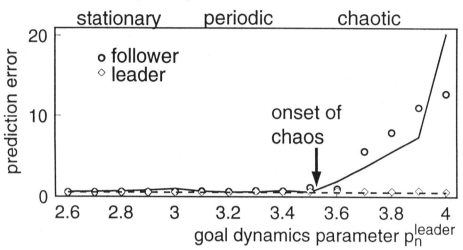

FIGURE 9 The prediction error of the leader and follower in the leader-follower configuration, as a function of the goal dynamics of the leader. The circles represent the results of our numerical experiments; the continuous line the analytic result. The parameter values are $\sigma_D = 0.006$, $N_n^1 = N_n^2 = 100$, $K_n^1 = K_n^2 = 10$, and $p_n = 4.0$.

agent entrains the environment to a period-four dynamics by using a period-four driving force, the other agent cannot disentrain the environment to a period-two dynamics with a period-two driving force, since a period-two entrainment would be disturbed by the period-four driving force of the first agent. However, if the second agent tries to entrain the environment to a period-eight dynamics, this may be stable if the period-four driving force of the first agent is taken into account by the second agent. Another situation arises if the goal dynamics of the follower is chaotic. It is then difficult to compensate the periodic driving force of the leader and the environment does not entrain with the follower during the trial period. An overview of situations in which it is advantageous for the follower to become active is given in Figure 10. In region A the follower has a more complicated dynamics than the leader. Therefore, it can entrain the environment successfully and does not increase its prediction error by becoming active. In region B the goal dynamics of the follower is simple compared to that of the leader. The prediction error increases significantly if the follower becomes active. In region C the goal dynamics of the follower is chaotic and usually does not lead to entrainment of the environment. Therefore, it is not advantageous for the follower to switch on its control in regions B or C. In the region D the goal dynamics of the leader is chaotic. This leads to very large prediction errors for the follower and makes control for the follower almost

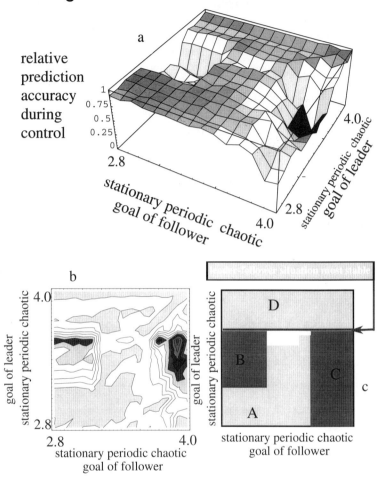

Two Agents: Leader-Follower situation

FIGURE 10　An overview of situations in which it is advantageous for the follower to become active. Depicted is the ratio between the five-step prediction error in the trial period of an active follower and mean prediction error of a passive follower as a function of the parameter of the goal dynamics of leader and follower. The surface plot (a) and the contour plot (b) are numerical results; (c) illustrates the different regions schematically. The parameters used in the simulation are $\sigma_D = .006, p_n = 4, K_n^1 = K_n^2 = 10$, and $N_n^1 = N_n^2 = 100$.

always advantageous. If the leader chooses a goal dynamics at the edge of chaos, i.e., $p_n^1 \approx 3.56$, the probability that the follower would become active is minimized, since regions B and C contain the entire range of p_n^2 values. Therefore, the leader follower-relation is most stable for a goal dynamics at the edge of chaos.

If we assume that challenge of the follower is successful if the complexity of its goal dynamics is larger than that of the leader, the lifetime of the leader-follower relation increases as p_n^1 increases. However, as soon as the goal dynamics of the leader becomes chaotic the prediction error of the follower rises sharply. This shortens the time span between two trial periods of the follower and makes it much more likely that the follower challenges the leader successfully. Therefore, the lifetime of the leader-follower configuration has a maximum for parameters which are close to $p_n^1 \approx 3.56$, the edge of chaos. As may be seen in Figure 11, our numerical studies indicate that **leader-follower configuration** at the edge of chaos is the most stable among all configurations including those where both agents are active, or both are passive.

Since a leader-follower configuration with a goal dynamics at the edge of chaos possesses a long lifetime, this configuration may be considered as an emerging structure. Despite the fact that no social aspects are included in the system, since both agents only seek to optimize their own prediction error, a structure with social aspects emerges, in which one of the agents takes the lead and the other agent follows the moves of the leader.

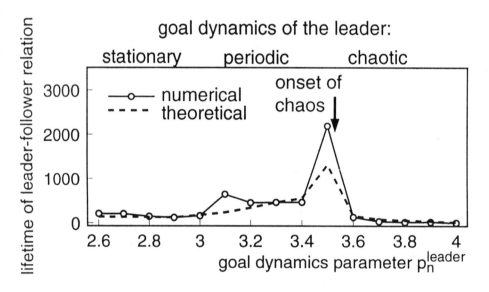

FIGURE 11   The lifetime of the leader-follower configuration as a function of the parameter of the goal dynamics of the leader, when the parameter of the goal dynamics of the follower is randomly chosen in the interval $2.6 < p_n^{c,1} < 4$. The continuous line represents the numerical result, the dashed line, the analytic result. The parameters are $\sigma_D = .006, p_n = 4, K_n^1 = K_n^2 = 10$, and $N_n^1 = N_n^2 = 100$.

3.3 TWO ACTIVE AGENTS

As noted above, the second agent will not remain passive, no matter how well it predicts its environment. It will, in time, switch on its controls in an attempt to improve. The system dynamics, for the general case of two active agents, with different goal dynamics, is complicated. If both agents use a control dynamics with a positive exponent, it leads to hyperchaos. While it is difficult to estimate analytically the prediction errors and configuration lifetimes, we observe in our numerical experiments that the resulting system dynamics settles into a state in which the environmental dynamics follows very closely one of the two imposed goal dynamics, in general the one which is more complicated (see Figure 12). Thus, if the goal dynamics of one agent is a stationary state, while the goal dynamics of the other agent

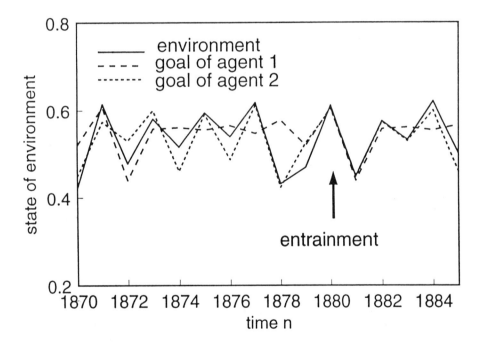

FIGURE 12 Numerical simulation of an environment subject to active agents which possess similar goal dynamics. Most of the time, the environment is disentrained; however it locks with the two agents when they utilize similar goal dynamics, such as at time $n = 1880$. The parameters are $\sigma_D = 0.006, N_n^1 = N_n^2 = 100, K_n^1 = K_n^2 = 10, p_n = 4.0$, and $p_n^1 = p_n^2 = 3.8$.

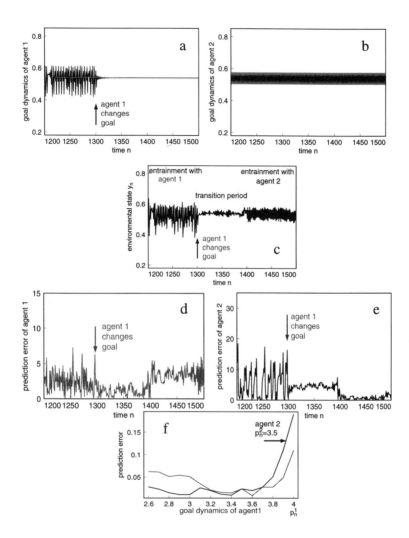

FIGURE 13 Two typical situations involving active agents. One agent has a chaotic goal dynamics up to time $n = 1300$ and then switches to a stationary goal (a), whereas the other agent has a periodic goal dynamics (b) for the whole time span. In (c) we show the response of the environment. Up to time $n = 1300$ it is entrained by the chaotic dynamics of the first agent. The first agent is the winner. When this agent switches to a stationary goal, the environment becomes entrained by the other agent after a transition period of 100 steps, which equals the rationality of the agent. In (d) and (e) we show the five-step prediction error of the agents. (f): This plot shows the five-step prediction error of both active agents as a function of p_n^1, for a fixed goal dynamics $p_n^2 = 3.5$ of one agent. We see that the agent with the larger p_n value, i.e., the more complex goal dynamics, is the winner. The parameters are $\sigma_D = 0.006$, $N_n^1 = N_n^2 = 100$, $K_n^1 = K_n^2 = 10$, and $p_n = 4.0$.

is a period-two dynamics, the environmental dynamics would follow the period-two dynamics. If one agent uses a highly chaotic goal dynamics, for example $p_n^1 = 4$ and the other agent a less chaotic goal dynamics, for example $p_n^2 = 3.6$, then the environment follows the first agent's goal. This "competition principle" would suggest that the agent with the more complicated goal dynamics has a good chance to outperform his competitor, as illustrated in Figure 13. We also observe that the driving force of the unsuccessful competitor tends to approach a constant, even if its goal dynamics is not stationary.

In this winner-loser configuration, the prediction error of the successful competitor is the same as for an active agent without any competitor, whereas the prediction error of the unsuccessful competitor is significantly larger. The difference between the two prediction errors can be estimated by the average difference of the goal dynamics of the two agents.

Since the unsuccessful competitor can usually improve his prediction error by switching his control off, the lifetime of the winner-loser configuration depends only on time span between two trial periods of the loser, and is determined by the prediction error of the unsuccessful competitor. The loser may be considered to be maladapted since his goal dynamics is too simple. In this case the maladapted agent would lose the competition. Preliminary results indicate that other types of maladaptation, such as suboptimal rationality or suboptimal complexity, may also lead an agent to lose the competition.

Our preliminary studies also indicate that time span required before the winner-loser configuration emerges from a state where both agents are active increases with the Lyapunov exponent of the goal dynamics of the successful competitor. This observation suggests that an agent that has a slightly more complicated goal dynamics than his competitor has the best chance for a rapid improvement of his prediction error.

3.4 ADAPTATION TO THE EDGE OF CHAOS

A scenario for transitions between successive configurations is presented in Figure 14. It takes the following form:

- First configuration: **two passive agents**; second configuration: one agent switches on its control and becomes the leader; the prediction error of both agents typically goes down. One thus arrives at the **leader-follower configuration**.

- While comparatively stable, the leader-follower configuration will not persist because as noted earlier, no matter how low the prediction error of the follower, it will eventually switch on its control. Most of the time this will not initially increase its prediction error. However the prediction error of the leader will increase; the leader will then respond by adjusting his controls, reducing his prediction error, but increasing that of the "follower." The follower will in

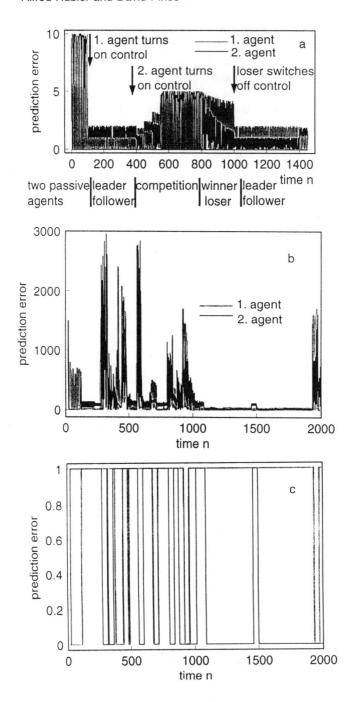

FIGURE 14   (a) This schematic figure shows a typical sequence of events; (continued)

FIGURE 14   (b) the results of a numerical simulation for the parameter values $\sigma_D = 0.006, N_n^1 = N_n^2 = 100, K_n^1 = K_n^2 = 10, p_n = 4.0, p^1 = 3.5, p^2 = 2.8, p^{\text{control}} = 0.0002$; and (c) the dynamics of the control coefficients which indicate whether the agents are active or passive.

turn challenge the leader, who loses his leadership role. This chain of events is full blown **competition** which can continue for a long time. Eventually there emerges a clear **winner** (the agent with the more complicated goal dynamics) and a clear **loser**.

- The **winner-loser** configuration is, of course not stable, since on the average, the loser, who has the larger prediction error, will switch off his control, returning the system to

- the **leader-follower** situation in which the loser significantly improves his prediction error.

In the course of the competition each agent will find, by trial and error, that when it improves a goal dynamics with a complexity greater than that of the apparent leader, its prediction powers improve as it assumes a leadership role. On the other hand, the prediction error of the "new" follower increases sharply when the goal dynamics of the "new" leader is chaotic; this condition will lead it, in turn, to shorten the time between two successive trial periods, and make it likely that the "new" follower becomes active after a short period of time. Eventually the leader will use a goal dynamics at the edge of chaos. This configuration is the most stable one, since the follower no longer find it advantageous to switch on his control. This is illustrated in Figure 15.

3.5 DISCOVERING OTHER AGENTS

It is natural to ask whether one agent can detect the presence of a second. If the second is passive, the answer is obviously no. If, however, that other agent is active, the answer is yes. The strategy for detection is to switch on a control: if the control works, there is no other active agent present. If, however, the environment cannot be entrained to a simple goal dynamics, it is likely that another agent is already entraining the environment. Our adaptive agents use this information in the sense that if they detect the other agent and therefore experience an increase of their prediction error during the trial period, they become passive. However, if they do not detect another agent, i.e., improve their prediction error during the trial period, they become active themselves.

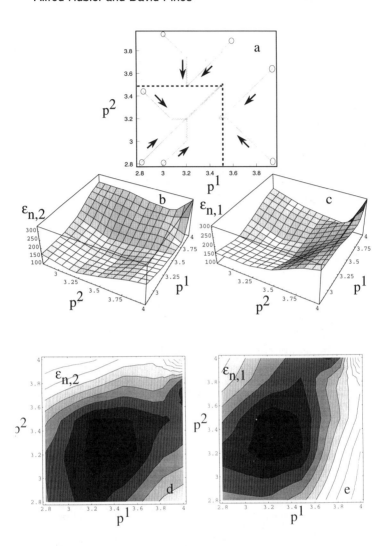

FIGURE 15   (a) The evolution of the "goal dynamics" parameters, as revealed by numerical experiments with different initial conditions, denoted by a circle. The agents are seen to adapt to the edge of the chaotic regime, indicated here by a dashed line. (b) and (c) are surface plots of the the prediction error of the two agents as a function of the parameters of the goal dynamics. (d) and (e) are contour plots of (b) and (c). The parameters are $\sigma_D = 0.006, N_n^1 = N_n^2 = 100, K_n^1 = K_n^2 = 10, p_n = 4.0, c^{\text{control}} = 0.0002$, and $c^{\text{goal}} = 10^{-5}$.

4. SUMMARY

The adaptive agents we have studied, although following a comparatively simple strategy, are able to identify regularities, generalize and compress observed data by using different sets of schemata and explore strategies to change their environment. We find that quantitative measures of adaptation in a complex environment, such as complexity, or the learning rate of the agents, approach limiting values that depend on the rate of change of the environment and the noise level in the environment. We hope it is possible to use these measures to test theoretical predictions experimentally in physical systems, such as phase-locked loops, as well as in economic systems. In our examination of the coevolution of two agents we observe the emergence of leader-follower configurations, in which the leader entrains the environment to a weakly chaotic dynamics. This suggests that a primary goal, such as a small prediction error, or a large return for an investment in an airline company in a highly competitive market, may cause secondary goals which are chaotic: the chaotic goal dynamics for the environment in this study or a chaotic dynamics of the pricing of products such as air fares.

It will be interesting to explore the relationship of the tendency of the agents to move to the edge of chaos with the general idea of adaptation to the edge of chaos as discussed by Langton,[22] Packard,[27] Kauffman,[20] and Crutchfield.[5]

It is also interesting to speculate on the applicability to real-world situations of some of the results we have obtained from our toy model. For example, we have seen that active adaptation, using control to improve prediction, is under most circumstances the preferred strategy. This finding seems in accord with experience, whether one is analyzing the way an infant develops predictive power by exercising control of its immediate environment (parents) through cries or smiles, or analyzing the behavior of two interacting adults. Consider, too, the attempt by traders on a stock market in the early minutes after the opening, to exercise control and improve their short-term predictive powers, by carrying out a series of trades designed to probe, actively, likely subsequent trading patterns on that day.

To cite another example; we have seen that in active competition between agents, the agent with the more complicated strategy will win. This finding accords with experience on the political scene. It helps suggest why, for example, the Serbs have proved so successful to date in pursuing their strategy of ethnic cleansing; their strategy may be regarded as one involving a series of controlled experiments which enable them to predict the UN response.

Although the system that we study is simple, it makes possible a quantitative comparison between numerical and analytical results. We hope that some of our findings for simple adaptive agents in a chaotic environment are applicable to the behavior of real adaptive agents in complicated economic and/or biochemical systems. Of course, to extend our approach to economic systems it is important to incorporate into our model the cost of constructing a model, exercising control, etc. Still, even at the present level, our approach would seem to provide insight into the success of technical trading systems. Technical trading systems usually use time bars to describe the spread of values of a time series, whereas in physics and engineering the variance is commonly used for that purpose. It can be shown[15] that for chaotic time series time bars are maximum likelihood estimates of the spreading of the data in contrast to the variance. To the extent that economic time series are chaotic, this could explain why it is advantageous to use time bars for analyzing their behavior.

We intend in the future to extend in a number of ways the numerical experiments presented here. We plan to study the competition between agents of markedly different adaptive capacities (as manifested both in the ability to model the environment and to control it) and to extend our approach to many interacting agents in order to examine possible collective behavior. While we have seen that the outcome of the present simulations of the interaction between two agents appears to lead to either a "win-win" or a "win-lose" situation, we anticipate that the actions of a powerful maladaptive agent can lead to a "lose-lose" situation, and it will be interesting to specify the conditions under which this comes about. As our program is currently written, no matter how well an agent is doing, it will, over time, seek to improve its predictive powers by changing its strategy, which means that all the configurations we have considered are metastable. We therefore plan to modify our innovation paradigm, Eq. (5), by introducing a threshold for change; we expect that the resulting "happy agent" configuration may lead, in some circumstances, to stable "win-win" configurations. Finally, in order to make more direct contact with economics, we intend to introduce both a cost of computation and a cost of control in our numerical experiments.

In another direction, we call the attention of the reader to a closely related set of independent numerical experiments carried out by Kaneko and Suzuki[19] on a model for the evolution of the complex song of a bird. They use a simple logistic map for the song dynamics and consider a two person game between competing "birds." Their agent "birds" adapt to one another, exercising control through their songs. Kaneko and Suzuki find that the dynamics of the complex song evolves toward the edge of chaos. It will be interesting to explore the relationship between their simulations and our own, and to see to what extent a complicated environment might play a role in that evolution.

ACKNOWLEDGMENTS

We wish to thank John Miller for his helpful advice in the early stages of our specification of this model as well as his cautions concerning its immediate applicability to economics, and Gottfried Mayer-Kress for advice on the development of our graphic displays and for stimulating discussions on these and related topics. We thank Bill Fulkerson for a critical reading of a preliminary version of this manuscript, and a number of helpful suggestions. This work was begun at the Santa Fe Institute with support from a Robert Maxwell Professorship, and has been subsequently supported both by the Santa Fe Institute and by the Center for Complex Systems Research at the Beckman Institute of the University of Illinois at Urbana Champaign; we thank both institutions for their support. The present version of our manuscript has profited from the informal remarks of our fellow participants at the Integrative Themes Workshop, whom we thank for their advice and encouragement.

A. APPENDIX

In this appendix, addressed to the mathematically inclined reader, we spell out some relevant details of our numerical experiments.

A.1 THE ENVIRONMENTAL DYNAMICS

The environmental dynamics (Eq. (2)) is given by a logistic map dynamics. The dynamics of the logistic map converges to a nonzero fixed point for $1 \leq p \leq 3$, labeled as stationary in our plots; it converges to period-2 cycle for $3 \leq p \leq 3.449449\ldots$, labeled as periodic, and to more and more complicated period-2^{β} cycles, $\beta = 2, 3, \ldots$ for $3.449449\ldots \leq p \leq p^{\text{crit}}$. Above $p^{\text{crit}} = 3.569946\ldots$ the attractor of the map is chaotic, except in special "periodic windows."[16] In our plots this last parameter region is labeled as chaotic. The Lyapunov exponent of the map, defined by

$$\lambda^e = \lim_{N \to \infty} \frac{1}{N} \sum_{n=1}^{\infty} \ln \left| \frac{df(y)}{dy} \right|_{y=y_n} \Big|, \tag{10}$$

may be positive for $p_{1,n} > p_{1,n}^{\text{critical}}$ and is negative otherwise; here $f(y) = py(1-y)$.

A.2 MODELING

An agent develope a model, f_n^{model}, of the environmental dynamics for single- and multiple-step predictions of the environment

$$\tilde{y}_{n,m}(\tilde{y}_{n,m-1}) = f_n^{\text{model}}(\tilde{y}_{n,m-1}) \tag{11}$$

where $\tilde{y}_{n,0} = y_n$ and where $m = 1, 2, 3\ldots$ counts the number of steps. This is based on the assumption that a good predictor for the observed events is a good predictor for future events.

The first step of the modeling process is for the agent to represent the observed events in a state space (see Figure 2). In the simulations, we restrict the attention of each agent to events that lie within a region of interest, a range of events such that $y_n^{\min,\mathrm{I}} \le y \le y_n^{\max,\mathrm{I}}$. $y_n^{\min,\mathrm{I}}$ and $y_n^{\max,\mathrm{I}}$ are the boundaries of the region of interest at time step n. In **modeling** the environmental dynamics each agent uses those N_n events which are most recent and in which the initial state is in the region of interest. In our numerical examples the region of interest is usually slightly larger than the region where events have been observed during the first 1000 time steps of each simulation. Further we introduce a equidistant grid $y_{j,n}^{\mathrm{grid}}$, $j=1,..,N_G$, $N_G \gg N_{n_m}$ and estimate the events $(y_{j,n}^{\mathrm{grid}}, f_{j,n}^{\mathrm{grid}})$ at these grid points through linear interpolation[15] (see Figure 2). The interpolation represents a generalization of the observed events, since the agent is guessing the behavior of the environment for those states where no observations are available.

The relation between $y_{j,n}^{\mathrm{grid}}$ and $f_{j,}^{\mathrm{grid}}$ is represented by a Fourier series:

$$f_n^{\mathrm{model}}(y_n) \equiv \sum_{k=1}^{N_G} p_{k,n}^{\mathrm{model}} \sin\left(\frac{\pi}{y_n^{\max,\mathrm{I}} - y_n^{\min,\mathrm{I}}} k(y_n - y_n^{\min,\mathrm{I}})\right)$$
$$+ d_{0,n}^{\mathrm{model}} + d_{1,n}^{\mathrm{model}}(y_n - y_n^{\min,\mathrm{I}}) \tag{12}$$

where $d_{0,n}^{\mathrm{model}}$ and $d_{1,n}^{\mathrm{model}}$ are parameters of the Fourier analysis chosen to improve the convergence of the Fourier series. Unless we specify otherwise, we assume that the observed events are almost homogeneously distributed in the region of interest, and that interpolation errors are small compared to statistical errors. In this case the standard deviation σ_{p_k} of the Fourier coefficients is given by $\sigma_{p_k} = \sigma_D \sqrt{2/N_n}$.

The last step of the modeling process is to compress the information which is contained in the generalized observed events by segregating information from random noise. The values of the model parameters $p_{k,n}^{\mathrm{model}}$ are determined by a least-squares fit which minimizes the prediction error by minimizing the difference between the generalized events at zero noise level and a model for single-step predictions (Eq.(12)) for the observed events. If we assume that the observed states are homogeneously distributed in the region of interest, that the additive noise is uncorrelated and that the rate of change of the environment is small, the fit problem has a unique solution and the optimal parameters $p_{k,n}^{\mathrm{model}}$ are given given by the Fourier coefficients $\tilde{p}_{k,n}^{\mathrm{model}}$ of $f_{j,n}^{\mathrm{grid}}$,[15] or are equal to zero, if the Fourier coefficient is smaller than its standard deviation σ_{p_k}. An analogous procedure, pruning, is followed in neural nets, where it is found to improve their performance.[12]

The environmental dynamics is a parabolic function with only one parameter. Therefore, a Tschebycheff series or a Legendre series would converge even faster than the Fourier series since these are also polynomial. However we intentionally program the agents to use a Fourier series in order to illustrate the point that an exact match between the set of models (schemata) and environmental dynamics is not necessary.

The Fourier coefficients of many continuous, piecewise linear functions converge parabolically[3], i.e., $\tilde{p}_{k,n}^{\mathrm{model}} \approx \tilde{p}_{c,n}^{\mathrm{model}}/k^2$, with an appropriate choice of $d_{0,n}^{\mathrm{model}}$ and

$d_{1,n}^{\text{model}}$ and converge linearly otherwise i.e., $\tilde{p}_{k,n}^{\text{model}} \approx \tilde{p}_{c,n}^{\text{model}}/k$, where $\tilde{p}_{c,n}^{\text{model}}$ is a number. For parabolic convergence an estimate of K_n^{opt} is given by

$$K_n^{\text{opt}} = \left(\frac{\tilde{p}_{c,n}^{\text{model}} \sqrt{N_n}}{\sqrt{2}\sigma_D} \right)^{1/2} \tag{13}$$

The prediction error for an agent with complexity K_n can be estimated by writing

$$\epsilon_{n,1} = 1 + K_n \frac{1}{N_n} + \sum_k \frac{(\tilde{p}_{k,n}^{\text{model}})^2}{2\sigma_D^2} \delta_{k,n} + \sum_k \left(\frac{p_{k,n}^{\text{model}}\sigma_p}{p_n \sigma_D} \right)^2 \frac{N_n}{6} \tag{14}$$

$$\epsilon_{n,m+1} = \epsilon_{n,m} \exp\left(2\lambda^e\right) \tag{15}$$

where $m = 1, 2, \ldots$ and where $\delta_{k,n} = 1$ if $p_{k,n}^{\text{model}} = 0$ and $\delta_{k,n} = 0$ otherwise.

The last term of Eq. (14) increases with the rate of change of the environment σ_p and N_n. Since the first term in Eq. (14) decreases with N_n, the prediction error $\epsilon_{n,l}$ is minimal for

$$N_n = N_n^{\text{opt}} = \left(\frac{6K_n^{\text{opt}}}{\sum_k \left(\frac{p_{k,n}^{\text{model}}}{p_n} \right)^2} \right)^{\frac{1}{2}} \frac{\sigma_D}{\sigma_p} \tag{16}$$

as depicted in Figure 3. Agents with optimal rationality and optimal complexity possess a prediction error,

$$\epsilon_{n,1}^{\text{opt}} = 1 + 2 \left(\frac{K_n^{\text{opt}} \sum_k (\frac{p_{k,n}^{\text{model}}}{p_n^2})}{6} \right)^{1/2} \frac{\sigma_p}{\sigma_D} + \frac{(\tilde{p}_{k,n}^{\text{model}})^2}{2\sigma_D^2} \delta_{k,n} . \tag{17}$$

The noise level in the environment and the rate of change of the environment determine the optimal rationality of the agent. If the Fourier series converges parabolically the optimal complexity of the agent with optimal rationality is:

$$K_n^{\text{opt}} = \left(\frac{3(\tilde{p}_{c,n}^{\text{model}})^4}{2 \sum_k (\frac{p_{k,n}^{\text{model}}}{p_n})^2 \sigma_D^2 \sigma_p^2} \right)^{1/7} \tag{18}$$

A.3 CONTROL

Model-based control (Eqs. (8), (9)) does not require immediate feedback from the environment in contrast to closed-loop controls.[21] We assume that the agents need a minimum period of time $\Delta n_{\min} > 1$ to evaluate an observation at time n and to predict the next state of the environment. Further, we assume the agent needs only insignificantly more time to compute multiple step predictions, such as $\Delta n = \Delta n_{\min} + \gamma m$ where $\gamma < 1$ and m counts the number of steps. If $m > 1 - \gamma/\Delta n_{\min}$ an agent can, in principle, predict future states of the environment. In a chaotic environment the prediction error rises exponentially with m and makes predictions for large m impossible. We will show later that control makes it possible for the agents to improve the quality of multiple step predictions significantly, and enables agents to predict the future even if Δn and γ are not small. In addition, it can be shown that controls are surprisingly stable against random noise, even if the parameters of the model differ from the exact parameter settings.[4] However, it seems to be difficult to estimate the exact boundary of the basin of attraction for a given control.[17,18] Therefore, we focus our discussion on the situation where the control is known to be stable,[25] i.e., where the control coefficient:

$$\lambda_n^c = \lim_{N \to \infty} \frac{1}{N} \sum_{n=1}^{\infty} \ln \left| \frac{df(y)}{dy} \right|_{y=x_n} \right|$$ (19)

is less than zero. For a logistic-map dynamics all controls which have a goal dynamics in the interval $(3/8, 5/8)$ are stable.[26] We therefore use in all our numerical investigations a goal dynamics which is in this interval, such as a logistic map dynamics which is scaled onto this interval

$$g(x_n) = \frac{z_{n+1}}{4} + \frac{3}{8}$$ (20)

where $z_{n+1} = p_n^{\text{goal}} z_n (1 - z_n)$, $0 \le p_n^{\text{goal}} \le 4$, and $0 \le z_n \le 1$. The goal dynamics may simultaneously have several attractors, with the same states but different phases. For example, period-two limit cycles may at some time n be in the upper state or in the lower state. In order to achieve entrainment, each agent tries to pick that phase which is as close as possible to the phase of the environmental system, i.e., it memorizes the minimum $y^{\min}$ value of the environmental dynamics and the minimum value of its goal dynamics $z_n^{\min}$. As soon as the environmental dynamics comes close to $y^{\min}$ the agent resets its goal dynamics to $z_n = z^{\min}$. The parameter p_n^{goal} determines whether the Lyapunov exponent of the goal dynamics

$$\lambda_n^g = \lim_{N \to \infty} \frac{1}{N} \sum_{n=1}^{\infty} \ln \left| \frac{dg(x)}{dx} \right|_{x=x_n} \right| \ .$$ (21)

is positive or negative, i.e., whether the attractor of the goal dynamics is chaotic or periodic.

If an agent is successful in controlling the environment, the prediction error is given by:

$$\epsilon_{n,1} = 1 + \exp\left(2\lambda^c\right) + K_n \frac{1}{N_n} + \sum_k \frac{(\tilde{p}_{k,n}^{\text{model}})^2}{2\sigma_D^2} \delta_{k,n}$$

$$+ \sum_k \left(\frac{p_{k,n}^{\text{model}} \sigma_p}{p_n \sigma_D}\right)^2 \frac{N_n}{6}, \tag{22}$$

$$\epsilon_{n,m+1} = \epsilon_{n,m} \exp\left(2\lambda^c\right) + \epsilon_{n,1}. \tag{23}$$

Equation (22) shows that the single-step prediction error increases slightly as agents attempt to control the environment; however, the multiple step error may be significantly reduced. If the uncontrolled environmental dynamics is chaotic, i.e., $\lambda_n^e > 0$ the prediction error increases exponentially with the number of steps m, whereas in the case of a stable control, i.e., $\lambda_n^c < 0$ the prediction error does not increase beyond a certain value, even if the goal dynamics is chaotic, i.e., $\lambda_n^c < 0$. Therefore, if the agent tries to improve his multiple-step predictions, it is advantageous for him to control a chaotic environment. If the control is very stable, i.e., $\lambda_n^c \ll -1$, the prediction error for a controlled environment is the same for single-step predictions and multiple-step predictions. In Figure 7 we depict the results of numerical experiments for different types of control.

A.4 SYSTEMS WITH TWO AGENTS

We assume that the agents have the same region of interest but may have different goal dynamics, and different models of the environmental dynamics. To distinguish between the parameters of the two agents, we attach a superscript $a = 1, 2$ to the model parameters $p_n^{\text{model},a}$, the rationality N_n^a, the complexity, K_n^a, the parameter of the goal dynamics p_n^a, the ON/OFF switch of the control, C_n^a, the Lyapunov exponent of the goal dynamics $\lambda_n^{g,a}$, the control coefficient $\lambda_n^{c,a}$, the prediction error $\epsilon_{n,m}^a$ and the control force F_N^a. The dynamics of the environmental system is then given by:

$$y_{n+1} = p_n y_n (1 - y_n) + F_n^1 + F_n^2 + F_n^{\sigma_D}, \tag{24}$$

$$p_n = p_n + F_n^{\sigma_p}. \tag{25}$$

In the following we discuss the prediction errors of the two agents for the situation where both are passive, i.e., both have their control switched off, both are active, i.e., both have their control switched on, and the leader follower situation, in which one is active (leader) and one is passive (follower). We study the lifetime of those structures, i.e., the number of time steps between the start and end of such a configuration, where trial periods do not count as the end of a configuration if this configuration reemerges after the trial period. Unless we specify otherwise, we assume that both agents have a rationality and complexity which would be optimal for single-agent systems.

As noted in the text, if **both** agents are **passive** the prediction error of each agent is the same as for single passive agents. For a chaotic environmental dynamics the prediction error is usually very large compared to the leader follower situation. Therefore, this configuration typically ends as soon as one agent becomes active during the trial period.

If **both** agents are **active**, with one emerging as a successful competitor, the other as an unsuccessful competitor, a rough estimate of the prediction error is given by:

$$\epsilon^1_{n,1} = 1 + \exp\left(2\lambda^{c,1}\right) + K_n \frac{1}{N_n} + \sum_k \frac{(\tilde{p}^{\text{model},1}_{k,n})^2}{2\sigma^2_D} \delta_{k,n}$$

$$+ \sum_k \left(\frac{p^{\text{model},1}_{k,n} \sigma_p}{p_n \sigma_D}\right)^2 \frac{N_n}{6}, \tag{26}$$

$$\epsilon^1_{n,m+1} = \epsilon^1_{n,m} \exp\left(2\lambda^{c,1}\right) + \epsilon^1_{n,1}, \tag{27}$$

for the successful competitor (winner), which is labeled here as the first agent, and by

$$\epsilon^2_{n,m} \approx \frac{1}{N_n} \sum_{i=n}^{n+N_n} \left(\frac{x^1_i - x^2_i}{\sigma_D}\right)^2 + \epsilon^1_{n,m+1} \tag{28}$$

for the unsuccessful competitor (loser).

The prediction error of the winner is the same as for a single active agent, whereas the prediction error of the loser is generally dominated by a term that measures the average difference between the two goal dynamics. The lifetime $L_{n,m}$ of this configuration can be estimated from the equation,

$$L_{n,m} = \frac{c}{\epsilon^2_{n,m}}. \tag{29}$$

In the **leader-follower situation**, where one agent is active and one agent is passive, the prediction error can be estimated by:

$$\epsilon^1_{n,1} = 1 + \exp\left(2\lambda^{c,1}\right) + K_n \frac{1}{N_n} + \sum_k \frac{(\tilde{p}^{\text{model},1}_{k,n})^2}{2\sigma^2_D} \delta_{k,n}$$

$$+ \sum_k \left(\frac{p^{\text{model},1}_{k,n} \sigma_p}{p_n} \sigma_D\right)^2 \frac{N_n}{6}, \tag{30}$$

$$\epsilon^1_{n,m+1} = \epsilon^1_{n,m} \exp\left(2\lambda^{c,1}\right) + \epsilon^1_{n,1}, \tag{31}$$

for the leader, and by

$$\epsilon_{n,1}^2 = 1 + K_n \frac{1}{N_n} + \sum_k \frac{(\tilde{p}_{k,n}^{\text{model},2})^2}{2\sigma_D^2} \delta_{k,n} + \sum_k \left(\frac{p_{k,n}^{\text{model},2} \sigma_p}{p_n} \sigma_D \right)^2 \frac{N_n}{6}, \quad (32)$$

$$\epsilon_{n,m+1}^2 = \epsilon_{n,m}^2 \exp\left(2\lambda^{g,1}\right) + \epsilon_{n,1}^2, \quad (33)$$

for the follower. The lifetime of the leader-follower configuration can be estimated by:

$$L_{n,m} = \begin{cases} \frac{c}{\epsilon_{n,m}^2} \frac{p^{\max} - p^{crit} + p_n^1}{p^{\max} - p^{\min}} & \text{for} \qquad p_n^1 < p^{crit} \\ \frac{c}{\epsilon_{n,m}^2} & \text{otherwise} \end{cases} \quad (34)$$

where $p^{\min}$ and $p^{\max}$ are the boundaries of the parameter range. The second term in Eq. (34) decreases for larger p_n^1 whereas the first term increases sharply when the goal dynamics becomes chaotic at $p_n^1 = 3.56$. Therefore, the lifetime has a maximum close to $p_n^1 = 3.56$ the edge of chaos (see Figure 11). This means that if the leader chooses a simple periodic goal dynamics, it is very likely that he will be successfully challenged by the follower, since the probability is high that the goal dynamics of the follower is more complicated than the goal dynamics of the leader. However, if the goal dynamics of the leader is highly chaotic, than the follower will also challenge the leader often since the follower's prediction error is poor. Equation (34) is only a rough estimate since it does not account for the periodic windows[16] of the logistic map dynamics, which are important at low noise levels.[24]

The lifetime of a configuration with two passive agents may be estimated by:

$$L_{n,m} = \frac{c^2}{\epsilon_{n,m}^1 \epsilon_{n,m}^2} \quad (35)$$

where the prediction errors are the same as of single passive agents. For a chaotic environment, passive agents have a large prediction error. Therefore, the lifetime of a configuration with two passive agents is quite short.

REFERENCES

1. Breeden, J., F. Dinkelacker, and A. Hübler. "Noise in the Modeling and Control of Dynamical Systems." *Phys. Rev. A* **42** (1990): 5827–5836.
2. Breeden, J., and A. Hübler. "Reconstructing Equations of Motion from Experimental Data with Hidden Variables." *Phys. Rev. A* **42** (1990): 5817–5826.
3. Bronshtein, I. N., and K. A. Semendyayev. *Handbook of Mathematics*, edited by K. A. Hirsch, 581–591. New York: Van Nostrand Reinhold, 1985.

4. Chang, K., S. Kodogeorgiou, A. Hübler, and E. A. Jackson. "General Resonance Spectroscopy." *Physica D* **51** (1991): 99–108.

5. Crutchfield, J. P., and K. Young, "Inferring Statistical Complexity." *Phys. Rev. Lett.* **63** (1989): 105–108.

6. de Sousa Vieira, M., and A. J. Lichtenberg. "On Sychronization of Regular and Chaotic Systems." Memorandum No. UCB/ERL M92/72, University of California at Berkeley, 1992. To appear in *Phys.Rev. E*.

7. Endo, T., and L. O. Chua. "Synchronization of Chaos in Phase-Locked Loops." Memorandum No. UCB/ERL M91/59, University of California at Berkeley, 1991.

8. Eisenhammer, T., A. Hübler, T. Geisel, and E. Lüscher. "Scaling Behavior of the Maximum Energy Exchange between Coupled Anharmonic Oscillators." *Phys. Rev. A* **41** (1990): 3332–3342.

9. Feigenbaum, M. "Quantitative Universality for a Class of Nonlinear Transformations." *J. Stat. Phys.* **19** (1978): 25–52.

10. Gell-Mann, M. "Complex Adaptive Systems." This Volume.

11. Grassberger, P., and I. Procaccia. "Characterization of Strange Attractors." *Phys. Rev. Lett.* **50** (1983): 346–349.

12. Hertz, J., A. Krogh, and R. G. Palmer, eds. *Introduction to the Theory of Neural Computation.* Santa Fe Institute Studies in the Science of Complexity, Lect. Vol. I. Redwood City, CA: Addison-Wesley, 1991.

13. Hübler, A. "Modeling and Control of Nonlinear Systems." Ph.D. Thesis, Technical University of Munich, Germany, 1987.

14. Hübler, A., and E. Lüscher. "Resonant Stimulation and Control of Nonlinear Oscillators." *Naturwissenschaften* **76** (1989): 67–69.

15. Hübler, A., and D. Pines. In preparation.

16. Jackson, E. A. *Perspectives of Nonlinear Dynamics*, Vol. 1, 142–225. Cambridge: Cambridge University Press, 1991.

17. Jackson, E. A. "Controls of Dynamic Flows with Attractors." *Phys. Rev. A* **44** (1991): 4839–4853.

18. Jackson, E. A., and S. Kodogeorgiou. "Entrainment and Migration Controls of 2-Dimensional Maps." *Physica D* **54** (1992): 253–265.

19. Kaneko, K., and J. Suzuki "Evolution to the Edge of Chaos in Imitation Game." Presentation at the Artificial Life Conference, Santa Fe, NM, June 1992, and preprint, 1992.

20. Kauffman, S. A., and R. G. Smith. "Adaptive Automata Based on Darwinian Selection." *Physica D* **22** (1988): 68–82.

21. Keefe, L. R. "Two Nonlinear Control Schemes Contrasted on a Hydrodynamic-Like Model." Nielsen Engineering and Research, Mountain View, CA. To appear in *Physics of Fluids A*, March 1993.

22. Langton, C. G. Ph.D. Thesis, University of Michigan, 1988.

23. Lorenz, E. N. "Deterministic Nonperiodic Flow." *J. Atmos. Sci.* **20** (1963): 130.

24. Mayer-Kress, G., and H. Haken. "The Influence of Noise on the Logistic Model." *J. Stat. Phys.* **26**, (1981): 149–171.

25. Merten, J., B. Wohlmuth, A. Hübler, and E. Lüscher. "Beschreibung und Steuerung des Getrieberauschen in einstufigen Getrieben." *Helv. Phys. Acta.* **61** (1988): 88–91.

26. Ohle, F., A. Hübler, and M. Welge. "Adaptive Control of Chaotic Systems." In *Proceedings of the Twelfth Turbulence Symposium*, edited by X. B. Reed, Jr., A11-1-A11-9, University of Missouri-Rolla, 1990.

27. Packard, N. H. "Adaptation Toward the Edge of Chaos." In *Dynamics Patterns in Complex Systems*, edited by J. A. S. Kelso, A. J. Mandell, and M. F. Schlesinger, 293–301. Singapore: World Scientific, 1988.

28. Poincaré, H. *Les Methodes Nouvelles de la Mechanique Celeste.* Paris: Gautier-Villars, 1892.

29. Schuster, H. G. *Deterministic Chaos*, p. 38. Weinheim: Physik-Verlag, 1984.

DISCUSSION

ANDERSON: I think you have here a theory which is very relevant to at least one important economic situation, namely, a two-person poker game; the leader-follower behavior, and so forth, remind me very much of strategies in poker.

HÜBLER: Actually, this is a pretty common management strategy where we just attach some numbers and labels on it. Because if you have a manager of a certain company, he needs to keep an overview of his environment. Therefore, he needs to introduce from time to time a new product, or something like this. But in addition, if he makes it too wild, he upsets the customers, or he upsets his employees. So you need to have, probably, something which is weakly chaotic. Now, this is known by managers. Some of the things we've added is we attached some numbers to that.

ANDERSON: I think there are certain marriages that are like this.

PINES: In fact, Phil, a fundamental question for our students of complex adaptive systems is this: take that two-person game, called a marriage, and ask does it function best when you're at the edge of chaos.

SIMMONS: It may not be best, but it certainly functions there!

KAUFFMAN: Alfred, I think this is just truly beautiful work, and I just wanted explicitly to draw a couple of connections that I was trying to point to in my own talk, when I was talking and pointing to you. One is the potential relationship between this and the story of coevolution to the edge of chaos, and these sort of coupled funny lattice models. And the second, I'll both state and then ask you a question about.

When I was trying to think about the bounded rationality issue, and—very much influenced by what you'd done—I finally thought about adaptive agents with no exogenous dynamics. In other words, remove the notion of an exogenous world out there. I'm just trying to do something, and so are you, and what I'm trying to do is build on the model that I have of you, and vice versa. Once you have bounded data, the bounded data now demands that you use a model of intermediate complexity, which, therefore, in turn, implies that the models that we have of one another will be inaccurate, which means that resulting dynamics between us could be chaotic, could be ordered. We could go to a mutually consistent state, with the third step then that if we're in a mutually consistent notion of rational expectations, we'll have more reliable data, and therefore build more complex models which will destabilize the system. Whereas, if we're in a chaotic regime, where I'm changing my model of you, and you of me, very often, we'll make simpler models and move back towards the ordered regime. It might also bring you right to the phase transition.

Do you think it could be used, or do you think that your techniques here could be used to construct some simulations, for example, of that kind of a coevolutionary process—but without an exogenous dynamics that everybody's trying to capture?

HÜBLER: Yes, I think that this is probably just a simplification of what we did. Probably you can kick out just environment. So, the surprising fact would be that maybe if you started from the situation where everything is stationary—this means one agent is making of the other agent a model, and tries to control the other agent, and everything is stationary—and, suddenly, you would find that the whole system of two agents moves to a state where they impose on each other a chaotic dynamics, and you would ask why are they doing that. And the reason for that would be that, if they impose a stationary state on each other, they lose the overview. They have large systematic errors in their modeling process. However, if they make a big, chaotic dynamics, they can reduce these systematic errors in the modeling process, and therefore they have better predictability. However, if they make it too wild, then they get in trouble with the Lyapunov exponent, and therefore they wouldn't do it; therefore, they evolve at the edge of chaos. So I think one could simplify this model to what you suggested.

LICHSTEIN: Just to follow up on Stu's comment: I'm not an expert on it, but I do talk to the people involved in derivative trading and trading in general. These are the guys who are not studying them, but trying to make money out of them. What they do talk about is the fact that their models work for a period of time, until the world reacts to them, and then they don't. But there are two interesting phenomena: One is that the spreads, or the margins, on which they're operating are continuously, over time, getting smaller, thinner. So that is a monotonic trend. The other one is, however, that they can play bigger and bigger games, so there's still an awful lot of money to be made. But the phenomenon you two are talking about, I think, is played out in the real world.

SIMMONS: Let me tell you what I think I understood of what you said, and then ask a question based upon that. You have a very nice model of a world in which there is a more or less exogenously imposed dynamic, with a chaotic environment, and you have two players who are seeking to make predictions and to control their environment. And you have very nice definitions of adaptation, optimal rationality, and the like. Now, I might imagine...not a real world, but maybe a slightly more realistic world, in which coevolution might take place, which is composed only of a collection of agents that are competing with each other, so that the environment is not this exogenously imposed, chaotic system, but is, in fact, nothing but the other agents. And I'm wondering if your work has any light to shed on that.

HÜBLER: Okay, I think this is a simplification, and I think something very similar happened. Because the first thing, what the agents are doing, these two agents: they kill the natural chaotic dynamics, and impose their own chaotic dynamics, in order to make it predictable. So I think this is a very good suggestion, and it probably will simplify the whole system a lot. I think you can remove it and get the same results; I'm almost certain. You don't need the environment. Because the agents kill it anyway.

GELL-MANN: I would like to make a related remark, which is that you can have adaptation, even in the limit of not having a changing time series for the environment. If you're trying to predict it, for example—or also react to it—then, if you're a long way from solving the problem (say, playing chess, not tic-tac-toe), then you can have adaptation without bothering to change the time series.

The other thing is that it's rather special, I think, to have control, as opposed to behavior with consequences that may approximate control under certain conditions. For example, as they said, in marriage actual control is rare. So in that respect, this is specialized, and more general systems—most of the things that people have considered at SFI—involve behavior where there is some effect, but not necessarily control.

HÜBLER: Okay. You are right, that this is just a special case. Because you could do the following: You could say, "I make a Taylor expansion of this function here." And what you control is the constant parameter. So if you do a Taylor expansion of this—perhaps you have something like $a_0 + a_1 y + a_2 y^2 + \ldots$ and so forth. What this additive driving force is doing, it is controlling the first parameter of this Taylor series. But of course, you could try to control the other parameters, too, and this one probably would not call control, but one would modify the behavior of the other agent. So you are right, this is a special case of modifying the behavior of the other agent

Peter Schuster

Institut für Molekulare Biotechnologie, Beutenbergstr. 11, O-6900 Jena, Germany; Institut für Theoretische Chemie der Universität Wien, Währingerstr. 17, A-1090 Wien, Austria; and Santa Fe Institute, 1660 Old Pecos Trail, Santa Fe, NM 87501, USA

How do RNA Molecules and Viruses Explore Their Worlds?

Abstract: RNA molecules in replication assays, viroids, and RNA viruses are considered as adaptive systems of minimal complexity. They use the Darwinian trial-and-error mechanism for optimization. In order to compete successfully with the defense mechanisms of the host, the viruses must tune efficiency and speed of optimization as high as possible by maximizing the tolerable fraction of mutants. Hence, the error rates lie very close to the critical threshold values.

Value landscapes, understood as mappings from sequence space into the real numbers, are computed for RNA secondary structures. They are highly complex potential functions and have to be explored in optimization of phenotypic properties. Correlation lengths of free energies were computed for different chain lengths of the RNA molecules. Different base-pairing alphabets (**AUGC, GC, AU**, etc.) were considered. Free-energy landscapes derived from two-letter alphabets have substantially smaller correlation lengths than those of their four-letter analogues. Accordingly, they have more local maxima and optimization is more difficult.

Complexity: Metaphors, Models, and Reality
Eds. G. Cowan, D. Pines, and D. Meltzer, SFI Studies in the
Sciences of Complexity, Proc. Vol. XIX, Addison-Wesley, 1994

The statistical investigation of properties derived from secondary structures is extended to the structures themselves. Secondary structures are converted into equivalent trees. A tree distance obtained by tree editing is used as a measure of the distance between structures. The tree distance induces a metric on the space of all structures called the shape space. Probability density surfaces cast most of the information stored in the mapping from sequences into shapes into a simpler three-dimensional object. They are computed for RNA secondary structures, and correlation lengths are derived from them. Structures formed by two letter **GC** sequences were found to be much more sensitive to mutation than their four-letter (**AUGC**) analogues.

RNA shape space was investigated by inverse folding, which is tantamount to searching for sequences forming the same secondary structure. The probability distribution of structures is sharply peaked: there are relatively few common structures and many rare structures. Sequences that fold into common structures are distributed randomly in sequence space. Inspection of structure-probability densities and the data on inverse folding show that almost all common structures are found a small neighborhood of any random sequence. There is an extended neutral network in shape space: almost any neutral sequence can be reached from almost any other neutral sequence in steps of Hamming distance, one or two without ever changing shape. Consequences of these results for evolutionary optimization and molecular biotechnology are discussed.

MINIMAL STRATEGIES FOR ADAPTATION AND LEARNING

Adaptation is often used in different and conflicting meanings. Herefore, we ought to be precise in what sense this term will be used. All systems respond to changes in the environment. A system at equilibrium, for example, commonly relaxes to another equilibrium state when it is perturbed. The new equilibrium is completely determined by the changed environmental conditions. Steady states and dynamical systems that are oscillating or chaotic will show analogous *passive* responses to environmental change. Complex adaptive systems reply *actively*, and changes caused by the active replies are added on top of the passive responses. An active reply manifests itself in a nontrivial modification of the adaptive system. Complex adaptive systems, in general, have many degrees of freedom. Active replies need neither be reproducible in detail, nor invertible, since essentially the same changes in properties and functions can be achieved in many different microscopic ways. Do complex adaptive systems learn? The answer is "yes" if we understand learning as a combination of exploration of the environment and improvement of performance through adaptive change. Such a working definition of learning will certainly appear too simpleminded in the eyes of behavioral biologists and psychologists. It has the advantage, however, of being very general and it allows us to find primitive forms of learning in systems that are sufficiently simple, making them accessible to analysis by the conventional methods of physics and chemistry.

What are the minimal requirements for adaptation? A very simple, if not the simplest, active reply to the environment consists of trial and error adaptation according to Darwin's principle of multiplication, variation, and selection. Needless to say, there is a rich variety of other, more complex adaptive responses, but we shall not be concerned with them here. Minimal requirements for Darwinian adaptation are easily stated:

- a population of objects that are capable of replication (and heredity),
- a sufficiently large set of variants of these objects,
- occasionally occurring inheritable replication errors leading to variants, and
- selection through restricted proliferation caused by limited resources.

All four requirements can be met by replication of nucleic acid molecules in cell-free media. Thus, the Darwinian mechanism of evolution is no privilege of cellular life. It can be observed equally well in test tubes or other suitable chemical reaction setups.

Variation, in essence, is based on two different principles: mutation and recombination. The first case represents a deviation in genotype of the offspring from the parent. It is commonly caused by erroneous replication. Recombination uses two parental genotypes and produces a mixed offspring by partial exchange of genetic information. Here we shall be concerned only with variation through mutation (see also the next section). The (sub)set of accessible variants has to be sufficiently large, preferentially of combinatorial complexity, in order to guarantee a rich reservoir of forms from which selection can choose. Nucleic acids are heteropolymers and, accordingly, the set of variants is indeed of combinatorial complexity (there are as many as κ^{ν} different polymer sequences for κ classes of monomers with sequence length ν; $\kappa = 4$ for natural nucleic acids).

Adaptive selection requires nothing but nonvanishing differential fitness between variants. Manfred Eigen has shown in his pioneering study[6] how differential fitness can be expressed by kinetic rate constants of replication, mutation, and degradation. It is not difficult to compute expressions for differential fitness provided the replication mechanism and the selection constraints are precisely known. Changes in genotypes are not always consequences of adaptations. In cases of vanishing differential fitness, sequences may vary by random genetic drift. From mere observation of a change in the sequence, it is impossible to tell whether a mutation is adaptive or neutral. Motoo Kimura developed a stochastic model of population dynamics that casts random drift into precise and quantitative testable terms. It is the basis of his neutral theory of evolution.[30]

We can try now to give an answer to the initial question concerning molecules and learning. Individual molecules do not learn. Replicating molecules, however, form time series or genealogies. The ultimately surviving molecular types are modified along their genealogies in such a way that their performance improves monotonously (at least in the average over several generations). Thus, populations of replicating molecules do indeed learn, and the mechanism of learning is Darwin's trial-and-error principle. Learning in this sense is based on heredity and, consequently, the template mechanism of nucleic acid replication is essential.

MOLECULAR EVOLUTION

The simplest entities that do actually fulfill all four requirements listed above are RNA molecules. They are known to replicate *in vitro* in cell-free assays that contain activated monomers, a specific enzyme-catalyzing replication (a replicase), and a suitable RNA molecule as template. Sol Spiegelman and co-workers were the first to show that selection,[37,44] and adaptation to changes in environmental conditions,[31] do indeed occur in assays of this kind. In order to point at the analogy between molecular evolution and conventional evolutionary biology, we identify the sequence of bases (the monomeric units of the RNA molecule chosen from the four classes **A**, **U**, **G**, and **U**) with the genotype. Accordingly, and in agreement with Sol Spiegelman's view, the spatial structure of a RNA molecule is considered as its phenotype. An adaptation to the environment implies a change in the RNA sequence and, like other genotypes, RNA molecules carry a record on previous adaptations. This record may be deciphered by the same techniques as applied in molecular phylogeny of DNA organisms.[13,32] Because of the much higher mutation rates the use of statistical geometry[10,12] is essential. Such investigations give, for example, important insights into the epidemiology of RNA viruses.

The capability for molecular replication is found with both classes of nucleic acid molecules, RNA and DNA. The principle is template copying by means of complementation of a single strand to a double helical structure. Expressed in thermodynamic terms complementarity is the strong energetic preference for $(G \equiv C)$ and $(A = U)$ base pairs in the geometry of the double helix of nucleic acids that leads to an (almost) exclusive formation of these Watson-Crick-type base pairs. Boolean logic is very rare in chemistry and requires strong nonlinearities that have their origins in special structures (like here), or in special dynamics (for example, in proofreading mechanisms). Template chemistry is not restricted to nucleic acid molecules; other examples of complementary template interactions were found recently. They represent candidates for unnatural molecular replication systems.[38]

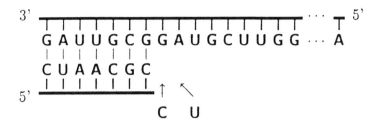

Point mutations: **GAUUG|C|GGAU** $\Longrightarrow$ **GAUUG|U|GGAU**

Insertions: **GAU|UGC|GGAU** $\Longrightarrow$ **GAU|UGCUGC|GGAU**

Deletions: **GAU|UGC|GGAU** $\Longrightarrow$ **GAU|GGAU**

FIGURE 1   Base complementarity as the principle of RNA replication and the three main classes of mutations.

Replication of nucleic acids in nature (Figure 1) makes use of Watson-Crick-type base pairing as occurring in double helical regions of nucleic acids. The simplest molecular replication mechanism is found in nature with single-stranded RNA viruses. They replicate in host cells using only one enzyme. Double helix formation is used only as an intermediate stage: the template and the newly synthesized strand are separated during the course of replication, and two RNA molecules (called plus and minus strands) are present after termination of the process. This replication mechanism has been studied in great detail.[2,4] It is visualized best as an analogue of the conventional photographic (positive–negative) copying mechanism.

The phenotypes of RNA molecules (Figure 2) are their three-dimensional structures which are formed spontaneously by folding the sequences in proper aqueous solutions (including appropriate concentrations of structure stabilizing divalent cations, ionic strength, pH, and temperature). The major driving force for structure formation is Watson-Crick base pairing ($\mathbf{G{\equiv}C}$, $\mathbf{A{=}U}$) mediated by partial intramolecular complementarity of sequences. In addition, $\mathbf{G{-}U}$ base pairs occur as well. Other intermolecular forces and the interaction with the aqueous solvent

Variation

↓

GCGGAUUUAG · · · GCACCA   **Genotype**:   *Nucleotide Sequence*

‖

⇓

Unfolding of the Genotype

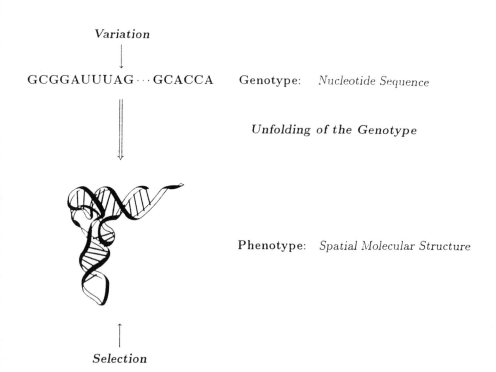

Phenotype: *Spatial Molecular Structure*

↑

Selection

FIGURE 2   Molecular genotypes and phenotypes in RNA evolution experiments.

shape the spatial structure of an RNA molecule. Small- and medium-sized RNA molecules (with chain lengths $\nu < 200$) form equilibrium structures that are independent of the mechanism of folding and, thus, are completely determined by the sequence and the environmental conditions. Structure formation of large RNA molecules appears to be controlled by the kinetics of folding, too. How does the three-dimensional structure of an RNA molecule determine its fitness in evolution experiments? The replicase recognizes the RNA template by its structure: only a subset of RNA structures binds to the enzyme strongly enough for the initiation of replication. Small RNA species $(50 > \nu > 25)$ that are replicated by the enzyme were isolated recently.[5] They share characteristic structural elements, a free 3-foot end and a double helical region at the 5-foot end of the RNA sequence, and minimal requirements for the binding of RNA molecules to the replicase are derived from these data. All efficiently replicating RNA molecules have rich internal structure along the sequence which is apparently required to separate the plus and the minus strands during the process of replication. Speed and efficiency of several reaction steps of the replication process depend strongly on structural details of the enzyme, and both RNA molecules, template and product. In general, all rate and

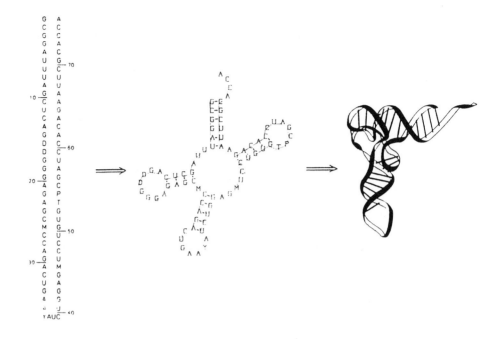

FIGURE 3   Folding of an RNA sequence into its spatial structure. The process is partitioned into two phases: in the first phase only the Watson-Crick-type base pairs are formed (which constitute the major fraction of the free energy), and in the second phase the actual spatial structure is built by folding the planar graph into a three-dimensional object. The example shown here is phenylalanyl-transfer-RNA (t-RNA[phe]) whose spatial structure is known from X-ray crystallography.

equilibrium constants of the replication process, and hence, also the overall rate of RNA synthesis, are functions of the three-dimensional structure. It determines, therefore, the fitness of an RNA molecule in molecular evolution experiments as the conventional phenotype does in organismic biology.

Formation of spatial structures of RNA molecules is commonly partitioned into two steps (Figure 3). The first step comprises conventional base-pair formation (**AU, GC, GU**) and yields the so-called secondary structure of the RNA molecule. The secondary structure, in essence, is a listing of the base pairs which can be represented as a two-dimensional, unknotted graph. In the second step the base pairing pattern is converted into the three-dimensional structure. There are several reasons for considering the secondary structure as a crude first approximation to the spatial structure of the RNA, for example:

■   conventional base pairing and base-pair stacking cover the major part of the free energy of folding,

- secondary structures are used successfully in the interpretation of RNA function and reactivity, and
- secondary structures are conserved in evolutionary phylogeny.

Two features of RNA secondary structures will be important later on. Secondary structures are discrete by definition (two bases either form a base pair or they don't), and secondary structures are composed of largely independent structural elements. There are several classes of substructures commonly characterized as stacks, loops, joints, and free ends.

In molecular evolution (and apart from recombination in all biology), the basis of variation is simply the limited accuracy of replication. Replication errors or mutations produce RNA sequences which differ from the parental template sequence. A change in the sequence will commonly cause a change in the structure. Mutations thus provide the genetic reservoir from which better adapted variants are chosen. They fall into three classes (Figure 1):

- point mutations leading to single-base exchanges at constant chain length,
- insertions in which the daughter sequence contains part of the parental sequence twice (or more often), and
- deletions in which part of the parental sequence is omitted in the daughter sequence.

In reality, the frequencies of all classes of mutations are phenotype-dependent in the sense that the molecular structure of the RNA determines position-dependent mutation rates. Position dependence of point mutations is much weaker than that of insertions and deletions and, therefore, a uniform error model of point mutations was conceived and successfully applied to replication and mutation of RNA molecules in $vitro$, of viroids, and of RNA viruses in $vivo$.[6,7,10,11] The model assumes that the error rate per (newly incorporated) base and replication event, p, is independent of the position in the sequence and of the nature of the point mutation (transition or transversion). It allows to express all mutation probabilities in terms of only three parameters, the copying fidelity $q = 1-p$, the mutant class h_{ki} (this is the number of positions in which the daughter sequence $\mathbf{I}_\kappa$ differs from the parental sequence $\mathbf{I}_i$ which is commonly called the Hamming distance between the two sequences), and the chain length ν:

$$Q_{ki} = p^{h_{ki}} q^{\nu - h_{ki}} = q^\nu \left(\frac{1-q}{q} \right)^{h_{ki}} = q^\nu \epsilon^{h_{ki}} \tag{1}$$

with $\epsilon = (1-q)/q = p/(1-p)$. No such model can be justified for insertions and deletions since they are directly related to the secondary structure of the RNA molecule. They occur very frequently at the beginning or at the end of hairpin loops. A plausible molecular interpretation postulates the jumping of the replicase at these positions on the template in the forward direction (deletion) or in the backward direction (insertion).

Selection experiments in molecular evolution were mainly carried out in well-stirred homogeneous solutions. The material consumed by replication is renewed either discontinuously, as in the serial transfer experiments,[44] or continuously, as in elaborate flow reactors.[23,24] Statistical removal of RNA molecules produced in excess restricts the populations size to essentially a constant (in the discontinuous serial-transfer technique, the population size is constant on the average). A new selection technique was introduced by John McCaskill and co-workers.[1] In their molecular evolution experiments they use capillaries that contain a medium suitable for RNA replication. RNA is injected and a wave front spreads through the medium. The front velocity of the traveling wave increases with the replication rate and, hence, faster replicating mutants are selected by the wave propagation mechanism. The time axis of conventional evolution experiments is mapped onto a spatial axis by this new technique.

Recalling RNA secondary structures and their importance for the replication process, we recognize an important difficulty with recombination in evolution of RNA molecules. Unless the sites of recombination are directly related to the secondary structure of the template (for example, the sites have to lie in single-stranded stretches outside hairpin loops), the probability to produce viable descendants is very low. Recombination in RNA-based genomes is practically unknown in nature except in cases where special mechanisms are available. The RNA virus *influenza A* is mentioned as a typical example: it has a split genome consisting of eight independent RNA molecules. Recombination occurs through simultaneous infection of a cell by two (or more) virus particles. Individual RNA molecules replicate independently and are randomly packed into the virus particles of the daughter generation. Genes from different parental viruses can be recombined in the virus progeny.

ADAPTATION AND THE EDGE OF DISORDER

The first experimental evidence that replicating ensembles of RNA molecules show adaptation of their structures to changes in the environment was provided by Sol Spiegelman and co-workers.[31] The medium for reproduction was deteriorated by the addition of a heterocyclic dye (ethidium bromide in the particular case) that intercalates between Watson-Crick base pairs and, thus, interferes with template-induced replication. Then the most frequently occurring genotypes (commonly called "master sequences") were isolated from populations and analyzed in the course of serial transfer experiments. A three-error mutant of the original master sequence was found to replicate faster in the new medium. More recent investigations of replication in the presence of RNA-cleaving enzymes ("ribonucleases": these are enzymes that cleave RNA specifically in single-stranded regions after certain bases) have shown directly how the structure of RNA molecules is adjusted in order to cope with detrimental changes in the environment.[46] Mutants appear and are enriched in the populations that compensate for the change in the environment by having fewer cleavage sites than the original master sequence, or none at all. In cases where the number of cleavage sites cannot be reduced to zero, the strand, which still carries cleavage sites, has a shorter mean lifetime in the stationary replicating ensemble than its complementary strand, which is usually resistant to ribonuclease.

Adaptation, however, is not restricted to the optimization of molecular structures. Mutation rates can also be subjected to change. According to Eq. (1) such a change can be caused by a variation of the error rate p as well as by a change in the chain length ν. Let us first consider the role of replication errors in population dynamics. Replication errors lead to new molecular species whose replication

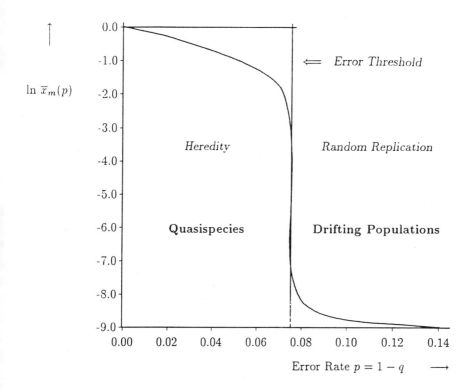

FIGURE 4   Evolution at the error threshold of replication. The fraction of the most frequent species in the population, called the **master sequence**, is denoted by $x_m(p)$. It becomes very small at the error threshold. Accordingly, the total fraction of all mutants, $1 - x_m(p)$, approaches one at the critical mutation rate.

efficiency is evaluated by the selection mechanism. The higher the error rate, the more mutations occur and the more viable mutants appear in the population. The stationary mutant distribution has been characterized as "quasi-species" since it represents the genetic reservoir of asexually replicating populations. An increase in the error rate thus leads to a broader spectrum of mutants and makes evolutionary optimization faster and more efficient in the sense that populations are less likely caught in local fitness optima. There is, however, a critical error threshold[6,7,10,11,47]: if the error rate p exceeds this critical limit, heredity breaks down; populations drift in the sense that new RNA sequences are formed steadily and old ones disappear; and no evolutionary optimization according to Darwin's principle is possible (Figure 4). Survival in variable environments demands sufficiently fast adaptation, and species with tunable error rates will adjust their quasi-species to meet the environmental challenge. In constant environments, on the other hand, such species will tune their error rates to the lowest possible values in order to maximize fitness.

Viruses are confronted with extremely fast changing environments since their hosts developed a variety of defense mechanisms ranging from the restriction enzymes of bacteria to the immune system of mammals and man. RNA viruses have been studied extensively. Their multiplication is executed by enzymes that do not allow large-scale variations of replication accuracies $(q = 1 - p)$. RNA viruses vary the error rate by changing the length of their genomes, and adjust the RNA chain lengths to optimal values that correspond to maximal chain lengths[6,7]:

$$\nu_{\max} \approx \frac{\ln \sigma}{p} . \tag{2}$$

Herein $\sigma \leq 1$ is the superiority of the master sequence in the stationary population. The superiority expresses differential fitness between the master sequence and the average of the remaining population. In the limit $\lim \sigma \rightarrow 1$ we are dealing with "neutral evolution."[30] Experimental analysis of several RNA virus populations has shown that almost all chain lengths are adjusted to yield error rates close to the threshold value. Thus, RNA viruses adapt to their environments by driving optimization efficiency towards the maximum.

Thus, adaptation in rapidly changing environments drives populations towards the error thresholds that are tantamount to the border between order (expressed in terms of quasi-species) and disorder (in the form of drifting populations). Populations of RNA viruses were found to evolve indeed at the edge of disorder. Can we generalize this result to other organisms? If Van Valen's "Red Queens hypothesis"[36,48] is met by actual ecosystems, coevolution of species would require each species to evolve at its maximum speed in order to compete successfully with evolutionary changes of the other species in the same ecosystem even under constant physical conditions, and the catchphrase "life is evolution at the edge of disorder" would be precisely to the point. We should keep in mind, however, that there are also "living fossils" in nature that have undergone hardly any macroscopically detectable change in hundreds of million years of life under apparently constant conditions. Under such constant environments, optimization would favor small error rates since then the fitness of a given variant increases with increasing replication accuracy.

COMBINATORY LANDSCAPES AND COMBINATORY MAPS

The efficiency of evolutionary adaptation, in essence, depends on the relation between genotypes and phenotypes. Sewall Wright[53] introduced the notion of a fitness landscapes in order to illustrate evolution as an hill-climbing process on an (presumably) exceedingly complex surface. Recently the concept of fitness landscapes saw a revival in biology.[27,29,33,34] We shall try here to verify and analyze the abstract concept of fitness landscapes and more general mappings from genotype space into phenotype space for the simplest example known at present, the evolution of RNA molecules.

Firstly, we define a distance between genotypes, in our case RNA sequences, such that they can be ordered in a natural way. Apparently this is the minimal number of mutations required to convert one sequence into another. The restriction

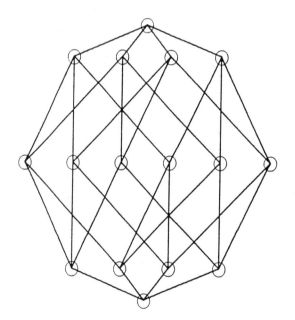

FIGURE 5 The sequence space of binary (**AU** or **GC**) sequences of chain length $\nu = 4$. This is a discrete vector space. Every circle represents a single sequence of four letters. All pairs of sequences with Hamming distance $d_h = 1$ (these are pairs of sequences that differ in one position) are connected by a straight line. The geometric object obtained is a hypercube in four-dimensional space and, hence, all positions (and all sequences) are topologically equivalent.

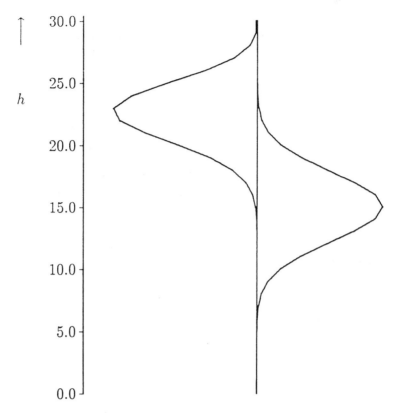

FIGURE 6   Distributions of sequences into error classes, $P(h)$, with the Hamming distance h from the reference. The curve on the left-hand side refers to four-letter (**AUGC** or **GC**) sequences ($\kappa = 2$). The chain length is $\nu = 30$ in both cases.

to point mutations (which are also predominant in nature) is equivalent to the usage of the Hamming distance (d_h).[20] It counts the number of positions in which two aligned sequences differ, and represents the minimum number of point mutations that are required to convert one sequence into another. Moreover, the Hamming distance d_h induces a metric on the abstract space of all sequences commonly called the **sequence space**. The concept of sequence space was first used in coding theory.[20,21] It was suggested for application to proteins[35] and nucleic acids.[40] Later on it was expressed in quantitative terms for nucleic acids with different base-pairing alphabets.[7,10,11,47] The space of binary sequences (**AU**, **GC**) of chain length $\nu=4$ is shown in Figure 5 as an example. The sequence space of binary sequences is simply the hypercube of dimension ν when ν is the chain length of the sequences.

The sequence spaces of four-letter sequences are more complex objects.[12,47] The distribution of sequences around a reference sequence is described by

$$P(h) = \frac{(\kappa - 1)^h}{\kappa^\nu} \binom{\nu}{h} \tag{3}$$

where κ is the number of the letters in the alphabet (as before) and h is the index of the error class expressed in Hamming distance. In Figure 6 the distribution of sequences in sequence space is shown for binary and four-letter sequences of chain length $\nu = 30$. According to Eq. (3) the distribution of binary sequences is just the (normalized) binomial distribution. In the case of four-letter sequences the maximum of the distribution is shifted towards higher error classes. An important difference to which we shall refer later concerns the number of sequences that differ in all positions from the reference: there is only a single ν-error mutant—the complementary sequence—of a binary sequence, whereas we have 3^ν ν-error mutants of a four-letter sequence.

Sequences are of combinatory complexity and, therefore, we shall use the term "combinatory" for objects that are built on sequences spaces. A combinatory landscape Λ assigns scalar values to sequences and can be understood, therefore, as a mapping from a discrete metric vector space, the sequence space $\mathcal{X}$ (with the Hamming distance d_h as metric), into the real numbers:

$$\Lambda : \quad (\mathcal{X}, d_h) \implies (\mathbf{R}^1) . \tag{4}$$

In molecular biology these scalar values (f_k)—for example, free energies, replication rate constants, or fitness values—are properties and, thus, functions of the RNA structures $(\mathbf{S}_k)$, which in turn are functions of the RNA sequences $(\mathbf{I}_\kappa)$:

$$\mathbf{I}_\kappa \implies \mathbf{S}_k = \mathcal{S}(\mathbf{I}_\kappa) ; \quad f_k = \mathcal{F}(\mathbf{S}_k) = \mathcal{F}\left(\mathcal{S}(\mathbf{U}_\kappa)\right) . \tag{5}$$

Here $\mathcal{S}(.)$ stands for the folding process, and $\mathcal{F}(.)$ symbolizes the evaluation of the structure.

Combinatory landscapes may be characterized and investigated by the conventional techniques from mathematical statistics.[11,16,18,51,52] In particular, an autocorrelation function of the landscape is computed from the following equation

$$\varrho_f(h) = 1 - \frac{\langle |\Delta f(h)|^2 \rangle}{\langle |\Delta f|^2 \rangle} . \tag{6}$$

Mean square averages of the differences $\Delta f = f_k - f_j$ are taken here over all sequences in sequence space $(\langle |\Delta f|^2 \rangle)$, or over all sequences in the error class h of the reference sequence $(\langle |\Delta f(h)|^2 \rangle)$, i.e., over all sequences at Hamming distance h from the reference. The autocorrelation functions can be approximated by exponential functions, and correlation lengths (ℓ_f) are estimated from the relation:

$\ln(\varrho_f(\ell_f)) = -1$. Correlation lengths of landscapes are an appropriate measure for their complexity. On the average there will be one local optimum in a ball with the correlation length as radius. Thus, the smaller the correlation length is, the harder is the corresponding optimization problem on the average.[11]

Computation of autocorrelation functions for RNA-based landscapes requires the knowledge of several hundred thousand values derived from RNA structures.

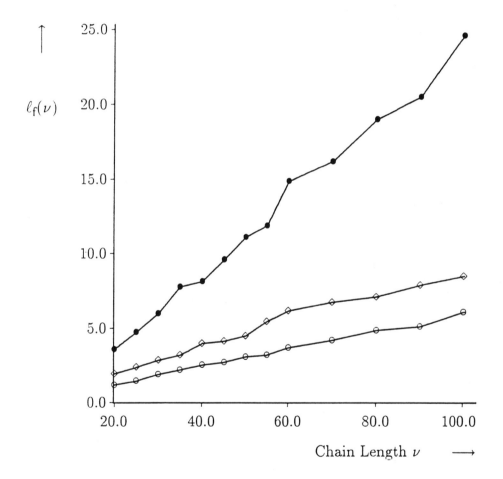

FIGURE 7   Correlation lengths of free energies (l_f) of RNA molecules in their most stable secondary structures as functions of the chain length ν. Values are shown for binary pure **GC** sequences (◇), for binary pure **AU** sequences (○), and for natural **AUGC** sequences (•). The correlation lengths are computed from $(\ln \varrho_f(h), h)$-plots by means of a least root mean square deviation fit.

These data are not available at present, either through experimental measurement or through computation of the three-dimensinal structures (which are highly time consuming and unreliable by the current algorithms). The restriction to secondary structures as a crude approximation to real structures, however, renders computation possible. We conceived a fast-folding algorithm that was originally used for realistic computer simulations of evolutionary optimization of RNA structures.[14,15,41] It is a variant of the frequently used dynamic programming method by Zuker.[54] The free energy landscape is considered as an example. Correlation lengths of free energies (ℓ_f) are essentially linear functions of of the chain length ν (Figure 7). The base-pairing alphabet has remarkably strong influence on the correlation length: the ℓ_f-values for **AUGC** sequences are much longer than the corresponding values for pure **GC** or pure **AU** sequences. The optimization of properties of RNA molecules derived from the natural **AUGC** alphabet is, therefore, much easier than that of RNA molecules, which consist predominantly of **G** and **C** or **A** and **U**, respectively.

The structure of the RNA molecule acts here as mediator between the sequence and the (scalar) property that is considered in the landscape. The relation between sequences and structures thus seems to be basic to all landscapes. A straightforward statistical analysis of RNA secondary structures was performed recently.[18] Structures are nonscalar objects, and there is no "structure landscape." Nevertheless, RNA secondary structures form a metric space ($\mathcal{Y}$) that we characterize as **shape space**. It is defined as the set of all RNA secondary structures formed by all sequences of chain length ν derived from a given base-pairing alphabet. The notion of shape space was used previously in theoretical immunology for the set of all structures presented by all possible antigens.[39,43] In order to find a measure for the relationship between RNA secondary structures the structures are converted into equivalent trees.[18] A "tree distance" d_t is defined which can be evaluated by tree editing as the minimal costs required to convert one tree into the other. The tree distance induces a metric on the shape space. RNA folding thus can be understood as a mapping from one metric space into another, in particular, from sequence space into shape space. The support is of combinatory complexity, and the corresponding map

$$\Phi : \quad (\mathcal{X}, d_h) \implies (\mathcal{Y}, d_t) \tag{7}$$

is an example of a combinatory map. We remark that the map is not invertible: a structure is uniquely assigned to every sequence, but several sequences may be mapped onto the same structure (see the next section).

An autocorrelation function of structures based on tree distances (d_t) is computed from the equation

$$\varrho_t(h) = 1 - \frac{\langle d_t^2(h) \rangle}{\langle d_t^2 \rangle} . \tag{8}$$

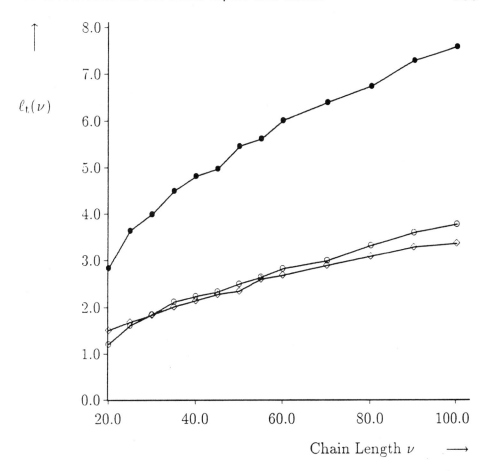

FIGURE 8 Correlation lengths of structures (l_t) of RNA molecules in their most stable secondary structures as functions of the chain length ν. Values are shown for binary pure **GC** sequences ($\diamond$), for binary pure **AU** sequences ($\circ$), and for natural **AUGC** sequences ($\bullet$). The correlation lengths are computed from $(\ln \varrho_t(h), h)$-plots by means of a least root mean square deviation fit.

As in the case of landscapes, mean square averages are taken here over all sequences in sequence space $(\langle d_t^2 \rangle)$, or over all sequences in the error class h of the reference sequence $(\langle d_t^2(h) \rangle$, i.e., over all sequences at Hamming distance h from the reference). The autocorrelation function is approximated by an exponential function, and a correlation length (l_t) is estimated from the relation: $\ln(\varrho_t(l_t)) = -1$. The correlation length of tree distances l_t is a measure of the average stability of structures against mutations. The shorter the correlation length, the more likely is a structural change occurring as a consequence of mutation. The correlation length thus measures stability against mutation. In Figure (8), correlation lengths of RNA structures are plotted against the chain length. An almost linear increase is observed. Substantial differences are found in the correlation lengths derived from different base-pairing alphabets. In particular, the structures of natural (**AUGC**) sequences are much more stable against mutation than pure **GC** sequences or pure **AU** sequences. The analogous results on free energies reported above are thus a straight consequence of the sequence-structure mapping (and not of the relation between structures and free energies). The observation of high sensitivity of structures derived from **GC** sequences against mutation is in agreement with structural data obtained for ribosomal RNA.[49] It provides also a plausible explanation for the use of two base pairs in nature: optimization in an RNA world with only one base pair would be very hard, and the base-pairing probability in sequences with three base pairs is too low to form thermodynamically stable structures for short chain lengths. The choice of two base pairs thus appears to be a compromise between stability against mutation and thermodynamic stability.

THE SHAPE SPACE OF RNA STRUCTURES

The number of structures that are acceptable as minimum free-energy secondary structures of RNA molecules can be computed from combinatorics of structural elements.[22,42,45,50] Thermodynamic stability considerations exclude single base pairs as stacks and hairpin loops with less than three bases. Then the number of structures can be approximated by $1.485 \times \nu^{-3/2}(1.849)^{\nu}$. This number is much smaller than the number of natural (**AUGC**) sequences, 4^{ν}. It is also smaller than the number of binary sequences, 2^{ν}. Accordingly several or many sequences fold into the same secondary structure and, hence, the mapping from sequence space into shape space cannot be invertible. In order to search for regularities of this map, we shall try several statistical approaches.

The information contained in the mapping from sequence space into shape space is condensed into a two-dimensional, conditional probability density surface called structure density surface (SDS):

$$S(t,h) = \text{Prob}\left(d_t = t \mid d_h = h\right). \tag{9}$$

The structure density surface expresses the probability that the secondary structures of two randomly chosen sequences have a structure distance t provided their Hamming distance is h. Two examples of a structure density surfaces for natural **AUGC** sequences and for pure **GC** sequences of chain length $\nu = 100$ is shown

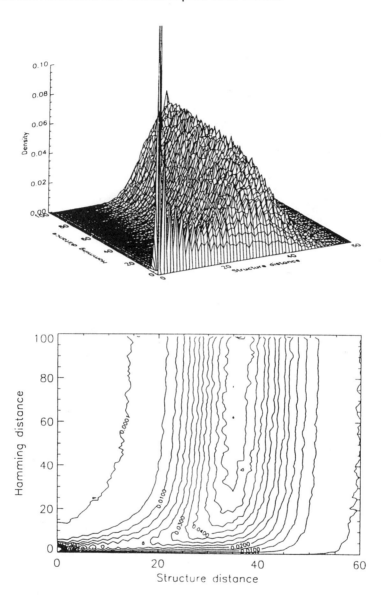

FIGURE 9 The structure density surface $S(t, h)$ of natural **AUGC** sequences of chain length $\nu = 100$. The density surface (upper part) is shown together with a contour plot (lower pat). In order to dispense from confusing details the contour lines were smoothed. In this computation a sample of 1000 reference sequences was used which amounts to a total sample size of 10^6 individual RNA foldings.

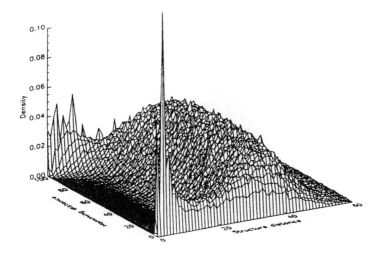

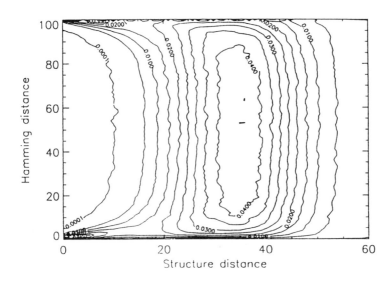

FIGURE 10 The structure density surface $S(t, h)$ of binary $\mathbf{GC}$ sequences of chain length $\nu = 100$. The density surface (upper part) is shown together with a contour plot (lower part). In order to dispense from confusing details the contour lines were smoothed. In this computation a sample of 1000 reference sequences was used which amounts to a total sample size of 10^6 individual RNA foldings.

in Figures 9 and 10. The SDS of **AUGC** sequences has an overall shape that corresponds to one half of a horseshoe and has rugged details superimposed on it. The contour plot illustrates an important property of the structure density surface: at short Hamming distances ($1 \leq \nu < 16$) the probability density changes strongly with increasing Hamming distance, but further away from the reference sequence ($16 < \nu < 100$) this probability density is essentially independent of the Hamming distance h. The first part reflects the local features of sequence-structure relations. Up to a Hamming distance of $h = 16$ there is still some memory of the reference sequence. Then, at larger Hamming distances the structure density surface contains exclusively global information that is independent of the reference. The critical Hamming distance at which we observe the change from the local to the global probability distribution is a little more than twice the correlation length of the tree distance ($\approx 2.1 \times \ell_t$).

The second example, the SDS of binary **GC** sequences, is horseshoe-like and exhibits approximate mirror-plane symmetry. The sequence at distance ν from the reference is compatible with the secondary structure of the reference. This means it could, in principle, form the same structure, although it need not be the minimum free energy structure of the complementary sequence. As a matter of fact, complementary sequences fold into the same or into closely related structures. Again we observe a change from local to global features of the SDS at a Hamming distance that is about $2.4 \times \ell_t$. The SDS computed for binary **AU** sequences is not substantially different from that for the **GC** case.

In order to gain more information on the relation between sequences and structures, an inverse folding algorithm that determines the sequences that share the same minimum free energy secondary structure was conceived and applied to a variety of different structures.[42] The frequency distribution of structures has a very sharp peak: relatively few structures are very common, many structures are rare and play no statistically significant role. The results obtained show in addition that sequences folding into the same secondary structure are, in essence, randomly distributed. In Figure 11 we present four examples of sequences folding into arbitrarily chosen secondary structures. Practically identical results were obtained for other structures, in particular for the secondary structure of phenylalanyl-t-RNA. Since there are relatively few common structures and the sequences folding into the same structure are randomly distributed in sequence space, all common structures are found in relatively small patches of sequence space. For natural **AUGC** sequences of chain length $\nu = 100$, a sphere of radius $h = 16$ (in Hamming distance) is sufficient to yield the global distribution of structure distances. In this case we can expect all common structures to be found in such spheres in sequence space. There are as many as 6.2×10^{25} sequences in such a ball. Although this number is large, it is nothing compared to the total number of sequences of this chain length: $4^{100} \approx 1.6 \times 10^{60}$. In Figure 12 we show a sketch of the mapping from sequence space into shape space that accounts for the results obtained with the four-letter sequences.

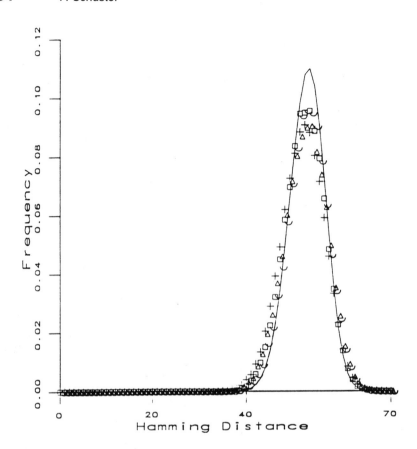

FIGURE 11   The distribution of Hamming distances between sequences folding into the same secondary structure as compared to the analogous distribution in a sample of random sequences (full line). Four examples are shown. The target structures were obtained by folding four arbitrarily chosen sequences.

In order to complement this illustration of the RNA shape space, we use a computer experiment that allows an estimate of the degree of selective neutrality (two sequences are considered neutral here if the fold into the same secondary structure). As indicated in Figure 13 we search for "neutral paths" through sequence space. The Hamming distance from the reference increases monotonously along such a neutral path but the structure remains unchanged. A neutral path ends when no further neutral sequence is found in the neighborhood of the last sequence. The length ℓ of a path is the Hamming distance between the reference sequence and the last sequence. Clearly, a neutral path cannot be longer than the chain length ν ($\ell \leq \nu$). The length distribution of neutral path in the sequence space of RNA

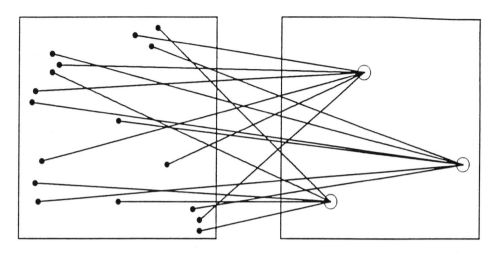

Sequence Space   Shape Space

FIGURE 12   A sketch of the mapping from sequence space into shape space.

molecules of chain length $\nu = 100$ is shown in Figure 14 for the three base-pairing alphabets considered here. We observe substantial differences. It is remarkable that about 20% of the neutral paths in the shape space of natural **AUGC** sequences have the maximum length. They lead through the whole sequence space to a sequence that is different in all positions from the reference and shares its structure. In shape spaces derived from binary sequences, almost no neutral path reaches the complementary sequence. This is certainly a consequence of the symmetry of the binomial distribution: there are very few sequences in the error classes $\nu-1$, $\nu-2$, etc., and it is unlikely that we find one among them that folds precisely into the same structure as the reference sequence. Comparing the two binary alphabets we find that the **AU** shape space sustains longer neutral paths than the **GC** shape space. This is clearly a consequence of the strength of base pairing: **GC** base pairs are stronger and, thus, are able to close smaller loops and create thereby a greater variety of equilibrium structures than their **AU** analogues.[16,18]

The combination of information derived from Figures 9, 11, and 14 provides insight into the structure of the shape space of RNA secondary structures, which may be cast into four statements:

- the frequency distribution of structures is sharply peaked (there are comparatively few common structures and many rare ones),

Mutant Class:

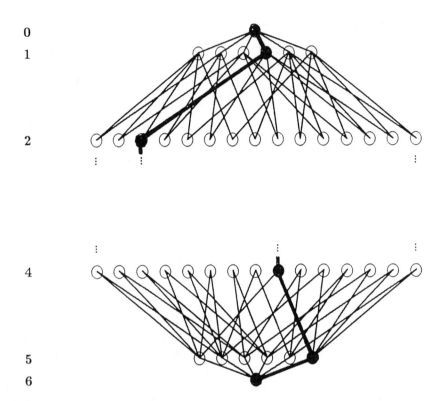

FIGURE 13 Percolation of the sequence space by neutral networks. The example shows the space (or subspace) of binary (AU or GC) sequences with chain length $\nu = 6$. The emphasized path corresponds to one complete neutral trajectory that connects sequences folding into identical structures through the entire sequence space.

- sequences folding into one and the same (common) structure are distributed randomly in sequence space,
- all common structures are formed by sequences within (relatively) small neighborhoods of any random sequence, and
- the shape space contains extended "neutral networks" joining sequences with identical structures (a large fraction of neutral path leads from the initial sequence through the entire sequence space to a final sequence on the opposite side or in an error class that is close to it).

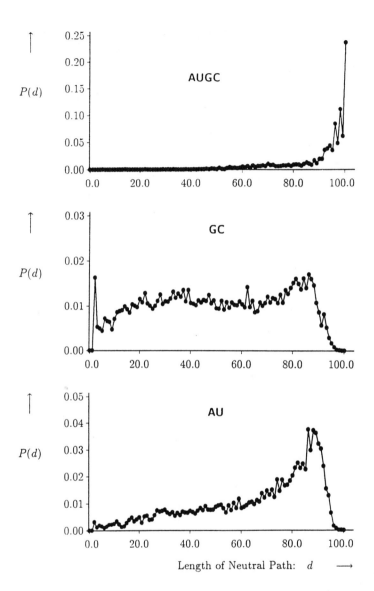

FIGURE 14 Length distribution of neutral paths in the shape spaces derived from **AUGC, GC**, or **AU** sequences of chain length $\nu = 100$. A neutral path connects pairs of sequences with identical structures and Hamming distance $d_h = 1$ (single-base exchange) or $d_h = 2$ (base pair exchange). The Hamming distance to the reference sequence is monotonously increasing along the path.

The structure of shape space is highly relevant for evolutionary optimization in nature too. It provides a firm answer to the old probability argument against the possibility of successful adaptive evolution. How should nature find a given biopolymer by trial and error when the chance to guess it is as low as $1 : \kappa^{\nu}$? Previously given answers[6,7,11] can be supported and extended by precise data on the RNA shape space. The numbers of sequences that have to be searched to find adequate solutions in adaptive evolution are many orders of magnitude smaller than those guessed on naive statistical grounds. In the absence of selective differences, populations drift readily through sequence space, since long neutral paths are common. This is in essence what is predicted by the "neutral theory of evolution,"[30] and what is often observed in molecular phylogeny by sequence comparisons of different organisms.

EVOLUTIONARY BIOTECHNOLOGY

The application of RNA-based molecular adaptive systems to solve problems in biotechnology by Darwin's selection principle was proposed in 1984 by Eigen and Gardiner.[8] Somewhat later a similar suggestion was made by Kauffman[26] for large-scale screening of proteins based on recombinant DNA techniques and selection methods. Meanwhile, many research groups started to apply evolutionary concepts to produce biomolecules with new properties (for two recent reviews see Joyce[25] and Kauffman[28]). At present molecular evolution seems to give birth to a novel branch of applied biosciences.

The essence of evolutionary biotechnology is sketched in Figure 15. Experiments are carried out on the level of populations of molecules. Replication of nucleic acid molecules is used as the amplification factor. Variation is introduced into the populations either by artificially increased mutation rates or by partial randomization of RNA or DNA sequences. The synthesis of oligonucleotides with random sequences has become routine by now. The essential trick of this new technique is to encode the desired properties and functions into the selection constraint.[25,28] Evolutionary biotechnology also provides new challenges for the design of high-tech equipment which is required to carry out massively parallel experiments under precisely controlled conditions.[24,46]

The results on RNA shape space derived in the previous section suggest straightforward strategies in the search for new RNA structures. It is of little advantage to start from natural or other preselected sequences since any random sequence would serve equally well as the starting molecule for the selection cycles shown in Figure 15. Any common secondary structure with suitable properties is accessible within a few selection cycles. Since the secondary structure is only a crude first approximation to the actual three-dimensional structure, fine tuning of properties by choosing from a variety of molecules sharing the same secondary structure will be necessary. In order to achieve this goal it is advantageous to adopt alternations of selection cycles with low and high error rates. At low error rates the population performs a search in the vicinity of the current master sequence (the most common sequence, which usually is the fittest sequence as well). If no RNA molecule with satisfactory properties is found, a change to high error rate is adequate. The population then spreads along the neutral network to other regions of sequence space that are explored in detail after tuning the error rate to low again.

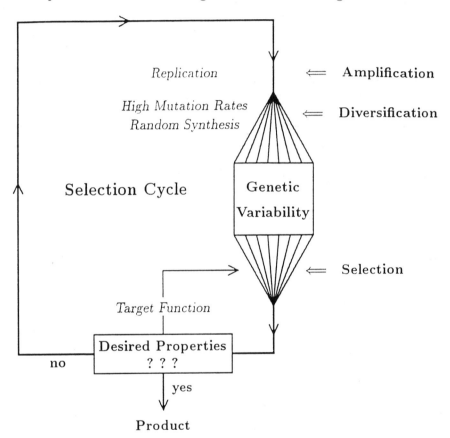

FIGURE 15 A sketch of the selection technique in evolutionary biotechnology.

MOLECULAR EVOLUTION AND COMPLEX ADAPTIVE SYSTEMS

Molecular evolution of RNA *in vitro* provides presumable the simplest molecular adaptive system that can be studied by experiment as well as by extensive computer simulation. It can be investigated by the conventional analytical tools of physics and physical chemistry and, thus, allows us to gain deep insights into the principle underlying adaptation and learning in its crudest form where it becomes intimately related to Darwin's concept of evolution. In this sense molecular evolution finds its place somewhere at the base of the hierarchical tree of complex adaptive systems as sketched in the research program of complexity.[19]

The study of molecular evolution, however, leads us into another relevant issue of the origin of complexity. The genotype-phenotype relation, which is fundamental to all evolutionary phenomena, is reduced here to its essence: **sequence** and **structure**. The emergence of properties and functions becomes tantamount to the formation of shapes in physical space that are encoded in the sequences of monomers. RNA shape space restricted to secondary structures provides a toy system with a firm physical basis close to the real world since secondary structures are first approximations to the three-dimensional structures. At the same time they are simple enough to be studied by rigorous mathematical and statistical tools as well as by computer simulation. RNA secondary structures demonstrate in straightforward manner how simple rules lead to complex phenomena. These rules, in essence, are little more than the base-pairing rules (**A=U**, **G≡C**), and some thermodynamic or kinetic criterion (minimum free energy, for example, that this could be replaced by a "maximum number of base pairs" criterion without changing any of the basic features). The rules, harmless as they look at a first glance, carry their enormous power by being **nonlocal**: any base can pair with (almost) any base, regardless of whether they are close by or far apart on the sequence. Therefore, all kinds of small and large loops can be formed and a great variety of different structures may be created by folding the sequence.

Considering the mapping from (RNA) sequence space into shape space (Figure 12), we can hardly escape the feeling that we are dealing with a kind of random mapping from a larger into a smaller discrete metric space. Complexity, not fully understood, will often make the impression of randomness. What distinguishes real complexity, and life, from arbitrariness are the regularities that are hidden beyond the confusing surface. One major challenge for future research on complex systems in general, as well as for investigations on biopolymers in particular, is to discover these not yet known regularities.

ACKNOWLEDGMENTS

The statistical analysis of the RNA shape space presented here is joint work with Dr. Walter Fontana, Dr. Peter F. Stadler, and others. It will be described in detail elsewhere. Financial support by the Austrian Fonds zur Förderung der Wissenschaftlichen Forschun (Projects S 5305-PHY and P 8526-MOB) is gratefully acknowledged. The Institut für Molekulare Biotechnologie, Jena, Germany, is sponsored by core funding from the Thüringische Ministerium für Wissenschaft und Kunst (Erfurt) and the Bunderministerium für Forschung und Technologie, Bonn, Germany.

REFERENCES

1. Bauer, G. J., J. S. McCaskill, and H. Otten. "Traveling Waves of *in Vitro* Evolving RNA." *Proc. Natl. Acad. Sci. USA* **86** (1989): 7937–7941.
2. Biebricher, C. K., M. Eigen, and W. C. Gardiner, Jr. "Kinetics of RNA Replication." *Biochemistry* **22** (1983): 2544–2559.
3. Biebricher, C. K., M. Eigen, and W. C. Gardiner, Jr. "Kinetics of RNA Replication: Plus-Minus Asymmetry and Double-Strand Formation." *Biochemistry* **23** (1984): 3186–3194.
4. Biebricher, C. K., M. Eigen, and W. C. Gardiner, Jr. "Kinetics of RNA Replication: Competition and Selection Among Self-Replicating RNA Species." *Biochemistry* **24** (1985): 6550–6560.
5. Biebricher, C. K., and R. Luce. "Sequence Analysis of RNA Generated *de novo* by Qβ Replicase." Preprint, 1992.
6. Eigen, M. "Self-Organization of Matter and the Evolution of Biological Macromolecules." *Naturwissenschaften* **58** (1971): 465–523.
7. Eigen, M., and P. Schuster. "The Hypercycle. A Principle of Natural Self-Organization. Part A: Emergence of the Hypercycle." *Naturwissenschaften* **64** (1977): 541–565.
8. Eigen, M., and W. Gardiner. "Evolutionary Molecular Engineering Based on RNA Replication." *Pure & Appl. Chem.* **56** (1984): 967–978.
9. Eigen, M., J. McCaskill, and P. Schuster. "The Molecular Quasispecies—An Abridged Account." *J. Phys. Chem.* **92** (1988): 6881–6891.
10. Eigen, M., R. Winkler-Oswatitsch, and A. Dress. "Statistical Geometry in Sequence Space: A Method of Quantitative Comparative Sequence Analysis." *Proc. Natl. Acad. Sci. USA* **85** (1988): 5913–5917.
11. Eigen, M., J. McCaskill, and P. Schuster. "The Molecular Quasispecies." *Adv. Chem. Phys.* **75** (1989): 149–263.

12. Eigen, M., and R. Winkler-Oswatitsch. "Statistical Geometry on Sequence Space." In *Methods in Enzymology*, edited by R. F. Dolittle, Vol. 183, 505–530. New York: Academic Press, 1990.

13. Findley, A. M., S. P. McGlynn, and G. L. Findley. *The Geometry of Genetics*. New York: John Wiley & Sons, 1989.

14. Fontana, W., and P. Schuster. "A Computer Model of Evolutionary Optimization." *Biophys. Chem.* **26** (1987): 123–147.

15. Fontana, W., W. Schnabl, and P. Schuster. "Physical Aspects of Evolutionary Optimization and Adaptation." *Phys. Rev. A* **40** (1989): 3301–3321.

16. Fontana, W., T. Griesmacher, W. Schnabl, P. F. Stadler, and P. Schuster. "Statistics of Landscapes Based on Free Energies, Replication and Degradation Rate Constants of RNA Secondary Structures." *Mh. Chem.* **22** (1991): 795–819.

17. Fontana, W., D. A. M. Konings, P. F. Stadler, and P. Schuster. "Statistics of RNA Secondary Structures." Preprint 90-02-008, Santa Fe Institute, 1990. To appear in *Biopolymers*, 1993.

18. Fontana, W., P. F. Stadler, E. G. Bornberg-Bauer, T. Griesmacher, I. L. Hofacker, M. Tacker, P. Tarazona, E. D. Weinberger, and P. Schuster. "RNA Folding and Combinatory Landscapes." *Phys. Rev. E* (1993): in press.

19. Gell-Mann, M. "Complexity and Complex Adaptive Systems." In *The Evolution of Human Languages*, edited by J. A. Hawkins and M. Gell-Mann, Santa Fe Institute Studies in the Sciences of Complexity, Proc.Vol. X, 3–18. Redwood City, CA: Addison-Wesley, 1992.

20. Hamming, R. W. "Error Detecting and Error Correcting Codes." *Bell Syst. Tech. J.* **29** (1950): 147–160.

21. Hamming, R. W. *Coding and Information Theory*. Englewood FCell fs, NJ: Prentice Hall, 1980 and 1986.

22. Hofacker, I. L., P. Schuster, and P. F. Stadler. "Combinatorics of RNA Secondary Structures." *SIAM J. Disc. Math.*, submitted, 1992.

23. Husimi, Y., K. Nishgaki, Y. Kinoshita, and T. Tanaka. "Cellstat—A Continuous Culture System of a Bacteriophage for the Study of the Mutation Rate and the Selection Process at the DNA Level." *Rev. Sci. Instrum.* **53** (1982): 517–522.

24. Husimi, Y., and H.-C. Keweloh. "Continuous Culture of Bacteriophage Qβ Using a Cellstat with a Bubble Wall-Growth Scraper." *Rev. Sci. Instrum.* **58** (1987): 1109–1111.

25. Joyce, G. F. "Directed Molecular Evolution." *Sci. Am.* **267** (**6**) (1992): 48–55.

26. Kauffman, S. A. "Autocatalytic Sets of Proteins." *J. Theor. Biol.* **119** (1986): 1–24.

27. Kauffman, S. A. "Adaptation on Rugged Fitness Landscapes." In *Lectures in the Sciences of Complexity*, edited by D. Stein, Sante Fe Institute Studies in the Sciences of Complexity, Lect. Vol. I, 527–618. Redwood City, CA: Addison Wesley, 1989.

28. Kauffman, S. A. "Applied Molecular Evolution." *J. Theor. Biol.* **157** (1992): 1–7.

29. Kauffman, S. A., and S. Levin. "Towards a General Theory of Adaptive Walks on Rugged Landscapes." *J. Theor. Biol.* **128** (1987): 11–45.

30. Kimura, M. *The Neutral Theory of Molecular Evolution.* Cambridge: Cambridge University Press, 1983.

31. Kramer, F. R., D. R. Mills, P. E. Cole, T. Nishihara, and S. Spiegelman. "Evolution *in vitro*: Sequence and Phenotype of a Mutant RNA Resistant to Ethidium Bromide." *J. Mol. Biol.* **89** (1974): 719–736.

32. Li, W.-H., and D. Graur. *Fundamentals of Molecular Evolution.* Sunderland, MA: Sinauer, 1991.

33. Macken, C. A., and A. S. Perelson. "Protein Evolution on Rugged Landscapes." *Proc. Natl. Acad. Sci. USA* **86** (1989): 6191–6195.

34. Macken, C. A., P. S. Hagan, and A. S. Perelson. "Evolutionary Walks on Rugged Landscapes." *SIAM J. Appl. Math.* **51** (1991): 799–827.

35. Maynard-Smith, J. "Natural Selection and the Concept of a Protein Space." *Nature* **225** (1970): 563–564.

36. Maynard-Smith, J. *Evolutionary Genetics.* New York: Oxford University Press, 1989.

37. Mills, D. R., R. L. Peterson, and S. Spiegelman. "An Extracellular Darwinian Experiment with a Self-Duplicating Nucleic Acid Molecule." *Proc. Natl. Acad. Sci. USA* **58** (1967): 217–224.

38. Orgel, L. E. "Molecular Replication." *Nature* **358** (1992): 203–209.

39. Perelson, G. A., and G. F.Oster. "Theoretical Studies on Clonal Selection: Minimal Antibody Repertoire Size and Reliability of Self-Non-Self Discrimination." *J. Theor. Biol.* **81** (1979): 645–670.

40. Rechenberg, I. *Evolutionsstrategie.* Stuttgart-Bad Canstatt, Germany: Frommann-Holzboog, 1973.

41. Schuster, P. "Complex Optimization in an Artificial RNA World." In *Artificial Life II*, edited by C. G. Langton, C. Taylor, J. D. Farmer, and S. Rasmussen, Santa Fe Institute Studies in the Sciences of Complexity, Proc. Vol. X, 277–291. Redwood City, CA: Addison-Wesley,1992.

42. Schuster, P., W. Fontana, P. F. Stadler, and I. L. Hofacker. "How to Search for RNA Structures." Preprint, 1993.

43. Segel, L. A., and A. S. Perelson. "Computations in Shape Space: A New Approach to Immune Network Theory." In *Theoretical Immunology, Part Two*, edited by A. S. Perelson, Santa Fe Institute Studies in the Sciences of Complexity, Proc. Vol. III, 321–343. Redwood City, CA: Addison-Wesley, 1988.

44. Spiegelman, S. "An Approach to the Experimental Analysis of Precellular Evolution." *Qtr. Rev. Biophys.* **4** (1971): 213–253.

45. Stein, P. R., and M. S. Waterman. "On Some Sequences Generalizing the Catalan and Motzkin Numbers." *J. Discr. Math.* **26** (1978): 261–272.

46. Strunk, G. "Automatized Evolution Experiments *in vitro* and Natural Selection Under Controlled Conditions by Means of the Serial Transfer Technique." Doctoral Thesis, Universität Braunschweig, Germany, 1993.

47. Swetina, J., and P. Schuster. "Self-Replication with Errors—A Model for Polynucleotide Replication." *Biophys. Chem.* **16** (1982): 329–345.

48. Van Valen, L. "A New Evolutionary Law." *Evol. Theory* **1** (1973): 1–30.

49. Wakeman, J. A., and B. E. H. Maden. "28 s Ribosomal RNA in Vertebrates. Location of Large-Scale Features Revealed by Electron Microscopy in Relation to Other Features of the Sequences." *Biochem. J.* **258** (1989): 49–56.

50. Waterman, M. S. "Secondary Structures of Single-Stranded Nucleic Acids." *Adv. Math. Suppl. Studies* **1** (1978): 167–212.

51. Weinberger, E. D. "A More Rigorous Derivation of Some Results on Rugged Fitness Landscapes." *J. Theor. Biol.* **134** (1988): 125–129.

52. Weinberger, E.D. "Correlated and Uncorrelated Fitness Landscapes and How to Tell the Difference." *Biol. Cybern.* **63** (1990): 325–336.

53. Wright, S. "The Roles of Mutation, Inbreeding, Crossbreeding, and Selection in Evolution." *Proceedings of the Sixth International Congress on Genetics* **1** (1932): 356–366.

54. Zuker, M. "The Use of Dynamic Programming Algorithms in RNA Secondary Structure Prediction." In *Mathematical Methods for DNA Sequences*, edited by M. S. Waterman, 159–184. Boca Raton, FL: CRC Press, 1989.

DISCUSSION

LAPEDES: I'm finding the result that you see different correlation lengths as you go from a two-letter alphabet to a four-letter alphabet, and also that the resilience of structure increases, very intriguing. One might ask, what would happen, say, if you went to a six-letter alphabet. Could one show, for example, that four-letter alphabets are optimal, in some sense?

SCHUSTER: We did that and something more that allowed us to distinguish the separate effects of alphabets because otherwise the data would be somewhat confusing. In the natural four-letter alphabet we are not dealing simply with two base pairs, the two base pairs are of different strength and GU pairs are allowed as well. We looked, therefore, at a model alphabet with four bases forming two base pairs of equal and GC strength—with no analog to GU. This alphabet forms structures that are less stable against mutation than the natural AUGC structures. They are, of course, more stable than the GC-two-letter structures. To build structures from a four-letter-alphabet with base pairs of AU strength is difficult because of the weakness of the AU pairs. This is even more true for six-letter AU-alphabets.

Thus, we looked then at a six-letter alphabet with three base pairs of GC-strength. The trend going from two-letter to four-letter continues: the six-letter alphabet shows more resilience, but the difference between the two and four is much larger than between four and six. The take-home lesson is that you get some stability against mutation in going from four to six bases, but what you lose in structural stability more than balances this gain. Too many structures in the six-letter alphabet are unstable, you are often dealing with large loops, etc. Thus four letters do truly represent an optimal compromise between resilience and structural stability, and this optimum is further improved in nature by the additional effects of the AUGC system that I have just mentioned: AU pairs weaker than GC and the possibility to form GU pairs.

FONTANA: I guess there is a tradeoff between how smooth your landscape and structure. The smoother your landscape is, the less structures you have. That's why they're smoother. But, on the other hand, if you buy the argument that structure conveys function, obviously the smoother the landscape is, the less function you can implement. So somewhere, the graph pops up again here, where there is a tradeoff between ruggedness, and the amount of structure that you can construct with a given ruggedness.

ANDERSON: Why don't we just call it the Laffer curve, and be done with it?

SCHUSTER: Walter, one more comment. You said, "The more structure per molecule, the fewer distinct, individual structures you have." That's not precisely right. You have the same number of structures in the two-letter alphabet, and in the four- and the six-letter alphabet. However, in the four-letter alphabet you have more sequences, so the number of structures per sequence becomes smaller.

FONTANA: I'm really talking about the minimum free energy structures.

SCHUSTER: You have fewer, that's right.

FONTANA: That's what counts.

GELL-MANN: Minimum free energy.

SCHUSTER: Minimum-free-energy structures, you have fewer.

ANDERSON: I just wanted to make the point that this could easily be the reason why there are only seven thousand, or seventy thousand, or whatever number there are, of exons. That many is all you need to cover shape space. Or, at least, it certainly covers a lot more shape space than one would think.

SCHUSTER: I would guess there is a lot more history in the exons than is in this RNA molecule.

ANDERSON: Nonetheless, there is this factor.

SCHUSTER: Yes, but...

KAUFFMAN: Maybe not, though. It would be really interesting if whatever it is—seven thousand, or seventy thousand—whether or not they actually cover a very large fraction of whatever the shape space is. We need to clone them into antibody molecule V-regions, for example, and see what diversity...

GELL-MANN: The shape space you're talking about for the exons is what?

ANDERSON: Proteins.

GELL-MANN: You mean protein shape space, in that case? It's not like this.

KAUFFMAN: No, it is like that.

GELL-MANN: Well, different arrangement.

KAUFFMAN: So we need to see whether or not that 70,000, put in V-regions, for example, is a pretty good immune system.

PERELSON: That's also a very restrictive class of shapes, the ones that you get out of antibody V-regions. We're really interested in the all-possible-proteins...

SCHUSTER: If you don't have some assumptions, which are actually guided by your experimental situation which you want to describe, it's not so simple to construct a protein shape space. Because, you have to do discretization; you have to know how to do that; "what are the same structures" is not a trivial...

GOODWIN: Peter, I'm very intrigued by this result that you only need to search a small part of space to find any structure you want. It strikes me that it may be generalizable to morphologies, generally speaking. We translate that into the sort of problem that we're interested in, in relation to morphogenesis. It may well be that there are many domains of parameter space where you find the same structures, and that could be the generalization. That seems to me a very interesting conjecture.

SCHUSTER: One answer to that is just a guess but, since it has been raised in a conference here at the Santa Fe Institute, I would like to repeat it. It was Jane and David Richardson, at one of these meetings; we were talking about making synthetic proteins, and they said if you make a guess how a synthetic protein— for a secondary structure; you want to make a certain secondary structure—you make a guess of the sequence; what you get is always bad. But if you change, in a constructive way, only ten amino acid residues, to the appropriate amino acid—you get the structure you want. And that would mean—even if your guess would be a random sequence—if you have nearby-lying good sequences, and you know how to do the changes, that would reflect the same result. This is an empirical result that people say (they do synthetic proteins) that all the guesses are bad, but when you do the proper changes, and you know how to do it, you get reasonable structures.

LLOYD: In fact, a further comment on that. One of the mysteries of how evolution works is that if you're doing mutations in a very high-dimensional space—so, a random walk in a very high-dimensional space—it's very hard to explore the space. A random walk is a poor way of exploring high-dimensional spaces. Therefore the only way that evolution, by mutation, in high-dimensional space can work is if, indeed, there are good things that lie all around you—the riches abound.

The second comment I'd like to make has to do with the first part of your talk. And I suspect many other people here noticed how close the analogy was between this evolution of self-replicating strands of RNA and Tom Ray's digital organisms in his talk. Of course, Tom Ray designed his organisms to be like this, but it's amazing that, without Tom having done too much design, that they would exhibit so many of the same features. They're roughly the same length, for instance; they evolve similar mechanisms of destroying other organisms and taking advantage of the space, or, in this case, the material that they have. It's really quite remarkable that with such different support—in one case, a computer; in this case, a set of enzymatic reactions that perform replication—that you should get so many of the same phenomena.

SCHUSTER: My answer to that would be "What's common is copying." And in computer systems, copying is very simple. But nature had to look for a very long time until she found a molecule that she could copy. There I would see the difference between chemistry and computer models. If you have the copying as the basic mechanism, I guess, then you see generic features of systems that copy and make errors and compete.

PERELSON: Everything you've done so far, when you talked about structures, is based on the secondary structure. Do you have any indications whether it would generalize to tertiary structures (which would be a little more relevant to the protein...)?

SCHUSTER: With tertiary structures, first of all, if you want to do formal comparison, you run into the same problem with proteins. You have to define what are the same tertiary structures; you cannot use coordinates for that; you have to do the proper coarse graining. And we intend to do studies on coarse graining, first as an exercise on the two-dimensional structures, and then try to go to three-dimensional, discretized structures. For example, put the atoms on a lattice makes comparison much simpler.

RAY: My colleague here noted that the RNA viruses had evolved to the edge of chaos in adjusting their mutation rates, so he knew that you could possibly defeat these viruses by just slightly raising their mutation rates. Is that right?

SCHUSTER: If we can do so, that would make an excellent anti-viral property. We must not forget, however, that viruses use—at least in the case of bacterial viruses—parts of the host system for their replication. For example, in the replicase, of the Qb-virus only one subunit comes from the viral gene, and three subbunits come from *E. coli*. In order to make use of the error threshold, you would, for example, have to find a modified virus that enters the cell competitively with the wild type but, because of a sloppy virus, specific protein replicates badly or not at all, and thus prevents other viruses entering. It is not simple but also not hopelessly hard to design such a useful competitor to the wild type.

RAY: In my theory you take the blood out of the body and raise its temperature.

SCHUSTER: But that's what the body does by itself. If you have fever, it tries to do it like that.

RAY: Just raise it a little more.

GELL-MANN: That's just the kind of suggestion that makes one's blood boil!

James H. Brown
Department of Biology, University of New Mexico, Albuquerque, NM 87131

Complex Ecological Systems

Abstract: The complexity of ecological systems is rivaled only by such systems as the brain and the global economy. From the standpoint of the present meeting, ecological systems have at least five features that make them interesting. First, they are comprised of many parts; most contain hundreds to billions of individual organisms and tens to millions of species. Second, ecological systems are open systems that maintain themselves far from thermodynamic equilibrium by the uptake and transformation of energy and by the exchange of organisms and matter across their arbitrary boundaries. Third, ecological systems are adaptive, responding to changing environments both by behavioral adjustments of individuals and by Darwinian genetic changes in the attributes of populations. Fourth, ecological systems have irreversible histories, in part because all organisms are related to each other genetically in a hierarchic pattern of descent from a common ancestor. Fifth, ecological systems exhibit a rich variety of complex, nonlinear dynamics.

In the last three decades we have seen an emphasis on deductive, reductionist, and experimental approaches to study the structure and dynamics

Complexity: Metaphors, Models, and Reality
Eds. G. Cowan, D. Pines, and D. Meltzer, SFI Studies in the
Sciences of Complexity, Proc. Vol. XIX, Addison-Wesley, 1994 **419**

of ecological systems. Empirical studies tended to use small-scale, short-term manipulations of relatively simple systems to investigate the interactions among species and between organisms and their abiotic environment. Complementing this empirical approach was an effort to construct and test simple, deterministic mathematical models. Within the last few years, however, an increasing number of ecologists have begun to adopt much more inductive, holistic, and nonexperimental research programs and to focus explicitly on issues of complexity. This has been fueled in part by the practical need to address serious human-caused environmental problems, and in part by conceptual, methodological, and technological advances in the study of other complex systems.

I describe two examples of quite different approaches to the study of ecological complexity. One focuses on the properties of networks of interaction among many species that are connected by exchanges of energy, materials, and services. Some investigators have studied food webs and sought emergent general properties of the trophic relationships among species. Others have used natural or manipulative perturbation experiments to study the response of ecological communities to altered patterns of interaction caused by changes in the composition of species or by variation in the abiotic environment. The second example of a recent approach to ecological complexity is what I call macroecology. It is somewhat analogous to a statistical mechanics for ecological systems. It is concerned with identifying and explaining statistical patterns, such as distributions of body size and abundance, that are exhibited by large numbers of individuals and species. Both of these approaches seem to be producing tantalizingly exciting results, but it may be too soon to assess their prospects for making major advances.

INTRODUCTION

The central problems of ecology concern the relationships of individual organisms with their environment, the interactions and diversity of species, and the fluxes of energy and materials through ecosystems. All levels of ecological organization, from individual organisms to assemblages of species to ecosystems to the entire biosphere, are examples of complex systems with most, if not all, of the structural and dynamical properties being discussed at this meeting.

Lack of time and expertise forces me to limit my comments to just one aspect of ecological complexity: species diversity. I will organize my presentation into three parts. First, I will discuss some general features of species diversity. In this section, I will try both to give a feeling for the magnitude of diversity and complexity, and to show how some of the concepts being discussed here apply. I will then try to become more concrete by presenting two examples. One concerns the structural and dynamical properties of networks of interaction in assemblages of many species. The second concerns statistical properties of the groups of related species produced by evolution from a common ancestor.

SPECIES DIVERSITY AND COMPLEX SYSTEMS

The diversity and complexity of ecological systems is rivaled only by such systems as the brain and the global economy. From the standpoint of the present meeting, the ecological systems of coexisting, interacting species have at least four features that make them interesting.

MANY DIFFERENT PARTS

First, they contain many different parts. There are estimated to be between 5 million and 50 million species on earth. A number that emerges from several attempts to make a quantitative estimate is thirty million.[50,78] Each species is comprised of hundreds to billions of individuals. Small patches of habitat, such as the piñon-juniper woodland that surrounds us here in Santa Fe, New Mexico, are inhabited by thousands to tens of thousands of species of microbes, plants, and animals.

One striking thing about these numbers is their uncertainty. We have no idea how many species actually inhabit the entire earth or any of its major habitats. This may seem strange to someone who is not a biologist, but the job of sampling, sorting, and identifying all of the species would be an immense task. It would require training experts for some very diverse groups, such as mites, nematodes, and many kinds of microbes, whose taxonomy and systematics are poorly known.

It is a tautology to say that each species is unique, but the ways in which species differ from and resemble each other has profound implications for the organization of ecological systems. Each species is comprised of individual organisms that are more similar to each other than to the individuals of other species in genetic composition, morphology, physiology, and behavior. It is the gaps between species, primarily in genetic and morphological space, that enables us (and other species) to recognize species as fundamental, relatively discrete units of biological organization. These unique combinations of traits provide the "labels" that John Holland uses in his model systems.

One consequence of these differences among species is that each species plays a distinctive ecological role. The differences in morphology, physiology, and behavior are reflected in different requirements for resources, different tolerances for abiotic environmental conditions, and different kinds and strengths of interactions with other species. These environmental relationships comprise the ecological niche of each species. A consequence of its particular niche is that each species exhibits a unique pattern of variation in the abundance of its individuals over both time and space.

OPEN FAR FROM EQUILIBRIUM SYSTEMS

The second interesting feature of ecological systems is that they are open systems that are maintained far from thermodynamic equilibrium by metabolism, the uptake and transformation of energy and materials. The fundamental units are individual organisms, which are usually relatively discrete units. These are thermodynamically unlikely entities with complex substructures and well-defined boundaries, that mark the transition between the varying external environment and the homeostatic internal conditions. Across their boundaries, organisms maintain high concentrations of materials, such as carbon, nitrogen, water, phosphorus, and potassium, that are rare in the environment but essential for survival, growth, and reproduction.

The different kinds of organisms that live together in one place are called an ecological community. Although they do not have discrete boundaries, most ecological communities are complex systems that maintain themselves far from thermodynamic equilibrium.[63] The coexisting species are functionally interconnected in networks that reflect the exchange of energy, materials, and services. The different kinds of organisms acquire energy from sunlight (green plants) or each other (everything else except chemosynthetic microbes), help each other by trading such commodities as food for sex (pollination) or protection or inorganic nutrients for energy (fungus/plant mutualisms), and compete with each other for all of these resources.

The community of organisms, together with their nonliving environment, comprise the ecosystem. Recently, it has become increasingly apparent that the activities of organisms play major roles in the structure and function of ecosystems. For example, in terrestrial ecosystems, organisms contribute importantly to both soil formation and erosion. Similarly, in marine ecosystems, organisms are important agents in both deposition (e.g., coral reefs) and erosion of geological surfaces.

ADAPTATION

The third interesting feature of ecological systems is that they are adaptive. They can respond adaptively to environmental change by at least three mechanisms. First, individual organisms can detect and respond facultatively to changes in their immediate environment. Examples range from simple compensatory changes in function, such as the solar tracking in which plants keeping their leaves and/or flowers perpendicular to incoming sunlight, to the chemotactic movements of bacteria in gradients of beneficial or toxic substances, to the complex learned foraging behaviors of predatory mammals. The second mechanism of adaptation involves the selective movement of both individual organisms and species. Such immigration and emigration enables ecological communities to respond adaptively to spatial and temporal variation in the environment. Both this and the previous mechanism involve individual organisms responding adaptively to environmental change in ways that tend to increase their probability of surviving and reproducing. Both also tend to produce fairly rapid negative feedback in ecosystems, maintaining flows of energy and materials and preventing the local accumulation of unused resources. One exception is the sometimes destabilizing effect of immigration of a new species into a previously relatively closed ecosystem, such as an oceanic island, lake, or other isolated habitat.

The third mechanism of adaptation is natural selection. At least two levels of organization, individuals and species, exhibit a combination of Malthusian and Darwinian dynamics. They reproduce themselves by birth or speciation, potentially producing many more descendants than can survive. Because attributes are heritable, the inevitable deaths and extinctions tend to be selective, eliminating the less fit and resulting in descent with adaptive modification. In contrast to the two previous mechanisms of adaptation, natural selection requires time scales of generations to millennia, because it involves the death or differential reproduction of individuals or species.

IRREVERSIBILITY

A final feature of ecological systems is that they are effectively irreversible. Each individual organism is the unique result of the interaction between the genetic program and the environment during its development. Each species represents the unique outcome of similar interactions during its phylogenetic history. Both ontogenetic development and phylogenetic evolution are also sensitive to initial conditions.

The evolution of living things has involved the production of new structures and functions by modifying the characteristics of pre-existing organisms. A cumulative record of these changes is recorded in the molecules and more complex characteristics of contemporary organisms. This record serves as the basis for phylogenetic reconstruction or cladistics. In 1966, the German biologist Willi Hennig developed an algorithm for reconstructing the history of a lineage based on the irreversible changes in characters. Initially, phylogenetic reconstructions, called cladograms, were made using traditional morphological characters. More recently, however, the techniques of molecular biology have been adopted and now most cladograms are based on nucleic acid sequences.[5,75]

Although the characteristics of assemblages of species might seem to be more reversible than those of the species that comprise them, this is probably not the case. It may be true to some extent at the smallest scales, where immigration and local extinction (or emigration) could potentially shift species composition back and forth between alternative states. At larger spatial and temporal scales, however, the imprint of irreversible history is readily apparent. Speciation, long-distance colonization, extinction, and coevolution leave their indelible imprint on the composition of regional biotas. In so doing, they influence even the smallest scales by determining the pool of species from which local communities are assembled.

NONLINEAR DYNAMICS AND OTHER CHARACTERISTICS OF COMPLEX SYSTEMS

It is clear that the dynamics of species interactions are inherently nonlinear. This is caused in part by the fact that the different species operate on different temporal and spatial scales. This is an area that has received considerable attention from ecologists, although it is not one where I have any firsthand experience. Relatively simple mathematical models of two- and three-species systems often produce low-dimensional chaotic dynamics.[27,33] Empirical analyses of long-time series of simple interactions, such as pathogen-host relationships and the famous Hudson Bay Company record of lynx and hare furs, have been quite successful in reconstructing low-dimensional attractors.[61]

It is perhaps too early to say to what extent and which components of ecological systems exhibit some of the other features of complex adaptive systems discussed at this meeting. There are intriguing hints that some species assemblages may tend to be poised at some critical point between order and randomness. For years I have been intrigued by the fact that these assemblages seem to be characterized, on the one hand, by both the high degree of seemingly unpredictable individualism in the variation in the abundance of each species in space and time, and, on the other hand, by the apparently predictable but probabilistic structure exhibited by the assemblages as a whole. I will cite one example from my own research on desert rodents. On the one hand, there is enormous spatial and temporal variation in the presence or absence or the relative abundances of the particular species that coexist to comprise local communities.[13,14] On the other hand, these same sets of species exhibit several kinds of structure (ratios of body and tooth sizes, core-satellite patterns of distribution and abundance, and nested subset composition) that would not be possible if the coexisting species had been assembled at random from the regional species pool.[3,6,21,25,30,54,67]

Do these kinds of observations suggest that ecological communities tend to "evolve to the edge of chaos" or to exhibit "self-organized criticality"? The possibilities are exciting and warrant investigation.

NETWORKS OF INTERACTIONS

I will now illustrate some interesting features of complex ecological systems by considering in somewhat more detail two quite different approaches. The first is the study of structural and dynamical properties of networks of coexisting, interacting species.

SOME HISTORY

This is a subject with a long history in ecology. It begins with theoretical and empirical studies of the population dynamics of pairs of predator-prey or competing species.[26,72,77] In the 1960s MacArthur[42,43] and Levins[37,38] suggested that such a

INDIRECT INTERACTIONS

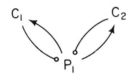

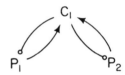

EXPLOITATIVE COMPETITION   "APPARENT COMPETITION"

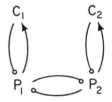

INDIRECT MUTUALISMS

FIGURE 1 Some simple networks of indirect interactions among three or four species. The P's and C's indicate species of producers at any trophic level and consumers at the next higher trophic level, respectively (e.g., plants and their herbivores, or herbivores and their predators); the arrows with points and circles indicate positive and negative effects, respectively. In exploitative competition, two consumers affect each other negatively by both depleting a common prey. In Holt's[29] apparent competition, two consumers can affect each other negatively if each contributes to maintaining a common predator species. There are a variety of indirect mutualisms in which two species can benefit each other by competing with a common third competitor or by feeding on competing prey.

pairwise approach might be extended to investigate the structure and dynamics of communities of many species. By the 1970s, however, the shortcomings of this approach were becoming apparent. Theoretical studies of the properties of simple networks of interaction among three or four species showed that simply adding one or two additional species could change the sign of the net interaction between a pair of species.[29,36] At about the same time, empirical studies were finding good examples of the simple kinds of indirect interactions (Figure 1).

Since the 1970s there has been a change in emphasis, away from studies that try to deduce the properties of communities by putting together species and their interactions from the bottom up, and toward top-down studies that focus on the emergent structural and dynamical properties of entire communities.

One important issue has been the relationship between species diversity and dynamical stability. MacArthur[42] had suggested that increasing the number of species tended to increase the stability of ecological communities. This was widely accepted, apparently because it fit the intuition of many empirical ecologists that some of the most species-rich ecological communities, such as those in tropical rain forests and coral reefs, were also some of the most stable ones. MacArthur's conjecture was dealt a severe blow when May[47] showed that random assemblages of species tended to become more unstable as they acquired more species, with instability defined in terms of magnitude of fluctuations and probability of extinction of component species.

Since diverse species do coexist to form relatively stable, persistent communities, it seems to follow that the species do not interact at random. Instead, the networks of interaction must have special structural and dynamical properties (Figure 2). May[47] suggested that the networks might be organized into parallel subnetworks, each of which tended to be relatively stable and insulated from interactions with species in other subnetworks. Lawlor[34] pointed out that interactions in all real communities are subject to the constraint that energy flows among species and is dissipated subject to the laws of thermodynamics. Nutritional relationships among species within communities began to receive increased attention.

FOOD WEBS: THEORY AND DATA

Ecologists use the term food web to refer to the network of trophic interactions among a set of species. The vast majority of studies of food webs share a number of features (Figure 3(a)). First, the food webs are just that: diagrams of which species eat which other ones. Thus, they indicate the flow of energy and material resources through a community. Second, food webs do not indicate the other kinds of interactions in addition to those between predator and prey, herbivore and plant, and parasite or pathogen and host. The networks do not indicate the services, such as protection from biological enemies or physical stress or transport of gametes, that

INTERACTION NETWORKS

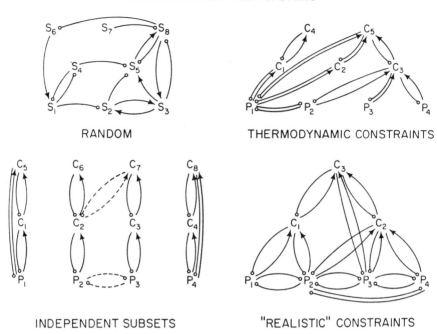

FIGURE 2 Hypothetical networks of interaction among species in ecological communities. Random assemblages of large numbers of species were shown by May[47] to have low probability of being dynamically stable (maintaining all of the species in the face of possible extinction). Compartmentalization of communities into independent subsets of linearly arranged food chains were suggested by May[47] and Pimm[57] to enhance stability. Thermodynamic constraints on energy flow, which tend to produce a food web of pyramidal shape, add an element of realism that was missing from many of the early models of many interacting species.[34] Simple rules based on empirical patterns of energy flow and trophic specialization were used to derive the organization of interactions reflecting "realistic constraints."

are often a critical element of mutualistic interactions. They also do not indicate the aggressive interference or pre-emption of space that can be important mechanisms of competitive interactions. Competition that occurs as a result of exploitation of common food resources is incorporated in the food webs as indirect interactions. Third, the interactions among species in food webs are usually characterized only qualitatively: as being either present or absent. Thus, all possible interactions among the species can be portrayed as a matrix, and entries of ones and zeros can be used to designate which species do or do not eat which other species. Fourth, the food webs contain no spatial or temporal information. Thus they fail to capture potentially important features of the dynamics and spatial structure of interactions.

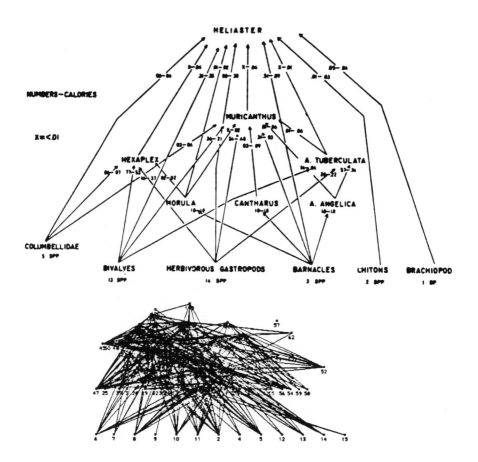

FIGURE 3   Two examples of empirically derived food webs. (a) The web developed by Paine[52] to characterize the important trophic interactions among species on rocky substrate in the intertidal zone of the northern Gulf of California. The top predator, the starfish *Heliaster kubinili,* was hypothesized to be a "keystone species." (b) The web drawn by Winemiller[79] to describe the strong interactions among the fish species in a swamp creek in the western llanos of Venezuela. Note the extreme difference in the degree of connectance between these two webs. This difference probably reflects more a difference between the two ecologists (and the objectives of their studies) than between the two communities in the organization of interactions.

Despite these simplifications the study of food webs has produced some intriguing empirical generalizations and related theory.[18,22,52,57,64,65,69] Compilations and analyses of the many food webs in the ecological literature suggest several statistical regularities. The number of trophic levels (links in a food chain), number of prey per predator, number of predators per prey, and measures of connectance (the extent to which species are linked to other species) all vary somewhat within and among communities, but the variation appears to be limited to a narrow range of values (about five for the first three parameters) in the vast majority of published food webs. This suggests that these networks may be organized according to a set of general and relatively simple principles.

One concern about these patterns, however, is that they may say more about the way that ecologists collect, analyze, and classify data, than about how nature is organized.[53] Of particular concern are the following: (1) qualitative/quantitative issues: whether all species that are consumed are represented by links, and, if not, how those links that are drawn reflect the importance of different species in the diet; (2) aggregation of species: often all species in the diet cannot be distinguished, and sometimes similar, "trophically equivalent" species are combined into one prey category; and (3) temporal and spatial variation: the area of space and length of time that are represented by the data. My own impression is that the more careful and complete the data, the more links tend to be found. This is illustrated by the webs for tropical fish communities assembled by Winemiller[79] and shown in Figure 3(b) (compare with Figure 3(a)). It is also illustrated by the gedanken study of preparing a food web for humans in any large city. The number of species consumed would be very large, at least one or two orders of magnitude greater than the value of about five prey species that is supposedly typical of most consumers.

This problem of empirical reliability of published food webs casts doubt on the value of the analytical and theoretical studies that have tried to treat these webs as quantitative data and to derive general principles of community organization. This does not necessarily indicate, however, that such general principles do not exist. I suspect that they do, and that they are reflected in statistical patterns derived from the published food webs. I also suspect, however, that further progress in elucidating these principles will come, not from analyzing ever larger numbers of qualitative webs, but from studies that focus on quantitative and dynamical aspects of all interactions among species, not just trophic relationships. Since it will be an immense task to assemble sufficiently detailed data for any real community, an alternative might be to analyze interaction networks in computer-simulated communities, such as elaborations on John Holland's *ECHO* or Tom Ray's *Tierra*.

PERTURBATION EXPERIMENTS

Another way to investigate the properties of networks of ecological interaction is to perturb a community by removing or adding one or more species, and monitor the responses as the network is reorganized. I will give two examples of studies where species have been removed, and then briefly consider the case of invading exotic species.

Edward Boyer,[4] a doctoral student at the University of Arizona, studied the effects of the natural disappearance of a top predator from the northern Gulf of California. The large starfish, *Heliaster kubinili*, was a major consumer of the sessile organisms (animals and algae) that compete for space on the intertidal rocky shore. In 1978, *Heliaster* was decimated by the outbreak of a microbial pathogen and nearly went extinct throughout its geographic range in the northern Gulf.[23] Boyer, who had been censusing the organisms on rocky surfaces before the disappearance of *Heliaster*, continued his study to obtain a record of the changes caused by this natural perturbation.

Figure 3(a), from a classic paper by Robert Paine,[52] illustrates the food web for the rocky intertidal where Boyer did his work. Paine suggested that *Heliaster*, as the top predator in this food web, is a "keystone" species that regulated the abundance and utilization of space by the other species in this communities. This would predict that when *Heliaster* disappeared, the barnacles, which are the superior competitors, should increase and monopolize a larger share of space. In the short term, one to two years after the extirpation of *Heliaster*, this was just what Boyer observed (Figure 4). In the longer term, however, even though *Heliaster* was still absent, the dominance of barnacles decreased again because other predators increased their consumption of barnacles. Boyer's study showed that the division of space among species on the rocky shore obeys a power law. Although this relationship was disrupted temporarily when *Heliaster* first disappeared, it was reestablished rapidly even though *Heliaster* remained absent. And it is also noteworthy that the relative dominance of particular species (indicated by their rank, not shown explicitly in Figure 4) was different before and after the perturbation.

The second example of response of a community to removal of species comes from my own experimental studies in the Chihuahuan Desert. In 1977 we initiated a number of experimental manipulations of 0.25-ha plots of desert shrubland.[8] Perhaps the most interesting results were those caused by the removal (by selective fencing and live-trapping) of a "guild" of three similar species of seed-eating kangaroo rats.[9] This perturbation initiated a series of changes during the subsequent 12 years. Within the first 2–4 years, species that interact directly with the kangaroo rats showed large changes in abundances: smaller rodent species that compete for food with kangaroo rats increased several fold,[15,51] and some large-seeded plant species, whose seeds are preferentially consumed by kangaroo rats, increased several orders of magnitude.[8,32] Within a period of 4–8 years, species that interact indirectly with kangaroo rats changed: small-seeded plants that compete with the large-seeded species decreased[8,60] and, in a similar experiment,[32] a specific fungal parasite of one of the large-seeded plant species increased, ultimately causing a decrease in its host. Finally, after 8–12 years the plots where kangaroo rats had been removed exhibited dramatic changes in vegetation, from shrubs and annual forbs desert to grasses, and these plots were colonized by animal species, typical

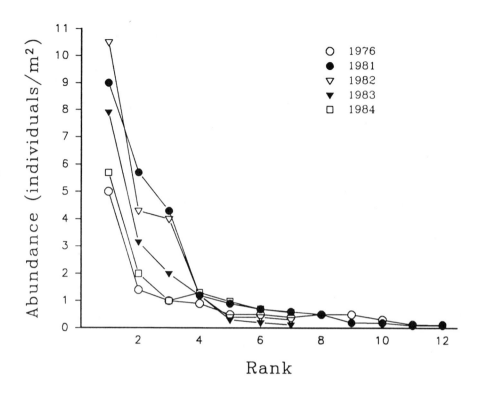

FIGURE 4 "Dominance-diversity" relationships showing the ranked relative abundances of species on rocky shore in the intertidal of the northern Gulf of California before and after the near extinction of the top predator, the starfish *Heliaster kubiniii*, in 1978. Note that immediately after the disappearance of the starfish, the community became dominated by a few species as predicted by Paine's keystone predator hypothesis but, as time passed and the starfish remained absent, the distribution of abundances among species returned the original values.[4]

of grassland and never before observed on the study site.[9,28] So, as in Boyer's study, removal of species from a community first had fairly straightforward direct and indirect effects on certain other species, but ultimately resulted in a complete reorganization of the relationships among the species, including the shifts in the plants and animals that hold most of the space and form the base of the food web.

It is interesting that in both of these examples, removal of key species caused large changes in abundance (and, in the desert community, local colonization and perhaps extinction) of species, but it did not result in any substantial decrease in overall species diversity. A similar picture could be painted for the consequences of the extirpation of the chestnut from deciduous forests in eastern North America following the outbreak of a fungal pathogen (introduced from Eurasia) in the first half of this century. This contrasts with many examples in which colonizing exotic species caused large decreases in diversity. I will just note that one difference between these systems is that the rocky intertidal, desert shrub, and deciduous forest communities where the species removal "experiments" occurred are large, open systems. In contrast, the impact of additional "experiments" involving colonizing exotics are most dramatic on small, relatively closed communities, such as islands and isolated lakes and springs.

STATISTICAL CHARACTERISTICS OF LARGE BIOTAS

The second approach to investigating complex ecological systems I will call macroecology, because of its similarity to macroevolution,[68] and to a lesser extent to macroeconomics. It could also be called statistical ecology, and there are similarities to statistical mechanics and other statistical approaches to physical systems.

SOME HISTORY

Statistical approaches to ecology have a long history. Seemingly nearly every time properties of diverse assemblages of species have been examined, intriguing statistical patterns appear. These patterns have long perplexed ecologists.[56] Some, such as the frequency distributions of abundance,[58,76] body sizes,[31,48] and areas of geographic ranges,[59,77] and the $-3/2$ power law of self-thinning in plants,[40,73,74,80] appear to be very general. Even in these cases, however, testable explanations have proven elusive.

Some of us continue to investigate these statistical patterns in the hope that they may hold clues to important aspects of the structure and dynamics of complex ecological systems. Most current studies take one of two different but complementary approaches. An almost completely ecological perspective has been adopted toward some of the patterns, such as the $-3/2$ thinning in plants and the distribution of abundance or space among species. These patterns reflect regularities in the way that resources are divided among the species that coexist in the same place at the same time. They have intriguing similarities to Pareto and related distributions of wealth in economic systems.[70] While, in a sense, these distributions have a history, they represent the self-assembly of interacting units by movements across the boundaries of an open system.

A much more evolutionary perspective has been adopted in studies of other patterns, such as the distributions of body sizes and areas of geographic ranges. These patterns have often been viewed as the products of cladogenetic evolution, the proliferation of species from a single common ancestor over evolutionary time. Each of these distributions reflects a unique history, a pedigree of ancestor descendent relationships (Figure 5). Thus, they provide a statistical representation of the

interaction, over evolutionary time, between the intrinsic constraints and potential-
ities of a lineage and influence of extrinsic environmental limits and opportunities
on its diversification.

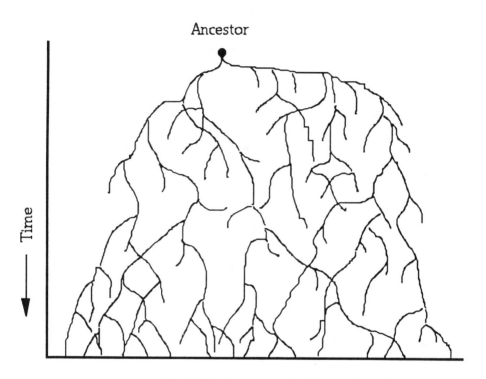

Variable (e.g. morphology, diet ...)

FIGURE 5   The pattern of descent of a trait, such as body size in a hypothetical
lineage descended from a common ancestor. The lines indicate the trajectory of the
trait value over evolutionary time, with branches and terminals showing speciation
and extinction events, respectively. It may be possible to reconstruct much of such
history for some real organisms, using data from the fossil record to measure the
trait in different species at different times in the past and using molecular and
other characteristics of living and extinct organisms to determine the phylogenetic
relationships among the species.

BODY SIZE

To illustrate some of the recent advances and future promise of the statistical macroecological approach, I will consider the distribution of body sizes in large assemblages of related species, and then discuss some of the implications.

There are several reasons to focus on body:

1. variation in size is one of the most important components of biological diversity (body mass spans more than 20 orders of magnitude, 10^{-13} to 1^8 g from mycoplasma to whales[62]);
2. there are good, quantitative data on the sizes and associated characteristics of organisms of many different kinds;
3. many characteristics of organisms from the structure and function of individuals to the ecological and evolutionary dynamics of species vary with size, and many of these relationships can be characterized by so-called allometric equations of the form $T = cM^b$, where T is the trait of interest, c is a constant, M is individual body mass, and b is another constant; and
4. some size-related phenomena are so general that they must reflect fundamental constraints on structure, function, and diversity.

One of these general patterns is the characteristic frequency distribution of body size among the species in any large taxonomic group (Figure 6). While the actual sizes of different kinds of organisms obviously vary enormously, the qualitative shape of the frequency distribution, highly modal and right-skewed, even on a logarithmic scale, appears to be almost universal. It has been found in organisms from bacteria to plants to vertebrates.[2,31,48,71] I know of no exceptions.

The characteristic shape of these distributions reflects the result of the independent evolution of many different lineages in a wide variety of environments. What is there about the process of evolutionary diversification in response to ecological opportunity that predictably produces this outcome? There are two main hypotheses. The first is that as body size deceases, organisms inherently become more specialized, have more restrictive requirements, can divide resources more finely, and, hence, more species can exist. This hypothesis, first proposed by Hutchinson and MacArthur,[31] has recently been recast in terms of fractal geometry.[35,49] It can potentially account for the decrease in the number of species as size increases to the right of the mode in Figure 6. It cannot, however, account for the sharp decrease in the number of species as size decreases to the left of the mode. The greatest diversity is not in the very smallest species.

An alternative hypothesis has recently been proposed by our research group. We have developed a model for an optimal body size that is based on an explicit energetic definition of fitness.[12] This model assumes that the fitness of an individual can be defined as the rate that resources (energy, nutrients, and other essential

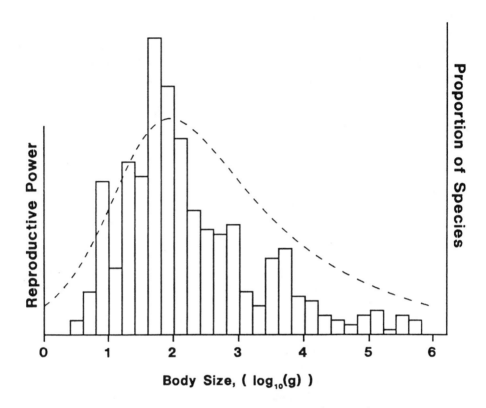

FIGURE 6 Histograms: frequency distribution of number of species with respect to logarithms of body mass for North American terrestrial mammals. Dashed line: The hump-shaped curve of reproductive power as a function of the logarithm of body mass predicted by a model that assumes that the allocation of energy to offspring is a two-stage process, involving first the uptake of energy from the environment and, second, the transformation of energy into offspring. Note that the model predicts an optimal body size of approximately 100 g and a distribution of reproductive power that closely matches the observed frequency distribution.[12]

commodities, such as water), in excess of those required for growth and maintenance of the individual, can be harvested from the environment and utilized for reproduction. Thus, we equate fitness with reproductive power, the rate of conversion of energy into useful work for reproduction.

Reproductive power is modeled as the consequence of two limiting rates, which are allometric functions of body mass, M: (1) the rate at which an individual can acquire resources from its environment, which scales as $M^{0.75}$, the same as individual metabolic rate, productivity, and growth rate[16,55,62]; and (2) the rate at which it can convert those resources into reproductive work, which scales as $M^{-0.25}$, the same as the rate of mass-specific metabolism and nearly all biological conversion processes.[16,55,62]

From these assumptions we derive a mathematical expression for the optimal size. The biological interpretation of this model is straightforward. The smallest individuals have a great capacity to convert resources into reproductive work, but they are limited by the rate of acquisition of resources for reproduction. They must spend most of their time foraging just to meet their high mass-specific maintenance metabolism. In contrast, large individuals have a great capacity to acquire resources, but they are constrained by the rate at which these can be converted into viable offspring.

This model does three things that are fundamentally new. First, it can predict the quantitative value of the optimum based on available (or obtainable) data on individual productivity and energy turn over. The allometric exponents, 0.75 and −0.25 are assumed to be the same for all organisms. The value of the optimum depends on taxon-specific constraints on structure and function that are reflected in two constants, which can be estimated from allometric equations for maximum productivity and maximum energy turn over, respectively. Estimating these constants for mammals using data by Peters,[55] the model predicts the distribution of reproductive power shown in Figure 6. Note that this closely matches the size-frequency distribution for mammals, and it predicts an optimum body mass of approximately 100 g.

Two kinds of data from insular mammal faunas suggest that the optimal size for mammals is indeed about 100 g (Figure 7): (1) populations of species that on continents are larger than this size tend to evolve dwarf insular races, whereas populations of smaller species tend to evolve giant insular races[39]; and (2) as the area of a land mass and the number of species present decrease, the range of sizes represented in the fauna also decreases so that, when there is only a single species present, it tends to be close to the optimal size.[45,46]

The second innovative feature of the BMT model is that the two-stage process of energy acquisition and allocation results in a hump-shaped distribution of fitness. This, in turn, predicts that the relationship between many life history and ecological traits and body size will also be hump-shaped. As organisms deviate from the optimal size and the rate that energy can be allocated to reproduction decreases, this requires compensatory adaptations that reduce mortality. Thus, the model can explain why hummingbirds and vespertilionid bats, among the smallest birds and mammals, have smaller clutch/litter sizes and much longer lifespans than their somewhat larger relatives and than predicted by the standard allometric equations.[17,24] The higher reproductive power of optimal-sized organisms should also enable them to become more abundant than their larger and smaller relatives. Several data sets of population densities in different kinds of animals are consistent with this prediction.[13,14,19,20,44] So far as we are aware, ours is the only model that offers a general explanation for these patterns.

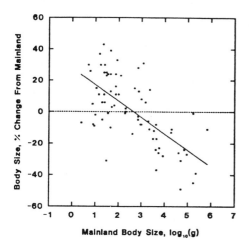

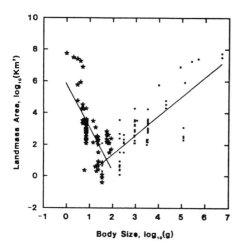

FIGURE 7 Two empirical patterns in the body sizes of mammals on islands and continents that suggest the existence of an optimal body size of approximately 100 g. Above: microevolutionary trends in insular populations of mammals, showing that populations above and below the optimum tend to evolve dwarf or giant insular forms, respectively. Below, maximum (squares) and minimum (stars) body sizes of terrestrial mammals inhabiting a large sample of islands and continents. Note that as the size of the land mass and the number of species present decreases, the range of body sizes tends to converge toward the optimal body mass.[12]

Third and most importantly, by developing an explicit energetic definition of fitness as reproductive power, our model suggests a unifying physical basis for investigating biological complexity. Evolutionary biologists and population geneticists have traditionally defined fitness in terms of the rate of production of viable offspring or as the rate of increase (or net rate of replacement) of a trait or genotype. This concept of fitness has been criticized for containing elements of circularity or tautology. Our formulation circumvents these problems and follows Boltzman,[1] Lotka,[41] and Schrödinger[66] in seeking a physical, thermodynamic, and energetic basis of fitness.

This leads me to one last insight. A common feature of many complex adaptive systems is that increased complexity evolves by evading the consequences of conservation laws through "innovations" that change the constraints on the system. This is true of biological systems, whether we are considering individual organisms or ecological communities of many species. On the one hand, the supply of energy and materials is limited. Organisms inherit from their ancestors structural and functional constraints that restrict their capacities to transform energy to do the work of acquiring, against organism-environment concentration gradients, the materials that they require to maintain homeostasis and produce offspring. On the other hand, these limits are not absolute. Evolutionary innovations can break preexisting constraints, enabling organisms to process more energy and acquire more materials. These advantageous innovations enable the systems that possess them to move farther from thermodynamic equilibrium and to develop greater diversity and complexity.

At least at the extremes, these innovations can have two kinds of effects. In one case, the breaking of constraints may enable one evolutionary lineage to increase its share of the existing energetic and material pie, increasing its abundance and distribution at the expense of other lineages. An example may be the origin and increase to dominance of angiosperm plants in the Cretaceous. The key innovations of the angiosperms, mutualistic relationships with animals to perform pollination (and, in many lineages, also to disperse seeds), enabled them to evolve species adapted and confined to the microenvironments that were available in spatially and temporally heterogeneous landscapes. This potentially gave them an enormous advantage and enabled them to replace the wind-pollinated gymnosperms as the dominant land plants. In the other case, a key innovation may enable a lineage to exploit a new source of energy and/or materials, thus increasing the size of the pie. An example would be the adaptations for acquiring and retaining water that enabled both plants and animals to first colonize the land. For the great majority of the history of life on earth, it was confined to the oceans. Key innovations that occurred in the Mesozoic allowed the colonization of land and led ultimately to the large buildup of biomass and organic diversity in terrestrial ecosystems. In both of these cases, key innovations broke existing constraints on the acquisition of energy and materials and resulted in the evolution of more diverse and complex systems.

SUMMARY

Ecological systems of many coexisting species provide excellent examples of complex adaptive systems; they possess most of the characteristics of systems being discussed here. These ecological systems are typically composed of enormous numbers of individual organisms belonging to many structurally and functionally distinct species. These are open systems that maintain themselves far from thermodynamic equilibrium by the exchange of energy and materials among organisms and between organisms and the abiotic environment. Ecological systems are constantly changing, and they have the capacity to respond adaptively to environmental change. Ecological systems are effective by irreversible and, as a consequence, they have a unique history. I discuss in some detail two promising approaches that ecologists have recently begun to use to try to understand the fundamental features of complex systems of many interacting species. The first is the analysis of structural and dynamical properties of food webs, which are networks of interaction among species based on the exchange of energy and materials. These networks may have interesting emergent properties of connectance among the species and reorganization in response to perturbation. The second approach is the analysis of the statistical distributions of properties, such as body size, abundance, and distribution in space and time, among the species in large biotas. Here, too, there appear to be patterns that emerge repeatedly in different kinds of organisms from different environments, suggesting that some relatively simple processes underlie the seemingly complex organization.

ACKNOWLEDGMENTS

So many individuals have influenced my thinking on ecological complexity that it is impossible to thank them all individually. Special thanks must go to Brian Maurer, Pablo Marquet, Mark Taper, Tim Keitt, Eric Toolson, Alan Johnson, Eric Scheider, and Astrid Kodric-Brown for many hours of stimulating discussion. Sarah Linehan helped to prepare the manuscript and Felisa Smith drew Figure 5. The National Science Foundation has supported much of my own research, most recently with grants BSR-8718138 and BSR-8807792.

REFERENCES

1. Boltzman, L. *Populare Schriften*. Leipzig: J. A. Barth, 1905.
2. Bonner, J. T. *The Evolution of Complexity by Means of Natural Selection*. Princeton: Princeton University Press, 1988.
3. Bowers, M. A., and J. H. Brown. "Body Size and Coexistence in Desert Rodents: Chance or Community Structure." *Ecology* **63** (1982): 391–400.
4. Boyer, E. H. "The Natural Disappearance of a Top Carnivore and Its Impact on an Intertidal Invertebrate Community: The Interplay of Temperature and Predation on Community Structure." Ph.D. Thesis, University of Arizona, 1987.

5. Brooks, D. L., amnd D. A. McLennan. *Phylogony, Ecology, and Behavior.* Chicago, IL: University of Chicago Press, 1991.

6. Brown, J. H. "On the Relationship Between Abundance and Distribution of Species." *Amer. Nat.* **124** (1984): 255–279.

7. Brown, J. H., and M. A. Bowers. "Patterns and Processes in Three Guilds of Terrestrial Vertebrates." In *Ecological Communities Conceptual Issues and the Evidence,* edited by D. R. Strong, Jr., et al., 282–296. Princeton: Princeton University Press, 1984.

8. Brown, J. H., D. W. Davidson, J. C. Munger, and R. S. Inouye. "Experimental Community Ecology: The Desert Granivore System." In *Community Ecology,* edited by J. Diamond and T. J. Case, 41–61. New York: Harper & Row, 1986.

9. Brown, J. H., and E. J. Heske. "Control of a Desert-Grassland Transition by a Keystone Rodent Guild." *Science* **250** (1990): 1705–1707.

10. Brown, J. H., and M. A. Kurzius. "Composition of Desert Rodent Faunas: Combinations of Coexisting Species." *Annales Zoologici Fennici* **24** (1987): 227–237.

11. Brown, J. H., and M. A. Kurzius. "Spatial and Temporal Variation in Guilds in North American Desert Rodents." In *Ecology of Small Mammal Communities,* edited by Z. Abramsky, B. J. Fox, D. W. Morris, and M. R. Willig, 71–90. Austin: Texas Technical University, 1989.

12. Brown, J. H., P. A. Marquet, and M. L. Taper. "Evolution of Body Size: Consequences of an Energetic Definition of Fitness." *Amer. Nat.* **142** (1993): 573–584.

13. Brown, J. H., and B. A. Maurer. "Evolution of Species Assemblages: Effects of Energetic Constraints and Species Dynamics on the Diversification of North American Avifauna." *Amer. Nat.* **130** (1987): 1–17.

14. Brown, J. H., and B. A. Maurer. "Macroecology: The Division of Food and Space Among Species on Continents." *Science* **243** (1989): 1145–1150.

15. Brown, J. H., and J. C. Munger. "Experimental Manipulation of a Desert Rodent Community: Food Addition and Species Removal." *Ecology* **66** (1985): 1545–11563.

16. Calder, W. A., III. *Size, Function, and Life History.* Cambridge: Harvard University Press, 1984.

17. Calder, W. A., III. "Avian Longevity and Aging." In *Genetic Effects on Aging II,* edited by D. E. Harrison, 204. Caldwell, NJ: Telford Press, 1989.

18. Cohen J. E. "Food Webs and Niche Space." Princeton: Princeton University Press, 1978.

19. Damuth, J. "Population Density and Body Size in Mammals." *Nature* **290** (1981): 699–700.

20. Damuth, J. "Interspecific Allometry of Population Density in Mammals and Other Mammals: The Independence of Body Mass and Population Energy-use." *Biol. J. Linnean Soc.* **31** (1987): 193–246.

21. Dayan, T., and D. Simberloff. "Morphological Relationships Among Coexisting Heteromyids: An Incisive Dental Character." *Amer. Nat.* (in press).
22. DeAngelis, D. L. *Dynamics of Nutrient of Cycling and Food Webs*. London: Chapman & Hall, 1991.
23. Dungan, M. L., T. E. Miller, and D. A. Thomsom. "Catastrophic Decline of a Top Carnivore in the Gulf of California." *Science* **185** (1982): 1058–1060.
24. Findley, J. S. *Bats*. Cambridge: Cambridge University Press, in press.
25. Fox, B. J., and J. H. Brown. "Assembly Rules for Functional Groups in North American Desert Rodent." *Oikos* (in press).
26. Gause, G. F. *The Struggle for Existence*. Baltimore: Williams & Wilkins, 1934.
27. Hastings, A., and T. Powell. "Chaos in a Three-Species Food Chain." *Ecology* **72** (1991): 896–903.
28. Heske, E. J., J. H. Brown, and S. Mistry. "Long-Term Experimental Study of a Chihuahuan Desert Rodent Community: 13 Years of Competition." *Oecologia* (in press).
29. Holt, R. D. "Predation, Apparent Competition, and the Structure of Prey Communities." *Theor. Popul. Biol.* **12** (1977): 197–229.
30. Hopf, F. A., and J. H. Brown. "The Bull's-Eye Method for Testing Randomness in Ecological Communities." *Ecology* **67** (1986): 1139–1155.
31. Hutchinson, G. E., and R. H. MacArthur. "A Theoretical Model of Size Distributions Among Species of Animals." *Amer. Nat.* **93** (1959): 117–125.
32. Inouye, R. S. "Interactions Among Unrelated Species: Granivorous Rodents, a Parasitic Fungus, and a Shared Prey Species." *Ocealogia* **49** (1981): 425–427.
33. Kot, M., and W. M. Schaffer. "The Effects of Seasonality on Discrete Models of Population Growth." *Theor. Pop. Biol.* **26** (1984): 340–360.
34. Lawlor, L. R. "Structure and Stability in Natural and Randomly-Constructed Competitive Communities." *Amer. Nat.* **116** (1980): 394–408.
35. Lawton, J. H. "Species Richness and Population Dynamics of Animal Assemblages. Patterns in Body Size: Abundance and Space." *Phil. Trans. Roy. Soc. London B* **330** (1990): 283–291.
36. Levine, S. H. "Competitive Interactions in Ecosystems." *Amer. Nat.* **110** (1976): 903–910.
37. Levins, R. *Evolution in Changing Environments*. Princeton: Princeton University Press, 1968.
38. Levins, R. "Evolution in Communities Near Equilibrium." In *Ecology and Evolution of Communities*, edited by M. L. Cody and J. M. Diamond, 16–50. Cambridge: Harvard University Press. 1975.
39. Lomolino, M. V. "Body Sizes of Mammals on Islands: The Island Rule Reexamined." *Amer. Nat.* **125** (1985): 310–316.
40. Lonsdale, W. M. "The Self-Thinning Rule: Dead or Alive?" *Ecology* **71** (1990): 1373–1388.

41. Lotka, A. J. *Principles of Physical Biology.* Baltimore: Williams & Wilkins, 1922.

42. MacArthur, R. H. "Patterns of Species Diversity." *Biol. Rev.* **40** (1965): 510–533.

43. MacArthur, R. H. "The Theory of the Niche." In *Population Biology and Evolution*, edited by R. C. Lewontin, 159-176. Syracuse: Syracuse University Press, 1968.

44. Marquet, P. A., S. N. Navarrete, and J. C. Castilla. "Scaling Population Density to Body Size in Rocky Intertidal Communities." *Science* **250** (1990): 1125–1127.

45. Marquet, P. A., and M. L. Taper. "Linking Patterns and Process Across Levels of Organization: From Individuals to Populations, Communities and Biotas." Unpublished manuscript.

46. Maurer, B. A., J. H. Brown, and R. D. Rusler. "The Micro and Macro in Body Size Evolution." *Evolution* **46** (1992): 939–953.

47. May, R. M. *Stability and Complexity in Model Ecosystems.* Princeton: Princeton University Press, 1973.

48. May, R. M. "The Dynamics and Diversity of Insect Faunas. " In *Diversity of Insect Faunas*, edited by L. A. Mound and N. Waloff, 188–204. Royal Ent. Soc. Symposium. London, Sept., 1977. Oxford: Blackwell, 1978.

49. May, R. M. "The Search for Patterns in the Balance of Nature: Advances and Retreats." *Ecology* **67** (1986): 1115–1126.

50. May, R. M. "How Many Species are There on Earth?" *Science* **241** (1988): 1441–1449.

51. Munger, J. C., and J. H. Brown. "Competition in Desert Rodents: An Experiment with Semipermeable Enclosures." *Science* **211** (1981): 510–512.

52. Paine, R. T. "Food Web Complexity and Species Diversity." *Amer. Nat.* **100** (1966): 65–75.

53. Paine, R. T. "Food Webs: Road Maps of Interactions or Grist for Theoretical Development?" *Ecology* **69** (1988): 1648–1654.

54. Patterson B. D., and J. H. Brown. "Regionally Nested Patterns of Species Composition in Granivorous Rodent Assemblages." *J. Biogeography* **18** (1991): 395–402.

55. Peters, R. H. *The Allometry of Growth and Reproduction.* Cambridge: Cambridge University Press, 1983.

56. Pielou, E. C. *Mathematical Ecology.* New York: Wiley, 1977.

57. Pimm, S. L. *Food Webs.* London: Chapman & Hall, 1982.

58. Preston, F. W. "The Canonical Distribution of Commonness and Rarity: Part I and II." *Ecology* **43** (1962): 185–215 and 410–432.

59. Rapoport, E. H. *Aerography: Geographical Strategies of Species.* New York: Pergamon Press, 1982.

60. Samson, D. A., T. E. Philippi, and D. W. Davidson. "Granivory and Competition as Determinants of Annual Plant Diversity in the Chihuahuan Desert." *Oikos* **65** (1992): 61–80.

61. Schaffer, W. M. "Order and Chaos in Ecological Systems." *Ecology* **66** (1985): 93–106.
62. Schmidt-Nielsen, K. *Scaling, Why is Animal Size So Important?* Cambridge: Cambridge University Press, 1984.
63. Schneider, E. D., and J. J. Kay. "Life as a Manifestation of the Second Law of Thermodynamics." *Comp. & Math.* (in press).
64. Schoener, T. W. "Food Webs From the Small to the Large. " *Ecology* **70** (1989): 1559–1589.
65. Schoenly, K., and J. E. Cohen. "Temporal Variation in Food Web Structure: 16 Empirical Cases." *Ecol. Monographs* **61** (1991): 267–298.
66. Schrodinger, E. *What is Life, Mind and Matter?* Cambridge: Cambridge University Press, 1947.
67. Simberloff, D., and W. Boecklen. "Santa Rosalia Reconsidered: Size Ratios and Competition." *Evolution* **35** (1981): 1206–1228.
68. Stanley, S. M. *Macroevolution: Pattern and Process.* San Francisco: Freeman, 1979.
69. Sugihara, G. "Graph Theory, Homology and Food Webs." *Proc. Symp. Appl. Math.* **30** (1984): 83–101.
70. Ulanowicz, R. E., and W. F. Wolff. "Ecosystem Flow Networks: Loaded Dice?" *Math. Biosci.* **103** (1991): 45–68.
71. Van Valen, L. "Body Size and Numbers of Plants and Animals." *Evolution* **27** (1973): 27–35.
72. Volterra, V. "Variations and Fluctuations of the Number of Individuals in Animal Species Living Together." *J. Cons. Perm. Int. Ent. Mer.* **3** (1926): 3–51.
73. White, J. "The Allometric Interpretation of the Self-Thinning Rule." *J. Theor. Biol.* **89** (1981): 475–500.
74. White, J., and J. L. Harper. "Correlated Changes in Plant Size and Number in Plant Populations." *J. Ecol.* **58** (1970): 467–485.
75. Wiley, E. O. *Phylogenetics: The Theory and Practice of Phylogenetic Systematics.* New York: Wiley, 1981.
76. Williams, C. B. *Patterns in the Balance of Nature.* New York: Academic Press, 1964.
77. Willis, J. C. *Age and Area.* Cambridge: Cambridge University Press, 1922.
78. Wilson, D. E. *Biodiversity.* Washington, DC: National Academy Press, 1988.
79. Winemiller, K. O. "Spatial and Temporal Variation in Tropical Fish Trophic Networks." *Ecol. Monographs* **60** (1990): 331–367.
80. Yoda, K., T. Kira, H. Ogawa, and K. Hozumi, "Self-Thinning in Over-Crowded Pure Stands Under Cultivated and Natural Conditions." *J. Biol. Osaka Cy. Univ.* **14** (1963): 107–129.

DISCUSSION

BAK: I'd like to return to your fascinating experiment, where you started with a Pareto distribution of species, and then you disturbed the system, and after a transient period, it returned to a Pareto, or power law, distribution of species. That's precisely the kind of behavior one would expect for self-organized critical system: It builds up to a critical system, where you have a power-law distribution; you can do whatever you want to it, and since this critical state is an attractor, it must eventually return to it—not the same state, but some other state belonging to the critical attractor. So I would take this kind of behavior as empirical evidence that we are talking about a critical phenomenon. And that's also in agreement with the general picture that you have scaling laws, and so on. But I find it very fascinating; I would like, myself, to work on it, to understand that in detail.

GELL-MANN: That was Boyer's project on the Sea of Cortez, right?

KAUFFMAN: This is a really important point: is it really consistent? Brian, what's the standard explanation for a Pareto distribution in economics?

ARTHUR: There is none.

BAK: There is a paper with myself and Scheinkman and Woodford of the University of Chicago, where we explain the occurrence, or we suggest that the Pareto distribution in economics is precisely because you have avalanches or chain reactions in a self-organized, critical system. I don't think there is any other explanation. That doesn't mean that it's right, but...

ARTHUR: Phil Anderson gave a very good explanation to me the other night, over drinks and so on. You have a tournament of players, and each loser pays the winner a dollar, and that gives you a Pareto distribution. Now imagine—and this leads to the following fuzzy thought—imagine that you've very few players. Then there's not that much difference in wealth, but you still get a Pareto distribution. Imagine you've a large land mass, with many players. Then you get a Pareto distribution, but scaled to have elephants at one end, and small organisms at the other end. So I think that there's a whole series of ideas here, that may be connecting.

LICHSTEIN: Bank size fits into that category; we're going through a dramatic change right now.

ARTHUR: City size, too.

BAK: Is that what is generally called a Zipf law?

GELL-MANN: Zipf's law is a particular power; it's the power near dx/x^2, so the integral goes like $1/x$. And that's good for populations of cities; it's good for word distributions; it's good for a tremendous variety of things—approximately.

KAUFFMAN: Jim, what's the explanation in ecology for it? This is a well-known distribution that people have been looking at for 50 years; what do people say about such Pareto distributions?

BROWN: The only person that I know of—and again, I hadn't intended to talk about that part, originally—the only person that I know of that's really talked about that in ecology is Bob Ulanowitz, theoretical ecologist at Maryland's Chesapeake Bay station. And he has at least talked about families of such distributions, when they appear in ecology—and I don't know if he's talked about the underlying basis of them. Now, one special class of that distribution is the three-halves thinning law, that you observe in pine plantations, both intra- and interspecifically; as a set of seedlings grow up, some grow big and dominate, and other ones get smaller. I didn't have Boyer's thesis with me, and so I can't recall what the slope of that is. But again, these are organisms competing for space in this particular case, in the rocky intertidal, and there might be reason for thinking that it looks something like a three-halves thinning relationship.

GELL-MANN: But in this work that was done on a Pareto law: was the power predicted? Or just that there would be a power law?

BAK: The power was nonuniversal; that depends on the kind of system that you have. But there is a power...

GELL-MANN: But the Zipf tendency is toward powers that lie in a certain range near -2 in the differential distribution, and as far as I know, that still remains unexplained.

GOODWIN: Do you know what happens within species? There's an interesting recent article by Graham Bell, in which he demonstrates that if you look at norms of reaction for genotypes of a particular species, you find an inconsistent match with the environment. In other words, some genotypes within that species change one way; others change another, as you change environmental variables. Now I wonder if there's any study to match the diversity within species, with diversity between species. In other words, is it the same strategy everywhere?

BROWN: I really don't know the answer to that, and I just haven't looked at that. The variation within most of these kinds of species that we're working with is so small, compared to the variation between species, that we've in general just sort of ignored it, on taking an average value for the species. I think if Marc Feldman were here, he would be a little bit upset with this sort of general fitness function, and he would say, "What about frequency dependence?" And I think that bears on your question, that you can have, in some cases, alternative strategies within species that depend on the relative abundance of the other type as well. But I think that you can move up one level, and if there are two different strategies—because one can eat A better than B, and the other can eat B better than A—they can still be maximizing some rate of conversion of energy and resource into offspring. So frequency-dependent systems don't seem to me to underlie some kind of a general fitness function of this sort.

GELL-MANN: Can you say why ecologists moved away, in the latest wave of fashion, from this r and K set of ideas?

BROWN: I guess they've moved away for a variety of reasons. In part I think it's just because life-history theory has got a lot more sophisticated, once you really get in...

GELL-MANN: You mean there's a spectrum, instead of two strategies?

BROWN: There's a whole spectrum, and it's not just a matter of producing a lot of offspring in a colonizing environment, or few offspring in a competitive environment; there are organisms that live in highly competitive environments that produce many, many offspring—have a very high juvenile mortality, low rates of recruitment into the adult classes, but the adults then live for long periods of time. Sequoia trees, for example, are hardly colonists, but over their lifetime they have produced millions of seeds. And so, I think that's a case of a generalization that was just sort of too general for the taste of most ecologists.

GELL-MANN: Oh, so it was the correlation with colonist, or non-colonist, that people have moved away from, not the classification into these two kinds of behavior.

BUSS: It was a statement about r and K in its initial formula, a statement about a particular kind of mortality schedule, and a particular kind of natality schedule, and those were only two instantiations of natality and mortality schedules. And it was very successful early on and, in fact, it spawned an enormous amount of research over a range of different kinds of natality and mortality schedules that have effectively displaced r and K selection, and...

GELL-MANN: So this is a wider distribution over mortality and natality schedules than they thought originally. I see.

HOLLAND: One comment. On these oversimple systems of mine, and on inadequate amount of runs, and all that kind of thing—nevertheless, in those runs where I have gone from just an initial primary producer and a single predator, and get this diversity of 20 or 30 species...by eye at least, I see something very similar to what you're talking about, and the other thing's that important is that these nets are far from random. The heredity that's in that net constrains it greatly in what goes on.

GELL-MANN: You mean the founder effects?

HOLLAND: Partially founder effects, but when you're three or four successive steps away, it's hard to call it still a founder effect (although it is, in a sense).

BROWN: This relates to both Murray's r and K question and your comment. One thing that looked like it was neat in your system is...there is one thing that's very clear about these life histories, and that is there's an enormous body-size dependence of those—the smaller things are, the faster they go, and the less space they use, and so forth. That seems to emerge from your model system; just as a sequence of things gets longer, the things tend to go slower. So I think that some of these scaling properties are really important, and I want to get that issue out on the table here.

HOLLAND: Yes, in the crudest sense there is this r and k...I found a very unusual phenomenon in several runs in which I had some very small individuals—that is, short chromosomes—that can reproduce very rapidly in my setup, and then some very sophisticated individuals. And, strangely enough, very few in the intermediate region. And I'm not at all sure what that's about.

BROWN: One thing that you observe if you look at these patterns, that appears to come up over and over again: I have that curved sort of fitness function as a function of body space. Well, it turns out that the very smallest birds and mammals have very low fecundities, and very long lifetimes. That is, hummingbirds, which weigh only two grams (the smallest birds), have two eggs per year, and they live 13 years in the wild, which is four times what you'd predict on the basis of the equation for all birds. There are two groups of very small mammals: shrews, which we don't know much about, and bats. The tiny little bats, which get down around two grams, have been observed to live 25 years in the wild, and they have one or two offspring per year. The intermediate things—the sparrows, and the mice, and the blackbirds, and so forth—have much, much higher rates of reproductive output. And there are a whole series of things that start falling into place once you adopt this kind of view.

GOODWIN: Did you do the reverse to the experiment where you allowed the kangaroo rats to come in again, and go back to where it was?

BROWN: We're waiting to do that when the system stops changing in interesting ways—and we're 15 years in, and it hasn't happened yet. These are quarter-football-field-sized patches of desert. If you want to have four replicates, and controls, and so forth, you don't have a lot of things to mess around with. We want to do exactly that, once interesting things stop happening, and they haven't stopped happening yet.

GELL-MANN: May I ask a question about that experiment? In such small plots, were you able to do much about the redistribution of birds? Can you say anything about that?

BROWN: Well, we're working on that this summer. Certain birds respond to the patchiness that's created by those manipulations, and there are two things. We didn't get money to study birds early on, so we don't know if birds increased initially when we removed the seed-eating rodents, and/or ants. We do know that in the longer run, birds decreased in abundance, and that appears to be related to this filling in with grass, and so forth. . .

GELL-MANN: But surely certain kinds. . .

BROWN: Well, the granivorous birds, which forage by looking for seeds on open patches of ground, and have to run along the ground, have decreased their use of these plots where kangaroo rats have been removed. And we also know that there's a species of bird that's specific to desert grassland, and within the last three years it started to reproduce on our plots—the territorial males are all centered over each one of these grassy. . .

GELL-MANN: What are those, Cassin's sparrow?

BROWN: Cassin's sparrow; you got it! We're working to quantify that this summer.

BAK: May I be allowed to put on record a prediction that you will not go back to your original state, but you will go back to another state where you'll have Pareto distribution. And I think those things are irreversible.

MOORE: When you have these exponents, like three-fourths. . .how much of that comes from the network properties, and how much of it comes from physical things, like the fact that these are three-dimensional creatures, and their surface area is like their mass to the two-thirds. I just read how the FDA and EPA just agreed to use three-fourths to scale the dosage from rats to humans, for instance.

BROWN: So far as we can tell, those are properties of individual organisms (those scaling things). That's our hypothesis. That's why I couldn't agree more, for example, with the call that Brian Goodwin and Leo Buss have for a theory of the organism. Because while we can account, I think, for these very sort-of-emergent properties, on the basis of the properties of the individual organisms that comprise these systems, there's no good explanation for why that power is three-fourths. All these bright people—biophysicists, and physiologists, and stuff—have been working on that problem for a hundred years, and there's no general explanation for why that's three-fourths.

GELL-MANN: That's not the only place in the world where the power in power laws is difficult to predict. Benoit Mandelbrot said that he spent a good part of his career showing the existence of power laws, and never trying to predict the power—and that was how he became a great man.

Kenneth J. Arrow
Department of Economics, Stanford University, Stanford, CA 94305

Beyond General Equilibrium (Abstract)

I present four alternative theoretical perspectives on the economy: (1) competitive general equilibrium theory (CGE); (2) economics of differential information; (3) increasing returns, imperfect competition, and growth; and (4) bounded rationality. These are by no means mutually exclusive; each clearly captures some aspect of reality. Yet they are in general inconsistent with each other.

1. CGE is still the only coherent account of the entire economy. It recognizes not only the present but also the future, even uncertainty about the future. In a sense the future (or at least anticipations of the future) influence the present, and therefore the rationality of expectation formation plays an important part. Though coherent, both the assumptions and the implications of the model are clearly false in many cases.

2. The last thirty years have seen the recognition that individuals have different information and that these differences have important implications for the workings of the economy. These observations have given great insight into the development of some institutions, but they have not been integrated into a coherent theory. Further, the demands on the rationality of economic agents are even greater than those of CGE.

3. That there are economies to greater scale has long been known. It is now clear that a major cause of scale economies is the possibility of acquiring information. Some economies are internal to firms, some external. Economies of scale are incompatible with perfect competition. They are increasingly being used to explain the persistence of growth in advanced economies and the slow rate of convergence about nations.

4. All preceding theories make great demands on the rationality of economic agents and their ability to predict. That is contrary to a good deal of evidence from psychological experiments. This provides an important agenda for economic research but not yet a definite theoretical structure.

DISCUSSION

ARTHUR: My problem with bounded rationality is why perfect rationality would come into play in the first place. Imagine a Martian coming to Earth and making certain observations. First, he looks at an electron in a magnetic field and he would see that it obeys certain equations. He wouldn't ask the electron why it is that in a field it decides to turn right. Next he watches an ant nest and he identifies the behavior of the individual ants and how they act together or not together. And then he watches the whole nest and makes empirical observations based on how the thing would grow as a function of time. Next, he would notice a consumer and find out how he decides whether to go to this supermarket or that one and he would observe certain patterns and that would be the input to his model. And then he analyzes an economy in terms of that model. And finally, he meets Ken Arrow. My problem is at what point in that process do thinking and rationality come in?

ARROW: I would suppose that even when you're dealing with animals you have evidence of purposive behavior, behavior that looks ahead from where it is now. Watch a dog try to go around an obstacle. He doesn't just go up to it and get his nose bashed in. He thinks ahead far enough to know that there's an obstacle there. You can see him exploring and deciding what the best way is to go around it. It would be reasonable to attribute to him some model with which he is trying to minimize injury to himself, trying to minimize the time that it takes to get to a certain point.

ARTHUR: How do you write that down in simple dynamical rules?

ARROW: The evidence for it is that he is walking back and forth. I watch his exploratory behavior. I can't interrogate him. There have been schools of economic thought that say the aim is to pile up lots and lots of evidence, important from the '20s to the '50s in which the view was very much that all of this introspection about rational behavior was nonsense. We just want lots of data on how prices move and how many telephone calls are made, how much pig iron is produced, and they were very useful for the collection of data, which were very useful. It didn't lead, so far as I know, to any conclusion of any kind. The idea is that verbal behavior, responses to questions, is a form of information just like watching the dog move back and forth. This is a world, remember, in which experimentation of real magnitude is not possible although there are people who experiment. So one source of our generalizations is that, practically speaking, you can only watch. You can say, "Does this magnet really move that particle? Let me move the magnet around and see what happens when I move it further away." That kind of experimentation is very difficult and, therefore, other kinds of information assume greater value. And one form of information is asking people, including yourself, what you would do in certain hypothetical situations. The other thing is defining what relevant variables you are looking at.

KAUFFMAN: As an outsider who has been trying to learn some economics, I'm struck by the fact that all of the usual concepts don't seem to work. General equilibrium doesn't seem to work because, if you have bread and butter and cheese, there's no mechanism that actually gets you to a market equilibrium in a guaranteed way. There's no dynamic to get you to Nash equilibria. With rational expectations, there's nothing that guarantees that you converge on the rationally expected result. I find myself wondering whether some of the ideas here that lead us toward the edge of chaos idea or Per's self-organized criticality might prove useful. Perhaps what we need is a body of theories that say not that the markets clear but that they come close to clearing and they don't go too far away from it. One would need a mechanism of balance there.

ANDERSON: I have two comments. One is that, starting from the fact that I think rationality is much more bounded than you can possibly believe, once you allow the nose of bounded rationality under the tent, you arrive at a theory that has no Lyapunov function, no fundamental theory of value, and then you ask yourself what is better, what are you optimizing, what are you getting to in the equations of motion. They become much more intractable in terms of any kind of equilibrium theory. This isn't an original remark but it seems to be one that doesn't seem to come into play in conventional economic theory. There may be no convergent equilibrium, the whole thing is a dynamic process. A second remark is that there is also unequal competition among individuals as well as firms. A fact that most of my economic friends seem to be concealing from me but is true in that wealth satisfies something like a Zipf's law which implies, in a sense, the accumulation of competitive advantage rather than equality among the agents. I'd like your comment on that.

ARROW: About the first part, it fits in with the first part of Stuart's question and is, of course, very important. I have been raising the problem of questions rather than of the models that attempt to answer them. Most of the models, as in most other fields, tend to be somehow equilibrium concepts. We say that there are forces on both sides and that they tend to come to equilibrium. Then the question is what makes them get to that equilibrium? Let me not pursue that simply because it's such a large topic. You raised a question about something I didn't discuss, income distribution. One of the reasons I didn't is that, in a sense, the CG model isn't so much wrong or right about income distribution as that it really has nothing useful to say. One answer, of course, can be given. It is implicit in equilibrium theory but it's not such a useful answer. People are different. They have particular talents at a given moment and that's why they command higher incomes. Wealth distribution is very unequal. Whether you think about its distribution as being stable or unstable, it's like whether it's half empty or half full. You can fit Pareto curves, something like that, the coefficients turn out to be fairly different but, if you start out with the belief that they can be anything, then indeed they fall into a narrow range. If you start with the belief that they are constant, then you say they're really pretty variable.

ANDERSON: Let me remind you that there is one advantage of having great wealth, which is concealing your wealth. The data are not trustworthy.

ARROW: Let me use the word income rather than wealth, partly because our date are mostly on income. Our wealth data are really very poor and people would argue that income, in some sense, is a more relevant concept. The trouble with CGE theory is that it only explains things by the fact that people are different and, of course, it adds to it that, if you have some wealth, you can add to it from investment. There is a cumulative tendency. But in the United States, this adds to the inequality but by no means dominates. Wealth has other cumulative aspects. You can come from a family that is better educated, has better connections, and so forth. This is an inadequate answer

WALDROP: I'm troubled by the notion that there is no theory of bounded rationality. I understood that Newell and Simon, starting in 1956, developed a very detailed theory of human problem solving which addressed exactly this question. It may be a good theory or a bad theory but it is a very detailed theory.

ARROW: I wouldn't say it's a theory about how actual human beings solve problems. It's a theory of how you ought to go about solving problems. They have rules, they've varied quite a bit over time. Bacon is the latest word there, I guess. It's true that they do argue that you can start from purely empirical data and get to generalizations by, essentially, taking the simplest kinds of generalizations and, it turns out, that goes very far, and they argue that science could have developed just by fitting data. If you think of science as human problem solving, among many philosophers of science it has been very strongly criticized. But, in any case, I'm not sure that the behavior of scientists is the behavior of average human beings.

WALDROP: I wasn't thinking of Bacon. I was thinking about a general problem-solving approach which they based on quite a lot of empirical data, listening to people talk as they solve problems.

ARROW: I cannot see how. Maybe it can be done, but if you take an ordinary business firm and ask how would you apply this to problem solving in your business firm, their conclusions, and they did some interesting empirical work in eonomics, by the way, or that school, Cyert and others, and mostly they concluded that people follow rules of thumb, and they had some remarkable successes, by the way. The trouble is that there was no theory, as far as I could see, as to how those rules of thumb were arrived at. They explain department store pricing. They ask people how they set their prices. Then they verified that prices were set according to those rules. Somebody did something very interesting. Ten years later they went back to the same department store, didn't ask them about new rules, they used the old rules, and they still got very good predictions. But where do those rules come from, why are they used and not other possible rules, these things have numerical parameters that are quite arbitrary, and that's why I said that I didn't think it was a real theory.

SIMMONS: This is the third time in the time I've been here that this issue has come up about the relation between stimulus-response and the ability to do look-ahead or anticipation or building models and it's important to know that there's an early theorem in automata theory that says that any look-ahead system that I build, if it is finite, can be reduced to a stimulus-response (S-R) system. Now that can always be done. The question is, should you? Does it help? So to say that I can explain all this in terms of stimulus-response or behaviorism is beside the point in terms of understanding and especially so if you ask Ken's question—where do the rules come from, where does the S-R come from—then you find, in general in my experience, that the system that does the anticipation, trying to get at that, is much more relevant than the system that's describing S-R. Since this has come up so many times. I think it's important that, at some point, we debate this further.

John Maynard Smith
University of Sussex at Brighton, School of Biological Sciences, Biology Building, Falmer,
Brighton BN1906, England

The Major Transitions in Evolution

As a geneticist, I think that things are alive if they have the properties of
multiplication, variation, and heredity, and if they don't have those properties they
are not. The reason is that if they do, they have a chance of evolving by natural
selection, and life is a consequence of such evolution; whereas if they do not, they
will not evolve. What I want to discuss, however, is the increase in complexity in
the course of evolution. The theory of natural selection does not say very much
about that. It does not predict that things should get more complicated: the most
you can say is that you would expect them to get better at doing whatever they
are doing right now, or at least not get strikingly worse at it. Further, empirically,
many organisms not only do not get more complicated, but do not change at all
with time: crocodiles today are not greatly different from crocodiles in the Jurassic.
So the fossil record shows that organisms do not necessarily change with time, let
alone become more complicated.

All the same, there is a sense in which oak trees or human beings or elephants
are more complicated than bacteria. I suggest that there have been a series of major
transitions in the way in which genetic material is transmitted between generations,
and that it is these transitions that have been responsible, at least in large part, for

TABLE 1 The Major Transitions in Evolution

STAGE	TRANSITION
1. Replicating molecules	Origin of compartments
2. Populations of molecules within compartments	Origin of chromosomes, genetic code, protein synthesis
3. Prokaryote stage, cells with single circular chromosome	Symbiotic origin of organelles, mitosis, linear chromosomes with multiple origins of replication
4. Eukaryote stage, nucleated cells	Origin of sex
5. Life history with meiosis and gamete fusion	Multicellular organization
6. Animals, plants, fungi	Animal societies with A. Castes
7A. Insect societies	B. language
7B. Human societies	

the increase in complexity. This way of looking at evolution did not originate with me. It is expressed very clearly in Leo Buss'[1] book, *The Evolution of Individuality*, and for all I know he has already talked about this. However, I suspect he probably has not, because, although he is brilliant at asking all the right questions, he is equally brilliant at giving all the wrong answers.

Let me start by giving a brief outline—and at this stage I agree closely with what Buss has said—of what the stages have been, and what were the major transitions. I want to give an account of what happened in evolution (see Table 1).

I think that the first objects with the properties that make natural selection possible were simple replicating molecules, presumably RNA or something very like it. The first major transition was that between a set of isolated replicating molecules to a situation in which these molecules were enclosed within compartments of some kind—I don't want to call them cells at this stage—each compartment containing a population of molecules. When such molecules were first enclosed within compartments, there were, presumably, many different kinds of molecule in each compartment, but each molecule replicated on its own, independently of the others. But when one reaches the prokaryote stage of evolution, these different molecules are linked together on a chromosome, so that when one is replicated, all are replicated. The transition from individual molecules to chromosomes is difficult to explain, because chromosomes will replicate more slowly than molecules, but it is, nevertheless, essential for further evolution.

From the origin of prokaryotes to the next major transition took about two thousand million years. It is not quite clear what they were waiting for, but they must have been waiting for something. The origin of eukaryotic cells—we are all eukaryotes—involved the origin, from cells without nuclei, of much larger nucleated cells. This transition involved a number of changes. The one that is most familiar, and most dramatic, is that it involved the swallowing by one cell of others of a different kind, and then the coexistence of the swallower and the swallowed—that is, of the outer cell and of an "endosymbiont." The first endosymbiont evolved into our mitochondria (organelles that enable us to use oxidation reactions to provide energy). A second such event gave rise to cells containing chloroplasts, which fix solar energy. In this way, prokaryote cells which had been capable of independent replication came together in a single, more complex individual.

However, the symbiotic origin of organelles may not have been the most important thing that happened in the origin of eukaryotes. The thing that really limits the degree of complexity that can be achieved by prokaryotes is this: when a prokaryote replicates its chromosome, it has a single origin of replication and a single terminus. The chromosome is circular: replication starts at one point, and proceeds in both directions round the circle, meeting at the terminus. Replication cannot start in two places. There is a technical reason for this, which arises from the way in which, when the cell divides, one copy of the chromosome is passed to each daughter cell: the mechanism that ensures that this happens would not work if there were several origins of replication. Of course, this is important. If, when the cell divides, one daughter cell receives two copies of the chromosome and the other no copies, the latter is a complete waste of material. The eukaryotes invented a completely new way of moving chromosomes about, and this enabled them to have many origins of replication. Until this happened, there was a sharp limit on how much DNA a cell could have. It takes a long time to replicate DNA—it takes about 40 minutes to replicate the chromosome of *E. coli*. This limits how much DNA a prokaryote can have: eukaryotes, in contrast, have many origins of replication, and therefore can have much more DNA per cell. This made possible—not necessary, but possible—the subsequent increase in complexity of the eukaryotes.

The next major transition, which followed soon after the origin of eukaryotes, was that from asexually reproducing cells to a life history in which reproduction was occasionally interrupted by sex. I hope you are all clear that sex is the opposite of reproduction: reproduction is one cell turning into two, and sex is two cells turning into one. It is because sex is an interruption of reproduction that it is hard for an evolutionary biologist to explain. A Darwinist has no difficulty in explaining its opposite, reproduction. All I will say now is that sex does have dramatic consequences for the way in which genetic information is transmitted between generations. In a sense, all the genetic material of a species forms a common population of genes, a "gene pool," that has a common future. Genes that are in different individuals today may produce copies that are in the same individual tomorrow. So one can think of all the genes of a species as constituting an evolving entity, rather than the genes in a single individual constituting such an entity. Calculations show that a sexual population can evolve more rapidly to meet changing circumstances: sex certainly confers advantages on the population, but it is less obvious that it does so on the individual organism.

The next major transition—and this is the one that Leo Buss primarily concentrates on—is that from a single cell to a multicellular organism, with a very large number of copies of the genetic information. The cells of our bodies are of many different kinds. When these differentiated cells divide, they tend to produce offspring cells like themselves, even though, usually, they have identical DNA. Thus, in tissue culture, fibroblast cells give rise to fibroblasts, epithelial cells to epithelial cells, and so on, even though the two types of cells have precisely the same DNA sequences in their nuclei. Thus, there is a second hereditary system operating in multicellular organisms with differentiated cells. The origin of multicellular organisms with many kinds of differentiated cells happened, not once, but independently on at least three occasions, giving rise to animals, plants, and fungi.

The final transition I want to mention is that between organisms that exist mainly as isolated individuals, or as mated pairs, to insect and human colonies, consisting of large numbers of individuals, playing different roles, and dependent on one another for survival. This has happened a number of times in the insects (once in the termites, and repeatedly in the hymenoptera). We are beginning to learn that it has happened in some other groups: for example, social spiders have evolved on at least three occasions. The individuals are small, but they construct a communal web higher than this room. It is now becoming clear that there is a division of labor between females that lay eggs, and others that run up and down the web collecting food. Finally, of course, there are human societies. These differ from all other animal societies in having a second "genetic material"—language—whereby information is transmitted between individuals. There is some cultural transmission of information in other animals, but it is quantitatively trivial in other animals, and it is predominant in humans. Thus, the origin of human language is another major transition in the way in which information is transmitted.

This picture of a series of transitions I stole from Leo Buss. Having stolen it, I have found a colleague, Eörs Szzthmáry, with whom I am writing a book which will attempt to provide answers to all the problems that arise.

The justification for discussing all these transitions in one chapter—or one book—is that they have common features and, therefore, may have common explanations. Again, in identifying the common features, I am still following the picture presented by Leo Buss. Essentially, what these transitions have in common is this. Before the transition, there were entities that were capable of independent reproduction. This is most obvious, perhaps, in the case of the symbiotic origin of the eukaryotes. Before the transition, the host cell, the ancestor of the mitochondrion, and the ancestor of the chloroplast, were all independently reproducing cells. After it, the three entities are part of a single unit, completely dependent on one another. The same thing is true of other transitions. The cells of your body are like cells which once were capable of independent reproduction; the members of an ant colony are like solitary insects, and so on. This raises a problem. Why didn't selection between independent replicators prevent the evolution of cooperation and interdependence? Why does not selection favoring selfish behavior disrupt integration? First, I will give some concrete examples of selection for selfishness. It turns out that, although the entities that make up the whole do usually cooperate, there are exceptions. First, consider what I call a "fair meiosis." When gametes (eggs or sperm) are made, if there is a pair of alleles at a locus, those two alleles have exactly equal chances of getting into the gamete: this is the basis of Mendel's laws. If a fair meiosis was an unbreakable rule, then there is nothing that a gene could do that was selfish. The only way in which a gene could improve its chances of propagation into the future would be by helping the survival and reproduction of the organism in which it found itself. But a fair meiosis is not universal: there are ways in which genes can and do cheat. The two main ways are, first, "meiotic drivel," which operates in meiosis, and causes a gene to be unfairly represented, relative to its allele, in the gametes, and second, "transposition," whereby a gene does not wait for its chromosome to replicate, but replicates independently, and jumps about. Almost always, meiotic drive genes, and transposable genes, reduce the fitness of the organism.

A second example of "selfish" behavior: your body only works well because your kidney cells stay in the kidney doing their job, your muscle cells stay in your muscles, and so on. But there are cells which do not obey the rules, but multiply out of control: they are called cancer cells, and will kill you. The good behavior of differentiated cells is not an unbreakable rule. Or consider sex. From the point of view of a female, there is an enormous short-term disadvantage to sex, at least in species like our own. A parthenogenetic female, producing eggs without fertilization, genetically identical to herself, in an otherwise sexual species, would, if other things were equal, produce twice as many offspring like herself as would a typical sexual female. Consequently, a mutation causing parthenogenesis would double in frequency every generation. So one would expect parthenogens to replace sexuals, even though the long-term effect on the survival of the species might be disastrous. I have little doubt that, if you go into the countryside near here, you will find lizards that are doing just that. There are, in these parts, lizard species consisting wholly of parthenogenetic females. In a sense, to go parthenogenetic is cheating: it is to the short-term advantage of the individual female, but not to the longer-term advantage of the population.

Perhaps the classic example of "altruism" is the behavior of the sterile castes in social insects. A worker bee that stings an intruder to the hive thereby kills itself. However, worker bees are not universally altruistic. They do sometimes lay eggs. Because of the peculiar genetic system of bees, they can only have sons, but they do have sons.

Are there any mechanisms that have ensured the preservation of cooperation? I think there are three such mechanisms. I do not claim that they are exhaustive, or that all are relevant to all the transitions. The most important one is the principle that when a new individual arises, whether it be a new colony, or new individual like myself—there are only one or a few copies of the genetic information. A striking fact, which we are all aware of, usually without wondering why it should be so, is that we develop from a single egg, from one cell. It's not the way an engineer would arrange for the production of children. It seems silly to go back to a completely undifferentiated cell, and give it all the work of differentiation to do over again. Why not make a little homunculus? Going back to one cell does have the consequence that, except for the rather rare event of somatic mutation, all the cells in the body are identical. So, from a gene's-eye point of view, if I were a gene sitting in a kidney cell, there would be no point whatever in leaving the kidney for the blood stream and traveling to the gonads, in the hope of entering a gamete, because exact copies of me will be present in the gametes anyway.

The importance of having only a few copies of the genetic material is relevant not only in the case of the single egg cell. Mitochondria are uniparentally inherited, usually from the female parent, and the numbers of mitochondria, at some point in the growth of the egg, is very small, so that typically all the mitochondria in an individual are genetically identical. If mitochondria were inherited from both parents, all the mitochondria in a cell would not be genetically identical, and there would be effective selection for "selfish" mitochondria, which replicated more rapidly, at the expense of doing what from the cell's point of view they ought to be doing, which is making ATP. As a final example, most insect colonies are founded by very few individuals, usually by a single mated female. There are exceptions, but this is the common situation. This does not mean, of course, that all the members of the colony are genetically identical, but they are at least genetically similar. The origin of my whole argument lies with W. D. Hamilton's[4] suggestion about the importance of genetic relatedness for the evolution of cooperative behavior.

The second mechanism is as follows. There is a degree of irreversibility about these transitions, which does not explain their origins, but does help to explain why, once they have occurred, they are very stable. For example, a cancer cell may in the short run produce more copies of itself, but it has no long-term future: even if cancer cells do reach the gonad, they are not going to pass to another generation. They will cause death. Although the ancestors of your body cells were once capable of independent existence, they no longer are. A worker bee does have the option of producing sons: it does not have the option of leaving the colony and reproducing on its own. Its whole biology is so bound up with the colony that it cuold no more live on its own than a kidney cell could live on its own. Or, consider parthenogenesis. If a lineage has been reproducing sexually for 800 million years, so many other things will have got bound up with sex that it is hard to unscramble it all. I once wrote an article in *Nature* saying that there were no reliably recorded cases of parthenogenesis in any mammal, but no one knew why. Several people wrote in, asking how I could be so ignorant not to know why mammals are never parthenogens. I did have some excuse—the relevant information had not then been published. The reason is in fact amusing, but contingent, and in a sense, trivial. It is that, in mammals, a few genes are "imprinted" so that, in the developing embryo, they remember whether they came from father or mother. In certain tissues,

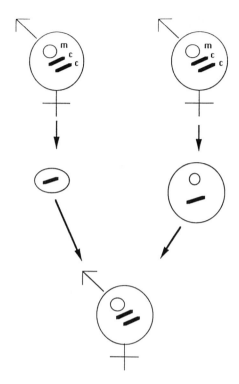

FIGURE 1   The inheritance of chromosomes (c) and mitochondria (m) in a hermaphrodite plant.

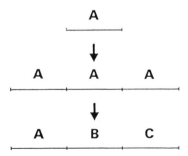

FIGURE 2   Increase in complexity by duplication followed by divergence.

only the father's gene is active, and in other tissues only the mother's gene is active. So if you do not have at least one father and one mother, there are some tissues in which neither gene is active. Since the genes are essential, you die. The point of this example is to illustrate how trivial the reasons may be why reversal—in this case from sex to parthenogenesis is impossible. But such examples can only help to explain the irreversibility of the transitions, once they have happened: they do not help to explain origins.

The final mechanism ensuring cooperation is the notion that there might be some kind of police force. This is clearly one reason why people pay taxes. I will give one example of "central control" from a plant (Figure 1). Most higher plants are hermaphrodite. The mitochondria are transmitted only in the egg, not in the pollen. Suppose you were a gene in a mitochondrion. It would pay you to suppress pollen production, so as to ensure that all available resources were devoted to female functions (growth of seeds). In some plants, common thyme is an example,[3] there are mitochondrial genes that do suppress male function. However, the spread of such selfish mitochondrial genes is prevented by genes in the nucleus that suppress these mitochondrial genes. There is a war between mitochondrial genes causing male sterility, and nuclear genes restoring male fertility. You can think of this, if you like, as the nuclear genes acting as a police force, preventing the mitochondrial genes acting selfishly.

To make further progress, it is useful to describe three ways in which complexity can increase. The simplest, though not the most interesting, way is by duplication and divergence (Figure 2). You start with one copy of the genetic information specifying A, and you duplicate it, to carry several copies of A. At this stage, you have not more information: you merely have some backup copies. If the copies then diverge, however, to specify A, B, and C, you have more information than you had when you started. Between the major transitions, this has been the main process whereby genetic information has increased. But I do not think it was important in the major transitions.

The second process is symbiosis (Figure 3). You start with a set of different, independently replicating entities. They may form an ecological relationship with one another: for example, they may be related as the elements of a hypercycle,[2] so that they support one another's growth. They are then enclosed in a compartment of some kind, so that, to some extent, they have a common future, and finally they are linked together, so that when one replicates, the others replicate. This almost certainly describes the origin of chromosomes. As explained earlier, symbiosis was important in the origin of eukaryotes, although in this case the mitochondrial and nuclear genes are not physically tied together so that if one replicates the other must.

The third process can be called epigenesis (Figure 4). A genetic message, $ABC...$, is replicated in the "germ line." In each generation, a large number of copies of it are produced, and transmitted to the body cells, but different genes

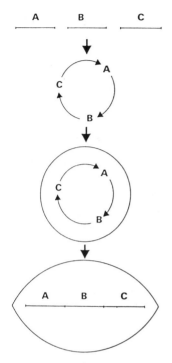

FIGURE 3   Increase in complexity by compartmentalization, followed by synchromnized replication.

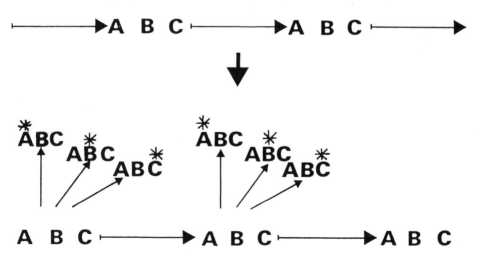

FIGURE 4   Increase in complexity by epigenesis, as exemplified by multicellular organisms and insect societies. * indicates genes that are switched on in particular cells.

are activated in different lineages. In this system, either there must be a "reset" button, so that, when a new individual is formed, all the genes are restored to their "totipotent" state, or there is an undifferentiated germ line. This is the process involved in multicellular organisms. One of the things that had to be invented in the Precambrian, before complex multicellular organisms could arise, was a "dual inheritance system," whereby differentiated cells give rise to daughter cells like themselves, without having to change the basic DNA sequence. Formally, insect societies are also like this, with the reproductive caste representing the germ line and the sterile castes the somatic cells.

Finally, I want to write a little more about symbiosis. Symbiosis can readily evolve either towards parasitism, or mutualism. Which path is actually taken *ought* to depend, it seems to me, on whether inheritance is direct or indirect (Figure 5). In direct inheritance, when the host produces a propagule, it contains a copy of its symbiont. In indirect inheritance, it doesn't, and the new individual has to swallow, or otherwise acquire, a symbiont. Direct inheritance one would expect to lead to cooperation, because a symbiont only has a future in so far as it keeps the host alive. Indirect inheritance one would expect more often to lead to parasitism. However, the facts are that, even in cases usually thought of as mutualistic, indirect inheritance is commoner than direct.[5]

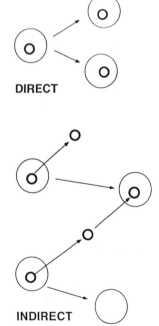

FIGURE 5   Direct and indirect inheritance of symbionts.

There are some cases of direct inheritance. For example, there are bacteria that live within the cells of aphids, without which the aphid cannot live (the bacteria synthesize vitamins which the aphid cannot make itself). The bacteria are transmitted in the egg. There is now molecular evidence that they have been vertically transmitted for tens of millions of years, without horizontal transfer. But this case is the exception. A more typical example is afforded by the luminous bacteria of fish. The fish depend on those bacteria for camouflage, for finding mates, and for finding prey. But their eggs do not contain the bacteria: the baby fish must find and swallow them. Indirect inheritance is also found in the mycorrhizal fungi of plants, the sulphur bacteria of animals in anoxic environments, algae in invertebrates, and in most lichens. There is a puzzle here. Of course, indirect inheritance often has led to parasitism, but there are examples of mutualism with indirect inheritance.

This chapter has mainly been a list of problems, with few answers. But it is the way biological complexity looks to me right now.

REFERENCES

1. Buss, L. *The Evolution of Individuality*. Princeton: Princeton University Press, 1987.
2. Eigen, M., and P. Schuster. *The Hypercycle*. Berlin: Springer-Verlag, 1979.
3. Gouyon, P., and D. Couvet. *The Evolution of Sex and Its Consequences*, edited by S. C. Stearns. Birkhauser, 1987.
4. Hamilton, W. D. "The Genetical Evolution of Social Behavior." *J. Theor. Bio.* **7** (1964): 1–32.
5. Margulis, M., and R. Fester. *Symbiosis as a Source of Evolutionary Novelty*. Boston: MIT Press, 1991.

DISCUSSION

GOODWIN: You excerpted a phrase, the same phrase that I did in a review, from Neil's book, but for totally different reasons. In relation to prokaryotes, you mentioned that they don't have a common gene pool, that they transfer genes horizontally so to speak. Isn't this as respectable a way of transmitting genes as sex or even more so?

MAYNARD SMITH: That's what I'm working on now. I'm working on the evolution of drug resistance in gonococcus and streptococcus and other bacteria. There are essentially three ways in which bacteria exchange genes. There's virus-mediated transduction, when a virus carries, I'm going to say "by mistake," a piece of genetic material from one bacterium to another. I do not think that process was selected because it transferred genes. It was an accident but it does have important consequences. The second is the way the plasmid has of getting from one host to another. Here, I don't think that evolved because it transferred bacterial genes although it occasionally does. The one that really interests me is the process of transformation because that does look as if it evolved to transfer genes. Bacteria like streptococcus and gonococcus, which go into transformation, have really quite complicated, evolved functions for carrying genes from cell to cell. They have really evolved functions for taking up DNA from the medium. I spent a lot of the last two days in Virginia debating about this. I don't know what the answer is right now. Do they do it because DNA is good to eat? DNA is good nourishing stuff. Or do they do it for the same reason that we go in for sex, i.e., for exchanging genes? We really don't know.

BROWN: What we've been talking about is a sort of dichotomy between the phenotype and the genotype or I'd prefer to say between metabolism and information. When you go through your major features of evolution, it seems to me that, of the advances in metabolism and phenotypic organization, some are advances and changes in the way that information is organized and processed. There's one other thing that fits in here: there's something interesting that goes on when you get to each new level that has both metabolism and information. There's a sort of feeding in of both of them, but when you have these little units, like individuals, you tend to have specialization of what I would call metabolism. That's what these things do but they share the information and yet they don't precisely share the information. This presents opportunities for them to do many interesting things and I wonder if, in some sense, this isn't representing another thing we've been talking about which is the conflict when these interesting complex systems sit at some kind of a boundary between these highly ordered states and states where cheating and complicated games are possible but still not representing a transition to completely disordered states.

MAYNARD SMITH: On the last point, I do think that the puzzle for a modern geneticist is explaining why our transposable elements and our viruses and the various intragenomic conflicts we have haven't totally destroyed us. When you think of all the ways DNA has of jumping around the place, it is very remarkable that we're here at all. That's another way of summarizing what I've been talking about, how do we prevent this kind of conflict? On the metabolism information thing, I'm an information first guy, you know that. Metabolism is boring and goes along for the ride. You can always tell a biologist's religion by whether he's a gene-first man like me or a protein-first man like Stu.

KAUFFMAN: Why complexity, John? One of the things that has been emerging into which we have the least insight is the question you started with: why do things bother to get complicated? What you've done largely in your talk is to tell us that when things do become complicated, you might need to solve the problem of making them stable, not why they bother to get complicated. These interactions have something to do with the things that economists talk about, namely the advantages of trade. The reasons things trade is that they both have something to gain. Somehow or other Adam Smith was telling us that, as things go on somehow or other, there are more and more things for which there's an advantage to trade and you do it either in an ecosystem or you decide to get together and do it in-house, sort of like the firm problem. Why do things bother to get complicated? Is it just entropic or is there some advantage?

MAYNARD SMITH: I think that the thing we must keep in mind is that the vast majority of evolutionary lineages do not get complicated. Only some do. The most boring thing I can say, and in my off moments I think that this is the only thing that can be said about the evolution of complexity, is the following: If there are quite complex things that can be efficient and quite simple things that can be efficient, then you're likely to find both. Since the origin of life, there has been no way to go but up. There's no intrinsic drive to get more complicated. That's clear. I'm inclined to think that you really can't say much more than that. If there exists a more complex way of life that would conserve our fitness, then some organism will hit on it and do it. But I agree that the analogy with trade is, in some way, illuminating and that a lot of these increases in complexity do come about through symbiosis, including the conquest of land by plants. which depended on a true symbiosis between plants and micorrhizae.

HOLLAND: Aren't arms races one of the things that drives complexity?

MAYNARD SMITH: They drive change but I'm not sure about complexity.

SCHUSTER: A short comment about symbiosis. We had a model, as you mentioned, where in one parameter you could find competition and, in another parameter, you could find this type of hypercyclic interaction. Two years ago Stu and I started to talk about coevolution and this discussion caused me to extend our models a little. There was one interesting thing that came out. In a certain parameter range where you have two attractors, one attractor corrresponds to the competitive state and the other attractor corresponds to the symbiotic state. In that case you needn't wait for an external parameter change; you just change concentrations a little bit or neutrality a little bit, and you can jump from one basin of attraction to the other. This is a straightforward mechanism. However, the parameter range in which this occurs is rare, restricted, so it's an event that would not inevitably occur but is, nevertheless, a case where you can really jump from competition to symbiosis.

MAYNARD SMITH: They are capable of independent replication?

SCHUSTER: Yes. However, in this case you can compare the difference between the competitive system and the symbiotic system and you see that in the competitive system there is a decrease in the nonconserved resources, the symbiotic system conserves more resources and this may be a way to explain that it pays to be symbiotic.

Nonadaptive Systems, Scaling, Self-Similarity, and Measures of Complexity

Erica Jen
CNLS, Los Alamos National Laboratory, Los Alamos, NM 87545

Cellular Automata:
Complex Nonadaptive Systems (Abstract)

Traditional mathematics offers few methods for building a comprehensive theory of adaptive dynamical systems. A broad research program to study such systems grows naturally out of studies on complex nonadaptive systems—prototypical examples are cellular automata and random Boolean networks—that appear to be analytically more tractable. Studies of these systems lead naturally to consideration of the geometry or "landscape" of the systems' parameter space and the effects of parameter changes on system behavior. Since adaptation can often be viewed as evolution of a system in parameter space toward "optimal" behavior (for example, a neural net in the process of learning may be viewed as evolving in cellular automaton parameter space), the theoretical understanding of complex nonadaptive systems provides a basis for the detailed analysis of systems undergoing evolutionary behavior.

Cellular automata (CA)—a class of mathematical systems characterized by discreteness, determinism, local interaction, and an inherently parallel form of evolution—are prototypical models for complex processes consisting of a large number of identical, simple, locally connected components. Examples of phenomena that have been modeled using CA include turbulent flow resulting from the collisions of fluid molecules, dendritic growth of crystals resulting from aggregation of atoms,

Complexity: Metaphors, Models, and Reality
Eds. G. Cowan, D. Pines, and D. Meltzer, SFI Studies in the
Sciences of Complexity, Proc. Vol. XIX, Addison-Wesley, 1994   **473**

and patterns of electrical activity in simple neural networks resulting from neuronal interactions.

Over the past decade, significant progress has been made in the understanding of CA, primarily in areas relating to their dynamical systems features, computation-theoretic properties, structure of rule space, solution of inverse problems, and relation between CA and continuous systems such as PDEs. On the basis of the theoretical and computational advances, it is clear that CA provide valuable insight into the behavior and analysis of general spatially extended dynamical systems, with the potential for answering fundamental questions such as dependence of behavior on system parameters and use of optimality criteria to induce evolutionary or adaptive behavior.

An example of a recently obtained result in the field is the development of a method—reminiscent of inverse scattering for soliton-bearing PDEs—for mapping certain nonlinear CA onto analytically tractable linear CA. Under evolution of the nonlinear automata, the lattice of CA sites "organizes" itself into multiple contiguous domains within which behavior is ordered and highly correlated. The domains are separated by ostensibly randomly propagating domain walls that may be interpreted physically as dislocations or, equivalently, as propagators of information. The solution method for these nonlinear automata maps them onto an exactly solvable linear "template" automaton with a closely related evolution function. Analysis of the nonlinear automaton is achieved by transforming the initial condition, allowing the transformed sequence to evolve under the linear automaton, and then inverting the transformation to reconstruct the original nonlinear system.

The implications of the work on nonlinear automata are several-fold. First, it clarifies the effects of nonlinear perturbations in the evolution function of a cellular automaton. Second, by providing an exact solution for systems conventionally viewed as near-random, the results provide a nonintuitive measure of the complexity both of the CA themselves and of the computation being implemented by the CA. Third, exact solvability permits analysis of an enormous range of dynamical characteristics for nonlinear systems, including fundamental features such as equilibrium behavior, information-theoretic characteristics, and transience length. Finally, the results are expected to play a critical role in the design of machine reconstruction techniques (as developed by Crutchfield) that attempt to construct from data the minimal graph representation of the system states attainable at each time step. The long-term goal is to provide an adaptive learning capability for these data analysis techniques that would enable them to detect higher-level processor structures, domain formation and annihilation, and information propagation in CA. Reconstruction techniques with this capability are expected to have wide applicability in the analysis of general spatially extended systems, including coupled lattice maps, discretizations of PDEs, and neural networks.

DISCUSSION

GOODWIN: In the problem that you're pointing to, we have the difficulty of understanding what is generic, what is robust. Do you see an alternative to the mathematical classification of generic properties? If we could study a parameter space and classify the extent of the domain for the different types of behaviors, that is an empirical approach, how close do you think that would be?

JEN: I'm not sure what the alternatives are. I think that, basically, you always use a model to generate predictions so that's the only way I know to classify these properties.

RASMUSSEN: We tried to look at a mapping, to try to identify some of the features of cellular automata based on a rule table and it's really difficult to do. If you can't have a rule table property on one axis and some other property on the other axis, what should we do then?

JEN: I believe in looking at the rule table. You should concentrate on looking at the genotype and how it leads to certain phenotypical behavior and not give up on that problem which is probably why I resist the idea of putting in adaptation right now because I don't feel that we're ready to do it yet.

QUESTION: What kind of adaptation do you mean?

JEN: Well, that's actually one of the questions: how do you put it in? The best place that I think and have tried is to change the rule depending on the dynamics. Depending on the state of the system then, you start to change your rules. Or you can do adaptation in the sense of defining a system function. You want a rule to perform a certain function like pattern recognition. You have a well-defined fitness function and you start mutating the rule in such a way as to move the cellular automaton rule toward something that performs the function better. So there are well-defined ways of doing the adaptation.

MOORE: I want to talk about undecidability. To me what an undecidability result does is to challenge you to classify as many classes as you can with a problem that is solvable. The other thing it tells you is that no finite classification will cover the whole state and I think that that's a very good result.

JEN: I like the decidability results because it tells you something. They are phrased in terms of the amount of compute time it takes to figure something out. This is very important because people talk about finite resources and it is often a way to tell if it is possible, given your finite resources, to figure something out.

MOORE: I agree with your emphasis on "let's see what we can solve." A lot of people have only a very slight understanding of undecidability and somehow view it as a sledge hammer that smashes the whole field to pieces.

Per Bak
Brookhaven National Laboratory, Department of Physics, Upton, NY 11973

Self-Organized Criticality:
A Holistic View of Nature

Abstract: "Self-Organized Criticality" (SOC) describes the tendency of large dynamical systems to drive themselves to a critical state with a wide range of length and time scales. The idea provides a unifying concept for large-scale behavior in systems with many degrees of freedom; it complements the concept of "chaos" wherein simple systems with a small number of degrees of system display quite complex behavior. The phenomenon seems to be quite universal; indeed, it has been looked for in such diverse areas as geophysics (earthquakes and volcanic activity), economics, biological evolution, condensed matter physics, and astrophysics (solar flares and quasars). It might be the underlying mechanism for the "1/f noise" emitted by many sources in Nature. Theoretical work on models displaying self-organized criticality, as well as a variety of applications will be discussed. Finally, some thoughts on the connection between the concepts of "complexity," "criticality," and "adaptability."

Complexity: Metaphors, Models, and Reality
Eds. G. Cowan, D. Pines, and D. Meltzer, SFI Studies in the
Sciences of Complexity, Proc. Vol. XIX, Addison-Wesley, 1994   **477**

1. INTRODUCTION

1.1 SCALE-FREE PHENOMENA IN NATURE

In Figure 1 I show the distribution of the magnitude of earthquakes in the New Madrid earthquake zone in the southeastern United States.[24] The logarithm of the number of earthquakes exceeding the magnitude M is plotted. The linear behavior is known as the Gutenberg-Richter law (G-R).[23] For each earthquake of magnitude 6 there are 10 earthquakes of magnitude 5, 100 earthquakes of magnitude 4, and so on. The magnitude can roughly be interpreted as the logarithm of the energy released during an earthquake. The linearity expresses the fact that the number of earthquakes for which a given energy is released decays as a power law in that energy, $N(E) = E^{-1-b}$. The linearity extends over five decades. If worldwide earthquakes are included, the linearity can be extended another couple of decades, with no apparent cutoff. This simple law is impressive in view of the complexity of the phenomenon.

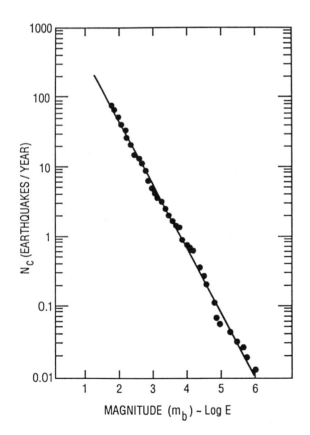

FIGURE 1 Cumulative distribution of earthquakes in the New Madrid zone during the period 1974–1983. The data were collected by Johnston and Nava.[24]

This is an example of a scale-free phenomenon: there is no answer to the question "how large is a typical earthquake?" Similar behavior has been observed elsewhere in Nature. Raup[35] has noted that biological evolution is intermittent with long periods of stasis interrupted by rapid extinction events. This behavior is related to Gould's "punctuated equilibrium."[21] There are many small events and very few large events, such as the extinction of the dinosaurs 50 million years ago, the great Permian extinction 200 million years ago, and the Cambrian explosion about 550 million years ago. The distribution of events follows a law similar to the Gutenberg-Richter law. Mandelbrot[29] has observed that the distribution of fluctuations of prices of cotton and railroad stocks are non-Gaussian with power-law tails, similar to the Gutenberg-Richter law. The fact that large catastrophic events appear at the tails of regular power-law distributions indicates that there is "nothing special" about those events, and that no external cataclysmic mechanism is needed to produce them.

Nature is full of objects, known as "fractals," that have spatial scale-free structure. These include the structure of the universe, mountain landscapes, clouds, coastlines, river basins, etc. A good deal of effort has been put into the geometrical characterization of these objects, but there has been practically no progress in understanding their dynamical origin. We have a tendency to overlook the dynamical nature of the problem of fractals: we often think of the universe, mountain landscapes, biology, and economics to be in a static equilibrium. If fractals are indeed the geometry of Nature, one must still understand how Nature produces them.

In addition to the fractal objects having features of all length scales, there is a ubiquitous phenomenon in Nature having temporal scale-free behavior known as one-over-f $(1/f)$ noise. $1/f$ noise has been found in places as diverse as sunspot intensity, traffic flow, voltage across resistors, and river flow. The signal has components of all frequencies, scaling as $1/f^{\phi}$, where the exponent is in the range $0.6 < \phi < 2$. The signal can be formally constructed by superimposing events of all durations: it has no characteristic time scale. Turbulence is a phenomenon with both temporal and spatial scaling aspects. Turbulent fluids have vortices on all length scales, and the temporal variations of the velocity at a given spot is intermittent, with large and small bursts of activity. But where does the scaling come from in all these systems?

1.2 SELF-ORGANIZED CRITICALITY

All the systems above are large dynamical systems with many interacting parts. The systems are also "dissipative." They are driven by constantly supplying "energy" from an exogenous source (tectonic plate motion in earthquakes, pumping energy in turbulence, labor and raw materials in economics, food and energy in biology, etc). The energy is eventually burned (dissipated) somewhere in the system.

It appears that we are dealing with a kind of critical phenomenon, because scale-free spatial and temporal behavior is the hallmark of systems at a critical point for a continuous phase transition. However, one usually has to tune some parameter, such as a temperature or a magnetic field, to a unique special value in order to achieve criticality. But for the natural phenomena above there is nobody to tune the parameters, so where does the criticality come from?

A few years ago we suggested[5] that slowly driven dynamical systems, with many degrees of freedom, naturally self-organize into a critical state, with avalanches of all sizes obeying power-law statistics. The critical state is an attractor for the dynamics: the system is unavoidably pulled towards the critical state for a wide range of initial conditions. Thus, in contrast to equilibrium physics, where criticality is the exceptional case, in nonequilibrium physics, criticality could be the typical state of matter.

The canonical metaphoric example is a simple pile of sand. Adding sand slowly to an existing heap will result in the slope increasing to a critical value. At that point there will be avalanches of all sizes. In the beginning, while the pile is flat, a local description in terms of individual grains is appropriate, but in the critical state where the interactions tie far-away parts of the system together, only a holistic description in terms of one sandpile will do. The picture was supported by numerical simulations on "sandpile automatons" which have since been studied by numerous authors.

The models are extremely simple: define an integer variable Z, representing the local height or slope of the pile, on a two-dimensional lattice. Increase Z (add sand) somewhere. Check if Z exceeds a critical value. If not, continue to increase Z somewhere. If yes, reduce Z by four units and send one unit to each of the four neighbors. Check if Z at any of the neighbor sites exceeds the critical value, and continue the process until the avalanche stops when there are no supercritical values of Z anywhere. Count the total number of topplings involved in the avalanche. Continue adding sand to generate more avalanches. Make a histogram of the number of avalanches of each size. The plot turns out to be a power law similar to the G-R law (Figure 1).

The most impressive analytical work is that of Dhar and coworkers.[17] They were able to construct exactly solvable models of self-organized criticality. For the standard sandpile model, they constructed an "Abelian algebra" for avalanche-generating operators, connecting the various configurations of the pile in the self-organized critical state. The properties of the algebra were used to calculate exactly the number of configurations of the pile belonging to the critical attractor. For a pedagogical discussion on "Abelian sandpiles," see the papers by Creutz.[13,14]

There have also been a number of experiments on real piles. Some early experiments failed to show the predicted behavior. In Figure 2 I show the results of recent experiments by Grumbacher et al.[22] They built small heaps on a scale, and monitored the distribution of avalanches of particles falling off the edges. In the figure is shown log-log plots of the normalized distribution function of avalanches. The experiments were performed using iron spheres (triangles) and glass spheres (circles) of the same size. In all cases a power-law distribution function was found.

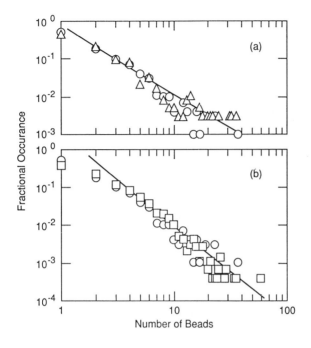

FIGURE 2 Distribution of avalanches in sandpiles (Grumbacher et al.[22]). Plot (a) compares iron spheres (triangles) with glass spheres (circles) of the same size. Plot (b) compares polystyrene (squares) with glass (circles) beads.

In the following sections, an overview will be given of phenomena in Nature which have been suggested as candidates of self-organized criticality. The final section contains a few thoughts of complexity, criticality, and adaptability.

2. EARTHQUAKES

2.1 SELF-ORGANIZED CRITICALITY IN BLOCK-SPRING MODELS

The idea of self-organized criticality (SOC), as applied to earthquakes, may be visualized as follows: think of the crust of the earth as a collection of tectonic plates, being squeezed very, very slowly into each other. In the beginning of our geological history, maybe, the stresses were small, and there would be no large ruptures or earthquakes. During millions of years, however, the system evolved into a kind of stationary state where the buildup of stress is balanced in average by the release of stress during earthquakes. Because of the long evolutionary process, the crust has "learned," by suitably arranging the building blocks at hand into a very balanced network of faults, valleys, mountains, oceans, and other geological structures, to respond critically to any initial rupture. The earthquake can be thought of as a critical chain reaction where the process is just barely able to continue.

The result of this self-organization process is in sharp contrast to any network of faults that one might set up by construction or engineering. Such networks would, with unit probability, not be critical, but either supercritical, causing a global explosion, or subcritical with only small events. We do not know how it all started, but that is not important for our arguments: the self-organized critical state is an attractor which will be reached eventually irrespectively of the initial conditions.

It makes no sense to separate the dynamics of the seismicity from the statics. It is not productive to think of earthquakes as being generated by a network of "preexisting faults." One can trivially explain the G-R law by assuming a fractal distribution of faults with a power-law distribution of characteristic fault sizes, but that leaves us with the equally difficult problem of explaining the dynamical origin of that distribution. What appears to be a static configuration of large faults in a human lifetime merely constitutes a snapshot of a slow ongoing geological process that has been hundreds of millions of years underway. During that period, faults have come and gone. The dynamics of the fault structure and the G-R law must be produced within a unified picture. In order to represent a realistic view of geophysics, the models must be robust, or adaptive, in the sense that if the physical properties were changed, or if noise were added, the system would reorganize during a transient period and become critical again. This is indeed the case for SOC models of earthquakes.

We want to study the simplest possible models which contain the essential physics of earthquakes. While there have been studies of three-dimensional crack-propagation models with slightly more realistic long-range redistribution of elastic forces following rupture, simple local models are probably more instructive, and certainly much more managéable to numerical and analytical study. We must emphasize that we do not think of the G-R law as originating from a single fault which must necessarily have a characteristic energy, and thus, no scale invariance. Our models are "toy" models supposed to illustrate the principle of *global* organization of the crust of the earth in a large area.

Consider a two-dimensional lattice of interacting blocks. The initial block structure represents a discretization of the space in much the same way as the lattice in lattice gauge theories of particle physics. The block size does not represent an intrinsic length scale in the problem. On each block, at sites (i, j) act a force $F_{i,j}$ in the general direction of motion in some fault region. In the beginning, $F_{i,j}$ may assume some random, small value. The initial state is not important for the long-term dynamics. Let the force increase uniformly by a infinitesimal amount per unit time; this simulates the slow driving by the tectonic plate motion. Eventually, the force at some site (i, j) must exceed a critical threshold value FC for rupture, which may be either uniform or random. The initial rupture is simulated by updating the forces at the critical site and the sites of the neighbors at $(i, j \pm 1)$ and $(i \pm 1, j)$:

$$F_{i,j} \rightarrow 0,$$
$$F_{nn} \rightarrow F_{nn} + \alpha F_{i,j}.$$

These equations represent the transfer of force to the neighbors. This may cause the neighbors to be unstable and a chain reaction to take place. This chain reaction is the earthquake. The equations are completely deterministic, with no external noise. We are not dealing with a noise-driven phenomenon; on the contrary, the physics turns out to be stable with respect to noise, i.e., noise is irrelevant. When the earthquake stops, the system is quiet until the force at some other location exceeds the critical value and a new event is initiated. The process continues again and again. One observes that for some time the earthquakes become bigger and bigger. When one is convinced that the system has self-organized into a stationary state, one might start measuring the energies of subsequent earthquakes as defined by the total number of rupture events following a single initial rupture. A histogram similar to that in Figure 1 for real earthquakes can be constructed.

This version was suggested by Olami, Christensen, and Feder,[32] who realized that the model could be directly related to earlier spring-block models. The value of α is directly related to the elastic parameters of the crust of the earth. For

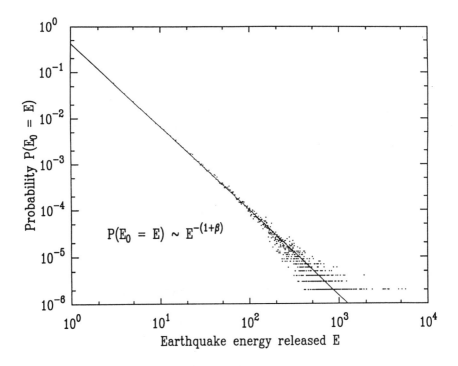

FIGURE 3   The energy distribution function for earthquakes produced by the Olami-Christensen-Feder model.[30] The histogram is the result of the statistics of half-a-million earthquakes on a square lattice of linear size 100 with $\alpha = 0.20$.

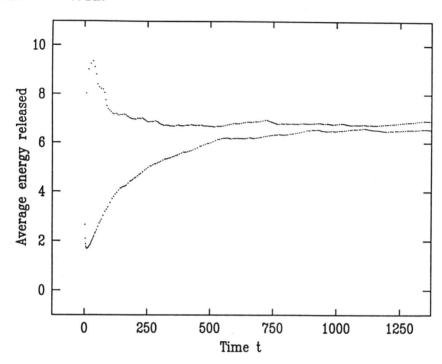

FIGURE 4   The average size of earthquakes as a function of time during the self-organization process (lower curve). The rise of the average is a measure of the rise of the correlation length in the system as a function of time. Notice that the initial rise is linear. The upper curve shows the average starting from a critical correlated configuration.

$\alpha = 1/4$ the force is conserved; i.e., the amount lost on the unstable site equals the total amount gained by the four neighbor sites. The criticality in this case prevails for values of α down to 0.05, with only 20% conservation. This came as a surprise since there was then a widespread belief that the lack of conservation would spontaneously generate a length scale, i.e., a "characteristic earthquake size." In fact, it seems that criticality occurs generically almost independent of the details of the toppling rule.

In Figure 3 I show the distribution of earthquakes for $\alpha = 0.20$. The straight line yields a b value of 0.8. The straight line indicates that the system has self-organized into the critical state. The slope depends on the degree of dissipation, $(1/4 - \alpha)$, so there is no universality of the exponent b in the nonconservative case. One should not look for unique b values in Nature. Indeed, different b values have been observed in different geographical areas.

Figure 4 illustrates the slow nature of the self-organization process. The running average of earthquake sizes, reflecting an upper cutoff for earthquakes, has been plotted vs. time, starting from a random uncorrelated configuration of forces. The average grows until it reaches a plateau limited only by the size of the system. This indicates self-organization into a critical state. Actually one can think of the curve as a measure of increasing complexity. We shall return to discuss the relations between large correlations and complexity in Section 7. The upper curve shows the running average starting from a state which has already had time to reach the steady state. The initial variations of the curves are statistical fluctuations.

The power-law distribution of earthquakes stems from the fractal nature of the SOC state, with correlated regions ranging over all length scales; those correlated regions, generated by the long-term dynamics, are the equivalent of the active faults or fault segments in real earthquakes. The fault structure changes on large geological time scales. More realistic long-range SOC models produce faults which topologically look much more like a real fractal-like arrangement of two-dimensional faults in a three-dimensional matrix.

2.2 ON EARTHQUAKES AND TURBULENCE

A liquid driven by imposing a velocity difference v over a length scale L undergoes a transition to a turbulent state. In the turbulent state, there are vortices of a large range of length scales. The energy is dissipated locally within a short length scale known as the Kolmogorov length. The dissipation is believed to occur very locally on a fractal set of zero measure.[29]

The physics of earthquakes can be described in a similar language. The crust in a fault region is driven by imposing a force or a strain over a large length L. In the stationary state, the energy is dissipated in narrow fault structures forming a fractal set. The spatio-temporal correlation functions for the two phenomena are quite similar,[5] although the time scales are vastly different. In both cases, the energy enters the system uniformly (zero wave vector) and leaves the system locally. The analogy has been explored in some detail by Kagan.[25]

Maybe it is useful to think of all self-organized critical processes as turbulent. Traditionally, turbulence has been thought of as being synonymous with the dynamics of the Navier Stokes equation, whatever that might be. It seems that it might be useful to think of Navier Stokes turbulence as a special case of something more general, giving the theorist more freedom to study simpler models. To that end we have constructed a simple forest fire model where energy is burned locally at fire fronts.[4] The fire, and the forests appear to evolve into a structure with temporal and spatial power-law correlations.

There is an analogous situation in equilibrium statistical mechanics: In the theory of equilibrium phase transitions, real progress was made only by solving simple toy models (Ising models) which contain essentially all the important physics without cluttering from irrelevant details. Where would we have been now, had we insisted on solving the quantum mechanics of the complicated electronic system of

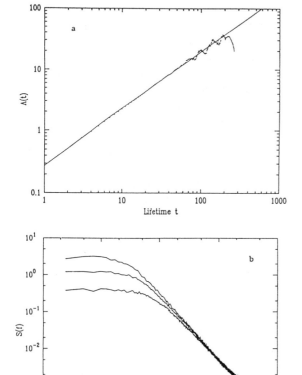

FIGURE 5 (a) Distribution of weighted duration of avalanches for $\alpha = 0.20$ for a system of size $L = 100$. The slope yields an exponent $\mu = 0.8$. (b) Power spectrum for systems with $\alpha = 0.2$ and different system sizes, $L = 45,\ 70,\ 100$. The slope, $\phi = 1.8$, is consistent with μ. The lower frequency cutoff scales with system size, while the upper cutoff is constant.

Iron in order to understand the critical properties of its magnetic-phase transition? The study of simple "caricature" models seems to be the only way open to us for understanding complex phenomena at this point.

3. 1/F NOISE

We have suggested that 1/f noise can be thought of as a signal formed by super-imposing the avalanches occurring in the self-organized critical state.[1,2] In order to illustrate how this works, Christensen et al.[10] introduced the weighted lifetime distribution of avalanches

$$\Lambda(t) = \sum s^2 P(S = s, T = t)$$

where $P(S,T)$ is the probability that an avalanche of size S has a duration of T time steps. They showed that if Λ has the scaling behavior $\Lambda(t) = t^{\mu}$, then the power spectrum $S(f)$ becomes

$$S(f) = f^{-(\mu+1)}, \text{ for } -1 < \mu < 1.$$

In Figure 5(a) I show the distribution of Λ for $\alpha = 0.20$ for the earthquake model above.[10] The slope of the straight line gives $\mu = 0.8$. In Figure 5(b) is shown a direct measurement of the power spectrum; an exponent $\phi = 1.75$ was found from the slope of the log-log plot, in reasonable agreement with the value 1.8 expected from the lifetime-distribution function defined above.

In Nature, values of the exponent of the $1/f$ noise in the interval 0.6–2.0 have been reported. A value of 1 corresponds to $\mu = 0$; this particular value of μ is obtained for $\alpha \approx 0.11$ in our model. The exponent depends on the parameters of

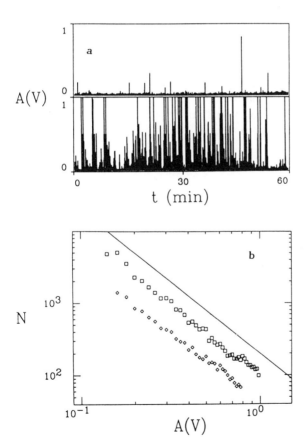

FIGURE 6 (a) Recorded samples of acoustic activity near Stromboli measured at two locations. (b) Distribution of bursts at the two sites (Diodati et al.[19]).

the model; thus, it is not universal as exponents for equilibrium critical phenomena usually are.

4. OTHER APPLICATIONS IN GEOPHYSICS AND ASTROPHYSICS

4.1 VOLCANIC ACTIVITY

Volcanic activity is intermittent, with events of all sizes, just like earthquakes. Diodati et al.[19] have measured bursts of acoustic emission in the area around Stromboli in Italy. They placed piezoelectric sensors coupled to the free ends of steel rods tightly cemented into rock-drill holes. They observed the distribution function for intervals between the bursts, and the distribution of burst amplitudes, E. In Figure 6 I show the latter distribution. Scaling behavior $N(E) = E^{-\delta}$, with $\delta \approx 2$ was observed.

4.2 CLOUD FORMATION

Clouds are fractal, with no typical size, suggesting an underlying critical dynamics. Nagel and Raschke[30] have derived a cellular automaton model which turns out to have a strong resemblance to percolation-based growth models. They studied the projections of the perimeter of clusters and found what they call a realistic value of the fractal dimension.

4.3 SOLAR FLARES

Observations show that the distribution of solar flare hard x-ray bursts is a power law with exponent 1.8. The power law spans over five orders of magnitude.[16] Lu and Hamilton[28] have suggested that this power-law dependence is a consequence of the coronal magnetic field being in a self-organized critical state, and they constructed a simple model describing the numerous coupled magnetic reconnection events of magnetic field lines (Figure 7).

5. APPLICATIONS IN CONDENSED MATTER PHYSICS

5.1 PINNED FLUX LATTICES AND CHARGE DENSITY WAVES

Type II superconductors in a magnetic field trap magnetic flux lines by pinning them to impurities. Pla and Nori,[34] Tang,[37] and Vinokur et al.[38] have suggested that the dynamics of the flux lines is self-organized critical, with collective bursts of flux line motion. One would like to observe those bursts directly; several experiments are in progress.

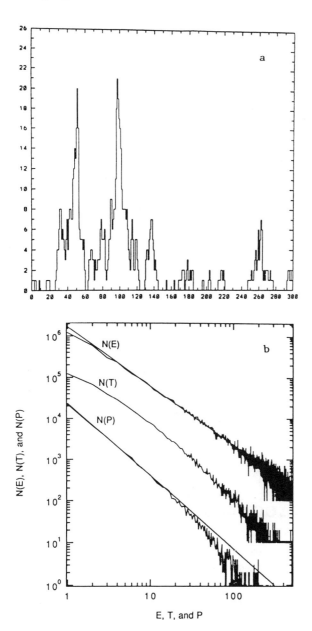

FIGURE 7 (a) Solar flare activity vs. time. (b) Distribution of energy release, peak activity, and duration of x-ray bursts (Lu and Hamilton[28]).

In charge-density wave systems there is a depinning transition vs. the applied voltage where the charge density wave starts sliding. Peinke et al.[33] have measured the sliding just above the critical voltage. They observed that the motion is not uniform, but intermittent. They measured the distribution of charges involved in these bursts, and also found a power-law distribution. When the voltage V is increased beyond the critical value V_c, there is a current which scales as $I = (V - V)^{-\beta}$. Several experimentalists have measured the exponent. Sandpile models give $\beta = 0.7$, and experiments seem to be in fair agreement, although many different, and mostly larger, values have been reported. This exponent is related to the other exponent characterizing avalanche distributions, etc., through scaling relations, similar to those found in equilibrium critical phenomena.

5.2 IONIZATION BREAKDOWN IN SEMICONDUCTORS AND LIGHTNING

Clauss et al.[11] have studied doped p-germanium near the onset of impact ionization breakdown. The power-law distribution of the breakdown events was interpreted as supporting a model of self-organized criticality. Nori[31] has suggested that lightning is a similar self-organized critical breakdown phenomena; the distribution of the charge released during lightning events indeed seems to follow a power law.

5.3 BARKHAUSEN NOISE

The magnetization of a ferromagnet is determined by its magnetic domain distribution and the response of the domains to applied magnetic fields. The changes in magnetization in a magnetic field is an intermittent phenomenon, involving complicated correlated domain flips. This noise behavior is known as the Barkhausen effect. The changes in magnetization can be detected as voltage pulses in a pick-up coil near the specimen. Cote and Meisel[12] have measured the distribution of the size of the voltage pulses and (you guessed it) found a power-law distribution.

6. APPLICATIONS IN BIOLOGY AND ECONOMICS

Economics and biology deal with "large dissipative dynamical systems." In economics, various agents (consumers, producers, economists...) interact. Raw materials and energy are consumed (dissipated) in the process. In biology, many species evolve together forming ecological systems.

6.1 BIOLOGY

Kauffman has constructed models, known as NKC models[26] of coevolving species. The fitness of the individual species is a function of the genetic code, represented by a binary string (10011100101111...). The fitness, as function of the genetic code, can be thought of as a rough landscape. The species try to improve their fitness (climb a local top) by mutating single genes at a time. The fitness also depends on the genetic code of the coevolving species. The landscape is changing all the time, to a degree that depends on the strength of the interactions between species. We have a collection of "interacting dancing landscapes." In a physicist's language, the model is a collection of spin glasses with asymmetric interactions.

If the coupling between the landscapes is strong, the climbing is a Sisyphean effort, where one would not be able to climb the top before the top disappears and new ones arise. If the interaction is weak, one will typically be able to climb only a local shallow top, but not be able to reach the global optimum. There is a phase transition separating the two states which are often called "frozen" and "chaotic," respectively.[7] The optimal result is supposed to occur if the system is precisely critical. The individuals will try to tune their complexity to that point. For species coevolving in the critical state the impact of single random mutations will be coevolutionary avalanches of all sizes, as observed by Raup.[35] Unfortunately, we have not been able yet to demonstrate this scenario in a satisfactory way, but there are many indications, so we are hopeful.

In the critical state, the individual species would interact to form a single, highly coherent biology, in the same way that individual grains of sand conspire to form a single pile of sand. Self-organized criticality may be the underlying dynamical principle behind Lovelock's "Gaia" hypothesis,[27] in which life on earth form a single organism, far out of equilibrium. In the SOC state the environment conspires against you to the extent that you are barely able to hold on by your fingernails—you may be wiped out by the smallest disturbance occurring anywhere in the system.

The closest that we have come is a simulation on the "Game of Life," a cellular automaton mimicking a society of interacting individuals. Using the language above the model can be thought of as a collection of species, each with one gene that can exist in two different states, interacting with eight other neighbor species. Indeed, the model self-organizes to the critical state, or very close to the critical state, when driven by random mutations.[4] It is not clear whether this is a robust scenario, or is due to an unlikely coincidence. Very recently, Ray[36] has simulated a system of reproducing organisms living in a computer memory. He measured the variation of the distribution of species and found a power law like the Gutenberg-Richter law. Thus, Ray's "life" seems to operate at the critical point. The idea that evolution works at a critical point has subsequently been taken up by several authors. Some of them have developed a new phraseology, calling it "evolution to the edge of chaos," etc., but the fundamental idea is the same. (Very recently, a simple model of evolution which actually self-organizes to a critical point with punctuated equilibria has been constructed.[8])

6.2 ECONOMICS

We have constructed a model of interacting economics agents.[7] The agents buy supplies, produce, and sell. When receiving orders, they either sell (if they have goods in stock), or produce (when they have not). In order to be able to produce, they order supplies from agents further down the chain. They produce extra units which they stockpile. The model is driven by consumer demand.

The model can be mapped on to one of the models of self-organized criticality which was solved by Dhar.[17,18] The response to a single demand is an avalanche with a power-law distribution. The aggregate production from many individual demands is a non-Gaussian Pareto-Levy distribution, precisely as proposed by Mandelbrot.[29] This suggests that large fluctuations in economics might be understood as a self-organized critical phenomenon. However, the model is way too simplified to be realistic: more work on better models is needed.

---

7. ON CRITICALITY, COMPLEXITY, AND ADAPTABILITY

What is complexity? Certainly ordered systems where every point in time and space looks like every other point are not complex. Also, it does not make sense to talk about complexity when the system is random, and each point in space is complete, uncorrelated with any other point. Zhang[39] has come up with what looks to be a useful definition of complexity: the system must have information on all length and time scales. The complexity is the integrated information, $-\sum p \log p$ of the system over all possible coarse-grainings in time or space. Using Zhang's formula, the Ising model of equilibrium phase transitions is most complex at the critical point, where there are clusters of all sizes. In short: Complexity is Criticality. Self-organized criticality might thus be interpreted as Nature's drive towards maximum complexity. As Conway tried to make the Game of Life complex, he inadvertently at the same time made it critical!

Then what is adaptability of a complex system? Since "purpose" and "rationality," and thus "learning," and "adaptability" do not really exist in deterministic dynamical system, the question should really be: which are the features of complex systems that an outside observer might interpret as adaptability? The only systematic way to define an "organism" is as a part of the system which is dynamically connected, in the sense that if part of the organism is affected, the whole organism is affected. An organism lives and dies as a whole. Because of the hierarchical nature of the critical or complex system, with interactions over all scales, we can arbitrarily define what we mean by a unit: In a biological system, one can choose either a single cell, a single individual, such as an ant, the ant's nest, or the ant as a species, as the adaptive unit. In a human social system, one might choose an individual, a family, a company, or a country as the unit. No unit at any level has the right to claim priority status. For each choice of unit there is a time scale.

No matter what our choice is, there are very many degrees of freedom inside the unit, with smaller length scale, and there are very many degrees of freedom outside the unit. What we observe as "adaptability" is the dynamics (with avalanches, etc.) of the degrees of freedom forming that unit, allowing the system to remain an integrated part of the critical system. The self-organization process, allowing the individual units to barely survive under different conditions, i.e., the robustness of the critical state, is probably what one would interpret as adaptability or resiliency. Fitness is a synonym for self-consistent integration into a highly integrated complex or critical state by any part of the system. We like to see ourselves as the center of everything. However, while we like to interpret ourselves as being well adapted and superbly fit, an outside observer might interpret us simply as a self-consistent, functionally integrated, part of an evolving complex system. The only unique description of biology might be as a whole, the Gaia theory, just as the only consistent way of describing the sand dynamics is in terms of one pile.

ACKNOWLEDGMENT

Supported by the U.S. Department of Energy under contract DE-AC02-76-CH00016.

REFERENCES

1. Bak, P., C. Tang, and K. Wiesenfeld. "Self-Organized Criticality: An Explanation of 1/f Noise." *Phys. Rev. Lett.* **59** (1987): 381–384.
2. Bak, P., C. Tang, and K. Wiesenfield. "Self-Organized Criticality." *Phys. Rev. A* **38** (1988): 364–374.
3. Bak, P., K. Chen, and M. Creutz. "Self-Organized Criticality in the Game of Life." *Nature* **342** (1989): 780.
4. Bak, P., K. Chen, and C. Tang. "A Forest Fire Model and Some Thoughts on Turbulence." *Phys. Lett.* **147** (1990): 297–299.
5. Bak P., and K. Chen. "Self-Organized Criticality." *Sci. Am.*, **246 (1)** (1991): 46–53.
6. Bak, P., K. Chen, J. Scheinkman, and M. Woodford. "Self-Organized Criticality and Fluctuations in Economics." Working Paper 92-04-018, Santa Fe Institute, 1992; *Richerche Economiche* **47** (1993): 3–30.
7. Bak, P., H. Flyvbjerg, and B. Lautrup. "Co-Evolution in a Random Fitness Landscape." *Phys, Rev. A* **46** (1992): 6724-6730.
8. Bak, P., and K. Sneppen. "A State Model of Evolution." *Phys. Rev. Lett.* **71** (1993): 4083–4086.

9. Christensen, K. C., H. J. Jensen, and H. C. Fogedby. "Dynamical and Spatial Aspects of Sandpile Cellular Automata." *J. Stat. Phys.* **63** (1991): 653-681.

10. Christensen, K., Z. Olami, and P. Bak. "Deterministic 1/f Noise in Non-Conservative Models of Self-Organized Criticality." *Phys. Rev. Lett.* **68** (1992): 2417–2420.

11. Clauss, W., A. Kittel, U. Rau, J. Parisi, J. Peinke, and R. P. Huebener. "Self-Organized Critical Behavior in the Low-Temperature Impact Ionization Breakdown of p-Ge." *Europhys. Lett.* **12** (1990): 423–428.

12. Cote, P. J., and L. V. Meisel. "Self-Organized Criticality and the Barkhausen Effect." *Phys. Rev. Lett.* **67** (1991): 1334–1337.

13. Creutz, M. "Abelian Sandpiles." *Comp. Phys.* Mar/Apr (1991): 198–203.

14. Creutz, M. "On Self-Organized Criticality." *Nuc. Phys. B* (Proc. Suppl.) **20** (1992): 748–752.

15. Christensen, K. C., H. J. Jensen, and H. C. Fogedby. "Dynamical and Spatial Aspects of Sandpile Cellular Automata." *J. Stat. Phys.* **63** (1991): 653–681.

16. Dennis, B. R. "Solar Hand X-Ray Bursts." *Solar Phys.* **120** (1985): 465.

17. Dhar, D., and R. Ramaswamy. "Exactly Solved Model of Self-Organized Criticality." *Phys. Rev. Lett.* **63** (1986): 1659–1662.

18. Dhar, D. "Self-Organized Critical State of Sandpile Automaton Models." *Phys. Rev. Lett.* **64** (1990): 1613–1616.

19. Diodati, P., F. Marchesoni, and S. Piazza. "Acoustic Emission from Volcanic Rocks: An Example of Self-Organized Criticality." *Phys. Rev. Lett.* **67** (1991): 2239–2242.

20. Drossel, B. and F. Schwabl. "Self-Organized Criticality in a Forest-Fire Model." *Phys. Rev. Lett.* **69** (1992): 1629–1632.

21. Gould, S. J. "Punctuated Equilibria: The Tempo and Mode of Evolution Reconsidered." *Paleobio.* **3** (1977): 135–151.

22. Grumbacher, S. K., K. M. McEwen, D. A. Halvorson, D. T. Jacobs, and J. Lindler. "Self-Organized Criticality: An Experiment with Sand Piles." *Am. J. Phys.* **61** (1993): 329–335.

23. Gutenberg, B., and C. F. Richter. *Seismicity of the Earth and Associated Phenomena.* New York: Hafner, 1965.

24. Johnston, A. C., and S. J. Nava. "Recurrence Rates and Probability Distribution Estimates for the New Madrid Seismic Zone." *J. Geophys. Res. B* **90** (1985): 6737–6751.

25. Kagan, Y. "Seismicity: Turbulence of Solids." *Nonlinear Sci. Today* **2** (1992): 1–13.

26. Kauffman, S. A., and S. Johnsen. "Coevolution to the Edge of Chaos—Coupled Fitness Landscapes, Poised States, and Coevolutionary Avalanches." *J. Theor. Biol.* **149** (1991): 467–506.

27. Lovelock, J. E. *Gaia, A New Look at Life on Earth.* Oxford: Oxford University Press, 1979.

28. Lu, E. T., and R. J. Hamilton. "Avalanches and the Distribution of Solar Flares." *Astrophys. J. Lett.* **380** (1991): L89–L92.

29. Mandelbrot, B. *The Fractal Geometry of Nature*. San Fransisco: W. H. Freeman, 1982.
30. Nagel, K., and E. Raschke. "Self-Organizing Criticality in Cloud Formation." *Physica A* **202** (1992): 519–531.
31. Nori, F. "Lightning as a Self-Organized Critical Phenomenon." Preprint, University of Michigan, 1992.
32. Olami, Z., H. J. Feder, and K. Christensen. "Self-Organized Criticality in a Continuous, Nonconservative Cellular Automaton Modeling Earthquakes." *Phys. Rev. Lett.* **68** (1992): 1464–1247.
33. Peinke, J., A. Kittel, and J. Dumas. "Critical Dynamics of the Quasi-One-Dimensional Blue Bronze $K_{0.3}Mo_3O$ at Low Temperatures." *Europhys. Lett.* **18** (1992): 127–131.
34. Pla, O., and F. Nori. "Self-Organized Critical Behavior in Pinned Flux Lattices." *Phys. Rev. Lett.* **67** (1991): 919–922.
35. Raup, M. D. "Biological Extinction in Earth History." *Science* **251** (1986): 1530–1532.
36. Ray, T. S. "An Approach to the Synthesis of Life." In *Artificial Life II*, edited by C. G. Langton, C. Taylor, J. D. Farmer, and S. Rasmussen. Santa Fe Institute Studies in the Sciences of Complexity, Proc. Vol. X, 371–408. Redwood City, CA: Addison Wesley, 1992.
37. Tang, C. "SOC and the Bean Critical State." *Physica A* **194** (1993): 315–320.
38. Vinokur, V. M., M. V. Feigelman, and V. B. Geshkenbein. "Exact Solution for Flux Creep with Logarithmic U(j) Dependence: Self-Organized Critical State in High-T_c Superconductors." *Phys. Rev. Lett.* **67** (1991): 915–918.
39. Zhang, Y-C. *J. Phys. 1* (France) (1991): 971–975.

DISCUSSION

FONTANA: It seems very important to build up to the critical state by slow driving.

BAK: Yes, it is. Fast driving can destroy it. If you take a hammer and hit it, you can destroy anything. The time scale must be relevant to the system we are looking at. Earthquakes may be in minutes, geological processes in millions of years. Also with evolution. there are different time scales in the buildup.

MAYNARD SMITH: I don't trust myself to remark on what you said about adaptation because I think it's rubbish but the earlier part is interesting. There are actually two quite different ways in which I can visualize criticality building up in a biological system. One is at the ecological level. There could be the continuous appearance of new species, gradually building up the number of species until a critical point was reached; we do not know what determines that point but it undoubtedly exists. The situation would then be suddenly relieved by an extinction, like an avalanche.

BAK: And little ones go on all the time.

MAYNARD SMITH: Sure. You can have little ones. The other, which you might think about in doing your modeling, arises because there is a critical relationship between the mutation rate, that is to say, the error rate in replicating DNA and the size of the genome. There are, however, processes constantly increasing the genome size and, sometimes, increasing the mutation rate and I can imagine that at the individual level, looking not at mass extinctions but at the individual organisms. These two processes could lead to a critical point. What evidence we have suggests that individual organisms have mutation rates just about at the critical level.

BAK: Yes. Remember, I don't pretend that these solutions in biology are all right. These are just suggestions that biology may have critical points.

MAYNARD SMITH: But what you call adaptation is not what is meant by biologists. You should not use our word for something different. Call it something else.

[Editor's Note: despite more discussion, we did not resolve the proper word. See pp. 579–583.]

Melanie Mitchell,* James P. Crutchfield, and Peter T. Hraber***

*Santa Fe Institute, 1660 Old Pecos Trail, Suite A, Santa Fe, New Mexico, U.S.A. 87501; e-mail: mm@santafe.edu, pth@santafe.edu and **Physics Department, University of California, Berkeley, CA 94720; e-mail: chaos@gojira.berkeley.edu

Dynamics, Computation, and the "Edge of Chaos": A Re-Examination

Abstract: In this chapter we review previous work and present new work concerning the relationship between dynamical systems theory and computation. In particular, we review work by Langton[22] and Packard[29] on the relationship between dynamical behavior and computational capability in cellular automata (CA). We present results from an experiment similar to the one described by Packard,[29] which was cited as evidence for the hypothesis that rules capable of performing complex computations are most likely to be found at a phase transition between ordered and chaotic behavioral regimes for CA (the "edge of chaos"). Our experiment produced very different results from the original experiment, and we suggest that the interpretation of the original results is not correct. We conclude by discussing general issues related to dynamics, computation, and the "edge of chaos" in cellular automata.

Complexity: Metaphors, Models, and Reality
Eds. G. Cowan, D. Pines, and D. Meltzer, SFI Studies in the
Sciences of Complexity, Proc. Vol. XIX, Addison-Wesley, 1994   **497**

INTRODUCTION

A central goal of the sciences of complex systems is to understand the laws and mechanisms by which complicated, coherent global behavior can emerge from the collective activities of relatively simple, locally interacting components. Given the diversity of systems falling into this broad class, the discovery of any commonalities or "universal" laws underlying such systems will require very general theoretical frameworks. Two such frameworks are dynamical systems theory and the theory of computation. These have independently provided powerful tools for understanding and describing common properties of a wide range of complex systems.

Dynamical systems theory has developed as one of the main alternatives to analytic, closed-form, exact solutions of complex systems. Typically, a system is considered to be "solved" when one can write down a finite set of finite expressions that can be used to predict the state of the system at time t, given the state of the system at some initial time t_0. Using existing mathematical methods, such solutions are generally not possible for most complex systems of interest. The central contribution of dynamical systems theory to modern science is that exact solutions are not necessary for understanding and analyzing a nonlinear process. Instead of deriving exact single solutions, the emphasis of dynamical systems theory is on describing the geometrical and topological structure of ensembles of solutions. In other words, dynamical systems theory gives a geometric view of a process's structural elements, such as attractors, basins, and separatrices. It is thus distinguished from a purely probabilistic approach such as statistical mechanics, in which geometric structures are not considered. Dynamical systems theory also addresses the question of what structures are generic; that is, what behavior types are typical across the spectrum of complex systems.

In contrast to focusing on how geometric structures are constrained in a state space, computation theory focuses on how basic information-processing elements—storage, logical gates, stacks, queues, production rules, and the like—can be combined to effect a given information-processing task. As such, computation theory is a theory of organization and the functionality supported by organization. When adapted to analyze complex systems, it provides a framework for describing behaviors as computations of varying structure. For example, if the global mapping from initial to final states is considered as a computation, then the question is: what function is being computed by the global dynamics? Another range of examples concern limitations imposed by the equations of motion on information processing: can a given complex system be designed to emulate a universal Turing Machine? In contrast to this sort of engineering question, one is also interested in the intrinsic computational capability of a given complex system; that is, what information-processing structures are intrinsic in its behavior?[8,17]

Dynamical systems theory and computation theory have almost always been applied independently, but there have been some efforts to understand the relationship between the two—that is, the relationship between a system's ability for information processing and other measures of the system's dynamical behavior.

RELATIONSHIPS BETWEEN DYNAMICAL SYSTEMS THEORY AND COMPUTATION THEORY. Computation theory developed from the attempt to understand information-processing aspects of systems. A colloquial definition of "information processing" might be "the transformation of a given input to a desired output" but, in order to apply the notion of information processing to complex systems and to relate it to dynamical systems theory, the notion must be enriched to include the *production* of information as well as its storage, transmission, and logical manipulation. In addition, the engineering-based notion of "desired output" is not necessarily appropriate in this context; the focus here is often on the intrinsic information-processing capabilities of a dynamical system not subject to a particular computational goal.

Beginning with Kolmogorov's and Sinai's adaptation of Shannon's communication theory to mechanics in the late 1950s,[20,34] there has been a continuing effort to relate a nonlinear system's information-processing capability and its temporal behavior. One result is that a deterministic chaotic system can be viewed as a generator of information.[33] Another is that the complexity of predicting a chaotic system's behavior grows exponentially with time.[6] The descriptive complexity here, called the Kolmogorov-Chaitin complexity,[4,21] uses a universal Turing machine as the deterministic prediction machine. The relationship between the difficulty of prediction and dynamical randomness is simply summarized by the statement that the growth rate of the descriptive complexity is equal to the information production rate.[3] These results give a view of deterministic chaos that emphasizes the production of randomness and the resulting unpredictability. They are probably the earliest connections between dynamics and computation.

The question of what structures underlie information production in mechanical systems has received attention only more recently. The first and crudest property considered is the amount of memory a system employs in producing apparent randomness.[7,15] The idea is that an ideal random process uses no memory to produce its information—it simply flips a coin as needed. Similarly, a simple periodic process requires memory only in proportion to the length of the pattern it repeats. Within the memory-capacity view of dynamics, both these types of processes are simple—more precisely, they are simple to describe statistically. Between these extremes, though, lie the highly structured, complex processes that use both randomness and pattern storage to produce their behavior. Such processes are more complex to describe statistically than are ideal random or simple periodic processes. The trade-off between structure and randomness is common to much of science. The notion of statistical complexity[8] was introduced to measure this trade-off.

Computation theory is concerned with more than information and its production and storage. These elements are taken as given and, instead, the focus is on how their combinations yield more or less computational power. Understandably, there is a central dichotomy between machines with finite and infinite memory. On a finer scale, distinctions can be drawn among the ways in which infinite memory is organized—e.g., as a stack, a queue, or a parallel array. Given such considerations, the question of the intrinsic computational structure in a dynamical system becomes substantially more demanding than the initial emphasis on gross measures of information storage and production.

Several connections in this vein have been made recently. In the realm of continuous-state dynamical systems, Crutchfield and Young looked at the relationship between the dynamics and computational structure of discrete time series generated by the logistic map at different parameter settings.[8,9] They found that at the onset of chaos there is an abrupt jump in computational class of the time series, as measured by the formal language class required to describe the time series. In concert with Feigenbaum's renormalization group analysis of the onset of chaos,[12] this result demonstrated that a dynamical system's computational capability—in terms of the richness of behavior it produces—is qualitatively increased at a phase transition.

Rather than considering intrinsic computational structure, a number of "engineering" suggestions have been made that there exist physically plausible dynamical systems implementing Turing machines.[2,26,27] These studies provided explicit constructions for several types of dynamical systems. At this point, it is unclear whether the resulting computational systems are generic—i.e., likely to be constructable in other dynamical systems—and whether they are robust and reliable in information processing. In any case, it is clear that much work has been done to address a range of issues that relate continuous-state dynamics and computation. Many of the basic issues are now clear and there is a firm foundation for future work.

DYNAMICS AND COMPUTATION IN CELLULAR AUTOMATA. There has also been a good deal of study of dynamics and computation in discrete spatial systems called cellular automata (CA). In many ways, CA are more natural candidates for this study than continuous-state dynamical systems since they are completely discrete in space, in time, and in local state. There is no need to develop a theory of computation with real numbers. Unfortunately, something is lost in going to a completely discrete system. The analysis of CA behavior in conventional dynamical systems terms is problematic for just this reason. Defining the analogs of "sensitive dependence on initial conditions," "the production of information," "chaos," "instability," "attractor," "smooth variation of a parameter," "bifurcation," the "onset of chaos," and other basic elements of dynamical systems theory requires a good deal of care. Nonetheless, Wolfram introduced a dynamical classification of CA behavior closely allied to that of dynamical systems theory. He speculated that one of his four classes supports universal computation.[36] It is only recently, however, that CA behavior has been directly related to the basic elements of qualitative dynamics—the attractor-basin portrait.[17] This has lead to a reevaluation of CA behavior classification and, in particular, to a redefinition of the chaos and complexity apparent in the spatial patterns that CA generate.[10]

Subsequent to Wolfram's work, Langton studied the relationship between the "average" dynamical behavior of cellular automata and a particular statistic (λ) of a CA rule table.[22] He then hypothesized that "computationally capable" CA, and in particular, CA capable of universal computation, will have "critical" λ values corresponding to a phase transition between ordered and chaotic behavior. Packard experimentally tested this hypothesis by using a genetic algorithm (GA) to evolve CA to perform a particular complex computation.[29] He interpreted the results as showing that the GA tends to select rules close to "critical" λ regions—i.e., the "edge of chaos."

We now turn our discussion more specifically to issues related to λ, dynamical-behavior classes, and computation in CA. We then present experimental results and a theoretical discussion that suggest the interpretation given of the results by Packard[29] is not correct. Our experiments, however, show some interesting phenomena with respect to the GA evolution of CA, which we summarize here. Longer, more detailed descriptions of our experiments and results are given by Mitchell et al.[24,25]

CELLULAR AUTOMATA AND THE "EDGE OF CHAOS"

Cellular automata are one of the simplest frameworks in which issues related to complex systems, dynamics, and computation can be studied. CA have been used extensively as models of physical processes and as computational devices.[11,16,31,35,37] In its simplest form, a CA consists of a spatial lattice of *cells*, each of which, at time t, can be in one of k states. We denote the lattice size (i.e., number of cells) as N. A CA has a single fixed rule used to update each cell; this rule maps from the states in a neighborhood of cells—e.g., the states of a cell and its nearest neighbors—to a single state, which becomes the updated value for the cell in question. The lattice starts out with some initial configuration of cell states and, at each time step, the states of all cells in the lattice are synchronously updated. We use the term "state" to refer to the value of a single cell —e.g., 0 or 1—and the term "configuration" to mean the pattern of states over the entire lattice.

In this chapter we restrict our discussion to one-dimensional CA with $k = 2$. In a one-dimensional CA, the neighborhood of a cell includes the cell itself and some number r of neighbors on either side of the cell. All of the simulations described here are of CA with spatially periodic boundary conditions (i.e., the one-dimensional lattice is viewed as a circle, with the right neighbor of the rightmost cell being the leftmost cell, and vice versa).

The equations of motion ϕ for a CA are often expressed in the form of a *rule table*. This is a lookup table listing each of the neighborhood patterns and the state to which the central cell in that neighborhood is mapped. For example, the following is one possible rule table for a one-dimensional CA with $k = 2, r = 1$. Each possible neighborhood η is given along with the "output bit" $s = \phi(\eta)$ to which the central cell is updated.

η	000	001	010	011	100	101	110	111
s	0	0	0	1	0	1	1	1

In words, this rule says that for each neighborhood of three adjacent cells, the new state is decided by a majority vote among the three cells.

The notion of "computation" in CA can have several possible meanings,[24] but the most common meaning is that the CA performs some "useful" computational task. Here, the rule is interpreted as the "program," the initial configuration is interpreted as the "input," and the CA runs for some specified number of time steps or until it reaches some "goal" pattern—possibly a fixed-point pattern. The final pattern is interpreted as the "output." An example of this is using CA to perform image-processing tasks.[32]

Packard[29] discussed a particular $k = 2, r = 3$ rule, invented by Gacs, Kurdyumov, and Levin (GKL)[13] as part of their studies of reliable computation in CA. The GKL rule was not invented for any particular classification purpose, but it does have the property that, under the rule, most initial configurations with less than half 1's are eventually transformed to a configuration of all 0's, and most initial configurations with more than half 1's are transformed to a configuration of all 1's. The rule thus approximately computes whether the density of 1's in the initial configuration (which we denote as ρ) is above the threshold $\rho_c = 1/2$. When initial configurations are close to $\rho = 1/2$, the rule makes a significant number of classification errors.[24]

Packard was inspired by the GKL rule to use a GA to *evolve* a rule table to perform this "$\rho_c = 1/2$" task. If $\rho < 1/2$, then the CA should relax to a configuration of all 0's; otherwise, it should relax to a configuration of all 1's. This task can be considered to be a "complex" computation for a $k = 2, r = 3$ CA since the minimal amount of memory it requires increases with N; in other words, the required computation is spatially global and corresponds to the recognition of a nonregular language.[1] The global nature of the computation means that information must be transmitted over significant space-time distances (on the order of N) and this requires the cooperation of many local neighborhood operations.[24]

In dynamical terms, complex computation in a small-radius, binary-state CA requires significantly long transients and space-time correlation lengths. Langton hypothesized that such effects are most likely to be seen in a certain region of CA rule space as parameterized by λ.[22] For binary-state CA, λ is simply the fraction of 1's in the output bits of the rule table. For CA with $k > 2$, λ is defined as the fraction of "nonquiescent" states in rule table, where one state is arbitrarily chosen to be "quiescent," and all states obey a "strong quiescence" requirement.[22] Langton

---

[1] See Hopcroft and Ullman[19] for an introduction to formal-language classes in computation theory.

performed a number of Monte Carlo samples of two-dimensional CA, starting with $\lambda = 0$ and gradually increasing λ to $1 - 1/k$ (i.e., the most homogeneous to the most heterogeneous rule tables). Langton used various statistics such as single-site entropy, two-site mutual information, and transient length to classify CA "average" behavior at each λ value. The notion of "average behavior" was intended to capture the most likely behavior observed with a randomly chosen initial configuration for CA randomly selected in a fixed-λ subspace. These studies revealed some correlation between the various statistics and λ. The correlation is quite good for very low and very high λ values. However, for intermediate λ values in finite-state CA, there is a large degree of variation in behavior.

Langton claimed on the basis of these statistics that as λ is incremented from 0 to $[1 - 1/k]$, the average behavior of CA undergoes a "phase transition" from ordered (fixed point or limit cycle after some short transient period) to chaotic (apparently unpredictable after some short transient period). As λ reaches a "critical value" λ_c, the claim is that rules tend to have longer and longer transient phases. Additionally, Langton claimed that CA close to λ_c tend to exhibit long-lived, "complex"—nonperiodic, but nonrandom—patterns. Langton proposed that the λ_c regime roughly corresponds to Wolfram's Class 4 CA,[36] and hypothesized that CA capable of performing complex computations will most likely be found in this regime.

Analysis based on λ is one possible first step in understanding the structure of CA rule space and the relationship between dynamics and computation in CA. However, the claims summarized above rest on a number of problematic assumptions. One assumption is that in the global view of CA space, CA rule tables themselves are the appropriate loci of dynamical behavior. This is in stark contrast with the state space and the attractor-basin portrait approach of dynamical systems theory. The latter approach acknowledges the fact that behaviors in state space cannot be adequately parameterized by any function of the equations of motion, such as λ. Another assumption is that the underlying statistics being averaged (e.g., single-site entropy) converge. But many processes are known for which averages do not converge. Perhaps most problematic is the assumption that the selected statistics are uniquely associated with mechanisms that support useful computation.

Packard empirically determined rough values of λ_c for one-dimensional $k = 2, r = 3$ CA by looking at the *difference-pattern spreading rate* γ as a function of λ.[29] The spreading rate γ is a measure of unpredictability in spatio-temporal patterns and so is one possible measure of chaotic behavior.[28,36] It is analogous to, but not the same as, the Lyapunov exponent for continuous-state dynamical systems. In the case of CA it indicates the average propagation speed of information through space-time, though not the production rate of local information. At each λ a large number of rules was sampled and for each CA γ was estimated. The average γ over the selected CA was taken as the average spreading rate at the given λ. The results are reproduced in Figure 1(a). As can be seen, at low and high λ's, γ vanishes, indicating fixed-point or short-period behavior; at intermediate λ it is maximal,

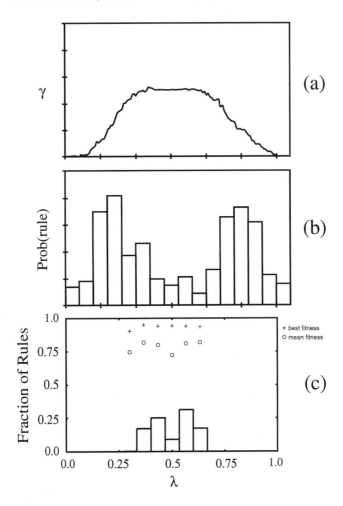

FIGURE 1   (a) The average difference-pattern spreading rate γ of a large number of randomly chosen $k = 2, r = 3$ CA, as a function of λ. (b) Results from the original experiment on GA evolution of CA for the $\rho_c = 1/2$ classification task. The histogram plots the frequencies of rules merged from the final generations (generation 100) of a number of runs. These populations evolved from initial populations uniformly distributed in λ. The histogram consists of 16 bins of width 0.0667. The bin above $\lambda = 1.0$ contains just those rules with $\lambda = 1.0$. Graphs (a) and (b) are adapted from Packard,[29] with the author's permission. No vertical scale was provided there. (c) Results from our experiment. The histogram plots the frequencies of rules merged from the final generations (generation 100) of 30 runs. These populations evolved from initial populations uniformly distributed in λ. Following Packard[29] the λ-axis is divided into 15 bins of length 0.0667 each. The rules with $\lambda = 1.0$ are included in the rightmost bin. The best (cross) and mean (circle) fitnesses are plotted for each bin. (The y-axis interval for fitnesses is also [0,1].)

indicating chaotic behavior; and in the transition or λ_c regions—centered about $\lambda \approx 0.25$ and $\lambda \approx 0.80$—it rises or falls gradually. While not shown in Figure 1(a), for most λ values γ's variance, like that of the statistics used by Langton, is high.

THE ORIGINAL EXPERIMENT

The empirical CA studies recounted above addressed only the relationship between λ and the dynamical behavior of CA as revealed by several statistics. Those studies did not correlate λ or behavior with an independent measure of computation. Packard[29] addressed this issue by using a genetic algorithm (GA)[14,18] to evolve CA rules to perform a particular computation. This experiment was meant to test two hypotheses: (1) CA rules able to perform complex computations are most likely to be found near λ_c values; and (2) when CA rules are evolved to perform a complex computation, evolution will tend to select rules near λ_c values.

Packard's experiment consisted of evolving binary-state one-dimensional CA with $r = 3$. The "complex computation" is the $\rho = 1/2$ task described above. A form of the genetic algorithm was applied to a population of rules represented as bit strings. To calculate the fitness of a string, the string was interpreted as the output bits of a rule table, and the resulting CA was run on a number of randomly chosen initial conditions. The fitness was a measure of the average classification performance of the CA over the initial conditions.

The result from this experiment are displayed in Figure 1(b). The histogram displays the observed frequency of rules in the GA population as a function of λ, with rules merged from a number of different runs with identical parameters but with different random number seeds. In the initial generation the rules were uniformly distributed over λ values. The graph (b) gives the final generation—in this case, after the GA has run for 100 generations. The rules cluster close to the two λ_c regions, as can be seen by comparison with the difference-pattern spreading rate plot (a). Note that each individual run produced rules at one or the other peak in graph (b), so when the runs were merged together, both peaks appear.[30] Packard interpreted these results as evidence for the hypothesis that, when an ability for complex computation is required, evolution tends to select rules near the transition to chaos. He argues, like Langton, that this result intuitively makes sense because "rules near the transition to chaos have the capability to selectively communicate information with complex structures in space time, thus enabling computation."[29]

OUR EXPERIMENT

We performed an experiment similar to Packard's. The CA rules in the population are represented as bit strings, each encoding the output bits of a rule table for $(k, r) = (2, 3)$. Thus, the length of each string is $128 = 2^{2r+1}$.

For a single run, the GA we used generated a random initial population of 100 rules (bit strings) with λ values uniformly distributed over [0,1]. Then it calculated the fitness of each rule in the population by a method to be described below. The population was then ranked by fitness and the 50 rules with lowest fitness were discarded. The 50 rules with highest fitness were copied directly into the next generation. To fill out the population 50 new rules were generated from pairs of parents selected at random from the current generation. Each pair of parents underwent a single-point crossover whose location was selected with uniform probability over the string. The resulting offspring were mutated at a number of sites chosen from a Poisson distribution with a mean of 3.8.

The fitness of a rule R is calculated as follows. R is run on 300 randomly chosen initial configurations on a lattice with $N = 149$. A new set of initial configurations is chosen each generation, and all rules in that generation are tested on it. The 300 initial configurations are uniformly distributed over densities in [0,1], with exactly half having $\rho < 1/2$ and exactly half having $\rho > 1/2$. R is run on each initial configuration for approximately 320 iterations; the actual number is chosen probabilistically to avoid overfitting. Note that 320 iterations is the measured maximum amount of time for the GKL CA to reach an invariant pattern over a large number of initial configurations on lattice size 149.

R's score on a given initial configuration is the fraction of "correct" bits in the final configuration. For example, if the initial configuration has $\rho > 1/2$, then R's score is the fraction of 1's in the final configuration. Thus, R gets partial credit for getting some of the bits correct. A rule generating random strings would therefore get a score of 0.5. R's fitness is then its average score over all 300 initial configurations. For more details and for justifications for these parameters, see Mitchell et al.[24]

The results of our experiment are given in Figure 1(c). This histogram displays the observed frequency of rules in the population at generation 100 as a function of λ, merged from 30 different runs with identical parameters but different random number seeds. The initial populations were each uniformly distributed over λ. The best and mean fitnesses of rules in each bin are also displayed.

There are a number of striking differences between Figures 1(b) and 1(c):

- In Figure 1(b), most of the rules in the final generations cluster in the λ_c regions defined by Figure 1(a). In particular, in Figure 1(b), approximately 66% of the mass of the distribution is in bins 3–5 and 12–14 combined (where bins are numbered 1–16 left to right). In Figure 1(c) these bins contain only 0.002% of the mass of the distribution (there are no rules in bins 3, 4, 12, 13, or 14, and there are only 5 rules in bin 5 out of a total of 3000 rules represented in the histogram).

- In Figure 1(b) there are rules in every bin. In Figure 1(c) there are rules only in the central six bins.

- In both histograms there are two peaks surrounding a central dip. As in the original experiment, in our experiment each individual run produced rules at one or the other peak, so when the runs were merged together, both peaks appear. In Figure 1(b), however, the two peaks are located roughly at bins 4 and 13 and thus are centered around $\lambda = 0.23$ and $\lambda = 0.83$, respectively. In Figure 1(c) the peaks are located roughly at bins 7 and 9 and thus are centered around $\lambda = 0.43$ and $\lambda = 0.57$, respectively. The ratio of the two peak spreads is thus approximately 4:1.

■ In Figure 1(b), the two highest bins are roughly five times as high as the central bin whereas in Figure 1(c) the two highest bins are roughly three times as high as the central bin.

In Figure 1(c) we also give an important calibration: the best and mean fitness of rules in each bin. The best fitnesses are all between 0.93 and 0.95, except the leftmost bin which has a best fitness of 0.90. Under this fitness function the GKL rule has fitness ≈ 0.98 on all lattice sizes; the GA never found a rule with fitness above 0.95 on lattice size 149, and the measured fitness of the best evolved rules was much worse on larger lattice sizes.[24] The fitnesses of the rules in Figure 1(b) were not given by Packard,[29] though none of those rules achieved the fitness of the GKL rule.[30]

DISCUSSION OF EXPERIMENTAL RESULTS
WHY DO THE RULES CLUSTER CLOSE TO $\lambda = 1/2$?

What accounts for these differences between Figures 1(b) and 1(c)? In particular, why did the evolved rules in our experiment tend to cluster close to $\lambda = 1/2$ rather than the two λ_c regions?

There are two reasons (discussed in detail below): (1) Good performance on the $\rho_c = 1/2$ task *requires* rules with λ close to 1/2, and (2) the GA operators of crossover and mutation intrinsically push any population close to $\lambda = 1/2$.

It can be shown that correct or nearly correct performance on the $\rho_c = 1/2$ task requires rules close to $\lambda = 1/2$. Intuitively, this is because the task is symmetric with respect to the exchange of 1's and 0's. Suppose, for example, a rule that carries out the $\rho_c = 1/2$ task has $\lambda < 1/2$. This implies that there are more neighborhoods in the rule table that map to output bit 0 than to output bit 1. This, in turn, means that there will be *some* initial configurations with $\rho > \rho_c$ on which the action of the rule will *decrease* the number of 1's. And this is the opposite of the desired action. However, if the rule acts to *decrease* the number of 1's on an initial configuration with $\rho > \rho_c$, it risks producing an intermediate configuration with $\rho < \rho_c$, which then would lead (under the original assumption that the rule carries out the task correctly) to a fixed point of all 0's, misclassifying the initial configuration. A similar argument holds in the other direction if the rule's λ value is greater than 1/2. This informal argument shows that a rule with $\lambda \neq 1/2$ will misclassify certain initial configurations. Generally, the further away the rule is from $\lambda = 1/2$, the more of such initial configurations there will be. Such rules may perform fairly well, classifying many initial configurations correctly or partially correctly. However, we expect any rule that performs reasonably well on this task—in the sense of being close to the GKL rule's 0.98 average fitness across lattice sizes—to have a λ value close to 1/2. This is one force pushing the GA population to $\lambda = 1/2$. We note that, not surprisingly, the GKL rule has $\lambda = 1/2$.

This analysis points to a problem with using this task as an evolutionary goal in order to test the hypothesis relating evolution, computation, and λ_c rules. As was shown in Figure 1(a), for $k = 2, r = 3$ CA the λ_c values occur at roughly 0.25 and 0.80. But for the ρ-classification tasks, the range of λ values required for good performance is simply a function of the task and, specifically, of ρ_c. For example, the underlying 0–1 exchange symmetry of the $\rho_c = 1/2$ task implies that if a CA exists to do the task at an acceptable performance level, then it has $\lambda \approx 1/2$. Even though this basic point does not directly invalidate the hypothesis concerning evolution to λ_c regions or claims about λ's correlation with *average* behavior, it presents problems with using ρ-classification tasks as a way to gain evidence about a generic relation between λ and computational capability. In our view, though, useful general hypotheses about evolution and computation should apply at least to computational tasks such as density classification.

A second force pushing rules to cluster close to $\lambda = 1/2$ is a "combinatorial drift" force, by which the random actions of crossover and mutation, apart from any selection force, tend to push the population towards $\lambda = 1/2$. The results of experiments measuring the relative effects of this force and the selection force in our experiment are given in another paper.

Our experimental results, along with the theoretical argument that the most successful rules for this task should have λ close to 1/2, lead us to conclude that it is not correct to interpret Figure 1(b) as evidence for the hypothesis that CA able to perform complex computations will most likely be found close to λ_c. This is an important conclusion, since Packard[29] is the only published experimental study directly linking λ with computational ability in CA.

In appreciating this, one must keep in mind that it has been known for some time that some CA, e.g., the Game of Life CA, are capable in principle of universal computation.[1] The Game of Life has $\lambda \approx \lambda_c$. Langton[23] demonstrated that another two-dimensional CA with $\lambda \approx \lambda_c$ is capable in principle of universal computation, using a construction similar to the proof of computation universality for the game of Life. However, as Langton points out, these particular constructions do not establish any necessary correlation between λ_c and the ability for complex, or even universal, computation.

As far as the GA results are concerned, we do not know what accounted for the differences between our results and those obtained in the original experiment. We speculate that the differences are due to additional mechanisms in the GA used in the original experiment that were not reported by Packard.[29] For example, the original experiment included a number of additional sources of randomness, such as the regular injection of new random rules at various λ values and a much higher mutation rate than that in our experiment.[30] These sources of randomness may have slowed the GA's search for high-fitness rules and prevented it from converging on rules close to $\lambda = 1/2$. Our experimental results and theoretical analysis give strong reason to believe that the clustering close to λ_c seen in Figure 1(b) is an artifact of mechanisms in the particular GA that was used rather than a result of any computational advantage conferred by the λ_c regions. We have also performed a wide range of additional experiments to test the robustness of our results. Not only have they held up, but these experiments have pointed to a number of mechanisms that control the interaction of evolution and computation.

WHAT CAUSES THE DIP AT $\lambda = 1/2$?

Aside from the many differences between Figure 1(b) and Figure 1(c), there is one rough similarity: the histogram shows two symmetrical peaks surrounding a central dip. We found that in our experiment this feature is due to a kind of symmetry breaking on the part of the GA; this symmetry breaking actually impedes the GA's ability to find a rule with performance at the level of the GKL rule. In short, the mechanism is the following. On each run, the best strategy found by the GA is one of two equally fit strategies:

STRATEGY 1. If the initial configuration contains a sufficiently large block of adjacent (or nearly adjacent) 1's, then increase the size of the block until the entire lattice consists of 1's. Otherwise, quickly relax to a configuration of all 0's.

STRATEGY 2. If the initial configuration contains a sufficiently large block of adjacent (or nearly adjacent) 0's, then increase the size of the block until the entire lattice consists of 0's. Otherwise, quickly relax to a configuration of all 1's.

These two strategies rely on local inhomogeneities in the initial configuration as indicators of ρ. Strategy 1 assumes that if there is a sufficiently large block of 1's initially, then the ρ is likely to be greater than $1/2$, and is otherwise likely to be less than $1/2$. Strategy 2 makes similar assumptions for sufficiently large blocks of 0's. Such strategies are vulnerable to a number of classification errors. For example, a rule might *create* a sufficiently sized block of 1's that was not present in an initial configuration with $\rho < 1/2$ and increase its size to yield an incorrect final configuration. But, as is explained by Mitchell et al.,[24] rules with $\lambda < 1/2$ (for Strategy 1) and rules with $\lambda > 1/2$ (for Strategy 2) are less vulnerable to such errors than are rules with $\lambda = 1/2$. A rule with $\lambda < 1/2$ maps more than half of the neighborhoods to 0, and thus, tends to decrease the initial ρ. Due to this it is less likely to *create* a sufficiently sized block of 1's from a low-density initial configuration.

The symmetry breaking involves deciding whether to increase blocks of 1's or blocks of 0's. The GKL rule is perfectly symmetric with respect to the increase of blocks of 1's and 0's. The GA, on the other hand, tends to discover one or the other strategy, and the one that is discovered first tends to take over the population, moving the population λ's to one or the other side of $1/2$.

The shape of the histogram in Figure 1(c) thus results from the combination of a number of forces: the selection and combinatorial drift forces described above push the population toward $\lambda = 1/2$, and the error-resisting forces just described push the population away from $\lambda = 1/2$. (Details of the epochs the GA undergoes in developing these strategies are described by Mitchell et al.[24])

It is important to understand how in general such symmetry breaking can impede an evolutionary process from finding optimal strategies. This is a subject we are currently investigating.

CONCLUSION

In this chapter we have reviewed some general ideas about the relationship between dynamical systems theory and the theory of computation. In particular, we have discussed in detail work by Langton and by Packard on the relation between dynamical behavior and computation in cellular automata. Langton investigated correlations between λ and CA behavior as measured by several statistics, and Packard's experiment was meant to directly test the hypothesis that computational ability is correlated with λ_c regions of CA rule space.

We have presented theoretical arguments and results from an experiment similar to Packard's. From these we conclude that the original interpretation of Packard's results is not correct. We believe that those original results were due to mechanisms in the particular GA used in Packard[29] rather than to intrinsic computational properties of λ_c CA.

The results presented here do not disprove the hypothesis that computational capability can be correlated with phase transitions in CA rule space.[2] Indeed, this general phenomena has already been noted for other dynamical systems, as noted in the introduction.[9] More generally, the computational capacity of evolving systems may very well require dynamical properties characteristic of phase transitions if they are to increase their complexity.

We have shown only that the published experimental support cited for hypotheses relating λ_c and computational capability in CA was not reproduced. One problem is that these hypotheses have not been rigorously formulated. If the hypotheses put forth by Langton[22] and Packard[29] are interpreted to mean that *any* rule performing complex computation (as exemplified by the $\rho = 1/2$ task) must be close to λ_c, then we have shown it to be false with our argument that correct

---

[2]There are some results concerning computation in CA and phase transitions. Individual CA have been known for some time to exhibit phase transitions with the requisite divergence of correlation length required for infinite memory capacity.[5]

performance on the $\rho = 1/2$ task requires $\lambda = 1/2$. If, instead, the hypotheses are concerned with generic, statistical properties of CA rule space—the "average" behavior of an "average" CA at a given λ—then the notion of "average behavior" must be better defined. Additionally more appropriate measures of dynamical behavior and computational capability must be formulated, and the notion of the "edge of chaos" must also be well defined.

Static parameters estimated directly from the equations of motion, as λ is from the CA rule table, are only the simplest first step at making such hypotheses and terms well-defined. λ and γ are excellent examples of the problems one encounters: their correlation with dynamical behavior is weak; they have far too much variance when viewed over CA space; and so on. What is needed is a more structural analysis that goes beyond measuring degrees of randomness and that allows one to detect the intrinsic computational capability in CA behavior. This need, all the more salient in light of the analysis given here, shows that there are problems with using any *particular* computational task to test statistical hypotheses relating λ to computational ability. Any particular task is likely to require CA with a particular range of λ values for good performance, and the particular range required is a function only of the particular task, not of intrinsic properties of regions of CA rule space.

Let us close by re-emphasizing that our studies do not preclude a future rigorous and useful definition of the phrase "edge of chaos" in the context of cellular automata. Nor do they preclude the discovery that it is associated with a CA's increased computational capability. Finally, they do not preclude adaptive systems moving to such dynamical regimes in order to take advantage of the intrinsic computational capability there. In fact, the present work is motivated by our interest in this last possibility. And the immediate result of that interest is this attempt to clarify the underling issues in the hope of facilitating new progress along these lines.

ACKNOWLEDGMENTS

This research was supported by the Santa Fe Institute, under the Core Research, Adaptive Computation and External Faculty Programs, and by the University of California, Berkeley, under contract AFOSR 91-0293. Thanks to Doyne Farmer, Jim Hanson, Erica Jen, Chris Langton, Wentian Li, Cris Moore, and Norman Packard for many helpful discussions and suggestions concerning this project. Thanks also to Emily Dickinson and Terry Jones for technical advice.

REFERENCES

1. Berlekamp, E., J. H. Conway, and R. Guy. *Winning Ways for Your Mathematical Plays.* New York, NY: Academic Press, 1982.
2. Blum, L., M. Shub, and S. Smale. "On a Theory of Computation over the Real Numbers." *Bull. AMS* **21** (1989): 1.
3. Brudno, A. A. "Entropy and the Complexity of the Trajectories of a Dynamical System." *Trans. Moscow Math. Soc.* **44** (1983): 127.
4. Chaitin, G. "On the Length of Programs for Computing Finite Binary Sequences." *J. ACM* **13** (1966): 145.
5. Creutz, M. "Deterministic Ising Dynamics." *Ann. Phys.* **167** (1986): 62.
6. Crutchfield, J. P., and N. H. Packard. "Symbolic Dynamics of One-Dimensional Maps: Entropies, Finite Precision, and Noise." *Intl. J. Theor. Phys.* **21** (1982): 433.
7. Crutchfield, J. P., and N. H. Packard. "Symbolic Dynamics of Noisy Chaos." *Physica D* **7** (1983): 201.
8. Crutchfield, J. P., and K. Young. "Inferring Statistical Complexity." *Phys. Rev. Lett.* **63** (1989): 105.
9. Crutchfield, J. P., and K. Young. "Computation at the Onset of Chaos." In *Complexity, Entropy, and the Physics of Information*, edited by W. H. Zurek. Santa Fe Institute Studies in the Sciences of Complexity, Proc. Vol. VIII, 223–269. Redwood City, CA: Addison-Wesley, 1990.
10. Crutchfield, J. P., and J. E. Hanson. "Turbulent Pattern Bases for Cellular Automata." Santa Fe Institute Report SFI-93-03-010, Santa Fe, New Mexico. Also *Physica D*: in press.
11. Farmer, D., T. Toffoli, and S. Wolfram, eds. *Cellular Automata: Proceedings of an Interdisciplinary Workshop.* Amsterdan: North Holland, 1984.
12. Feigenbaum, M. J. "Universal Behavior in Nonlinear Systems." *Physica D* **7** (1983): 16.
13. Gacs, P., G. L. Kurdyumov, and L. A. Levin. "One-Dimensional Uniform Arrays that Wash Out Finite Islands." *Probl. Peredachi. Inform.* **14** (1978): 92–98.
14. Goldberg, D. E. *Genetic Algorithms in Search, Optimization, and Machine Learning.* Reading, MA: Addison-Wesley, 1989.
15. Grassberger, P. "Toward a Quantitative Theory of Self-Generated Complexity." *Intl. J. Theor. Phys.* **25** (1986): 907.
16. Gutowitz, H. A., ed. *Cellular Automata.* Cambridge, MA: MIT Press, 1990.
17. Hanson, J. E., and J. P. Crutchfield. "The Attractor-Basin Portrait of a Cellular Automaton." *J. Stat. Phys.* **55(5/6)** (1992): 1415–1462.
18. Holland, J. H. *Adaptation in Natural and Artificial Systems*, 2nd ed. Cambridge, MA: MIT Press, 1992. (First edition, 1975).
19. Hopcroft, J. E., and J. D. Ullman. *Introduction to Automata Theory, Languages, and Computation.* Reading, MA: Addison-Wesley, 1979.

20. Kolmogorov, A. N. ""Entropy per Unit Time as a Metric Invariant of Automorphisms." *Dokl. Akad. Nauk. SSSR* **124** (1959): 754.

21. Kolmogorov, A. N. "Three Approaches to the Concept of the Amount of Information." *Prob. Info. Trans.* **1** (1965): 1.

22. Langton, C. G. "Computation at the Edge of Chaos: Phase Transitions and Emergent Computation." *Physica D* **42** (1990): 12–37.

23. Langton, C. G. "Computation at the Edge of Chaos: Phase Transitions and Emergent Computation." The University of Michigan, Ann Arbor, MI, 1991.

24. Mitchell, M., P. T. Hraber, and J. P. Crutchfield. "Revisiting the Edge of Chaos: Evolving Cellular Automata to Perform Computations." *Complex Systems* (1993): to appear.

25. Mitchell, M., J. P. Crutchfield, and P. T. Hraber. "Evolving Cellular Automata to Perform Computations: Mechanisms and Impediments." *Physica D* (1993): to appear.

26. Moore, C. "Unpredictability and Undecidability in Dynamical Systems." *Phys. Rev. Lett.* **64** (1990): 2354.

27. Omohundro, S. "Modelling Cellular Automata with Partial Differential Equations." *Physica D* **10** (1984): 128.

28. Packard, N. H. "Complexity of Growing Patterns in Cellular Automata." In *Dynamical Behavior of Automata: Theory and Applications*, edited by J. Demongeot, E. Goles, and M. Tchuente. New York: Academic Press, 1984.

29. Packard, N. H. "Adaptation Toward the Edge of Chaos." In *Dynamic Patterns in Complex Systems*, edited by J. A. S. Kelso, A. J. Mandell and M. F. Shlesinger, 293–301. Singapore: World Scientific, 1988.

30. Packard, N. H. Personal communication.

31. Preston, K., and M. Duff. *Modern Cellular Automata.* New York: Plenum, 1984.

32. Rosenfeld, A. "Parallel Image Processing Using Cellular Arrays." *Computer* **16** (1983): 14.

33. Shaw, R. "Strange Attractors, Chaotic Behavior, and Information Flow." *Z. Naturforsh.* **36a** (1981): 80.

34. Sinai, Ja. G. "On the Notion of Entropy of a Dynamical System." *Dokl. Akad. Nauk. SSSR* **124** (1959): 768.

35. Toffoli, T., and N. Margolus. *Cellular Automata Machines: A New Environment For Modeling.* Cambridge, MA: MIT Press, 1987.

36. Wolfram, S. "Universality and Complexity in Cellular Automata." *Physica D* **10** (1984): 1–35.

37. Wolfram, S., ed. *Theory and Applications of Cellular Automata.* Singapore: World Scientific, 1986.

James P. Crutchfield
Physics Department, University of California, Berkeley, California 94720

Is Anything Ever New?
Considering Emergence

Abstract: This brief essay reviews an approach to defining and then detecting the emergence of complexity in nonlinear processes. It is, in fact, a synopsis of "The Calculi of Emergence: Computation, Dynamics, and Induction."[8] [hereafter JC] that leaves out the technical details in an attempt to clarify the motivations behind the approach.

The central puzzle addressed is how we as scientists—or, for that matter, how adaptive agents evolving in populations—ever "discover" anything new in our worlds, when it appears that all we can describe is expressed in the language of our current understanding. One resolution—hierarchical machine reconstruction—is proposed. Along the way, complexity metrics for detecting structure and quantifying emergence, along with an analysis of the constraints on the dynamics of innovation, are outlined. The approach turns on a synthesis of tools from dynamical systems, computation, and inductive inference.

Complexity: Metaphors, Models, and Reality
Eds. G. Cowan, D. Pines, and D. Meltzer, SFI Studies in the
Sciences of Complexity, Proc. Vol. XIX, Addison-Wesley, 1994   **515**

1. EMERGENT?

Some of the most engaging and perplexing natural phenomena are those in which highly structured collective behavior emerges over time from the interaction of simple subsystems. Flocks of birds flying in lockstep formation and schools of fish swimming in coherent array abruptly turn together with no leader guiding the group.[32] Ants form complex societies whose survival derives from specialized laborers, unguided by a central director.[19] Optimal pricing of goods in an economy appears to arise from agents obeying the local rules of commerce.[9] Even in less manifestly complicated systems emergent global information processing plays a key role. The human perception of color in a small region of a scene, for example, can depend on the color composition of the entire scene, not just on the spectral response of spatially localized retinal detectors.[23,42] Similarly, the perception of shape can be enhanced by global topological properties, such as whether or not curves are opened or closed.[22]

How does global coordination emerge in these processes? Are common mechanisms guiding the emergence across these diverse phenomena?

Emergence is generally understood to be a process that leads to the appearance of structure not directly described by the defining constraints and instantaneous forces that control a system. Over time "something new" appears at scales not directly specified by the equations of motion. An emergent feature also cannot be explicitly represented in the initial and boundary conditions. In short, a feature emerges when the underlying system puts some effort into its creation.

These observations form an intuitive definition of emergence. For it to be useful, however, one must specify what the "something" is and how it is "new." Otherwise, the notion has little or no content, since almost any time-dependent system would exhibit emergent features.

1.1 PATTERN!

One recent and initially baffling example of emergence is deterministic chaos. In this, deterministic equations of motion lead over time to apparently unpredictable behavior. When confronted with chaos, one question immediately demands an answer—Where in the determinism did the randomness come from? The answer is that the effective dynamic, which maps from initial conditions to states at a later time, becomes so complicated that an observer can neither measure the system accurately enough nor compute with sufficient power, to predict the future behavior when given an initial condition. The emergence of disorder here is the product of both the complicated behavior of nonlinear dynamical systems and the limitations of the observer.[4]

Consider instead an example in which order arises from disorder. In a self-avoiding random walk in two-dimensions the step-by-step behavior of a particle is specified directly in stochastic equations of motion: at each time it moves one step in a random direction, except the one it just came from. The result, after some period of time, is a path tracing out a self-similar set of positions in the plane. A "fractal" structure emerges from the largely disordered step-by-step motion.

Deterministic chaos and the self-avoiding random walk are two examples of the emergence of "pattern." The new feature in the first case is unpredictability; in the second, self-similarity. The "newness" in each case is only heightened by the fact that the emergent feature stands in direct opposition to the systems' defining character: complete determinism underlies chaos and near-complete stochasticity, the orderliness of self-similarity. But for whom has the emergence occurred? More particularly, to whom are the emergent features "new?" The state of a chaotic system always moves to a unique next state under the application of a deterministic function. Surely, the system state doesn't know its behavior is unpredictable. For the random walk, "fractalness" is not in the "eye" of the particle performing the local steps of the random walk, by definition. The newness in both cases is in the eye of an observer: the observer whose predictions fail or the analyst who notes that the feature of statistical self-similarity captures a commonality across length scales.

Such comments are rather straightforward, even trivial from one point of view, in these now-familiar cases. But there are many other phenomena that span a spectrum of novelty from "obvious" to "purposeful." The emergence of pattern is the primary theme, for example, in a wide range of phenomena that have come to be labeled "pattern formation." These include, to mention only a few, the convective rolls of Bénard and Couette fluid flows, the more complicated flow structures observed in weak turbulence,[37] the spiral waves and Turing patterns produced in oscillating chemical reactions,[30,39,43] the statistical order parameters describing phase transitions, the divergent correlations and long-lived fluctuations in critical phenomena,[1,2,36] and the forms appearing in biological morphogenesis.[25,38,39]

Although the behavior in these systems is readily described as "coherent," "self-organizing," and "emergent," the patterns that appear are detected by the observers and analysts themselves. The role of outside perception is evidenced by historical denials of patterns in the Belousov-Zhabotinsky reaction, of coherent structures in highly turbulent flows, and of the energy recurrence in anharmonic oscillator chains reported by Fermi, Pasta, and Ulam. Those experiments didn't suddenly start behaving differently once these key structures were appreciated by scientists. It is the observer or analyst who lends the teleological "self" to processes which otherwise simply "organize" according to the underlying dynamical constraints. Indeed, the detected patterns are often *assumed* implicitly by the analysts via the statistics selected to confirm the patterns' existence in experimental data. The obvious consequence is that "structure" goes unseen due to an observer's biases. In some fortunate cases, such as convection rolls, spiral waves, or solitons, the functional representations of "patterns" are shown to be consistent with mathematical models of the phenomena. But these models themselves rest on a host of theoretical assumptions. It is rarely, if ever, the case that the appropriate notion of pattern is extracted from the phenomenon itself using minimally biased discovery procedures.

Briefly stated, in the realm of pattern formation, "patterns" are guessed and then verified.

1.2 INTRINSIC EMERGENCE

For these reasons, pattern formation is insufficient to capture the essential aspect of the emergence of coordinated behavior and global information processing in, for example, flocking birds, schooling fish, ant colonies, and in color and shape perception. At some basic level, though, pattern formation must play a role. The problem is that the "newness" in the emergence of pattern is always referred outside the system to some observer that anticipates the structures via a fixed palette of possible regularities. By way of analogy with a communication channel, the observer is a receiver that already has the codebook in hand. Any signal sent down the channel that is not already decodable using it is essentially noise, a pattern unrecognized by the observer.

When a new state of matter emerges from a phase transition, for example, initially no one knows the governing "order parameter." This is a recurrent conundrum in condensed matter physics, since the order parameter is the foundation for analysis and, even, further experimentation. After an indeterminant amount of creative thought and mathematical invention, one is sometimes found and then verified as appropriately capturing measurable statistics. The physicists' codebook is extended in just this way.

In the emergence of coordinated behavior, though, there is a closure in which the patterns that emerge are important *within* the system. That is, those patterns take on their "newness" with respect to other structures in the underlying system. Since there is no external referent for novelty or pattern, we can refer to this process as "intrinsic" emergence. Competitive agents in an efficient capital market control their individual production-investment and stock-ownership strategies based on the optimal pricing that has emerged from their collective behavior. It is essential to the agents' resource allocation decisions that, through the market's collective behavior, prices emerge that are accurate signals "fully reflecting" all available information.

What is distinctive about intrinsic emergence is that the patterns formed confer additional functionality which supports global information processing. Recently, examples of this sort have fallen under the rubric of "emergent computation."[12] The approach here differs in that it is based on explicit methods of detecting computation embedded in nonlinear processes. More to the point, the hypothesis in the following is that during intrinsic emergence there is an increase in intrinsic computational capability, which can be capitalized on and so can lend additional functionality.

In summary, three notions will be distinguished:

1. The intuitive definition of emergence: "something new appears";
2. Pattern formation: an observer identifies "organization" in a dynamical system; and
3. Intrinsic emergence: the system itself capitalizes on patterns that appear.

2. WHAT'S IN A MODEL?

In moving from the initial intuitive definition of emergence to the more concrete notion of pattern formation and ending with intrinsic emergence, it became clear that the essential novelty involved had to be referred to some evaluating entity. The relationship between novelty and its evaluation can be made explicit by thinking always of some observer that builds a model of a process from a series of measurements. At the level of the intuitive definition of emergence, the observer is that which recognizes the "something" and evaluates its "newness." In pattern formation, the observer is the scientist that uses prior concepts—e.g., "spiral" or "vortex"—to detect structure in experimental data and so to verify or falsify their applicability to the phenomenon. Of the three, this case is probably the most familiarly appreciated in terms of an "observer" and its "model." Intrinsic emergence is more subtle. The closure of "newness" evaluation pushes the observer inside the system. This requires in turn that intrinsic emergence be defined in terms of the "models" embedded in the observer. The observer in this view is a subprocess of the entire system. In particular, it is one that has the requisite information processing capability with which to take advantage of the emergent patterns.

"Model" is being used here in a sense that is somewhat more generous than found in daily scientific practice. There it often refers to an explicit representation—an analog—of a system under study. Here models will be seen in addition as existing implicitly in the dynamics and behavior of a process. Rather than being able to point to (say) an agent's model of its environment, one may have to excavate the "model." To do this, one might infer that an agent's responses are in co-relation with its environment, that an agent has memory of the past, that the agent can make decisions, and so on. Thus, "model" here is more "behavioral" than "cognitive."

3. THE MODELING DILEMMA

The utility of this view of intrinsic emergence depends on answering a basic question: How does an observer understand the structure of natural processes? This includes both the scientist studying nature and an organism trying to predict aspects of its environment in order to survive. The answer requires stepping back to the level of pattern formation.

A key modeling dichotomy that runs throughout all of science is that between order and randomness. Imagine a scientist in the laboratory confronted after days of hard work with the results of a recent experiment—summarized prosaically as a simple numerical recording of instrument responses. The question arises: What fraction of the particular numerical value of each datum confirms or denies the hypothesis being tested and how much is essentially irrelevant information, just "noise" or "error"?

This dichotomy is probably clearest within science, but it is not restricted to it. In many ways, this caricature of scientific investigation gives a framework for understanding the necessary balance between order and randomness that appears whenever there is an "observer" trying to detect structure or pattern in its environment. The general puzzle of discovery then is: Which part of a measurement series does an observer ascribe to "randomness" and which part to "order" and "predictability"? Aren't we all in our daily activities to one extent or another "scientists" trying to ferret out the usable from the unusable information in our lives?

Given this basic dichotomy, one can then ask: How does an observer actually make the distinction? The answer requires understanding how an observer models data—that is, the method by which elements in a representation, a "model," are justified in terms of given data.

A fundamental point is that *any* act of modeling makes a distinction between data that are accounted for—the ordered part—and data that are not described—the apparently random part. This distinction might be a null one: for example, for either completely predictable or ideally random (unstructured) sources, the data are explained by one descriptive extreme or the other. Nature is seldom so simple. It appears that natural processes are an amalgam of randomness and order. In our view it is the organization of the interplay between order and randomness that makes nature "complex." A complex process then differs from a "complicated" process, a large system consisting of very many components, subsystems, degrees of freedom, and so on. A complicated system—such as an ideal gas—need not be complex, in the sense used here. The ideal gas has no structure. Its microscopic dynamics are accounted for by randomness.

Experimental data is often described by a whole range of candidate models that are statistically and structurally consistent with the given data set. One important variation over this range of possible "explanations" is where each candidate draws the randomness-order distinction. That is, the models vary in the regularity captured and in the apparent error each induces.

It turns out that a balance between order and randomness can be reached and used to define a "best" model for a given data set. The balance is given by minimizing the model's size while minimizing the amount of apparent randomness. The first part is a version of Occam's dictum: causes should not be multiplied beyond necessity. The second part is a basic tenet of science: obtain the best prediction of nature. Neither component of this balance can be minimized alone; otherwise, absurd "best" models would be selected. Minimizing the model size alone leads to huge error, since the smallest (null) model captures no regularities; minimizing the error alone produces a huge model, which is simply the data themselves and manifestly not a useful encapsulation of what happened in the laboratory. So both model size and the induced error must be minimized together in selecting a "best" model. Typically, the sum of the model size and the error are minimized.[5,21,33,34,41]

From the viewpoint of scientific methodology the key element missing in this story of what to do with data is how to measure structure or regularity. (A particular notion of structure based on computation will be introduced shortly.) Just how structure is measured determines where the order-randomness dichotomy is set. This particular problem can be solved in principle: we take the size of the candidate model as the measure of structure. Then the size of the "best" model is a measure of the data's intrinsic structure. If we believe the data are a faithful representation of the raw behavior of the underlying process, this then translates into a measure of structure in the natural phenomenon originally studied.

Not surprisingly, this does not really solve the problem of quantifying structure. In fact, it simply elevates it to a higher level of abstraction. Measuring structure as the length of the description of the "best" model assumes one has chosen a language in which to describe models. The catch is that this representation choice builds in its own biases. In a given language some regularities can be compactly described, while in others the same regularities can be quite baroquely expressed. Change the language and the same regularities could require more or less description. And so, lacking prior God-given knowledge of the appropriate language for nature, a measure of structure in terms of the description length would seem to be arbitrary.

And so we are left with a deep puzzle, one that precedes measuring structure: How is structure discovered in the first place? If the scientist knows beforehand the appropriate representation for an experiment's possible behaviors, then the amount of that kind of structure can be extracted from the data as outlined above. In this case, the prior knowledge about the structure is verified by the data if a compact, predictive model results. But what if it is not verified? What if the hypothesized structure is simply not appropriate? The "best" model could be huge or, worse, appear upon closer and closer analysis to diverge in size. The latter situation is clearly not tolerable. An infinite model is impractical to manipulate. These situations indicate that the behavior is so new as to not fit (finitely) into current understanding. Then what do we do?

This is the problem of "innovation." How can an observer ever break out of inadequate model classes and discover appropriate ones? How can incorrect assumptions be changed? How is anything new ever discovered, if it must always be expressed in the current language?

If the problem of innovation can be solved, then, as all of the preceding development indicated, there is a framework which specifies how to be quantitative in detecting and measuring structure.

4. WHERE IS SCIENCE NOW?

Contemporary physics does not have the tools to address the problems of innovation, the discovery of patterns, or even the practice of modeling itself, since there are no physical principles that define and dictate how to measure natural structure. It is no surprise, though, that physics does have the tools for detecting and measuring complete order—equilibria and fixed point or periodic behavior—and ideal randomness—via temperature and thermodynamic entropy or, in dynamical contexts, via the Shannon entropy rate and Kolmogorov complexity.

For example, a physicist can analyze the dynamics of a box of gas and measure the degree of disorder in the molecular motion with temperature and the disorganization of the observed macroscopic state in terms of the multiplicity of associated microstates, that is, with the thermodynamic entropy. But the physicist has no analogous tools for deducing what mechanisms in the system maintain the disorder.

Then again, the raw production of information is just one aspect of a natural system's behavior. There are other important contributors to how nature produces patterns, such as how much memory of past behavior is required and how that memory is organized to support the production of information. Information processing in natural systems is a key attribute of their behavior and also how science comes to understand the underlying mechanisms.

The situation is a bit worse than a lack of attention to structure. Physics does not yet have a systematic approach to analyzing the complex information architectures embedded in patterns and processes that occur between order and randomness. This is, however, what is most needed to detect and quantify structure in nature.

The theories of phase transitions and, in particular, critical phenomena do provide mathematical hints at how natural processes balance order and randomness in that they study systems balancing different thermodynamic phases. Roughly speaking, one can think of crystalline ice as the ordered regime and of liquid water as the (relatively) disordered regime of the same type of matter (H_2O). At the phase transition, when both phases coexist, the overall state is more complex than either pure phase. What these theories provide is a set of coarse tools that describe large-scale statistical properties. What they lack are the additional, more detailed probes that would reveal, for example, the architecture of information processing embedded in those states, namely, the structure of those complex thermodynamic states. In fact, modern nonequilibrium thermodynamics can now describe the dominance of collective "modes" that give rise to the complex states found close to certain phase transitions.[16,28] What is still needed, though, is a definition of structure and a way to detect and to measure it. This would then allow us to analyze, model, and predict complex systems at the "emergent" scales.

5. A COMPUTATIONAL VIEW OF NATURE

One recent approach is to adapt ideas from the theory of discrete computation, which has developed measures of information processing structure.[6] Computation theory defines the notion of a "machine"—a device for encoding the structures in discrete processes. It has been argued that, due to the inherent limitations of scientific instruments, all an observer can know of a process in nature is a discrete-time, discrete-space series of measurements. Fortunately, this is precisely the kind of thing—strings of discrete symbols, a "formal" language—that computation theory analyzes for structure.

How does this apply to nature? Given a discrete series of measurements from a process, a machine can be constructed that is the best description or predictor of this discrete time series. The structure of this machine can be said to be the best approximation to the original process's information-processing structure, using the model size and apparent error minimization method discussed above. Once we have reconstructed the machine, we can say that we understand the structure of the process.

But what kind of structure is it? Has machine reconstruction discovered patterns in the data? Computation theory answers such questions in terms of the different classes of machines it distinguishes. There are machine classes with finite memory, those with infinite one-way stack memory, those with first-in first-out queue memory, and those with infinite random access memory, among others. When applied to the study of nature, these machine classes reveal important distinctions among natural processes. In particular, the computationally distinct classes correspond to different types of pattern or regularity.

Given this framework, one talks about the structure of the original process in terms of the complexity of the reconstructed machine. This is a more useful notion of complexity than measures of randomness, such as the Kolmogorov complexity, since it indicates the degree to which information is processed in the system, which accords more closely to our intuitions about what complexity should mean. Perhaps more importantly, the reconstructed machine describes *how* the information is processed. That is, the architecture of the machines themselves represents the organization of the information processing, that is, the intrinsic computation. The reconstructed machine is a model of the mechanisms by which the natural process manipulates information.

6. COMPUTATIONAL MECHANICS: BEYOND STATISTICS, TOWARD STRUCTURE

In JC[8] I review how a machine can be reconstructed from a series of discrete measurements of a process. Such a reconstruction is a way that an observer can model its environment. In the context of biological evolution, for example, it is clear that to survive agents must detect regularities in their environment. The degree to which an agent can model its environment in this way depends on its own computational resources and on what machine class or language it implicitly is restricted to or explicitly chooses when making a model. In JC[8] is also shown how an agent can jump out of its original assumptions about the model class and, by induction, can leap to a new model class which is a much better way of understanding its environment. This is a formalization of what is colloquially called "innovation." The inductive leap itself follows a hierarchical version of machine reconstruction.

The overall goal, then, concerns how to detect structures in the environment—how to form an "internal model"—and also how to come up with true innovations to that internal model. There are applications of this approach to time series analysis and other areas, but the main goal is not engineering but scientific: to understand how structure in nature can be detected and measured and, for that matter, discovered in the first place as wholly new innovations in one's assumed representation.

What is new in this approach? Computation theorists generally have not applied the existing structure metrics to natural processes. They have mostly limited their research to analyzing scaling properties of computational problems; in particular, to how difficulty scales in certain information processing tasks. A second aspect computation theory has dealt with little, if at all, is measuring structure in stochastic processes. Stochastic processes, though, are seen throughout nature and must be addressed at the most basic level of a theory of modeling nature. The domain of computation theory—pure discreteness, uncorrupted by noise—is thus only a partial solution. Indeed, the order-randomness dichotomy indicates that the interpretation of any experimental data has an intrinsic probabilistic component which is induced by the observer's choice of representation. As a consequence probabilistic computation must be included in any structural description of nature. A third aspect computation theory has considered very little is measuring structure in processes that are extended in space. A fourth aspect it has not dealt with traditionally is measuring structure in continuous-state processes. If computation theory is to form the foundation of a physics of structure, it must be extended in at least these three ways. These extensions have engaged a number of workers in dynamical systems recently, but there is much still to do.[3,6,7,17,27,29,44]

7. THE CALCULI OF EMERGENCE

In JC[8] I focus on temporal information processing and the first two extensions—probabilistic and spatial computation—assuming that the observer is looking at a series of measurements of a continuous-state system whose states an instrument has discretized. The phrase "calculi of emergence" in its title emphasizes the tools required to address the problems that intrinsic emergence raises. The tools are (i) dynamical systems theory with its emphasis on the role of time and on the geometric structures underlying the increase in complexity during a system's time evolution, (ii) the notions of mechanism and structure inherent in computation theory, and (iii) inductive inference as a statistical framework in which to detect and innovate new representations.

First, In JC^8 I define a complexity metric that is a measure of structure in the way discussed above. This is called "statistical complexity," and it measures the structure of the minimal machine reconstructed from observations of a given process in terms of the machine's size. Second, it describes an algorithm—ϵ-machine reconstruction—for reconstructing the machine, given an assumed model class. Third, it describes an algorithm for innovation—called "hierarchical machine reconstruction"—in which an agent can inductively jump to a new model class. Roughly speaking, hierarchical machine reconstruction detects regularities in a *series* of increasingly accurate models. The inductive jump to a higher computational level occurs by taking those regularities as the new representation. In the bulk of JC^8 I analyze several examples in which these general ideas are put into practice to determine the intrinsic computation in continuous-state dynamical systems, recurrent hidden Markov models, and cellular automata. It concludes with a summary of the implications of this approach to detecting and understanding structure in nature.

The goal throughout is a more refined appreciation of what "emergence" is, both when new computational structure appears over time and when agents with improved computational and modeling ability evolve. The interplay between computation, dynamics, and induction emphasizes a trinity of conceptual tools required for studying the emergence of complexity; presumably this is a setting that has a good chance of providing empirical application.

8. DISCOVERY VERSUS EMERGENCE

The arguments and development turn on distinguishing several different levels of interpretation: (i) a system behaves, (ii) that behavior is modeled, (iii) an observer detects regularities and builds a model based on prior knowledge, (iv) a collection of agents model each other and their environment, and (v) scientists create artificial universes and try to detect the change in computational capability by constructing their own models of the emergent structures. It is all too easy to conflate two or more of these levels, leading to confusion or, worse, to subtle statements seeming vacuous.

It is helpful to draw a distinction between discovery and emergence. The level of pattern formation and the modeling framework of computational mechanics concern discovery. Above, it was suggested that innovation based on hierarchical machine reconstruction is one type of discovery, in the sense that new regularities across increasingly accurate models are detected and then taken as a new basis for representation. Discovery, though, is not the same thing as emergence, which at a minimum is dynamical: over time, or over generations in an evolutionary system, something new appears. Discovery, in this sense, is atemporal: the change in state and increased knowledge of the observer are not the focus of the analysis activity; the products of model fitting and statistical parameter estimation are.

In contrast, emergence concerns the *process* of discovery. Moreover, intrinsic emergence puts the subjective aspects of discovery *into* the system under study. In short, emergence pushes the semantic stack down one level. In this view analyzing emergence is more objective than analyzing pattern formation in that detecting emergence requires modeling the dynamics of discovery, not just implementing a discovery procedure.

The arguments to this point can be recapitulated by an operational definition of emergence. A process undergoes emergence if at some time the architecture of information processing has changed in such a way that a distinct and more powerful level of intrinsic computation has appeared that was not present in earlier conditions.

It seems, upon reflection, that our intuitive notion of emergence is not captured by the "intuitive definition" given in the first section. Nor is it captured by the somewhat refined notion of pattern formation. "Emergence" is meaningless unless it is defined within the context of processes themselves; the only well-defined notion of emergence would seem to be intrinsic emergence. Why? Simply because emergence defined without this closure leads to an infinite regress of observers detecting patterns of observers detecting patterns.... This is not a satisfactory definition, since it is not finite. The regress must be folded into the system; it must be immanent in the dynamics. When this happens, complexity and structure are no longer referred outside, no longer relative and arbitrary; they take on internal meaning and functionality.

---

9. EVOLUTIONARY MECHANICS

Where in science might a theory of intrinsic emergence find application? Are there scientific problems that at least would be clarified by the computational view of nature outlined here?

In several ways the contemporary debate on the dominant mechanisms operating in biological evolution seems ripe. Is anything ever new in the biological realm? The empirical evidence is interpreted as a resounding "yes." It is often heard that organisms today are more complex than in earlier epochs. But how did this emergence of complexity occur? Taking a long view, at present there appear to be three schools of thought on what the guiding mechanisms are in Darwinian evolution that produce the present diversity of biological structure and that are largely responsible for the alteration of those structures.

Modern evolutionary theory continues to be governed by Darwin's view of the natural selection of individuals that reproduce with variation. This view emphasizes the role of fitness selection in determining which biological organisms appear. But there are really two camps: the Selectionists, who are Darwin's faithful heirs now cognizant of genetics, and the Historicists, who espouse a more anarchistic view.

The Selectionists hold that structure in the biological world is due primarily to the fitness-based selection of individuals in populations whose diversity is maintained by genetic variation.[24] In a sense, genetic variation is a destabilizing mechanism that provides the raw diversity of structure. Natural selection then is a stabilizing dynamic that acts on the expression of that variation. It provides order by culling individuals based on their relative fitness. This view identifies a source of new structures and a mechanism for altering one form into another. The adaptiveness accumulated via selection is the dominant mechanism guiding the appearance of structure.

The second, anarchistic camp consists of the Historicists who hold fast to the Darwinian mechanisms of selection and variation, but emphasize the accidental determinants of biological form.[15,26] What distinguishes this position from the Selectionists is the claim that major changes in structure can be and have been nonadaptive. While these changes have had the largest effect on the forms of present-day life, at the time they occurred they conferred no survival advantage. Furthermore, today's existing structures needn't be adaptive. They reflect instead an accidental history. One consequence is that a comparative study of parallel earths would reveal very different collections of life forms on each. Like the Selectionists, the Historicists have a theory of transformation. But it is one that is manifestly capricious or, at least, highly stochastic with few or no causal constraints. For this process of change to work, the space of biological structures must be populated with a high fraction which are functional.

Lastly, there are the Structuralists whose goal is to elucidate the "principles of organization" that guide the appearance of biological structure. They contend that energetic, mechanical, biomolecular, and morphogenetic constraints limit the infinite range of possible biological form.[10,11,14,20,38,40] The constraints result in a relatively small set of structure archetypes. These are something like the Platonic solids in that they pre-exist, before any evolution takes place. Natural selection then plays the role of choosing between these "structural attractors" and possibly fine-tuning their adaptiveness. Darwinian evolution serves, at best, to fill the waiting attractors or not depending on historical happenstance. Structuralists offer up a seemingly testable claim about the ergodicity of evolutionary processes: given an ensemble of earths, life would have evolved to a similar collection of biological structures.

The Structuralist tenets are at least consistent with modern thermodynamics.[16,28] In large open systems, energy enters at low entropy and is dissipated. Open systems organize largely due to the reduction in the number of active degrees of freedom caused by the dissipation. Not all behaviors or spatial configurations can be supported. The result is a limitation of the collective modes, cooperative behaviors, and coherent structures that an open system can express. The Structuralist view is a natural interpretation of the many basic constraints on behavior and pattern indicated by physics and chemistry. For example, the structures formed in open systems such as turbulent fluid flows, oscillating chemical reactions, and morphogenetic systems are the product of this type of macroscopic pattern formation. Thus, open systems offer up a limited palette of structures to selection. The more limited the palette, the larger the role for "principles of organization" in guiding the emergence of life as we know it.

What is one to think of these conflicting theories of the emergence of biological structure? In light of the preceding sections there are several impressions that the debate leaves an outsider with.

1. Natural selection's culling of genetic variation provides the Selectionists with a theory of transformation. But the approach does not provide a theory of structure. Taking the theory at face value, in principle one can estimate the time it takes a *given* organism to change. But what is the mean time under the evolutionary dynamic and under the appropriate environmental pressures for a hand to appear on a fish? To answer this, one needs a measure of the structure concerned and of the functionality it does or does not confer.

2. The Historicists also have a theory of transformation, but they offer neither a theory of structure, nor, apparently, a justification for a high fraction of functionality over the space of structures. Perhaps more disconcerting, though, in touting the dominance of historical accident, the Historicists advocate an antiscientific position. This is not to say that isolated incidents do not play a role; they certainly do. But it is important to keep in mind that the event of a meteor crashing into the earth is extra-evolutionary. The explanation of its occurrence is neither the domain of evolutionary theory, nor is its occurrence likely ever to be explained by the principles of dynamics: it just happened, a consequence of particular initial conditions. Such accidents impose constraints; they are not an explanation of the biological response.

3. In complementary fashion, the Structuralists do not offer a theory of transformation. Neither do they, despite claims for the primacy of organization in evolutionary processes, provide a theory of structure itself. In particular, the structure archetypes are neither analyzed in terms of their internal components, nor in terms of system-referred functionality. Considering these lacks, the Structuralist hope for "deep laws" underlying biological organization is highly reminiscent of Chomsky's decades-long search for "deep structures" as linguistic universals, without a theory of cognition. The ultimate failure of this search[18] suggests a reconsideration of fundamentals rather than optimistic forecasts of Structuralist progress.

The overwhelming impression this debate leaves, though, is that there is a crying need for a theory of biological structure and a qualitative dynamical theory of its emergence.[13] In short, the tensions between the positions are those (i) between the order induced by survival dynamics and the novelty of individual function and (ii) between the disorder of genetic variation and the order of developmental processes. Is it just an historical coincidence that the structuralist-selectionist dichotomy appears analogous to that between order and randomness in the realm of modeling? The main problem, at least to an outsider, does not reduce to showing that one or the other view is correct. Each employs compelling arguments and often empirical data as a starting point. Rather, the task facing us reduces to developing a synthetic theory that balances the tensions between the viewpoints. Ironically, evolutionary processes themselves seem to do just this sort of balancing, dynamically.

The computational mechanics of nonlinear processes can be construed as a theory of structure. Pattern and structure are articulated in terms of various types of machine classes. The overall mandate is to provide both a qualitative and a quantitative analysis of natural information processing architectures. If computational mechanics is a theory of structure, then innovation via hierarchical machine reconstruction is a computation-theoretic approach to the transformation of structure. It suggests one mechanism with which to study what drives and what constrains the appearance of novelty. The next step, of course, would be to fold hierarchical machine reconstruction into the system itself, resulting in a dynamics of innovation, the study of which might be called "evolutionary mechanics."

10. THE MECHANICS

By way of summarizing the main points, let's question the central assumption of this approach to emergence.

Why talk about "mechanics"? Aren't mechanical systems lifeless, merely the sum of their parts? One reason is simply that scientific explanations must be given in terms of mechanisms. Explanations and scientific theories without an explicit hypothesis of the underlying causes—the mechanisms—are neither explanations nor theories, since they cannot claim to entail falsifiable predictions.[31] Another, more constructive reason is that modern mathematics and physics have made great strides this century in extending the range of Newtonian mechanics to ever more complex processes. When computation is combined with this, one has in hand a greatly enriched notion of mechanism.

It might seem implausible that an abstract "evolutionary mechanics" would have anything to contribute to (say) biological evolution. A high-level view at least suggests a fundamental, if indirect, role. By making a careful accounting of where the observer and system-under-study are located in various theories of natural phenomena, a certain regularity appears which can be summarized by a hierarchy of mechanics. The following list is given in the order of increasing attention to the context of observation and modeling in a classical universe. The first two are already part of science proper; the second two indicate how computation and innovation build on them.

1. Deterministic mechanics (dynamical systems theory): The very notions of cause and mechanism are defined in terms of state space structures. This is Einstein's level: the observer is entirely outside the system-under-study.

2. Statistical mechanics (probability theory): Statistical mechanics is engendered by deterministic mechanics largely due to the emergence of irreducible uncertainty. This occurs for any number of reasons. First, deterministic mechanical systems can be very large, too large in fact to be usefully described in complete detail. Summarizing the coarse, macroscopic properties is the only manageable goal. The calculus for managing the discarded information is probability theory. Second, deterministic nonlinear systems can be chaotic, communicating unseen and uncontrollable microscopic information to affect observable behavior.[35] Both of these reasons lead to the necessity of using probabilistic summaries of deterministic behavior to collapse out the irrelevant and accentuate the useful.

3. Computational mechanics (theory of structure for statistical mechanics): As discussed at some length, it is not enough to say that a system is random or ordered. What is important is how these two elements, and others, interact to produce complex systems. The information processing mechanisms distinguished by computation theory give a (partial) basis for being more objective about detecting structure, quantifying complexity, and the modeling activity itself.

4. Evolutionary mechanics (dynamical theory of innovation): As noted above, evolutionary mechanics concerns how genuine novelty occurs. This is the first level at which emergence takes on its intrinsic aspect. Building on the previous levels, the goal is to delineate the constraints guiding and the forces driving the emergence of complexity.

A typical first question about this hierarchy is "Where is quantum mechanics?" The list just given assumes a classical physical universe. Therefore, quantum mechanics is not listed despite its undeniable importance. It would appear, however, either as the most basic mechanics, preceding deterministic mechanics, or at the level of statistical mechanics, since that is the level at which probability first appears. In a literal sense, quantum mechanics is a theory of the deterministic dynamics of complex "probabilities" that can interfere over spacetime. The interference leads to new phenomena, but the goals of and manipulations used in quantum mechanics are not so different from that found in stochastic processes and so statistical mechanics. My own prejudice in these issues will be resolved once a theory of measurement of nonlinear processes is complete. There are several difficulties that lie in the way. The effect of measurement distortion can be profound, for example, leading to irreducible indeterminacy in completely deterministic systems.[7]

So is anything ever new? I would answer "most definitely," and so reject Parmenidean changelessness in favor of novelty emerging, according to Heraclitus, from an "attunement of opposite tensions." With careful attention to the location of the observer and the system-under-study, with detailed accounting of intrinsic computation, with quantitative measures of complexity, we can analyze the patterns, structures, and novel information processing architectures that emerge in nonlinear processes. In this way, we demonstrate that something new has appeared.

11. ACKNOWLEDGMENTS

Many thanks are due to Melanie Mitchell for a critique of "The Calculi of Emergence: Computation, Dynamics, and Induction."[8] that led to the present essay. The author is also indebted to Dan McShea for his thoughts on the evolutionary trichotomy. Comments from Lisa Borland, Don Glaser, Jim Hanson, Blake LeBaron, and Dan Upper are gratefully acknowledged. This work was supported in part by AFOSR 91-0293 and ONR N00014-92-J-4024.

REFERENCES

1. Bak, P., and K. Chen. "Self-Organized Criticality." *Physica A* **163** (1990): 403–409.
2. Binney, J. J., N. J. Dowrick, A. J. Fisher, and M. E. J. Newman. *The Theory of Critical Phenomena.* Oxford: Oxford University Press, 1992.
3. Blum, L., M. Shub, and S. Smale. "On a Theory of Computation Over The Real Numbers." *Bull. AMS* **21** (1989): 1.
4. Crutchfield, J. P., N. H. Packard, J. D. Farmer, and R. S. Shaw. "Chaos." *Sci. Am.* **255** (1986): 46.
5. Crutchfield, J. P., and B. S. McNamara. "Equations of Motion from a Data Series." *Complex Systems* **1** (1987): 417.
6. Crutchfield, J. P., and K. Young. "Inferring Statistical Complexity." *Phys. Rev. Let.* **63** (1989): 105.
7. Crutchfield, J. P. "Unreconstructible At Any Radius." *Phys. Lett. A* **171** (1992): 52–60.
8. Crutchfield, J. P. "The Calculi of Emergence: Computation, Dynamics, and Induction." *Physica D* (1994): in press.
9. Fama, E. F. "Efficient Capital Markets II." *J. Finance* **46** (1991): 1575–1617.
10. Fontana, W., and L. Buss. "'The Arrival of the Fittest': Toward a Theory of Biological Organization." *Bull. Math. Biol.* **56** (1994): 1–64.

11. Fontana, W., and L. Buss. "What Would be Conserved If The Tape Were Played Twice?" *Proc. Nat. Acad. Sci.* **91** (1994): 757–761.

12. Forrest, S. "Emergent Computation: Self-organizing, Collective, and Cooperative Behavior in Natural and Artificial Computing Networks: Introduction to the Proceedings of the Ninth Annual CNLS Conference." *Physica D* **42** (1990): 1–11.

13. Goodwin, B., and P. Sanders, eds. *Theoretical Biology: Epigenetic and Evolutionary Order from Complex Systems.* Baltimore, MD: Johns Hopkins University Press, 1992.

14. Goodwin, B. "Evolution and The Generative Order." In *Theoretical Biology: Epigenetic and Evolutionary Order from Complex Systems,* edited by, B. Goodwin and P. Sanders, 89–100. Baltimore, MD: Johns Hopkins University Press, 1992.

15. Gould, S. J. *Wonderful Life.* New York: Norton, 1989.

16. Haken, H. *Synergetics, An Introduction,* 3rd edition. Berlin: Springer-Verlag, 1983.

17. Hanson, J. E., and J. P. Crutchfield. "The Attractor-Basin Portrait of a Cellular Automaton." *J. Stat. Phys.* **66** (1992): 1415.

18. Harris, R. A. *The Linguistic Wars.* New York: Oxford University Press, 1993.

19. Holldobler, B., and E. O. Wilson. *The Ants.* Cambridge, MA: Belknap Press of Harvard University Press, 1990.

20. Kauffman, S. A. *Origins of Order: Self-Organization and Selection in Evolution.* New York: Oxford University Press, 1993.

21. Kemeny, J. G. "The Use of Simplicity in Induction." *Phil. Rev.* **62** (1953): 391.

22. Kovacs, I., and B. Julesz. "A Closed Curve is Much More Than an Incomplete One—Effect of Closure in Figure Ground Segmentation." *Proc. Nat. Acad. Sci.* **90** (1993): 7495–7497.

23. Land, E. "The Retinex." *Am. Scientist* **52** (1964): 247–264.

24. Maynard Smith, J. *Evolutionary Genetics.* Oxford: Oxford University Press, 1989.

25. Meinhardt, H. *Models of Biological Pattern Formation.* London: Academic Press, 1982.

26. Monod, J. *Chance and Necessity: An Essay on the Natural Philosophy of Modern Biology.* New York: Vintage Books, 1971.

27. Moore, C. "Real-Valued, Continuous-Time Computers: A Model of Analog Computation, Part I." Technical Report 93-04-018, Santa Fe Institute, 1993.

28. Nicolis, G., and I. Prigogine. *Self-Organization in Nonequilibrium Systems.* New York: Wiley, 1977.

29. Nordahl, M. G. "Formal Languages and Finite Cellular Automata." *Complex Systems* **3** (1989): 63.

30. Ouyang, Q., and H. L. Swinney. "Transition From a Uniform State to Hexagonal and Striped Turing Patterns." *Nature* **352** (1991): 610–612.

31. Popper, K. R. *The Logic of Scientific Inquiry.* New York: Basic Books, 1959.

32. Reynolds, C. W. "Flocks, Herds, and Schools: A Distributed Behavioral Model." *Computer Graphics* **21** (1987): 25–34.
33. Rissanen, J. "Modeling by Shortest Data Description." *Automatica* **14** (1978): 462.
34. Rissanen, J. *Stochastic Complexity in Statistical Inquiry.* Singapore: World Scientific, 1989.
35. Shaw, R. "Strange Attractors, Chaotic Behavior, and Information Flow." *Z. Naturforsh.* **36a** (1981): 80.
36. Stanley, H. E. *Introduction to Phase Transitions and Critical Phenomena.* Oxford: Oxford University Press, 1971.
37. Swinney, H. L., and J. P. Gollub, eds. *Hydrodynamic Instabilities and the Transition to Turbulence,* Berlin: Springer-Verlag, 1981.
38. Thompson, D. W. *On Growth and Form.* Cambridge: Cambridge University Press, 1917.
39. Turing, A. M. "The Chemical Basis of Morphogenesis." *Trans. Roy. Soc., Series B* **237** (1952): 5.
40. Waddington, C. H. *The Strategy of the Genes.* London: Allen and Unwin, 1957.
41. Wallace, C. S., and D. M. Boulton. "An Information Measure for Classification." *Comput. J.* **11** (1968): 185.
42. Wandell, B. A. "Color Appearance: The Effects of Illumination and Spatial Pattern." *Proc. Nat. Acad. Sci.* **10** (1993): 2458–2470.
43. Winfree, A. T. *The Geometry of Biological Time.* Berlin: Springer-Verlag, 1980.
44. Wolfram, S. "Computation Theory of Cellular Automata." *Comm. Math. Phys.* **96** (1984): 15.

DISCUSSION

BAK: I'd like to advertise a definition of complexity by Zhang; he's a Chinese-Italian-Swiss physicist, known from the KPZ models, and other things. What he does is that he takes his string of bits, say in a three- or four-dimensional space-time, and he does all possible coarse-graining, and he calculates the entropy, or the information, on each coarse-graining, and integrates over all of them. That gives you a measure that does precisely what you want, that gives you something that has low complexity if it's disordered and low complexity if it's ordered, and high complexity if it's precisely critical. And one can think about what happens in evolution: that is sort of the quantity that will increase. He does all possible coarse-graining, and calculates some $p \log p$, you know, the information of that and integrates that summed over all coarse grainings.

GELL-MANN: All equally, all coarse grainings? Treated by what metric?

BAK: Yes. I don't now whether it's on a logarithmic scale; I'll have to look that up.

CRUTCHFIELD: Well, how does he deal with the uniform distribution? I presume the framework addresses arbitrarily small ϵ, since the entropy of the uniform continuous distribution is infinite.

BAK: Yeah. There's going to be some lower cutoff.
    You can precisely take your "strength," and feed into it, and you'll find precisely that when you have some critical configuration, you'll have the highest measure of complexity.
    What he did specifically is to take the configuration of the two-dimensional Ising model at any temperature, and found that it has the maximum complexity precisely at the critical point. So I'd like to ask, suppose you do the same, take the configuration of the Ising model at the critical point in two dimensions, and plop into your formalism, what did you get?

CRUTCHFIELD: Well, this goes somewhat indirectly. The complexity of the model is related to the mutual information. It sounds like he's computing a mutual information; you'll have to write out the exact expression.

BAK: Yes, he's calculating some kind of a sum over the correlation of all length scales.

CRUTCHFIELD: If Zhang is summing the Shannon entropy over all coarse-grainings then the resulting quantity sounds similar, if not identical, to a quantity that Norman Packard and I defined a decade ago, the total excess entropy. Rob Shaw explored this further and coined the phrase "stored information" for it. Which reminds me of some other more recent work related to your comments that he did in the late 1980s, which remains unpublished. Shaw looked at one type of mutual information within 2-D Ising model spin configurations as a function of temperature, and showed that it was maximized at the phase transition. In any phase transition with infinite correlations, the various mutual informations will be large since the systems exhibit the largest memory effects there.

LLOYD: But that also implies that your complexity is bounded below by mutual information.

CRUTCHFIELD: Yes. The statistical complexity is bounded below by the mutual information. Of course, you have to specify what probes you've chosen before defining the mutual information. Typically, two probes separated in space or time, as appropriate, have been used for estimating mutual information. If you use all possible probes and their combinations as a basis for mutual information then you obtain the total excess entropy. It is useful to note that, although the statistical complexity is defined and estimated with the same ease (or difficulty) as the excess entropy, it is a better indicator of the intrinsic computation in a process than either excess entropy or the more restricted types of mutual information.

BAK: But this is certainly much more complex; this is a more complete description of it.

CRUTCHFIELD: No, actually I would argue that Zhang's approach and the excess entropy are not complete if you don't acknowledge your computational models: the fact that there's nondeterminism; the fact that you have transition probabilities; the fact that you have memories organized as a stack. In fact, this is one of the points I wanted to make but I didn't have a chance to. Simply looking at systems which go through phase transitions will provide very good examples to test this new approach to detecting structure.

BAK: You don't have to do that; you can take any string of bits and feed into it.

CRUTCHFIELD: That is also the starting point of our procedure. But the main point is that at the onset of chaos, just because some complexity is maximized, that's not enough. It's also not enough to just give the critical exponents. Just because you have an infinite correlation, that's not enough. The memory, the correlation, is organized in very specific ways to do different types of computation, and that's a distinguishing feature. By demonstrating the mechanism, the machine, you can refine the universality class.

BAK: But what I'm saying is there's a deeper level of describing that.

CRUTCHFIELD: I'm saying that Zhang's measure, the way you've defined it, misses a number of computationally distinct properties. Computation theory is vastly more sophisticated in its understanding of structure than is contemporary statistical mechanics; I include in the latter the use of Shannon entropy, mutual information, and excess entropy.

KAUFFMAN: If you took that specific case, for example, presumably you would wind up saying that at the critical state for the spin glass, for the Ising model, if you're going to make a finite computational model of it, you might be driven to one of your higher classes?

CRUTCHFIELD: Yes. I think the Ising system would be a great example for that. Creutz has invented a two-dimensional Ising automaton; the suggestion then is to analyze that model and try to figure out its intrinsic computational properties and so how to define a measure of complexity. I didn't talk about evolving automata to recognize cellular automata patterns, but in that context there is a measure of space-time complexity. Jim Hanson and I generalized the approach I outlined in my talk to apply to spatio-temporal systems like cellular automata. The paper appeared in the *Journal of Statistical Physics* in March.

KAUFFMAN: Why shouldn't one start higher up in the hierarchy, and build a simpler model with field parameters that give up; you could sort of say, "Ah, the world's a little bit undeterministic."

CRUTCHFIELD: Right, that's a very good point. Why not start with a universal Turing machine? Perhaps I didn't state explicitly, but will now, why the approach we've taken starts at the bottom of the computational hierarchy. Bennett, Kolmogorov, and many others start in effect at the top with universal deterministic Turing machines. There's a very good reason that you don't want to do that. Namely, the cost—the computational cost for an observer – grows, on average as you move up the hierarchy. In fact, this comment was made this morning by someone. It becomes more difficult to answer questions using more sophisticated model classes, since they provide a richer descriptive framework. And that's the reason why we start at the lowest level, and have the available information drive the innovation process up the hierarchy if this is needed to find a finite model.

KAUFFMAN: What did you mean by cost of computation?

CRUTCHFIELD: If you want to make better and better approximations of the simple, nondeterministic two-state stochastic process mentioned in the talk, your internal model keeps growing. That's costing you. That means, if you're a biological organism, you're having to devote physical resources, metabolic resources, to using that model to make predictions.

KAUFFMAN: Then you're saying it's cheap to jump levels. You're saying that you're saving costs because your state's growing towards infinity so it's jumped up one level. But that doesn't answer the question why it shouldn't start at a higher level, yet.

CRUTCHFIELD: Well, there's an intermediate point here. The point is that the innovation in that example was the discovery of a counter register. Now, I can't make this argument in full generality yet, but I imagine that in any implementation that takes physical degrees of freedom and couples them together nonlinearly, it's easier to construct a counter than it is to make an arbitrary-sized finite automaton. I can't give you a biological or physical example, I can't design a set of differential equations for which that's true, since the appropriate real-valued computation theory isn't available yet. But a counter register is a more compact representation of the information processing needed to compute the measure over the sequences generated by that simple nondeterministic process. That is why I argue that an agent build of physical components would gain by innovating a counter register, that is, go beyond its limited, initial choice of modeling class.

KAUFFMAN: But that's the worry. Then why don't I start with the richer class of systems, higher up in your hierarchy, and work my way down?

CRUTCHFIELD: To be clearer about the point I just made, I'm going to have to give you more details. The short summary is that if you start out using a counter machine class, you'll have so biased yourself in terms of specific computational capabilities that it will be difficult to find the correct particular model: it might have been better to choose a queue machine class.

KAUFFMAN: But I would have been able to do it on a lower level?

CRUTCHFIELD: Starting at the lower level, you learn what is in the data, refining your assumptions only when the data justify it and in such a way that the innovation of which higher level to use is not biased.

KAUFFMAN: So that the higher level system misleads you with respect to getting an even simpler model at the lower level, and that's another cost of computation.

CRUTCHFIELD: Right. And you might think, "Well, counters are more general." Of course, but the model space is much bigger, in a sense. The computational hierarchy is organized such that, as you move up, the languages, the models, are more expressive, the search space is vastly larger higher up.

KAUFFMAN: So finding a good one is tougher.

CRUTCHFIELD: Right. So hierarchical machine reconstruction is a kind of bootstrap solution to this problem.

GELL-MANN: What's wrong with quantum mechanics?

CRUTCHFIELD: Oh, the potshot in the Whitehead quote. I'm sorry, but that's another talk about how deterministic measurements of deterministic processes can produce irreducible uncertainty even on the smallest space-time scales. This result leads me to question our interpretation of the mechanisms responsible for quantum phenonema.

ANDERSON: Quantum mechanics still has a lot of suprises.

CRUTCHFIELD: I'd have to switch gears here to address this properly. The bottom line is that I'm a diehard determinist still fascinated by the complexity of the classical universe.

General Discussion

John H. Holland, Chairperson
Division of Computer Science and Engineering, University of Michigan, Ann Arbor, MI 48109

Review and Remarks on Applications

HOLLAND: We need first to define what is an application. To me the notion of what's an application depends on what I think an explanation is. The definition of application depends on the prior definition of explanation. There are two broad notions of explanation. One is based on an explicit presentation in terms of tables. The other is an explicit presentation in terms of generators or algorithms.

If I go into an area that is relatively neutral to us, psychology, then I can trace a couple of lines of development. On one side, where psychology is striving to be a hard science, people hit on a very nice measure called reaction time. So you get a whole bunch of measurements on how long it takes to run a maze under different circumstances and you get a bunch of functions that look like power laws, so you get the notion of measuring the learning ability of something by how rapidly the time to run a maze decreases. So now I've got classic learning curves. That, for a long time, was the psychologist's definition of learning ability. Unfortunately, it turns out that under that definition of learning, rats are somewhat better than people so you might feel that you've missed some component of what you think you mean by learning. You offer strong rewards in both cases. Intermittent reinforcement is better than continuous reinforcement and you offer just enough to keep the subjects from starving in both cases.

Complexity: Metaphors, Models, and Reality
Eds. G. Cowan, D. Pines, and D. Meltzer, SFI Studies in the
Sciences of Complexity, Proc. Vol. XIX, Addison-Wesley, 1994   **541**

So, in the realm of psychology, this notion of an explicit approach to explanation led to the notion of reaction times which led to various kinds of curves which spilled over into artificial intelligence and eventually led to expert systems in which you try to be explicit about all the rules. The other approach, which was much slower and has been very sporadic and this is only an example, is where we try to perceive patterns in terms of various kinds of algorithms, so we ask, "What are sufficient generators for this kind of behavior?" In AI this led very early on to a learning system called Samuels' Checker Player. That part of AI went very much in this direction and it's only in the last decade or less that the field, now called machine learning because that's what Samuels called it way back when, became an important part of artificial intelligence and what is now called cognitive science is the feedback from AI back into psychology.

We have these two approaches to explanation; however you take this notion of explanation, it will have lots to do with what you consider to be an application of your theory or approach. I want now to reiterate that simply because systems are formally equivalent, whether you meant it in the axiomatic sense or some other sense, does not imply that they are semantically equivalent. They are not equally useful in trying to explore the problem. Here I point to two things. First of all, if I have any finite state system, it can have all sorts of look-aheads, whatever I like; it can always be reduced to an stimulus-response (S-R) system which is a table or a set of rules. These are, in a sense, formally equivalent systems. But they are not equivalent in the sense of my ability to explore subjects like evolution and adaptation. Another formal result is the not-so-well-known theory of Godel's, which is the following: Pick a theorem that you like, that's important, say, the Pythagorean theorem. If I look at the standard five axioms of Euclidean geometry, there is a shortest proof of the Pythagorean theorem. Godel's theorem, or meta-theorem, states that I can always find a formally equivalent axiom system for which the Pythagorean theorem is moved out as far as I like, say, for a million axioms. But if the number is large enough, I may not be able to apply Godel's theorem for lack of time. This is simply to illustrate the notion that formal equivalence does not mean semantic equivalence.

QUESTION: What about different computation systems?

HOLLAND: If we ask about a computationally universal system, like the Game of Life, I can get into conditions where particles move through and do not settle down, and the system remains perpetually novel. That's something we can discuss later.

We talked about hierarchy. Let me give you an illustrative example of hierarchy from classifier systems. Here we have a rule: I will respond to any bit string that starts with a one with action one. I don't care about hash marks. There's another rule here which responds to the same subset of this set of signals. It starts with a one but makes an additional requirement for a zero in the next position and responds, then, with action two. These rules, in a certain sense, are not consistent with one another and they undergo competition and the competition is based on how specific the condition is. The more specific the condition is, the more likely it is to win the competition. The more information it uses, the more it enforces the system, so to speak. Then, what I find is that I have what would be called a default rule, which is that the rule I apply in the absence of further information is the first but, if I get additional information, then the second rule takes over and displaces the first.

The point to be made is this: if I generate messages or signals at random, just as a test case, and ask what proportion of the signals are going to statisfy the first condition, the answer is one-half. If I ask about the second one, it drops to one-eighth. Obviously, one can imagine a hierarchy of rules like this. Which rule should I be constructing early on? When this system is just starting out, which rule does it make sense to construct? If I take a statistician's view that a certain number of samples is needed to gain some confidence in the rule, then clearly it takes four times as long to have the same level of confidence in rule two as in rule one. So, if I want a system to evolve gracefully, what that system should be doing is starting with rules that give it a bit of statistical edge. Those are the ones that will prevail first because I have to be lucky to hit one of the more specific rules if I don't have a sample.

So, if you watch a bunch of such systems over a period of time, some are using rules that are almost useless and are quickly displaced by systems that find simple, coarse-grained regularities. As the system goes on, it begins to use new rules, with recombination and mutation, to get more specific rules, progressively finer grained and more specifically defined, and begins to build a default hierarchy. That way, if you look at these systems, their capactiy expands in a relatively graceful way. You don't throw away the defaults. You try, instead, to find exceptions and refinements. Those are the basic ground points.

I want now to say two things. About interdisciplinary applications: I would claim, and I've seen this often enough in my own university, that once you start to get compartmentalized on these views that are essentially cross-disciplinary, you wind up with things that can often be very misleading. If we're talking about something like sustainable farming, it seems natural from one point of view that I will, as a self-interested agent, try to find those in a sustainable way. In fact, it doesn't happen. Why doesn't it happen? Because I have to look over into another discipline in order to see that, in fact, the whaling ships are not owned by the captain, they're owned by capitalists and these capitalists want to maximize the return on their investment and they know that capital can migrate very easily. So they're going to get the highest rate of return they can until the whales are all gone and then reinvest in another industry. It's not quite as easy as that but, if you take things like that into account, you're going to get a different picture than if you say that self-interested agents will not kill off all the whales. It's just not true. So one of the reasons that computer models are helpful is that we can pursue some of these interactions and still have relatively rigorous definitions and I'm really pointing to this kind of thing.

Another issue: As long as we're trying to build and understand computer models and trying to apply them, it is very important that we put a substantial effort into these systems, starting with cellular automata and going to models of complex systems that are multiagent, with nonlinear interactions, distributed, and all the rest of the litany, it's very important that we have a good way to interact with these things. I can't do the kind of thing that was such a big problem when we had computers early on with batch processing where, at the end of the day, out comes this stack of papers and I try to flip through it to try to find out where the interesting events occurred. What we need, I would claim, is something that's like a flight simulator. There are two reasons for this. One is that, if I've got a decent flight simulator and an experienced pilot, the pilot can go in there and try to control and, if that plane doesn't act appropriately under situations with which that pilot is familiar, then I have my reality check and I know that the flight simulator isn't very good even though that might be a very difficult thing to check in terms of the equations or the program. If the flight simulator doesn't do natural things under natural actions, then I know there's a problem with the simulator. On the other hand, once I've got the simulator, I can do things I can't possibly do with the real equipment. I can take the simulated plane and take it to where I have absolutely no control over any of the control systems and yet, as in an actual incident where the pilot managed, by changing the thrust on the engines, to save most of the lives of the people aboard. I can practice on a flight simulator things that can't be done with a real plane.

These two things are really important advantages. My claim would be that if we do some calibrations, some applications of our models of complex adaptive systems and have a decent interface, we can start doing something we almost can't do at all now and that is to let policy makers and their cohorts do flight simulation on some of these environmental scenarios and some of the other things we're interested in. But we won't do it unless it's interactive and we don't yet have any such interfaces. This problem will grow in spades when we start doing a lot of parallel computing.

Danny has some systems that you can interact with but, compared to SimCity, they're quite primitive. What you want to do in the policy arena is to make these things accessible to the policy makers, not just the model designers. You want them to give you reality checks without necessarily understanding the equations or the programs.

This is all prologue to a discussion of applications and what people think an application is when we're talking about complex adaptive systems.

RASMUSSEN: You're missing an explicit model which is not really algorithmic. When you use neural networks or time series analysis and are building a black box, you have an algorithm but not a fundamental law. So the kind of explanation you can get is not really a table but something in between having a law of physics and a table. This is where most of the explosive advances have occurred, using these black boxes.

HOLLAND: Let me turn this back to you. What would you say about the richness of an application and what you consider to be an application?

RASMUSSEN: Doyne and Norman are interested in finding structure in some time series. They don't need to understand all the intricate interactions going on in the marketplace but they are able, hopefully, to say something intelligent about future behavior. Isn't that an application?

HOLLAND: They ain't rich yet. I know that people talk about doing these things but I don't know of many actual demonstrated applications.

RASMUSSEN: There are many ways in which these networks have been used for industrial applications, to control processes, the steering of tankers. . .

HOLLAND: It's very interesting when you get behind the claims on these things. You usually find out that they've not actually been used. They've been used on a simulation of the process. They have not been used to control the real process. The only real application I know is where they have used this to assemble boxes on a production line.

RASMUSSEN: I remember a nonlinear application involving the stretching of optical fibers.

GELL-MANN: Was there really a model, even implicitly, or was it just a cybernetic control system?

RASMUSSEN: But wouldn't you say that these black box models are applications?

HOLLAND: What counts as an application? If I'm running a simulation and given a proof of principle, is that an application? Or do I have to put it out there on the machine that's drawing the optical fibers?

EPSTEIN: I had the impression that in sonar tasks, they have deployed neural nets.

HOLLAND: I know some experimental work but I don't know that anybody has used it for IFF or something like that. I'm not sure that's necessary but I want feedback from you as to what counts as an application.

SIMMONS: How about insight critical to the solution of some scientific problem by building a widget?

HOLLAND: That's why I said it's critical to say what counts as an explanation.

LICHSTEIN: The vocabulary that's in use in the world of commercial software, like driving ATMs, is exactly as you've described it. We look for table-driven systems and the tables become, in the minds of the users, the application. And the system that was used to generate the tables is called an application generator.

LLOYD: I wouldn't want adaptive ATMs.

LICHSTEIN: When you go up and give your name, we immediately speak to you in your language. We have ATMs, under the same operating conditions, that speak fourteen languages. That's an adaptive ATM being carried out exactly as described here.

HOLLAND: The point you raise, Seth—and we can get back to Per's notion—as to what constitutes adaptation, if we don't agree on that, we won't agree on application.

ARTHUR: What do you mean by adaptation, adaptation to what?

HOLLAND: Yes, when you say adaptive ATM, the point is, adaptive for what? If it's for language, that's great. And a lot of effort is going into where they place the window so they design the macros instead of you. When you get back to questions like adaptation, we must ask, do we need to describe a goal in order to talk about adaptation? Or can we take Per's view and say it either is or isn't? My decision is going to have a lot to do with how I use that system, what I expect of it in a particular context, and what I mean by the notion of application. I don't know really what it means to talk about applications of various complex adaptive systems.

PINES: I think that adaptation has got to involve a goal. At least I'd like to see that we all agree that we define adaptation to include the goal.

ARTHUR: So what is the problem that biology is solving when it's adaptive?

EPSTEIN: One thing you can do, without mentioning any goal is to use some kind of simulator to at least alert people to the existence of counterintuitive failure modes, counterintuitive thresholds, also to dynamical behavior that they weren't aware of and that they should be sensitive to. If you had things like Sim-X's or you have agents who interact and coevolve in accordance with parameters that are set by the pilot, the trainee, then that person can become more sensitive and informed to the fact that things can evolve in ways that are hard to anticipate, that are sensitive to initial conditions, and that there may be failure modes that were not anticipated.

HOLLAND: To try to reformulate this question, I would consider the flight simulator that alerts to me to these things to be an application. I don't know whether others would agree or not.

BUSS: Watching this meeting develop, clearly in the organization there were sets of speakers who would identify themselves as complexity theorists and then there would be groups of people who would identify themselves as economists or biologists or whatever. When I saw the schedule for this afternoon that said applications, I sort of assumed that it meant that it was not complexity theory but what the complexity theorists could tell people who identified themselves with one discipline or another. Which is, of course, a different point than the definition of application that you're moving us toward.

MOORE: You talk about flight simulators. I can see their usefulness for policy makers in a situation like when S&L's were relieved of liability and it might have served a purpose in a good model to simulate these agents that said, "Hey, we've been relieved of liability; let's go completely off the wall. But there's another level of modeling of human beings that I find not only abhorrent but completely unbelievable that it could ever be useful or accurate. I'm talking about people who model revolutions and political changes and even arms races, to some extent. I've never found that these models tell you anything about what human beings do that isn't already obvious and, if it isn't already obvious, it's usually wrong. I'm not an expert on these models and I will never be, but I think that, when you talk about flight simulators for policy makers, you have to pause for a second and ask what the limits of that are going to be. If you want to find out how people's air-conditioning use changes with temperature and add that to a global warming model as feedback, I think that's great. However, if you say you can model things like the formation of neighborhood organizations to start community banks to invest in insulation for low income homes, I'm not going to believe any mathematical model.

QUESTION: Do you include computer models?

MOORE: Yes. When you can use computers to write sonnets, then I'll believe you can do it.

HOLLAND: Can you tell us on what basis you make this extrapolation? It's a little like questioning why is there a science of psychology?

MOORE: Psychology is not a science that emphasizes predictive models.

HOLLAND: I think you'd find a lot of disagreement with people in my particular department of psychology.

MOORE: Of course, economists claim to have predictive models, too. Part of my conviction is that the human system is an inherently self-reflective system and that whenever there are models or patterns, human beings try to break or exploit them and I think that you can get computers to do that in a toy way which will give you some of these features and maybe look the same as human systems. But I don't think it will have any actual predictive power. You can probably write a computer program that can write science fiction stories about the future, but there's no chance that it actually will happen.

EPSTEIN: But nobody is claiming that. I'm not.

PINES: What John and many of us had in mind at the Santa Fe Institute is the potential of combining two types of technology with parallel processing technology, it begins to be possible to put together a simulator in which you know what went in and how it was wired and then someone can play that game and see what comes out with that set of rules. You can then change the rules and connections and see what comes out as a result of the changes. In the process you go way beyond the idea of an expert who testifies to a committee that if you do this, that will happen. You let the congressional person play the game and see what is going to happen. That is still a long way from modeling human behavior but it's a first step toward improving policy analysis.

EPSTEIN: John, you represent applications all the time at SFI. Can you give us a couple of applications of GAs and are these problems where it would help to have something like SimCity?

HOLLAND: I will give you an example of something that is claimed to be an application of GA. I am very cautious about all of this but here's an example that's clean. This was work done at GE for a high-bypass gas turbine for use in an airplane. They started with a designer using an expert system. They got the first design in two months. Then they got a simulated annealing revision of that design in another two days which was 50% better. Then they used a GA on that for another day and got another 50% improvement. This was all on a simulator for a gas turbine, nobody constructed the gas turbine, there are no gas turbines that look like this gas turbine. But this was the simulator that they use to help construct real gas turbines. That was an application of GA. You ask about the flight simulator. You're talking about something that is, for me, at least two generations ago. I haven't done much with them, except for the classifier systems, for over a decade. Classifier systems are more recent and I don't know of any real applications or even pseudo-applications. Finally, we get to things like ECHO. The flight simulator, this interface that allows you to do interactive things, is something that I would advocate for anybody who wants to use a highly parallel computer to explore complex adaptive systems. It hasn't got a lot to do with my particular stuff except that when you're dealing with agents that interact in nonlinear ways and are adapting, I think you need an interface like this to really exploit the power of the computer.

ARTHUR: You're talking about generic problems? Do you see this activity as an engineering problem?

HOLLAND: Certainly not.

KAUFFMAN: Let me describe a quite different model which I have been fooling around with after talking with Brian all these years. We model an evolving economy in which goods and services are symbol strings and strings act on strings to make strings like goods act on goods and become tools. The interesting thing here is that when you define a grammar where strings act on strings to make strings, in that model world it's formally undecidable that a specific string or good will ultimately be produced in the model economy that you're living in. Another way of saying that is that, therefore, if you change the grammar, you'll be in some different world. If you look at the unfolding of this world, it is ripe with historical accidents so that there may be statistical laws of the type that Per is talking about. But if you run the tape over and over again, as is said about evolution, you get quite different things each time. I suspect that this is true of the world that we are living in.

HOLLAND: How does this bear on Chris's point?

KAUFFMAN: It bears on it because whether we are talking about human agency or not, if the world that we are living in has formally undecidable properties, the only way to find out how it will work is to live it. Then any detailed simulation we do will wind up being wrong although we may still get statistically interesting things. So that for a system that's focused in that kind of way, namely nonstationary at all scales, which is the kind of thing Per discusses, we may not be able to simulate them in the same sort of engineering sense as we can when we're talking about systems that are closed, stationary types. What I get worried about with respect to the sustainability program is the danger of a certain kind of naivete that suggests we can intervene in some intelligent way with, ultimately, utterly catastrophic results.

HOLLAND: Is it better to intervene, as we always will, with such a model or without?

KAUFFMAN: I don't know.

EPSTEIN: I want to take up Stu's point and respond to Chris' point. I think that the term flight simulator is a bad choice. It invites this kind of predictive reading which is really wrong. People think you have made a prediction and things come out differently and this is very damaging. Somebody at Rand came up with the phrase "prosthetic imagination." (C. You mean "mental crutch.") I think we need something that blocks this other interpretation which we should avoid. The goal is not to predict things but essentially to jar people out of the sort of arrogant prediction that they can predict things. There are problems where all sorts of uninformed policy makers think they know what is going to happen. What you want is some sort of club with which you can beat some humility into people and show them that even the simplest rule will undergo a very complex evolution and you must be aware of that before claiming that this new weapon will do that, this new tax will do that. You don't know and we can almost prove that you don't know with these sorts of very simple demonstrations. Now, that said, there are some nonpredictive goals that models can pursue. They can suggest qualitative results and dynamical behaviors that are not foreseen and they can be used in testing the internal consistency of people's views. You might interview some person and you say, "Look, I'm just going to let you implicitly build a model and we'll let it play out according to the rules you think apply and then the response may be.... Holy mackerel, I didn't know it implied that; that's crazy," and then they go back and reexamine their assumptions. That's another way this enterprise can be useful. Finally, on prediction per se, I think you should hold off. It's a rhetorical question, what models can achieve. I think it's a little bit closed-minded to assert beforehand that a phenomenon can't be understood at this point. The history of science is full of examples where those claims are regularly refuted.

RASMUSSEN: What Chris said, probably precipitated by the term flight simulator, is that we should first specify what kinds of systems we want to build models of. There are systems which, at this time, we don't have a clue as to how to formalize.

EPSTEIN: And the dangerous thing is that all sorts of people think they have these models. So you go in with an explicit model and get all kinds of flak, which is fine, but the people who give you the flak go away with their own models, which are not explicated, and they think they know how things work. The main thing to combat is this unbelievable arrogance.

COWAN: John Steinbruner commented on the Moyer's program that we recorded last month that people do have models and there was a model about the Russians invading Western Europe which was totally unreal which, nevertheless, led to spending hundreds of billions of dollars on the basis of that model.

ARTHUR: There was such a plan.

COWAN: Of course, there was such a plan but it was unreal.

MOORE: Part of the reason I said what I said is that things like the sustainability program presuppose that solutions to the kinds of problems we face are for legislators and for people who make decisions and to somehow tweak the system. I think the solutions to these problems tends to come from below in many situations.

EPSTEIN: We agree. You know, one definition of a politician is a power drive with an applause meter for a brain. (It applies also to scientists. There are some applause meter scientists.)

PINES: I think that Josh has suggested the right new name for flight simulators, "humility injectors" or "club" for short.

Erica Jen, Chairperson
Los Alamos National Laboratory, Los Alamos, NM 87545

Interactions Between Theory, Models, and Observation

JEN: It was suggested yesterday that we organize our discussion today around the issue of to what extent people working on specific real problems actually find the concepts and techniques—that are enunciated by generalist interests in the problems of complexity and adaptation—[to be] useful. The usefulness of these concepts was in fact the explicit focus of Phil Anderson's talk, and the appropriateness and the usefulness of defining characteristic features of complex adaptive systems was certainly the implicit focus of Murray's talk. And I'm sure that by being here, we're all saying that we actually find these questions of intellectual interest; the question is to what extent are they, in fact, useful to people working on real problems.

It's obvious—and Jim's talk was a perfect example—that the people who are generalists who think about complexity and adaptation have drawn much of their inspiration from the specific empirical data coming from real systems. And so that direction, I think, is very clear. It's also clear that complex systems, and adaptation, and these ideas, have provided the tools for the solution of certain problems, such as protein secondary structure, that Alan Lapedes was talking about yesterday. It's not so clear to me—and maybe to some of you, also—that when Marc Feldman talks about, for instance, the construction of models for the transmission of cultural traits, or when Jim was talking just now about the ecology—it's not so clear whether the incorporation of complex interactions and adaptation is anything more than sort of a natural consequence of trying to refine and deepen one's understanding of a system.

And so what we're doing is posing the question: are the ideas that have been put forward by, say, Murray and Phil, and, say, Jim Crutchfield in talking about measures of complexity (and by various other people)—to what extent have they influenced our thinking when it comes to actually looking at a real problem? This may not be a well-posed problem, and it may just be that you have osmosis of ideas, and you can't really pinpoint things, and this philosophical sort of distinction may not be interesting. I think it would be interesting if people say—as, actually, I have heard some people say in private—that, in fact, they have had no influence on people's work.

GELL-MANN: But might have influence in the future.

JEN: We expect it to have great influence in the future. But if there are ways in which they have influenced people's thinking habits only, [it] would also be interesting.

GELL-MANN: Erica, couldn't we ask a slightly different question as well. Namely, now that one has heard some of these ideas, are they wrong? Are there ways in which they should be modified? Are there ways in which they should be made more sophisticated? And so on.

JEN: Or translated. I was thinking, actually, that since one of the focuses of the Institute in past years has been on the study of economics, and that may be an area in which the influence of these ideas is easier to articulate, maybe Jim Pelkey would be willing to say some things about it. But I would particularly like to have people like Peter Schuster, and Brian Goodwin, and Jim Brown, and Leo Buss—and maybe even Murray, talking about languages, how it's actually really changed your idea of the evolution of human languages (well, maybe that would be last). Would someone like to talk about economics, or do you feel that that's been covered in discussion?

ARTHUR: I'd be happy to say a few words.
First of all, I don't see economics as an applied field. I don't see studies of complexity as a kind of general field that feeds these applied fields, like biology and economics. I tend to see a to-and-fro; economics, for better or worse, is a highly theoretical field at the moment; biology is a field full of theories; physics similarly; and so on. And so, what we tend to discover in these fields is that there are themes, or issues—features among the theories we're coming up with that appear to have something in common with the theories and themes, and features, and that's just one. . .

JEN: Well, let me also just mention something that George Gumerman has been actually focusing on, which is that he would really like us to talk about the distinction between analogous and homologous phenomena; you can talk about common themes, and commonalities in behavior, but is that really representative of a common underlying mechanism, which you can then explore, or are these really analogous phenomena and we just point out the commonalities, and that's it? I don't know if that's a correct formulation of what you were trying to...

ARTHUR: I want to keep my remarks fairly short, because I'm sure a lot of people have things to say. So let me make two sets of remarks: One is that, as a number of people have said here, what we tend to be dealing with in the study of complex systems is systems that have a number of interacting elements; those elements can be in different behavioral modes, or different states, or there may be a number of states that elements can be in (that can point up and down; some sort of ferromagnet, or Ising model, or so on). The elements might be human agents, they may be cells, they may be monomers, or dimers, they may be neuronal cells, and so on. But these are the types of systems we tend to look at.

There's a certain amount of mutuality in the interactions and in some cases the behavioral states reinforce each other; in some cases they tend to mutually negate each other. That is, if you want, these elements like to do similar things some times, and they like to be different, other times. The way I look at it, I tend to see complex systems as ones that have a large measure of positive feedback, leavened by quite a lot of negative feedback thrown in. If that's the case, then, you get something like a spin glass, which consists very largely of positive feedbacks, or mutuality, or an ideal of cooperation among the elements—but again, throwing in negative feedbacks, which gives you a very high multiplicity of possible patterns, and the possibility of frustration, and so on.

When we were dealing with the economy, traditionally—you wanted a sort of before-and-after, so I'm going to give you two befores-and-afters. In economics, throughout the last hundred years or more, people have recognized that the economy consists of interacting elements; we call them agents. Economics, I believe, is a bit harder than physics, because the interacting agents have to look ahead and anticipate what the other elements might do, so they have expectations and strategy to consider. We set up a lot of the problems, over the last hundred years, as economists, to reflect mainly negative feedbacks where it counts on the margin. So near the solutions, we set things up so that if there's a lot of negative feedback (we call diminishing returns), that gave systems that had single, unique equilibria, and these solutions to economic problems were thereby predictable, and history didn't matter very much. When we stopped to consider complex systems along Santa Fe lines, we're talking about interactive elements—agents—where there might be quite significant measures of positive feedback. Good theorists have recognized this for decades. Marshall, even, in 1891, comments on increasing returns, or negative feedback, and says that it leads to a multiplicity of solutions, and says you have to be careful (that's almost a direct quote); If you have positive feedback and assume that the economy will reach only a static equilibrium, that's a false way of looking at things. He says the economy, under increasing returns (or, positive feedback), is basically organic. But he couldn't do anything about it because he didn't have the tools.

So the number one point is that the people I know of who've tried to incorporate what you might call positive feedback—or increasing returns, or nonconvexity, or nonlinearity, or whatever buzzwords you want to use—into economics, have all been highly influenced by condensed matter physics (I know I was, directly and indirectly), and by nonlinear dynamics. Because once you start to recognize that there are multiple metastable states, or multiple patterns that might form, or multiple equilibria (as we would call it in economics), there arises the question, "How is one equilibrium, or one possible solution to the problem, reached in a realization over time, rather than another?" Presumably, if you ran time again, small events—as a number of people have alluded to here: historical accidents—might steer you into a different solution.

So, when you start to incorporate these considerations into economics, several things have happened. Economics—from being a study of single, unique equilibria—has had to become a dynamic study of the selection of one pattern out of very many. So economics is changing at the moment, with much emphasis on Chris Langton's diagram; that is, this notion that historical events can lead you to one formation that might never be repeatable; that the economy is basically nonergodic; that the pattern that's selected is not necessarily the most efficient.

JEN: I don't mean to be pedantic about these things, but, in some sense, attributing that to Chris Langton is slightly peculiar.

ARTHUR: I mean, the diagram that he put in his talk.

JEN: Right, you're referring to the diagram but, since we are talking about sort of the lineage of ideas, and the influence of certain concepts—the concept of accidents, and coming into one equilibrium versus another, is not a new one.

ARTHUR: I didn't say it was a new one; I just referred to a diagram in somebody's talk.

All right, let me summarize.

Number One: As these ideas influence economics, and as people in economics think independently about such ideas, economics is changing from a study of unique, static equilibria, to a study of process; a study that must include dynamics, and must include the notion of how one pattern gets selected over another, and a study that includes nonergodicity, the possibility that the economy is not predictable, and that the economy may never reach an equilibrium. Ten years ago I asked Larry Kline—who is a very distinguished economist—"If I believe all your equations, Larry, how far would they be accurate in the future?" Larry said, "Oh, fifteen years, maybe to a couple of decimal places." [Tittering.] Nonsense; because I had started to think this way, but I just simply didn't believe it. We didn't have the tools, though, to refute that at the time.

JEN: You're talking about economics now being studied as a process, and what I'm really curious about is to hear from people who are working on problems that are inherently processes, like what George Gumerman works on...

GELL-MANN: Excuse me, Erica, could I ask Brian a question about what he just said? I know that you and John and Richard Palmer recently have done some very interesting work on interacting agents, each making a model of how the others will behave, and predicting, and so on—and that earlier you've done things along those lines as well—and I was wondering if you couldn't expand your remarks, briefly, to include these things. Because, that would address the information processing side of the issue—ignored so far in your remarks. You're treating these agents as some sort of system, and...

ARTHUR: The second major influence is that, and I wanted to say that. Up until recently, for reasons of mathematics, economists have interpreted the way people make decisions as using mathematical, deductive logic on well-defined problems. And, we know that that works very well if the problems are simple enough—I ask you to choose between something in the supermarket, and so on—however, if I put any of you in a situation where it's complicated, such as playing chess, we know that there are mathematically deductive solutions available that make great sense, and nobody's arrived at such; nobody I know of, nor has any computer been able to compute such solutions. Now, what I have learned from being exposed to the ideas of John Holland, mainly—but other people as well, including the sort of ideas coming out of psychology we were hearing yesterday—is that there is a way to go about how economics should operate in complicated situations. It's roughly this: we need to be inductive, rather than deductive.

We need to allow that people have internalized economic agents, are basically operating on the basis of internal models, or multiple hypotheses. A lot of people have realized that a long time ago. What I learned from John in particular is that those hypotheses could well be contradictory, and I learned this out of his classifier system. That is, I could be playing chess, and I could think, "My opponent's using such-and-such a defense—on the other hand, he may be using this defense, and on the other hand, he may have something else in mind." Now, those are three hypotheses, three internal models; I could deduce five moves, or ten moves, ahead, or seven moves ahead, on the basis of each of those, and make moves. But he's obviously not using all of his defenses at once. And so, what happens is that as time passes, those internal models (can I call them schemata?)...some of them are strengthened; that is, we get more confidence: "Ah, he is indeed using a Caro-Kann defense," or something like that. Others tend to get more and more refuted: "Oh, it appears that he's not doing this." But then as the game unfolds and progresses, there's a need to form further hypotheses and further internal models. So, to cut a long story short, I believe now—that is my main interest in the last three years (five years, since I came to Santa Fe)—immediately I came here, I dropped positive feedback, and thought about learning and adaptation. And the major problem in economics at the moment—and it's unsolved—is, how do economic agents operate under conditions of extreme complication (as opposed to complexity)? The answer that's appearing to me—and I heard echoes, by the way, in Stuart's talk the other day; other people would not view this as terribly novel, but it is new in economics— is that we form internal models, we have schemata, or we have hypotheses (and those are working hypotheses). And as we go, some of the hypotheses are strengthened, and thereby selected and acted upon; others are refuted, and die away. And so we live in this world of moving hypotheses, that are moving in and out of our minds, as we're dealing with unfamiliar, or complicated, situations.

PINES: I would suggest that a very useful thing to do before we start around again in responding to Erika's set of questions is to talk about the comparison of theory and observation to theory and experiment. It's not an accident, I think, that Brian didn't discuss this extensively because one problem, as I understand it, is the question of what body of data are out there that enable the new style economists, examplified by Brian and Per and José, to confront these new ideas and models with observations and determine whether they are doing a better job of explaining real phenomena.

ARTHUR: I would say that data from the stock market do not conform at all to standard theory but do correspond to the new style analysis.

KAUFFMAN: I was hoping that we would feel free to add to and amend the list, particularly in response to what Brian said. I think we missed a very crisp statement about the problem of bounded rationality which is obviously a core problem.

GELL-MANN: Henry Lichstein will spend a few minutes on a statement about banking.

COWAN: Brian, you mentioned that negative feedback dominated economic thinking and then you implied that now you have to use a large component of positive feedback compared to negative feedback. Can't you achieve what you are trying to achieve by not specifying the ratio of one to the other but just saying that there is positive feedback or must it be a very significant component?

ARTHUR: It turns out that some sectors of the economy operate mainly with negative feedback, like the production of coffee. In high technology virtually everything is positive feedback.

SIMMONS: I understand from what Brian says that, in being here at the Santa Fe Institute, his perspective on economics has changed. I also think I understand what in the former view of economics what would have been considered a solution to a problem. My question is what do you mean by a solution to the problem in the current paradigm?

JEN: We'll postpone a reponse to that until after the break. George, you have a question?

GUMERMAN: Chris Langton, the other day, talked about the "Cambrian explosion" and how we adapted it for archaeology. The photos on the wall represent a period in southwest history when we had a something like a cultural Cambrian explosion. We latched on to the concept and used it as an analogy or a metaphor, and it became useful for helping us to describe things but not necessarily to understand them. Years ago I used the term ecotone which is a biological term that means a transition region between two major environment zones. In an ecotone there are plant and animal species from both of the major environmental zones and there are also plant and animal species that are unique to that transition zone, the ecotone. I know that from using certain kinds of remote sensing imagery that there is a very distinct ecotone north of Phoenix, Arizona, between the Lower Sonoran and the Upper Sonoran life zones. I wanted to determine if it was a "cultural" ecotone as well, that is, could we see a mixture of cultures in this area and also some distinct cultural features. I used this biological term, ecotone, for cultural behavior and sure enough we found not only a mixture of cultures there but also distinctive features. But to this day I don't understand what shaped the cultural behavior in that area. The question is, was the human behavioral system analogous to the biological situation that resulted in an ecotone or are there basic underlying principles that facilitate similar behavior in both biological and cultural systems? What I haven't heard here and what I think needs to be addressed is the following question: Are we dealing with analogies or metaphors or are we looking at underlying principles?

JEN: We are mapping our areas for discussion later on. Chuck?

STEVENS: I'd like to ask two questions. The first question is: in neurobiology, there are what people call theories but they're not actually theories at all. What they are is a restatement of things we already know in a different language and there's no new result. Sometimes this can be useful but it's not a theory. I would be interested to hear the extent to which people think that complex systems theory has been a restatement or the extent to which, in all the various areas, there are results where we know something we didn't know before. The second thing is, and this is just my ignorance, that I don't have a clear idea of what the theory of complex systems is. I think I can recognize a complex system and I think I know some of the words we use to describe them but I don't know where you draw the line between, say, certain concepts in statistical mechanics and concepts or scale invariances, something that comes up a lot, in complexity. So I'm not quite sure what counts as parts of the theory of complex systems. I don't expect a precise answer. If you're going to do psychiatric diagnoses and you look at this big book, what they do is they have symptoms in three different categories and you have to have two in the first and one in the second and two more in one or the other. Maybe that's going to define a complex system, not any one category but several different categories. I can't comment on implications of a particular category until I know what counts.

EPSTEIN: I have one or two comments. I have to express some dismay that's built up over the last few days that most of the discussion would seem to suggest that the only social science that counts is economics. It seems to me there are lots of other social questions that are well posed and are worth thinking about. When do wars occur? When do revolutions occur? How do arms races work? Here's a case of military speciation, coevolution in which phenotypes and sizes are changing in response to one another. What about the huge social structures of religions and nations and the morphogenesis of ideological clusters, races, and so forth? We have examples of big aggregations like that and examples of where they collapsed, like the Soviet Union. These are not narrow economic questions. They are huge questions that are central to social theory, it seems to me, and I think that analogies to morphogenesis in other areas are potentially very powerful. I've been using the mathematical theory of epidemics and reaction and diffusion equations to try to think about revolutions, to try to think about slow and fast dynamics in these things. I don't know exactly how good the data are but it seems to me that we should try to develop theoretical formalisms that would generate the menu of gross qualitative behaviors of interest, up to equivalence classes that Brian was talking about. Let's get a crude thing that will generate the caricatures we're interested in and see if we can draw them so that they look exactly like the data tell us.

BROWN: Something about what Chuck Stevens said. I think that to make a lot of this useful to me, one of the things that needs to be done is to operationalize it. I know that this isn't easy but to struggle with these concepts so that they are framed in ways, being an empirical person, I can measure something with the data that I collect. I can collect lots of data. It varies in quality. But it's hard for me to get hold of some of these concepts because I don't know what they mean in terms of, precisely, the things I deal with.

MOORE: This is procedural. When we talk about people describing the impact of complex systems, whatever that is, on their field, I think one of the discussions that we haven't had yet is going in depth into, say, four or five ideas off that list of 20 and spend some time on those and have people say, "Well, I think that applies to my field," or it doesn't.

JEN: Yes, not only the 20 but also Phil Anderson's 8 and Murray's (I don't know how many). That would be a very useful thing to do.

BUSS: Let me say something that will probably drive every one to coffee. From the point of view of someone coming here that has with a specific set of problems, and listening to people from other fields who are primarily interested in solving particular problems in their fields, these people seem to share some similarities. They all seem to be interested in problems that are structurally similar in the following sense: They seem to involve problems with large numbers of heterogeneous players, in my systems gene cells and individuals, with the individual players nonlinearly coupled and where the system generates coherent macroscopic patterns. I think that clearly the homology and analogy issue is a central one but what brings us here is that we don't have answers to problems of that character. Scientists never have answers to problems of that character. We can't address homology or analogy until we've had some success. That's the first point I want to make. The second point is that, it seems to me, there is something of a taxonomy of these problems, and I see it as twofold. There is complexity at a given level, so understanding why Jim Brown can kick around three kangaroo rats and generate a graph from that is complexity at a given level. We have—what brings many of us back here—a variety of techniques that are good for asking those questions. When you look at John Holland's techniques and you look at a 15-year experiment, you really wish that there is something that can be worked out between these two guys. Clearly discovering the power of evolution and putting it into a computer is something useful for a lot of problems. The second part of the taxonomy has to do with the vertical level of complexity, the level that George Cowan focused our attention on at the beginning of the conference, getting from physics to the Boston Symphony. I think that we have far fewer techniques that are useful for that kind of problem. Walter's techniques are very powerful but that's one of a few classes of techniques. My final comment is that these are just techniques. In every one of the disciplines you have to ask the techniques the right questions. You're fundamentally limited there by endplay. You don't get it for free. If you ask the right questions, then you might get the right answers. And when you start getting the right answers, you have to ask data whether the answers are right. You can ask whether there are structural similarities in those answers that, in fact, imply the theory of complexity.

JEN:　　　　　　　Looking through the list of 20 points and adding in those that have been suggested later, we come back to the following questions: If we did have a good theory of complexity, would you care? Would it make any difference to you? If you did have a good way of constructing internal representations and testing whether agents have it, what difference does it make to you? Peter Schuster will talk about his views next.

SCHUSTER:　　　　You are asking a very complex question if one wants to go into the details. First of all, I think that a notion of complexity that can be cast into quantitative terms would be very useful in all fields and, in particular, in biology because there everybody is convinced that a human being is more complex than a virus. There is no problem if you compare things that are far apart in evolution. But when it comes to a turtle and a bird...

JEN: People here have been suggesting measures of complexity and do you find them useful to you in interpreting this problem?

SCHUSTER: There are those I have known before but, unfortunately, I missed the first two days and am unaware of all of the measures that were suggested.

JEN: But you know what they are designed to capture.

SCHUSTER: If I consider our system, then it is problem to see how one can use these quantitatively. They all apply in a qualitative way. This is not the first time this problem has come up. We started initially by applying conventional chemical methods to problems of molecular evolution and it turned out that we immediately raised a question that cannot be properly answered by this technique. And so we applied the concepts of irreversible dynamics which turns out to be something that you can cast in quantitative terms. But when we looked at systems, we ran into problems, one after another. In problems of molecular evolution, these techniques are not useful because you run into singularities, infinities, and so on. Later on, when we had developed already some ways to treat the problems, following stochastic processes and so on, we discovered a very nice parallelism, and this, I guess, is very important for developing notions of a general nature, between our kinetic systems and equilibrium properties of spin glasses. If you transform time to a spatial axis, you find that there is a nice approach and you see that the mathematics is the same, and that was a very fruitful cross-fertilization from our point of view because we saw the analogy to phase transitions where we could do analysis very quantitatively and see for this kind of value, landscape features; we can do phase transitions, we can evaluate all landscapes in that way. What I would really want to get from a general theory of complexity is a kind of a list of general phenomena and the techniques with which you can describe these general phenomena.

JEN: Is what Peter is doing and finding useful what you hoped he would find useful? Comments?

GELL-MANN: I'd like to add a comment. I referred to it very briefly in my talk. Jim Crutchfield discussed, and I also have discussed, a kind of effectiv ecomplexity, a measure of what we think of as complexity here, as opposed to algorithmic complexity which is simpler and not so relevant. Effective complexity is rather subjective or dependent on the situation, dependent on the complex adaptive system that's doing the observing, identifying the regularities, and so on. So, for example, for the grammar of a language, we said that the complexity is the length of the grammar book but the book might be written in a language that's similar to the language being described or very different, in which case the grammar might be long. There are certain cases where it looks as if effective complexity might be more intrinsic than in other cases. For example, if you could get at the internal grammar that a child registers in his head when it learns a native language, if we could read it out of the brain, that would be a much better measure in a certain sense, a less arbitrary measure of the complexity of the grammar than the length of a book. Likewise in biology if you take a genome and, particularly, must the exons, and look at the length of that, it gives a sort of intrinsic measure of complexity because you're actually getting right at the guts of what's happening rather than a description by a system that's not much involved.

SCHUSTER: If I may add to that, we, of course, looked at all of these different measures of complexity like algorithmic complexity but that doesn't help you in biology because it's missing something very important and that is time. Time is not a part of the algorithmic complexity and it matters whether you reach an optimum in a short time or a long time. The two aren't the same in evolution: to be fitter in 1,000 years is not the same as to be fitter in 10 years.

GELL-MANN: Can I say a word in answer to that? You can define measures of complexity that obey the arithmetic properties of information measures. There are also other measures, which are not information measures but very important, and they typically have something to do with time. Charlie Bennett's logical depth, for example, and various other kinds of depth. In general, one has to characterize a system by information measures, such as the length of a schema, and also non-information measures like logical depth.

MARTIN: One of the things I tried to do in my talk was to point to some variables of interest in psychology that I thought complex adaptive systems had something to say about, either because they had an analogy to make or because there was something homologous about these systems like psychological systems. I didn't address measures of complexity. But I think I could do that for psychology. I think this is particularly relevant to Jim Crutchfield's talk. In psychology one of the things that you'd like to know is can you find an ordinal scale of difficulty of learning in certain cases? How difficult is it to learn a certain class of things? How complicated is the environment in which learning has to occur? If you could quantify that in a way that corresponds to what we know empirically about how difficult it is for people to learn these things in these situations, that would be useful for psychologists. Jim Crutchfield's idea for finding, in effect, the machine that describes the environment as a way of measuring its complexity is an interesting one. There is some work by Rodette and Shapiro at MIT along similar lines that I've tried to apply to the case of the board game Mastermind. It's not a terribly complicated system but a somewhat interesting one from a human behavioral point of view. What you see is that, in fact, in Mastermind environments that show high diversity, according to Rodette and Shapiro's diversity measure, which is like Crutchfield's measure of what is the simplest machine that can describe the environment, there is a pretty good set of predictions about how humans should behave, which problems they'll find harder and easier to solve in this small domain. This is sort of a toy example but, I think, complexity, as a measure, can apply very well to certain psychological cases of that kind.

GELL-MANN: One more sentence that I think would illuminate what Ben just said. Charlie Bennett defined depth, roughly speaking, as the minimum number of steps needed to go from an average short program to the message string. There is an inverse quantity called crypticity, which is the depth of the program when you know the string. Crypticity is related to what you are asking, namely, for a very difficult learning problem, how hard is it to find the regularities. Peter Schuster asked how long it takes. The crypticity is a relevant measure, a non-information measure. If you see, for example, a fractal set which is derived from a trivial equation, how hard is it to go, how many steps do you need to go, from the observation of the fractal set to the trivial equation that generated it?

SIMMONS: Is that a well-posed question?

GELL-MANN: The way Charlie poses it, yes. I'm just rephrasing it.

JEN: I find it interesting that, in asking people to respond to the various concepts and techniques of complex systems, what they immediately seize on is measures of complexity and it's clear that it's because it's something that is close to being quantitative although Peter has expressed the difficulties in actually using it as a quantitative tool in his own field. I would hope that we would then look through the other things, for instance, that Murray was talking about, variability of schema, or the issue of fitness that Phil was talking about in terms of self-organized criticality and scaling. We should also talk about to what extent they correspond to techniques and tools that we actually find useful and what do we need to make them into things that are useful for our own problems.

ANDERSON: Throughout this discussion I'm going to be taking the negative view, which I hope is the point I took in the opening talk. I have yet to see a measure of complexity except, simply, information that really enlightened me in my search for some way that things work. I am, to some extent, on Peter's side. He sees the qualitative ideas of complexity as very often much more useful. A good example is the one that Murray gives, the grammar of a language. The grammar of primitive languages which are not used very much are much more complex than the grammar of English or Chinese which are used by billions of people. There must be some reason for that. Yet you can say whatever you want to say in English and probably say it more succinctly in Chinese. These sophisticated languages have a lot of compression. A measure of the complexity of these ancient languages is rather meaningless.

GOODWIN: I want to make some general comments about complexity in my area of work. There are two kinds of problems that we deal with because the problem of biological forms deals with morphology and behavior. In relation to morphology, my experience with complexity goes roughly as follows: Right from the beginning, it is a nonlinear problem, and nonlinear dynamics is a domain that I see as at the historical root of complexity. It's evolved from that but for me it's a kind of continuous evolution to develop a theory of organisms that can account for the existence of different types of morphology, different types of behavior, and that requires the understanding of a system that is hierarchically organized, with different levels of order. You have this relationship between hardware and software and they both require appropriate levels of complexity. They are somewhat distinct, one from the other, and they can be distinguished.

JEN: Both Phil and Jim talked about how nonlinear dynamics may have been involved from the beginning. What they meant by nonlinear dynamics is usually a very specific type of nonlinear dynamics, namely low-dimensional systems, and certainly I think we would all agree that complicated phenomena are inherently nonlinear and you are reinforcing that point. But is our approach to nonlinear phenomena one that you find useful in understanding the nonlinearities?

GOODWIN: Not in relation to specific problems of morphology because that is continuous with the traditional area of nonlinear dynamics. The Russian school of nonlinear dynamics was the root of the kind of tool we used in the study of morphogenesis, the generative theory of the organism with respect to pattern and form. But, now, when we come to behavior, that's where the transition begins to occur to a much more complicated notion of complexity and nonlinear dynamics. Let me give a brief example to illustrate that. Ant colony behavior is traditionally regarded from the sociobiological approach, the Wilson approach where specialized ants perform tasks determined by genotype and feeding regimes. Now that's a highly deterministic, centrally organized process. There's not much dynamic in that. What we find from the studies of people like Deborah Gordon and Cole and Frank, all experimentalists looking at ant colony dynamics, is a story quite different. The majority of ant species do not perform fixed tasks and the individuals in the colonies switch tasks. If you perturb the colony, you get a characteristic response pattern showing that there is a colony interdynamic totality. It's not broken down into a set of individually assigned tasks. They somehow communicate, interact, and balance the set of activities between foraging, patrolling, nest making, midden work, etc., so that there is switching going on. How are we to account for that? Furthermore, Cole's studies have shown that, if you look at individual members of a colony, their individual behaviors are chaotic. But the colony shows a well-defined periodicity with respect to activity. Now what we've done is to model that in terms of mobile cellular automata. We show that not only do you get temporal patterns emerging for a very simple dynamic but also spatial order which reflects the sort of patterns you see in an ant. The nest is organized in a particular way and that shows a quite close fit between what we get theoretically and what has been observed experimentally. This is where notions of complexity and chaos do begin to enter in relation to behavior. So I think that morphology and behavior have slightly different qualities of the dynamic that is required to explain them.

GELL-MANN: Do you really mean that the individual behavior is chaotic or do you just mean random?

JEN: It would be interesting to see how you react to what various people have been talking about in terms of individual agents constructing models of their ecology.

GOODWIN: Yes, because there's no model here, you notice; there is no dynamic program that tells them what to do . Everything is implicit.

BROWN: You really get the impression that the behavior of individual ants is very much limited by the number of neurons you can cram into that tiny little brain and that there is another emergent level of behavior and decision making that goes on at the level of the colony. Individual ants make decisions about what to pick up but what the colony forages for is very much a colony decision that's transmitted down to the individual workers as they experience different relative amounts of different foods in their environment.

GOODWIN: Furthermore, over periods of time, the colony can learn things that are greater than the lifetimes of the individual ants. You have a colony level of learning that is not directly reflected in what individuals are doing.

JEN: Do you need an understanding of when you have that type of behavior versus the type of things that Stuart has emphasized earlier?

CRUTCHFIELD: I don't understand why, when we talk about models, we have to think about there being explicit representation of models. What's the matter with a system whose behavioral dynamic, the whole thing, has just an instructionally meager implicit model.

EPSTEIN: But then you just have more data to explain. That's not an explanation. That's a replica of the behavior you're trying to explain.

KAUFFMAN: I agree with Brian. Morphogenesis doesn't seem to fit into the family of things we're talking about. It probably says something very deep about how evolution of robust forms can happen. But, Brian, I've tried to build models of genomic regulatory networks and there, I think, is the case where the images of complexity that we've been looking at may apply.

GOODWIN: This is a hardware/software question. I think you do need forms of complexity to study some packet of software instructions that are emergent.

WALDROP: There are two aspects of complexity I certainly get confused about and I sort of hear it around here. There's a measure of complexity where you look at something, at its structure, and wonder how complex is that. That's the sort of thing Jim has been getting at when he looks at how complex a machine he needs to encode some kind of bit string representing the structure of the thing he's looking at. But there's another kind of complexity where you ask what can it do. For example, if you have some schema in your head, like the grammar that a child has in his head, what is the computational power of that grammar, of the schema, of the device, and to me these are two different concepts of complexity. It's the difference between a program as "text" and a program as "executing." If that's a real distinction, it might be worth keeping in mind.

MARTIN: I think one of the bigggest problems in the study of language acquisition has been the mistake of trying to figure out what class of computational system can capture the grammar in a child's head, the differential language. It's been a complete boondoggle as far as I can tell. It's very much like what Walter has pointed out—the irrelevance of deciding whether things are decidable or undecidable. There's a question about what class of grammars you want to use. That ultimately only sidetracks the important issues in studying language acquisition. The important issue to me is to model the language used by the child as it's developing, not to model the implicit grammar that describes the language that will ultimately develop.

GELL-MANN: But if you study the various candidates, the schemata that the child is trying one after another...

MARTIN: That's a different question from trying to characterize the complexity of the class of grammar that the child is developing.

GELL-MANN: The actual trial ones presumably converge eventually.

MARTIN: Calling those grammars is a relaxation of the term grammar. I don't disagree with what you're saying.

GELL-MANN: The series of schemata that the child uses one after another eventually converge to the grammar.

HOLLAND: Murray, I think that they may, in some theoretical sense, eventually converge. But I can use lots of different kinds of generators and, if I only run them for a hundred years, then whether or not it belongs to this language group or that language group is irrelevant. I can get the same corpus up to a certain point, as long as I'm dealing with a finite corpus. If I'm not asking for the ultimate corpus that this set of rules will generate, if I'm only talking about some approximation or a limited corpus. In that case I can take a regular language or I can take a finite state grammar or I can take any one of these classes and wind up with that approximate corpus so to classify the generators is probably irrelevant to the problem. It's not going to give you much insight.

GELL-MANN: That may be but there's a slightly different point, that is to look (and, if possible, you would do it inside the brain) at the succession of child schemata and where they're going. I agree that to do it in infinite refinement as some people pretend they want to do it...

HOLLAND: We have no disagreement there. The disagreement is whether I should take the additional step of trying to classify that sequence as belonging to some grammar class, which probably doesn't offer much. Let me state it another way. I largely agree with Phil. I think the measures of complexity are largely irrelevant to the problems we have. If we look at the history of physics, Maxwell built models that were mechanical, like a bunch of gears turning around, to get some idea of how field theory worked, which isn't about gears at all. Well, what we're trying to do, in my opinion, is trying to build some models like that. In Maxwell's time there was a pretty good sense of how these gear trains worked and he put in idler gears to get the right conservation laws satisfied and so on. It had nothing to do directly with the differential equatons he was talking about but it gave him some insight and he used that insight to direct him to what kind of data would be useful, how to solve the equations, and so on. That qualitative kind of thing is, I think, terribly useful. Whatever we call it, analogy or metaphor, it guides us in how we think about the problem.

JEN: Regarding other features we have been talking about, like rugged landscapes, mechanisms for the generation of diversity and complexity, etc., can people respond to the question of whether these are also mostly regarded as being qualitatively useful as opposed to things that are worth trying to quantify?

PINES: Before we get to that, Erika, I suggest we talk about adaptation. We've been talking about complexity with various views expressed. If Phil and John are right, having said that a system is complex, a measure of complexity may or may not be relevant to understanding any particular system. I would like to raise the question, for consideration at that same level, what about adaptation? Adaptation, we agree, is essential. How can we go about defining it quantitatively and is it a useful concept for the particular systems that people are looking at? Is there some specified time scale that...etc., etc. That seems to me to be more important than rugged landscapes and qualitative features because it gets to the core of what we're trying to do.

MOORE: I do want to throw into the discussion of measures of complexity one well-defined context in which I think they are highly useful. If you have a formal model of the way wasps work and you can prove the theorem that puts an upper limit on the complexity in terms of what it can produce and then you can measure; according to that definition of a wasp's nest, you see it's bigger than that or it's outside that class, then you have a theorem in which the model that you suggested cannot create that type of nest. That's an example of where a measure of complexity is used to get an actual result.

GOODWIN: That was precisely the point I was going to suggest we come to because I don't personally find the concept of adaptive systems in the least useful and I'd like to expand on that. The ant colony that I just described has no notion of adaptation and yet the behavior that comes out is appropriate. In other words, we don't build in any concept or any measure of adaptation or optimization and what I suggest is whenever you use a measure of optimization, then you're forcing the system in a certain direction. We should eliminate all of those constraints and see what comes out of the dynamics. You see the solution emerge.

ANDERSON: The human brain is supposed to be a complex adaptive system and it consists of a lot of very independent units that are operating together because they're in contact. The dynamics of each individual unit is relatively simple. A neuron is, perhaps, as smart as an ant, maybe more, maybe less, but the combination of all those neurons works like the combination of all those ants and gives you something that's more complex or adaptive or whatever and my view of these questions is that the best thing you can do is to look at how things do it.

GOODWIN: Yes, but the question is what's the value of the concept of adaptation because that's a highly subjective assessment of what's going on; therefore, it's unnecessary to the analysis as far as I can see. I cannot see any way in which it functions as an essential component in the dynamic description.

KAUFFMAN: Would you say that in Peter's case of a replicating RNA molecule?

GOODWIN: Sure, Peter and I would entirely agree on that.

SCHUSTER: I would not exclude adaptation because I can see a distinction between a complex dynamic physical system like the B-C reaction which gives you beautiful waves, beautiful dynamics, but if you examine what's going on in that system, it's practically nothing. The chemical reactions at the end are the same as at the beginning. If you do the same with a set of replicating RNA molecules, at the end you get a molecule that is different from the previous ones. It has adapted to that condition where it can replicate faster.

GOODWIN: You're describing kinetics.

HOLLAND: You're talking about two different formalisms for describing something. I can describe processes by least action or by several different formal ways. Least action sounds as though there's something conscious going on but I have mathematics there which is not conscious. The point is: does the formalism help your intuition? Because I can come up with a half-dozen different formalisms.

To say that there's no adaptation there is just to say that you're using a formalism in which that's not a part of your construct but that's not to say, directly, that you can't build another formalism using the notion of adaptation that does help your intuition.

SCHUSTER: Here I don't agree. It's not a matter of the model I'm using, not a matter of the formalism. I put in here an RNA molecule, let it run, pick it up on the other side, and it's different.

HOLLAND: I would just say that I can often describe situations formally in a way that eliminates some primitive construct. It doesn't mean that the construct is not useful.

KAUFFMAN: I disagree with Brian, too. If you think about how flowers and insects meet and match one another, there is something to the particular stuff and the particular way that they are functionally integrated that is essential to their workings. I think that what we're missing is a theory dealing with functional integration between interacting entities and the approach that I suggest which bears on that question is Walter Fontana's lambda calculus or these autocatalytic polymer sets where, once you have kinds of molecules that are self-reproducing, what they share with one another, even if it's still hypothetical, can be food or toxin so that differentiating between poison and food is what I would want to call an adaptation. I think that, in that primitive sense, two autocatalytic sets coping with toxins that they're squirting at one another and learning what a toxin is and what a food is, these represent examples of functional integration, mutualism, and so on. So the real problem is that, outside of something like Walter's language, we don't have the start of a language to talk about functional integration and Walter's is the right framework to do that in.

BAK: Peter, I want to ask why are you so sure that when you do the B-B reaction, there being hundreds of species in this reaction, the system is not different at the end than at the beginning?

SCHUSTER: The set of chemical entities you have is stable.

ARTHUR: I agree with Brian completely. What you call functional integration, why can't you just call it self-consistency in a kinetic system? There was no problem to be solved when the whole thing started. The problem and its solution just coevolved kinetically.

ANDERSON: I want to say that there is something new that is introduced when Stu says it's food to an individual, and that's adaptation. That's one of the big issues here: what is an individual or a colony and why does it act in such a way as to perpetuate itself?

KAUFFMAN: What's an individual, what's yum, what's yuk. And you have to know all that to know what a schema is. I think that bacteria as a protein dealing with yuk or yum has a schema but it's not clear to me whether a protein floating out in a pot all by itself has a schema.

JEN: OK. Let's talk about banking.

LICHSTEIN: Let me describe a landscape and then you decide whether I've described a system that is exhibiting any of the characteristics that are of interest here.

The banking system until about 1965 was characterized by high degrees of geographical segregation at the level of cities and countries. It was also characterized by a fair amount of monopoly expressed in terms of limitations on price competition, and in terms of product. Very simply, we could not pay more than a certain amount of interest for deposits and we could not charge more than certain amounts of interest on loans and other products. And so we had a monopoly in terms of product and we had protection in terms of price. So we had geographical segmentation and even protection, monopoly in terms of product, and protection in terms of prices. Each one of these major characteristics in the the banking system has been completely eliminated over the last 25 years.

Let me go through what that has done. We have commercial banks, about 14,000 in the United States; investment banks, about 100 of any significance; and S&L's, which were a feature of the landscape for a while, about 4,000; insurance companies should be there, also. One of the most interesting segregations of products is that insurance services are provided by insurance companies and banking services are provided by banks. But if you uncover what they really are, they are doing similar things.

COWAN: Just say a word about what an investment bank does.

LICHSTEIN: An investment bank, in simple terms, provides access to capital markets. Capital markets are trade instruments for moving funds from people who have funds to people who need funds. This is intermediated by markets instead of intermediated through product. In 1965, if you went abroad, you would see a very wide range, from Canada with about 40 commercial banks, Japan at about 100 that they call city banks, all the way to France and the UK which had a wide variety. Now if you look at the landscape over time, each of the characteristics of the commercial banking environment changed dramatically. Increasing globality became a topic of conversation at conferences about 10 years ago. Now you don't hear it talked about because it has become a cliché. Every banker talked about the advent of technology. The ones that are around today are using technology, and the ones still only talking about it aren't here. I'll go through some very concrete examples in an attempt to get Brian to think about commercial banking.

ARTHUR: I do it all the time but it's a very sad affair.

LICHSTEIN: We can go into the Latin American debt problem, and it's fun to take a look at the S&L problem. The phrase "moral hazard" became important to us, meaning in loose terms that you're dealing with a customer who can affect the world in which he operates in such a way that it causes you great pain.

For example, in the early stages of the Latin American debt crisis, American banks had upwards of 100 billion dollars exposed to a set of countries that publicly said they couldn't pay it back. The countries could do things on the world stage that made it less likely that the banks would ever get paid back. This made it more likely, therefore, that we would come to terms and do a Brady plan kind of debt reduction. (The Brady plan involved substantial amounts of debt reduction.) That's on the LDC side.

With the S&L's we created the most incredible moral hazard structure where we said that the depositors are guaranteed to get their money back but there was no control over what the S&L's could do with the money, in the early days of the Reagan administration with Congress going along. There are four or five interesting phenomena associated with the S&L story and there are at least five or six books on what caused the S&L scandal. They each take one dimension and each one is correct. There was a tremendous amount of fraud and chicanery. There was a tremendous amount of moral hazard, where you just bet the bank and, if you won on the gamble, you made it in real estate or something else or, if you lost, the government took care of your debt. That was a real asymmetrical payoff situation.

COWAN: Was there anything different between what Maxwell did and what Keating did?

LICHSTEIN: They chose very different devices. The S&L's were run badly and the regulatory authorities allowed these crazies to make an accounting change which let them take the profit from selling mortgages in 1982 and amortize the losses over the next 12 or 14 years. And then somebody noticed that half of the S&L's in the country were losing money every single year. You can't operate that way. So they are going to disappear. Why did they get into that situation? They got into that situation because they were the other set of players when the price monopoly changed. If you remember back to those days, until 1981 when interest rates took off, the S&L's had a one-quarter percent advantage over commercial banks and, when commercial banks paid two and they paid two and a quarter, that was a big advantage. But when it went to five percent and five and a quarter percent, you'd be surprised how many people would leave their money in the S&L's instead of switching to banks. That was before money market funds and Eurodollars. Until about 1980, the price monopoly included this differential and allowed the S&L's to stay alive. The price cap dictate was removed when "globality," made possible by technology, made S&L's go out of business in a huge flash, costing all of us lots of money.

The change sequence on geography was that the United States, from the age of Jackson in 1836, was very anti-banks. So banks were limited to one city in almost every state and commercial banks, until the last fifty years, were only regulated by the states. In the last fifty years, first the Federal Reserve and then the office of the Comptroller of the Currency had more and more responsibility. But geographical segmentation meant that banks were in specific cities and rates were different in different cities. You could look in the paper and find that a Denver bank was paying more than a New York bank. When the differences became large enough, you turned to the mail and moved money around. So consumers showed a willingness to deal through the mail and with different cities. Then you had investment bankers doing very interesting things with brokered deposits, which further broke down the value of the city and state barriers to competition. The legislative initiatives for change followed, by about ten years, the advent of regional and countrywide competition. What it took to cause legislators to act were disasters, first in Texas and then the Northeast. The massive failures of some S&L's that required larger institutions, not in their home cities, to be able to take them over. It was disasters that forced the regulators to recognize what was really happening. First, city by city and state by state, then region by region. Now you have the exact phenomenon happening across continental boundaries. The Hong Kong and Shanghai Bank shifted its charter to London and provided a clearing bank.

So the United States has seen the S&L industry essentially go away. There are only about 2,100 of them now. They really no longer have a reason for existing.

The number of commercial banks has dropped from 14,000 to about 11,500, with an increase in concentration. The list of the top 20 banks is becoming more dominant in terms of size. One feature that I've yet to understand is that the top 50 or 100 banks have, within a factor of two, the same percentage total assets of banking around the world. The number of banks in the United States will probably continue to drop, to about five or six thousand. Technology is driving much of the reduction in forces. If you watch the BofA/Security Pacific merger in the West, or the Chemical Bank/Manufacturers Hanover merger in the East, what you see is a 15% reduction in cost the day they got their back offices integrated. We've all lived through this. None of us were alert enough to the process to give it a name but all of these phenomena are continuing. Commercial banks have tried to become investment banks, while the investment banks have tried to take over some of the commercial bank functions. Some of the banks have become boutiques, special product offerers. Products have disappeared from everybody's list as they become driven by markets and media technology, and there's a lot of change.

COWAN: The commercial banks had the same opportunities to inflate their profits as the S&L's had but they didn't. Am I wrong about that?

LICHSTEIN: Partly we did. But our regulators didn't let us do the same things.

COWAN: You mean you were better regulated?

LICHSTEIN: Well, we weren't so narrow in product. When that game was played in 1982 with mortgages, that was the beginning of the end. If you want one single thing that killed the S&L's, it was that game.

MOORE: This is just a general point but sociologically it's really interesting that for some people, complex systems has been a vindication of the free market and a vindication of the idea that things should be unregulated. For others, especially with the realization of the phenomenon of increasing returns and the formation of monopolies, etc., we have a denial of the fact that the free market works that well and that we might have to put in some well-defined regulations here and there. I haven't heard people go really head to head about this. The tension doesn't reach the surface but it's definitely there.

LICHSTEIN: I could give you a longer discussion on the fact that industry after industry, responding to the same issues of technology and globality, has seen a dramatic shrinkage in the number of players who are collectively making less money and whose monopoly powers are virtuallly impossible to exercise because of globality. I can take you through the aerospace industry, the air transport industry, every industry that I've looked at has been similarly affected.

ARTHUR: You can see this shrinkage and you can see it in other industries as well due to globality and technology but some of those industries have gone offshore. What are the chances that this will happen and we'll be banking with the Deutchbank or some foreign affilitate? A Japanese bank has a branch in California.

LICHSTEIN: The top ten American banks used to do a lot of work overseas. Citibank is the only one now with an extensive network and it is no longer in the top ten. The same is happening with Japanese banks. They are becoming less global as they have problems. So what you're going to see is fewer and fewer competitors who are worldwide in character. Then you'll have a lot of local players who know their customers.

PINES: Henry, in some of these things it seems that anyone who is a careful student of what is going on might have anticipated what happened. Were some of these aspects anticipated, discussed, and simply pushed aside? I ask it in the following context. Bob Maxfield and I were talking about what might be some of the products of the Santa Fe Institute. Bob said maybe in five or ten years you might be able to testify before a Congressional committee which is considering some disastrous regulation and say, "You're proposing this regulation. We can give you a handy-dandy model that contains the following assumptions." You can play the game. You can see what happens if you regulate this and that, granted these simulations may not have been available a decade ago, a certain amount of common sense indicates that it should have been clear. Was there any debate at that time suggesting these alternatives?

LICHSTEIN: We all think this way enough that we tried to do so. I gave a speech where I thought this through carefully and I have to admit that it amounted to "Therefore, what?" I couldn't come up with an answer. John Reed, chairman of Citicorp, made a presentation to Congress where he said, "Look what's happening in Texas." This was five years ago when we were trying to get nationwide banking. "Look what's happening in Texas, you can't segregate markets that way." And so he was implicitly using a model and trying to deal with it. But Congress ignored what we said.

SCHUSTER: One pressure in the pharmaceutical industry, in which I was working a little bit of the time, to spread worldwide was the response to local changes in regulations and laws governing production and so on. And it turns out that the multinational companies can simply respond to all these changes, like when several types of research become difficult here and there. Do you see in banking similar reasons for spreading all over the world, so that you escape local regulations?

LICHSTEIN: Yes, but we try not to put as sinister a tone to it as you do. The degree to which it's now practical to both move production and physical product is increasing all the time. What you do is move the concept and the design, things that are less mediated in physical terms. You can do it anywhere. It's done openly in offshore banking. There is a kind of Gresham's Law of banking. BCCI exploited it. The huge Cayman Island BCCI bank certainly isn't for local business.

COWAN: Do you see a strong ethical component in banking or do you think that's essentially irrelevant? In other words, what role does integrity play?

LICHSTEIN: We do everything we do as if it will show up in *The New York Times* the next day.

JEN: We need to consider what is the next well-defined problem that we should look at. Perhaps this is one. Should the round table discussions be structured around well-defined problems or should we structure the next two days in other ways?

HOLLAND: Murray has said, and I agree, that this group is much too large for effective discussion and that we're going to have to have some main working groups, with smaller numbers of people before you can have an effective interchange instead of the kind of polemic that this induces. I suggest that we try to do something like that when we have our discussion sessions.

COWAN: I thought the discussion groups in econ were not very useful. Perhaps I was in the wrong discussion group.

HOLLAND: The webs discussion group did, in fact, initiate several projects.

ARTHUR: I agree with John that maybe two or three groups would be useful.

COWAN: I'm not sure I agree. We're supposed to be mutually informing one another here and we're supposed to be sharpening our questions rather than taking action. I would guess that this sort of thing is mutually effective for that purpose.

David Pines, Chairperson
Physics Department, UIUC, 1110 W Green Street, Urbana, IL 61801

Search for Consensual Views

PINES: We're trying now to do several different things. To what extent is there any sort of consensus on what we've learned here? Also, are there issues which justify our splitting into smaller groups in order to consider these questions in more detail? We've asked again and again what are the common principles of the features that might underlie a study of the topic of complex adaptive systems. One possible cut for the following discussion is, based on what the outcome of the meeting might be, to give us a sense of where we're going: can we make an agreed-on list of some soluble model problems whose solution is relevant to understanding some general principles of some complex adaptive systems. I think we've heard one or two but we need to discuss whether they really work. Here we get to the idea that adaptation plays some role in CAS and that the environment plays some role in CAS and that, if we're talking about model problems that don't involve some explicit form of adaptation and do not involve a response to a changing environment, then we may be coming up with a very interesting set of model problems but they are not likely to be of general applicability. That's one general issue for discussion and, if we agree that adaptation and the environment are important in discussing CAS, then we can go on to discuss in small groups or in this larger group what might be some of the model problems which are currently being solved, which have been solved, or which might be solved that would cast light. By that I mean, to pursue John's point as to what constitutes an explanation, what constitutes a solution is either, conceivably in some simple cases, analytic solutions or, alternatively, computer simulations which anyone can repeat and arrive at the same answer.

GELL-MANN: There's a difference between an ideal set of examples and a primitive set of examples which, nevertheless, capture some of the features. Ideally, we would like numerous CAS's undergoing coevolution and so forth. The notion of CAS can exist in much simpler limits and one of them is a stationary time series for the environment. You could even have an exogenous fitness imposed. It's not what we want ultimately but it's a possible way to study CAS's because they still exist in the limit, provided that you're not near the solution of the problem, that it's not like a computer system for solving tic-tac-toe. That's really not adaptive because the problem is already solved. But if the problem is chess, which has not yet been solved, the problem can be very far from equilibrium and you can get a CAS even with the environment being a constant time series, which you don't want ultimately, even with an exogenous fitness which you don't want ultimately. So I think we should make the distinction between simple limits in which you still have a CAS and the more complex situations that we really want to study.

PINES: I had that clearly in mind and what I was trying to think of is can one imagine, for example, the program that Alfred Hubler and I have engaged on as being one candidate hydrogen atom for understanding CAS, by no means the only one, a primitive model but one which you can pursue through to the end since you can understand everything about it and you can go on from there.

GELL-MANN: I mentioned two objections I had to it both of which can be fixed. One of them is that, in a sense, it's too narrow because the environment is changing and that's an important part of it but you can still have a CAS without that. The second is that there's control of the noise, the chaos, which is perhaps too special. Very rarely does one have control of the environment rather than some fiddling around which influences the environment by feedback which is not control. Both of those can be fixed, however.

PINES: The second thing I thought we might try to do, so that you might have a notion of where we might try to go in the next 35 minutes, was to discuss how we would then, as a group, go about refining that list of twenty-odd questions, and do we do that by splitting into a small group that refines questions? I don't think that's a useful thing for us to do as a larger group. The final set of issues I hope we can get to are what John described as reality checks; namely, what kind of information is out there for specific systems which we agree might serve as specific examples of CAS to which we try to apply not only our concepts but for which we hope to develop new kinds of model solutions? With that preamble, John?

MAYNARD SMITH: I can start by saying that when I'm subject to jet lag, I have an unreasonably short fuse. Normally I am a rather amiable person. Secondly, I am a cynic. I am totally unconvinced that what you people are up to is in any way a sensible enterprise. I am not convinced that it is not a sensible enterprise but it's not the way I do science, it's not the way I've ever done science, you're trying to do it a different way, and I'm trying to find out what it's all about. What has stuck me over the years in relation to conversations between scientists from different disciplines, is that terrible troubles can arise over semantics. You guys obviously mean something by adaptive. I'm still trying to work out what it is. It's quite clear to me that it's not what I mean by adaptive. At least I don't think that it is. Perhaps in two days time, when I think about it, I can work it out. I don't intend to cause offense but I do think that biologists, in interacting with physicists, often find themselves completely infuriated by the fact that physicists seem to think that all biologists are ignorant and stupid, that they assume that we haven't actually thought about our problems, that we haven't actually got a theory of these things, and it can be very annoying.

GELL-MANN: Maybe you can list what you are reacting to.

MAYNARD SMITH: Well, the Per Bak talk. The first part I really liked and thought it might be applied to biology. That's true of the first two-thirds or so. He then started to make remarks about biology that showed he had a complete misunderstanding of what biologists mean...

GELL-MANN: But biologists always say this when they listen to his talks.
(Laughter)

MAYNARD SMITH: I've been listening to Stu for years and he makes me furious and he even has a medical degree.

ARTHUR: But at this meeting, we've even had an economist explaining jet engines.

MAYNARD SMITH: But biologists have thought about some of these problems and we may even have some ideas about some of them.

COWAN: I gather that you came here expecting modesty. Is that correct?

MAYNARD SMITH: On my own part, at least. We're going to have to watch semantics. I gather that you've been listening more to economists. But if biology is a CAS, you can have many thoughts that will be useful to us. They are very deep although there are many things you don't understand at all. We need help. But it takes small groups of people to hammer on these points and try to get the semantics right.

SIMMONS: I do think we should hear your definition of adaptation. It would be a great shame if it were only heard in a small group.

MAYNARD SMITH: Can I give an example of what I mean? What is biology adapted to? I don't think that biology has adapted to anything. No biologist would ask that question. Adaptation is not a property of an ecosystem, at least in my way of thinking about it. An adaptation is a property of an individual organism, not of an ecosystem, which makes that organism more likely to survive and reproduce. Semantically, we must only use the word "adaptive" if we think that the trait in question evolved by natural selection. I deliberately didn't use the word adaptation this morning because it's a very tricky word. But it's never a property of a population, let alone of an ecosystem. A heart is adapted for pumping blood around the body at the right speed, the right speed meaning to supply enough blood to make sure that we remain alive.

HOLLAND: Sometimes a distinction is made between an adaptation, which is the property of an organism, and the process of adaptation. How do you feel about the process of adaptation? Is there an evolutionary aspect related to the process?

MAYNARD SMITH: I am quite happy about that. There is a terminology, which is not universal in biology but which is actually quite useful. I prefer concrete examples rather than generalities. We spent a lot of time in Virginia the last few days arguing about whether the process of genetic transformation in bacteria was an adaptation. If so, what was it adapted to? Was it simply an accident, something that happened to bacteria not because of any natural selection at all. Or it could be an adaptation because DNA is good to eat, or it could be an adaptation for genetic transformation. In biology if we say that the function of some organism is to do x, we mean, not merely that the organism does do x, but that the organism evolved because it does x. Nautral selection favoring x has been responsible for the evolution of the organ. Thus, if we say that the function of transformation is to accelerate evolution, we mean, not only that it does, in fact, accelerate evolution but that that is why the transformation evolved. It would not be there had it not been for the fact that it helped evolution. If we merely say that it is a "consequence" of transformation that it helped evolution, then it's just there. We don't think that it is there because it helped evolution. So we spent a lot of time discussing whether it's a function or a consequence. Now I don't ask you to adopt those terms. But biologists would normally only think of something being an adaptation if it was, in the above sense, a function. The back of a horse is not adapted for people to ride on.

KAUFFMAN: I was trying to make the point the other day that there are severe difficulties with the assumption that we know what something has adapted to. In many cases, we simply don't know. It's not clear what fingerprints are for.

MAYNARD SMITH: Of course, we may not know the function of some organ. Usually it is obvious. Fingerprints aren't. There are plenty of things we don't know.

LLOYD: One of the most interesting points raised at this meeting, first by Peter Schuster and then by John Maynard Smith, is the point about adaptation toward adaptability. Peter made the point that some viruses of which you are aware have a mutation rate that is the maximum that still allows them to be viable. Similarly, John made the point that eukaryotes seem to have mutation rates that are as high as will still allow them to reproduce. This would be quite a remarkable general principle that what is being selected for is the ability to select faster. That the mutation rate, the rate of information processing, if you like, or the rate of exploration of an adaptive landscape is directly proportional to this.

MAYNARD SMITH: It's the right fact but the wrong interpretation. I think the reason for the high mutation rate of eukaryotes is that it costs a lot to keep the mutation rate down, you have to do a lot of proofreading and error correcting, so that it tends to drift up, so long as it doesn't matter. Not because it's good to have the high mutation rate but because it's not necessary to keep it down. At the same time there's a tendency for the message to get longer and longer, until these two reach the critical point where, if the mutation rate goes any higher, the animal can't survive. But I don't think the mutation rate has gone to that level because it helps the species evolve faster.

PINES: There may be settings in which that is true. If one took the economic model, for example, and if it really makes a difference as many people believe that an agent that is adaptive, that responds quickly to changes with the specific goal of improving its performance, if that agent survives longer and will reach a higher position within the economic environment, then Seth's point is well taken—that adapting toward adaptability might turn out to be a very good thing.

MAYNARD SMITH: You might argue that the virus mutation rate is another example.

SCHUSTER: If you consider viruses in mammals, for instance, where they have to survive the immune system, then a higher mutation rate is an advantage because, if it occurs in certain parts of the genome, it changes the antigenetic determinant and tends to fool the immune system. HIV seems to be particularly efficient in that. The problem is that at the same time the virus mutates in other parts that cannot afford mutation because of its conservative features. This seems to be reflected in the very low degree of viability in virus copies, for example, in influenza where one virus copy survives out of thirty. Twenty-nine copies are junk. They have mutations in places where they can't afford it. It seems to work in those places where you can interact with the immune system. In that case, I think it is a kind of adaptation. In the case of higher organisms I would be very hesitant to say that it has the same feature.

MAYNARD SMITH: You would use adaptation in the way that I just used it, then.

SCHUSTER: One thing is particularly important, as John pointed out: it is the individual replicating molecule that changes and then runs faster and, therefore, adapts to a situation where the rabbit runs faster.

KAUFFMAN: There is a puzzle about individuals, John. Suppose one were to take seriously the species selection argument. If one accepts the notion of species-level selection, then one can once again apply the notion of an adaptation to whatever the unit or level is. Would you agree with that use of the word?

MAYNARD SMITH: Absolutely. I think there is species-level selection. It's a default process. One tends to assume that it doesn't happen, but occasionally it does. I think it does in relation to sex, for example. So you can talk about a species-level adaptation, sure.

ARTHUR: Can you not talk about the adaptation of a whole group of species?

MAYNARD SMITH: Not if we're going to think of adaptation as the consequence of the process of natural selection. Ecosystems do not reproduce and do not compete with one another. We cannot assume that ecosystems have adaptations unless they are entities that multiply and have heredity.

ARTHUR: Suppose you had an ecosystem in some part of a bigger system and then something changed the system—for example, the kind of fish they were eating disappears, then the whole thing would change—you would get something different.

BROWN: It's true that at a lot of levels in biology; it may not be true beyond biology. It is a relatively precise Malthusian-Darwinian dynamic. We have units, the units reproduce, more are born than can survive, there is variation among the units, and, therefore, if there is differential survival and/or reproduction, we get modification. We can have units that do that, with cells, and cancer has been mentioned as an example, certainly from cells to individuals and up to species and, I think, cultural attributes can do that. Now, I guess the question is "If the differential reproduction is in some way related in some predictable mapping way to the environment, so that these things provide a better fit to the environment, then we call that adaptive evolution. I want to know two things. Is John willing to agree to that; and secondly, is there some reason why we should necessarily restrict our definition of adaptation in our discussion of these things to just that subset of phenomenologies?"

MAYNARD SMITH: I don't want to restrict the use of the word adaptation at all. It's just that we must understand each other. That's all. You use the word adaptation how you like. But it's important that we should be jolly clear how we are using it so we can't misunderstand each other.

PINES:　　　　　　　So you accept Jim's definition of adaptive evolution? That's okay.

MAYNARD SMITH:　　Sure.

LLOYD:　　　　　　I have another question for John. Earlier on, you said that the reason you didn't understand how sex had evolved—correct me if I'm wrong—is that although sex may be advantageous for the species as a whole, at the immediate level it may be disadvantageous for the individual. So it seems to me that you don't accept adaptation of the species as a whole. Is this why you regard this as a problem?

MAYNARD SMITH:　　You understood the problem perfectly. I am quite happy to regard sex as, in part, a species adaptation for just the reasons you described: the species abandons its own state. The reason I can't see it as a truly satisfactory explanation is the following: we find a variation within a species in reproducing sexually or asexually. If the only reason why sex is around is because it was good for the species as a whole, you can't have a situation where you go both ways because there is certainly genetic variability within a species as to whether you produce most of your seed one way or most another. There must be some shorter term force operating to keep something that operates every few generations, not over tens of thousands of generations; otherwise, there would be a tendency to stop producing their sexual flowers. There must be some short-term sexual advantage and people are doing experiments to find out what that sexual advantage might be. It probably has to do with resistance to viruses and things like that.

GELL-MANN:　　　So you don't find that explanation so bizarre, the one attributed to Hamilton.

MAYNARD SMITH:　　No, I don't find that bizarre at all.

GELL-MANN:　　　About outracing viruses or other parasites?

MAYNARD SMITH:　　No that's fine.

LLOYD:　　　　　　So outracing viruses is another example of trying to increase your genetic variability. Sexual variation would be another example of moving to much higher variability.

MAYNARD SMITH:　　Sure.

PINES:　　　　　　What about time scales for adaptation? Is that a question that one asks in a biological context? Do you speak of n generations where n is some number that is relevant to talking about classes of adaptive behavior?

MAYNARD SMITH: Well, we're probably inconsistent because sometimes one talks about organisms adapting to high temperature or high altitude where we don't mean genetic adaptation at all. We mean change within the lifetime of the indvidual by increasing the blood count or something of that kind. Adaptation during the lifetime of the individual is different. We're not too confident about rates of adaptation. One of the reasons why I work with bacteria nowadays is that I can do it in my lifetime. You can actually study the process whereas higher organisms tend to take thousands of years and are difficult to study.

GELL-MANN: If the explanation of the advantages of sex in terms of out-racing viruses and other parasites is acceptable, what was the resistance to it in Blacksburg, and elsewhere?

MAYNARD SMITH: The trouble is that we have more theories than we quite know what to do with, more candidate explanations. The main argument was between people who were taking a sort of outracing the environment line of argument, which is, on the whole, my own view although I don't feel deeply confident about it, and people who were taking the view that sex has to do with getting rid of damaged DNA in one way or another and there are a variety of models tending to show that if you reproduce sexually, it makes it possible to eliminate bad parts. You can make one good motor car out of two bad motor cars by taking the gear box out of one and so forth. So if you cross two organisms with different damaged genes, you can make one organism that works. That's the main rival explanation and the trouble is that they're both partly right.

SCHUSTER: That is precisely also the advantage for the short-term asexual reproduction because, if I already have a good car, there is no reason to take parts out of the other one.

MAYNARD SMITH: That's the Austrian argument. That's the problem. It's a difficult, complicated argument. It's really why I prefer the rapid change argument, the sharing genes argument.

PINES: Let me change to my moderator hat. Let us discuss the organization question for tomorrow.

KAUFFMAN: I propose that we do not do so immediately tomorrow morning but that we spend at least an hour or so with the group as a whole parsing our list and seeing how it assembles or falls apart.

COWAN: I think you might better say that you could spend an hour trying to select three items out of the list of twenty on which you want to spend the rest of your time. Maybe you can do it by vote. If you try to discuss that list extensively, you can spend the entire morning at it. It would probably be more profitable to trim it down very quickly.

KAUFFMAN: What I was trying to say is that if we look at that list, we'll see that they form overlapping families and we will winnow it down and it will fall apart pretty naturally.

COWAN: If you can do that very quickly, that would be an hour very well spent.

PINES: I have a certain skepticism as to whether, in a group of fifteen to twenty-five, we could in fact do it because it takes an hour for fifteen to twenty-five people each to get in his or her two minutes.

KAUFFMAN: Perhaps we could take a little longer but we need to see how the problems interrelate before we decide how we want to segment ourselves if we want to segment.

COWAN: I think that some one should walk in here tomorrow, or two or three people, and say, "My proposal for the short list is the following," and then we can debate that.

HOLLAND: That sounds like a good approach so we don't wander around so much.

GELL-MANN: As you said, George, it's not as though we have a list of twenty or so questions at the same hierarchical level and we're going to suggest dropping thirteen of them and leaving seven. It's a mishmash of overlap, redundancy, different hierarchical levels, contradictions...

COWAN: I really meant to say, "Find three summary statements which we can discuss."

SIMMONS: The list is presumably incomplete because it was prepared two days ago and we've had more thoughts.

PINES: We've probably already added about ten more questions to the list as subsequent speakers have said, "What I really want to know is this." I continue to think that, after the discussion in which I'd be happy to participate, we could still try to organize around the notion that some subgroup goes off and tries to refine the questions, some other subgroup goes off and talks about fundamental concepts and soluble problems, and some other subgroup discusses models and reality checks.

GELL-MANN: If we were to add to the list things that we've discussed since, one of them would surely have to do with this question of hierarchy, different stages, and the generation of those. We've heard about the generation of those by rather big jumps in biological evolution, jumps that involve a change in character, of organization. We've heard about that from a variety of people. Jim Crutchfield talked about it, Chris Moore talked about it, both of them from the point of view of Chomskyan hierarchies and mathematics (I won't say linguistics because I don't think they have much to do with linguistics, but mathematics, yes). We heard about it from Leo Buss and others in connection with biology, including John Maynard Smith to some extent. We heard about it from Walter Fontana in connection with a chemical model for the origin of life. In none of these cases did we have enough time to understand fully what was being said about hierarchies. I have the feeling that this is one of the most important subjects that we are discussing and I don't know how to deal with it in a comprehensive way.

COWAN: I'd like to go back to something you said, Murray. In response to Peter's insistence that time had somehow to be taken into consideration, you said, yes, not only do we need a measure of information but also a time relationship. Have I quoted you correctly?

GELL-MANN: Yes. Well, Seth included a number of those temporal measures in his list including Charlie Bennett's.

COWAN: Are you referring to the length of the computation?

GELL-MANN: Not just the length of any computation but the number of steps needed, for example, to go back from data to a simple scheme for generating the data. Time measures differ technically from information measures. Information measures are all subject to certain simple inequalities. The time measures do not obey the same inequalities.

COWAN: The reason I am suggesting that the time relationship be explored a bit more fully is that in the real world, when we discuss what we think of intuitively as a hierarchy of complexity, what we are doing is making big changes in time scale at each level. The fact that the time scale changes by orders of magnitude at each level may have something to do with how much computation is going on.

GELL-MANN: That's certainly true but there is also a much more elementary point things that look constant or look universal on one time scale are not when you look at them on another time scale. They are actually peculiar to an epoch. The same is true of things that are terrestrial constants but are cosmically just accidents, like nucleotides that are characteristic of life on Earth but may not be the same on other planets.

MAYNARD SMITH: Well, we're probably inconsistent because sometimes one talks about organisms adapting to high temperature or high altitude where we don't mean genetic adaptation at all. We mean change within the lifetime of the indvidual by increasing the blood count or something of that kind. Adaptation during the lifetime of the individual is different. We're not too confident about rates of adaptation. One of the reasons why I work with bacteria nowadays is that I can do it in my lifetime.

You can actually study the process whereas higher organisms tend to take thousands of years and are difficult to study.

George A. Cowan, Chairperson
Santa Fe Institute, 1660 Old Pecos Trail, Suite A, Santa Fe, NM 87501

What Are the Important Questions?

GELL-MANN: I'd like to discuss three topics. They are: (1) Stages of Operation of Complex Adaptive Systems, including:

- Information gathering and coarse graining;
- Identification of perceived regularities (as opposed to random or special components) and compression into a schema;
- Variation of schemata (including mutation or whatever is the mechanism of variation in the particular field);
- The use of the schema by the phenotype.
  Here we pay attention to Walter Fontana's warning that even when there is no physical distinction between the phenotype and the genotype we should still maintain the distinction between object and function. In feeding in data or simulated data, some random or special information is restored, yielding description, prediction, and behavior.
- A collection of complex adaptive systems coevolving.
  There are many other inclusion relations that we can consider here, many other ways in which complex adaptive systems form parts of other complex adaptive systems. For example, the mammalian immune system or the mammalian brain is a part of biology, produced by a biological evolution.

Some questions concerning the first topic are:

- At which stages do we encounter the order/disorder transitions?
- What do we mean by efficient adaptation in this transition region?
- When do we have attraction to the transition region?
- When do we have the transition region as a critical point with a power law distribution?
- When is it relevant to postulate all scales of space and time, giving the possibility of universal computation?
- Is there a possibility of especially high complexity in that domain and, if so, defined how?
- What is the appropriateness of the term chaos and its variation from case to case?
- What is the significance of the original cellular automaton example?

(2) Major Steps in Biological Evolution as Stages in Level of Organization.

Here we go from one place in the hierarchy to another; what do we mean by hierarchy? Here we can mention the Fontana-Buss duet, the so-called Chomsky hierarchies, the relevance of other hierarchies, and the connection with large steps in all sorts of other complex adaptive systems. People didn't always mention self-awareness near the end of that chain but probably it belongs there. They did mention changes like going from RNA (if that was the original stuff in the origin of life) to the invention of the organism and then from prokaryotes to eukaryotes.

SIMMONS: Does emergence of hierarchies belong in the stages of operation of a complex adaptive system or somewhere else?

GELL-MANN: We were saying that, in human thought processes, big jumps may be comparatively common compared to the usual jumps in evolution which usually use what is already available and make small changes. However, in biological evolution, as was pointed out by many people starting with Leo Buss, there are occasionally major changes and they are interpreted as changes in level of organization. The incorporation of endosymbionts into a cell as organelles is mentioned as one example and the development of metazoa is another one.

SIMMONS: Big jumps are, I think, a more frequent occurrence in economic evolution where a group of individuals get together and, from that, some small organization is formed which grows into a firm which...

GELL-MANN: In many kinds of complex adaptive systems you can have changes in levels taking place through aggregation.

QUESTION: ...What about punctuated equilibrium?

GELL-MANN: Some cases of punctuated equilibrium can be interpreted in terms of micro-mutations going on and on over a long period of time without producing major phenotypic changes. You can see that, in a certain sense, in Tom Ray's computer model, although there the physical distinction between genotype and phenotype is wiped out for simplicity. Nevertheless, as Walter pointed out, you have a distinction between the object as object and the object as function, a distinction that's still worth making. You have long periods during which nothing much is happening and suddenly there are very big changes in population. But this is an artifact, not a fundamental macro-mutation by any means. It's the result of small changes stored up.

BROWN: But there are lots of inventions, things that have appeared several times which have led to enormous explosions of diversity in different areas. I could go on and on.

GELL-MANN: I agree, when certain barriers are breached, there can be explosions of changes. So we can put here "biological evolution, e.g." Now how does Buss's theory operate relative to Chomsky hierarchies? They sound as though they might be generalized with profit so let me put in "other hierarchies." In other words, how do we think about the (various) kinds of mathematical models that people have dreamt up. And here I'll put what everybody has just been saying. "Connection with large steps in other kinds of complex adaptive systems." So this is the second question. And one can fill in other aspects. I probably haven't listed all the different ways in which we have discussed this issue.

(3) Identification and Compression of Regularities into a Schema

Jim Crutchfield presented a discussion; how can we generalize it? I would put here "and possible extensions and generalizations" and maybe comment from particular points of view. For example, those who work with neural nets may have some particular impression and also people who work with genetic algorithms. We should also discuss the general class to which these two examples belong, an area in which the Santa Fe Institute has not yet made the kind of progress that it may make. They are very different. Genetic algorithms are loosely modeled on the genetic aspects of biological evolution, not the developmental aspects which Brian Goodwin has worked on. Neural nets are loosely modeled on somebody's caricature of the human brain.

And, finally, we can mention here a certain type of complexity measured by the length of the schema. I believe that, in most popular and scientific discourse, this is what actually is meant when people say "complexity." In archaeology, for example, in discussing the complexity of a society, in linguistics in discussing the complexity of a grammar or phonology, or in biology in describing the complexity of some structure or organism. But it's a schema describing something as visual as viewed by some complex adaptive system. If it's the CAS under discussion that's doing the viewing, as in a genotype or an internal representation of the grammar of a language in the brain, then it's even more relevant than usual because the CAS is observing itself and the schema, the length of which one is discussing, is a kind of natural representation.

COWAN: Doyne Farmer addressed the comparison of neural nets with genetic algorithms early in the game and there is a useful SFI paper by him concerning these two methods.

GELL-MANN: I meant that the work on this topic is not yet as deep as what we can eventually produce. I think it is very important to characterize the class to which these two belong.

COWAN: May I suggest considering the possible role of time in the sense that Phil Anderson put it: if, in the real world, a process goes on and on without ever going anywhere so there is no definable end state, what can you say about the complexity of that process? What I am saying is that if the process is totally dynamic and just goes on and on, is it relevant to discuss complexity?

HOLLAND: Put it this way: there are certainly systems for which we can design models in which an attractor plays no role. The system simply goes on and gets increasingly complex as specified by some measure and there are no repetitions. The attractor, if it's there, is so far out that you're not going get there in the length of time that you look at it. How do we study such systems? Can we study such systems?

GELL-MANN: Yes.

HOLLAND: George was just saying that if you have a system like that, how do these questions apply? And also he was suggesting that that may be, in fact, the way the world operates.

SIMMONS: You mean, maybe, the Lindgren model?

HOLLAND: Or even Tom's model if he had a little more room.

BAK: That's only if there's no limit. In the limit, the system becomes bigger and bigger and it takes a longer and longer time to get to the attractor. You can only measure that asymptotically.

HOLLAND: But, Per, it is easy to design systems in which there's no limit.

BAK: No, if they're finite, there will be a limit.

HOLLAND: Sure, but it goes up as combinatorial factor so that the finiteness means nothing.

GELL-MANN: Can I put it in here in the following way: There is a contrast between the extremphile or tic-tac-toe case where things seem to go to a more or less stationary condition after a while and an open-ended evolutionary process. Is that what you meant, George? (Yes). It raises a separate set of questions, a very interesting set which we should explore further.

SCHUSTER: I should like to add one aspect which is only treated in part by the models I have heard discussed here and which I think is particularly relevant for biology and maybe also for other CAS. I refer to spatial diversity (occurring) in a way which we cannot yet treat appropriately. Most of the models either are poor in spatial dispersion or don't look at it at all. I think that, in the few examples we have where we can really see dynamics on a spatially dispersed system, the results are substantially different than those we find in spatially homogeneous systems or those in a few compartments.

GELL-MANN: We have heard in Chuck Stevens's discussion (and we've heard over the years from many people) about the relevance of geometrical considerations such as distance relationships in the human brain. Local regions have certain specializations communication is much greater nearby than very far away. Is that the kind of thing you mean?

SCHUSTER: No. That is not what we are looking at. We have a heterogeneous environment that provides us a kind of parallel development in many, many patches and this is very powerful.

GELL-MANN: As in ecology, for example.

SCHUSTER: Ecology is one example. I would call it spatial diversity.

GELL-MANN: But it's not completely diverse because you have recurrence.

SCHUSTER: It's not completely diverse.

GELL-MANN: So, let us add "spatial heterogeneity." I would say that's a separate topic, a very important one.

ARTHUR: A couple of things. I would add the topic that I and John Maynard Smith have talked about, although we came to opposite conclusions. I mean the theorem that as evolution takes place, things tend to become more complex.

GELL-MANN: You don't mean that, I'm sure. You mean that systems tend to permit greater complexity. Otherwise, every single thing would turn out to be more complex.

ARTHUR: No. I would include this item as a question. Do things tend to become more complicated as evolution proceeds?

GELL-MANN: But don't you mean it slightly differently? Don't higher levels of complexity arise as evolution proceeds?

KAUFFMAN: The question is why do things become complicated, and one of the answers is that it's an entropic exploration. Another is that maybe it's worthwhile and that's a very important question.

GELL-MANN: I understand that, but I want to introduce a small modification because some simple features are still present even though the maximum complexity has increased.

HOLLAND: If some elements of the system have more complexity, it doesn't necessarily mean that the system itself is more complex. Maybe the question should be "Does it have more complex components?" The two are not equivalent.

GELL-MANN: I'll write, "Does higher complexity tend to arise in the course of evolution?" Okay. (Yes)

ARTHUR: And I made a distinction between higher complexity in the sense of coevolutionary complexity with more interaction and more diversity and single entity evolution where the entity itself becomes more complicated or more sophisticated. It has to do with the part where you talk about cellular automata. The question is "Under what circumstances are coevolutionary systems always adapting to mutual changes in each other's elements?" If a change takes place in one element, that may or may not cause a cascade of adjustments in the other element. . .

GELL-MANN: May or may not? You mean that sometimes they are not coevolving.

ARTHUR: Yes. In a system that has a very strong attractor, say, all of the elements want to be at zero. If I change one of the elements to state one, the elements at state zero simply don't want to or find it difficult to. . .

GELL-MANN: So that's like a Nash equilibrium attractor as opposed to a coevolutionary attractor?

ARTHUR: No. It's like a zed-ordered system. In a highly chaotic system if you change one element, you may cause a complete cascade of changes. . .

GELL-MANN: Let me rephrase it. When asking whether there is an alternative attractor which is kind of a fixed point attractor like an equilibrium situation, as contrasted with an attractor. . .

ARTHUR: No. There is a spectrum. At one end of the spectrum, if I make a lot of changes in the elements, nothing happens to the other elements and everything settles down quickly and stays at the zed-ordered state.

GELL-MANN: Like a fixed point attractor?

ARTHUR: Yes. At the other end of the spectrum, I make a couple of changes and those changes percolate rapidly through the system and the system, which changes as a result, never settles down. Presumably in between, in the critical region, transience itself becomes a phenomenon and I believe this has something to do with the halting problem. It's something that Stuart has thought about. Does mutual adaptation go on forever?

KAUFFMAN: The question is whether or not there is a phase transition in a lot of coevolutionary systems and whether the edge of chaos, if I may use that phrase non-pejoratively, is in general the attractor of that dynamic?

COWAN: We've defined a group of questions which, obviously, are going to be discussed in greater detail when the subgroups convene. I would like, if we have a sufficient description of the problem, to call on David now to suggest a somewhat different set of items.

PINES: I'm not writing a book. I want to take a somewhat different cut through the wonderful set of topics that Murray has set forth. It seems to me that what he's done is to describe, if you like, not only a schema in terms of which he can write his book, but he has also described a research program for more or less the next century. Some of us can only anticipate contributing actively to that field for the next thirty years so I want to take a somewhat shorter term look at these aspects.

I want, first of all, to put more emphasis on adaptation, the point that Josh raised and that we started to discuss yesterday. I then want to talk about how we might make progress in identifying those elements of the general scheme that Murray put forth and how we can begin to answer some of the questions he raised in terms of either specific toy models or theories, the equivalent of some very elementary approach to complex adaptive systems which, however, can be worked through in considerable detail. And then I want to talk a little about reality checks.

I won't go through them again but wish to call John Maynard Smith's attention to the fact that not all physicists put adaptation in quite the same way as Per Bak did. This is the way that Alfred Hübler and I have approached the problem of trying to develop a quantitative description of adaptation in an evolving, chaotic environment. The reason for putting it down here is that I feel we haven't really focused on a number of these issues. We began to discuss different views of adaptation and had a lovely five minutes from John Maynard Smith yesterday on adaptation as viewed by the biologist. We are here trying to look not simply at biological systems but we're asking the question: Is it useful to look at adaptation in physical systems and, in particular, is it useful to look at adaptation in human beings, in the learning process in humans as a complex adaptive system? Is it useful to ask about the role that adaptation plays in all kinds of institutions starting with marriage, then the family, going on to social systems, political systems, to government, and to global organizations? Is that a useful idea? Are we now in a position to try to incorporate this way of thinking into a number of those problems and does the kind of approach that we're trying to develop for complex adaptive systems have some applicability there? So that's a general set of questions which are sort of subsumed on that transparency. As I said, I won't read them off but will consider them part of the record, which are here reproduced.

My second set of questions has to do with how many solved toy models are there out there at present that are relevant to the questions that Murray has raised, that we all have raised. Alfred and I have described one which is quite well worked out for a single agent, rather well worked out for two agents. It captures a number of the properties that Murray enumerated and I remind you of these. I'd like to see some discussion, before we break up, of a classification of other models which have been worked through from this same perspective. To what extent do they provide a quantitative description of what the system is doing, to what extent are dynamic details built in, to what extent is one dealing with a changing environment allowing for interaction between agents, and so forth. I just want to go away with a set of things that we should all read in this area and a set of things that we might think about. Put another way, what other model problems might come out of this discussion in which one has, in fact, some hope of arriving at a specific solution?

And then my final remark is that we should never lose sight of reality checks. A lot of what we've heard here from our friends in ecology, biology, biochemistry, molecular biology, etc, falls into that category but I'd really like to see more specific associations, such as "Oh, yes, in my field there is a specific set of observational data which supports this particular way of approaching a CAS" or "There is a set of things which are in clear contrast." Now, in my notes there are little bits and pieces of this from almost every speaker but it would surely be marvelous if I didn't have to try to read my handwriting in order to sort all of that out and we could collectively agree on some quite specific examples of reality checks.

COWAN: If we list the toy models we are presently aware of, what is your list?

PINES: Based on things I've heard here, there is a brief description by Per Bak of some specific calculations which have been done on a physical system concerning the role that rugged landscapes play. There are two beautiful papers that Per handed me dealing with some work that he has done on this topic. Then there is the artificial economy that Brian and John and Richard Palmer have worked on which falls into the class with a very well defined set of rules, a very clean set of results that any of us can go home and reproduce. I suspect that John may have a half-dozen other examples although when pushed for applications he wasn't totally forthcoming yesterday. But I would hope that John could add to that list and I don't know whether people would like to go around the table now and do that. It may not require a discussion group but it would surely be very interesting to know. Some of the things that Jim Crutchfield has done fall into this category but, again, I don't know how, in a quasi-taxonomic sense, to say Jim has dealt in a very clean way with this, this, and this. With respect to cellular automata, Erika made it very clear yesterday that there is a set of questions which have been addressed by CA and a set of questions which CA cannot even begin to address. It would likewise be interesting to revisit that subject.

COWAN: Does ECHO belong on that list?

HOLLAND: A point I wanted to get at yesterday is "what do we mean by application?" I honestly don't know what we mean by that. I would tend to say that ECHO is not an application but, if Jim Brown and I go ahead with this project using his data, then it would be an application.

COWAN: Now it's Stuart's turn.

KAUFFMAN: My list is remarkably congruent with Murray's. With his permission, I'd like to number his reasons.

1. This is "bounded rationality" which is a redescription of region 1 but I have a slightly different cut through it. What is the issue of schemata and compression?
2. This didn't come up in Murray's discussion of schemata but I want to point out something that John Holland has been telling us for a long time, which is that, in fact, if you think about his model, one often has parallel processing schemas in an immune network or in a classifier system or, in general, if you think of a cell with molecules acting on molecules and making things happen, schemas trigger schemas in some parallel network, and therefore, we have to talk about parallel processing webs of schemas and that actually leads you quickly to things like Boolean nets.
3. The whole thing by Jim Crutchfield and Holland, trying to point to the notion of hierarchical models and schema variation which is very much the same topic that's in here.
4. Something that might be called coevolution of schemas at the edge of chaos as an attractor. These are some additions to Murray's patch. Now add infinite vs. bounded rationality.

For Area 2, I'm going to make additions to your list. Is life critical?

SIMMONS: You are just saying that the question "Does coevolution occur at the boundary between order and disorder?" can be discussed in various contexts.

KAUFFMAN: Correct. There are a bunch of candidate concrete models to play with in this vicinity. Now here we have coevolution to the edge of chaos and we have clues and concrete models. John's ECHO model, being a coevolving model, smells of going to the edge of chaos. The Tierra model does it and, of course, the NK coupled landscapes that I've looked at which do likely go to the phase transition between order and chaos, precisely the point that Brian is making, and then Chris Lindgren's model...

ARTHUR: What is the evidence that these things are coevolving toward the edge of chaos?

KAUFFMAN: That's my whole point. The evidence in my model is that I explicitly showed that the phase transition looks like an attractor. (Comment: But it hasn't been done explicitly.) It hasn't been nailed down as much as it needs to be but it's one that we know how to go play with very explicitly. In the Tierra model the evidence is tentative but it has to do with the power law distribution of extinction events in the Tierra model which hints of being at a phase transition between order and chaos. ECHO is a model that I think John Holland can look at to see if he gets the same kind of phenomena that Tom Ray has gotten and then we have to ask whether that's evidence for criticality but these are concrete toy models to play with, to answer David's question. Then there is evolution to the edge of chaos. This gets to the notion that Peter Schuster and Manfred Eigen have worked on so much that has to do with the error catastrophe and phase transitions and spin glasses and so on. So spin glass, NK landscapes, and so on. All of the stories that we have been hearing about that viruses perhaps tune their mutation rates, or metazoa just happen to be at that mutation rate and population size that allows them to flow more or less over a landscape and it's a question...

WALDROP: What do you mean by error catastrophe?

KAUFFMAN: This is the Eigen-Schuster idea in their 1977 paper that, if you tune the mutation rate for finite-length RNA sequences, then at a low mutation rate, populations climb fitness peaks; as you increase the mutation rate, errors start to accumulate; and then, finally, you lose all information and smear out over the landscape. The conjecture is that you ought to be right at the melting point in order to explore landscapes well. Which is, of course, not so far from Alfred's notion that in order to get a good overview of the system you should be weakly chaotic.

BAK: Does it mean anything that you invariably translate disorder to chaotic? Do you mean disorder?

KAUFFMAN: I do but I've been calling it chaos for four years.

BAK: But it does little good to call it chaos. I have done it myself (Comment: But at least you regret it) and I regret it.

KAUFFMAN: I don't regret it at all and I think it's perfectly defensible in a number of cases. So this third one is "Do parallel processing networks get to the edge of chaos?"

SCHUSTER: It is really better to use a more general term than chaos so that, when we get to replication, we can talk about random replication which has nothing to do with chaos.

KAUFFMAN: I put in only "to the edge"—no further word. What I mean here is "melting." So this is the theme that asks whether parallel processing networks inevitably, in real cells and in real brains, get to the phase transition to carry out not just the capacity to adapt, but also to carry out complex calculations. There are numbers of models here to answer David's question again. There is Norm Packard's experiment, Chris's speculations, and there's the stuff I've been doing on Boolean games, where there's increasingly good numerical evidence that Boolean networks that play games with one another, can mutate, and change their internal structure and their internal logic, do evolve to the phase transition between order and disorder in order to play games well with one another. There are three other pieces that I want to put in explicitly, I think close to what Brian said: (1) the subcritical/supercritical boundary; (2) the notion of whether or not coevolution is, in some sense, driving diversity, in the Cambrian explosion or in technological evolution, so let's put down coevolution and diversity.

GELL-MANN: If you're going to ask a question like that, you might also ask if the removal of some obstacle is what's relevant in these explosions; that is, if evolution has been pushing against some barrier, when the barrier is gone, then there's a possibility of tremendous explosion. An analogue exists in island biogeography where, if a new island is reached, the possibilities of adaptive change are extraordinary. It may be that all of these explosions, including explosions in human thought, can be related, or many of them can be related, to the removal of some obstacle leading to the sudden opening up of new possibilities. I would add that as a possible additional point.

SCHUSTER: Can we vary the tenor in the following way? If a strong external stimulus appears, it need not be the removal of a barrier; for example, the appearance of oxygen in the atmosphere which was the stimulus for the appearance of eukaryotes. Or you can bring in some poison which you have to cope with.

BAK: Why do you call the appearance of oxygen an external factor? It's an internal feature.

SCHUSTER: It depends on your view, whether you look at the whole earth or you look at the affected organisms. If you look at the whole thing, then it's all internal.

HOLLAND: This is where we come back to the question of adaptation—where do you draw that line. The other thing that goes with that is the notion of multifunctionality. It's two things together, that something that's already there can serve a different or modified function with minor change and then the new opportunity.

SCHUSTER: You can say that the Earth develops but that organisms on the Earth adapt.

HOLLAND: Yes, yes. I think that's at the center of this argument.

KAUFFMAN: To finish my list. (3) There is a very intriguing possibility here that I would like to have some of us think about that's related to the criticality or "Is life critical" issue that I'm scared to say. The question is whether or not, at the critical state, for example, in fully developed turbulence, or at the critical state in lots of other systems, living systems actually have to use free energy to make things and is there something going on at the critical state that is doing something like optimizing the utilization of free energy or transport of energy or heat or something through the system? Is there some kind of optimizing system that characterizes this state that might serve as a physical underpinning for all of this? I talked about the subcritical/supercritical boundary. Peter and his colleagues took up a question I had whether blue bacteria and red bacteria that squirt stuff at one another for their mutual benefit come to an optimal ratio of squirting blue and red at one another which optimizes their growth. If I can summarize what Peter thinks, the answer is going to be "Yes, there is a regime in which the blue and red bacteria do that, even if they are just replicating RNA molecules. But curiously, the bet is that precisely at the point that the blue and red molecules optimize their mutual growth, they also optimize their utilization of a food resource that is sitting in the chemostat, optimization in the sense that they suck up the stuff that's good to eat and make the most of themselves. Now I think that there is an analogue between the ratio of blue and red that optimizes growth in bacteria and price, which is something I mentioned to Brian. My bet is that this is something that has been carried all the way up. It is related to why we see, prior to the emergence of life in prebiotic evolution, these quasi-catalytic swirls (feedback), that I'm trying to talk about (and so also is Walter Fontana). I'm wondering about the relevance of this question there and also for ecosystems. Ulanowitz talks about ecosystems almost having agents in them, having autocatalytic webs of transformation. The question is whether, if things are going to some critical state, but with conservation of mass energy, there is something that's going on in such systems that is being optimized in some way.

COWAN: Is there a set of experimental data that bears on that question? You can write down a set of equations for the reactions involved in all of these systems and you can calculate the free energies—so are there data that are useful in commenting on this point?

BAK: What can we go out there and measure and look at?

GELL-MANN: Free energy contains a term equal to various temperature times entropy, and in these systems one probably has to be very careful of the definition of entropy (Comment: because they're open systems)—it's not the openness that concerns me. There may be information and complexity terms in the entropy that are critical in this kind of system even though in most problems in physics and chemistry those terms are negligible. They may be very important here and, therefore, the definition of free energy may end up being the whole question.

KAUFFMAN: I don't know that I mean free energy either. I mean that these guys are optimizing something collectively. Life is optimizing the way in which it is gathering sunlight and spewing it forth into the stuff that we see out there as best it can and I think that at all levels you make units that collaborate with one another doing the best they can and the whole thing is critical.

SCHUSTER: Isn't there a clear answer to this problem? In order to make most of these models tractable, you have to assume irreversible reactions. The irreversible reactions take you out of any thermodynamic considerations. If you look, for instance, at free energy, then they tend to have instabilities in these models and so my comment is that either you go to the reversible reactions and then you have great problems to treat these systems properly because you end up with a mixture of complicated details that are relevant or you go to irreversibility but then thermodynamics is not the proper reference point.

KAUFFMAN: I agree and I put free energy in quotes because I know that it's an equilibrium concept. We' re talking about open, far-from-equilibrium systems that gather in energy and do things all the time they are processing information. Evidence is beginning to accumulate that suggests that criticality may be where these things go. Real physical systems do it in turbulence. What are the lessons that may bear upon whether ecosystems do it all with scaling up? What's happening?

COWAN: There is such a small amount of energy involved in information processing in most organisms compared to what is used for housekeeping chores, I'd like to know if the ratio can change significantly. If it does, it seems to me that this can be very important. If the organism learns to be a little more efficient and uses somewhat less housekeeping energy and somewhat more of the energy which then becomes available for processing information, that surplus could be poured into more offspring. Is that a subject that could affect your concept of "free energy"?

KAUFFMAN: Absolutely.

MAYNARD-SMITH: I've wondered for thirty years whether there might not be something that could be said about ecosystems as systems rather than individual species from a thermodynamic point of view, some principle of optimization or something that might offer a handle on their organization. I hope there is. I know perfectly well that my ignorance of thermodynamics precludes me from speaking about it here. That's one side of the problem. The other thing is that all papers so far published about biology containing the word thermodynamics are rubbish. I don't mean to say that they may not be true one day but I think that you have to proceed with great caution. Certainly the people who understand thermodynamics do this, which is why I'm keeping out.

HOLLAND: Peter's point of view should be reiterated. In these irreversible reactions, thermodynamics is largely irrelevant.

KAUFFMAN: The point should be the following: If we have evidence that adapting systems, in a variety of ways, go to some critical poised state between order and disorder, and particularly if that shows up in coevolving systems and if it shows up in models of the origin of life that may take one from organic molecules into self-reproducing collections of molecules and into ecosystems of collaborating self-reproducing molecules up into the biosphere, and if we should find that's critical at a lot of scales, it would be very nice if it turns out that there was something that somehow characterizes the situation that is like one can imagine characterizing in fully developed turbulence. I don't know what to call it and that's why I was so hesitant. . .

HOLLAND: I want to go back to George's point briefly. It seems to me that if you make this division which is just an intuitive division, I don't think you can ever make it precise, between housekeeping and information; nevertheless, there's a kind of flow tradeoff that is pretty interesting, which is much like the idea of a vacuum tube. That is, sometimes I can use what is a small energy requirement on information to make a big decrease in the housekeeping energy requirement and that kind of thing goes on a great deal, that kind of tradeoff where the system suddenly discovers by getting a bit more information and processing it at a relatively small cost, it changes its housekeeping requirement a great deal. This is the sort of thing we do in playing a game like chess. A bit of information can save a great deal of actual housekeeping and that must have a strong selection value.

COWAN: I think we have defined a very rich field of discussion and this can be continued later.

KAUFFMAN: Let me finish my list (3) which is remarkably similar to Murray's list (3). I call it emergence of hierarchies. It has Brian's request for a theory of the organism. It has implicit in it what. . .

MAYNARD SMITH: . . .I was talking about, the units of selection. In terms, what holds things together when there are replicators at different levels that might be selected on? It has the major transitions theme. It has "Does and why does complexity increase?" where I think that I join with Brian in suspecting that it does but I don't know why it ought to. And it has the local/global interpenetration. It's precisely because higher order units have dynamics among themselves that the higher order levels feed back and control what's going on at a lower level.

Let me say one more thing about this. If there is an analogy to price in bacteria coevolving with one another and becoming mutualists and exchanging products because there are advantages to trade and if it has a physical interpretation in the sense that the coevolving bugs that optimize how they exchange stuff so that they grow fastest become the ones that use their resources the fastest, then there is perhaps something that's going all the way up, some physical idea of the analogue of price and the utilization of resources that goes all the way up to economic systems. And that's what I'm trying to point out with all of that. Okay.

HÜBLER: I want to make a small comment on this last point about what is optimized in terms of physics. We started with simple physics systems. The simplest physics system is dendritic growth and we could show that it is possible to get stable, stationary dendritic fractures under certain circumstances. Now, on one side, those stable fractures are adaptive systems because, if you change environment, in this case the flow which you impose on the system, the system changes. In addition they are very similar to what Per Bak described in self-organized criticality because, if you look at the branch structure of this system, you can describe it in terms of self-organized criticality. But there is a third property. They optimize the exchange of energy or the flow of energy through the system. So you have a variation principle and self-organized criticality and adaptation at the same time. It turns out that this is a stable stationary state because it's a stable stationary dendritic structure. But you can do the same thing at a dynamic level. In this case you have to compare entrainment and the optimal energy exchange. It turns out that if you have two oscillators and they are not too different from one another, then you have entrainment that is like the optimal predictors of each other. But you can also show that, at the same time, there is an optimal exchange of energy. The energy flow between the two is optimized because they are at resonance.

BAK: That is very interesting. There is a person who also suggests that at the self-organized critical state you have a maximum flow of some conserved quantity.

HÜBLER: We could do this with the flow of heat or of a substance. If you look at the optimized density absorber so you have an absorber of fixed volume, you have diffusion equations surrounding the absorber and now you ask the optimal structure of this absorber. You find it is the dendritic structure very close to self-organized criticality and, in addition, you find that it is a stable, stationary structure from a physical point of view.

GOODWIN: I think it is clear that you don't have a measure of what is optimized.

BROWN: That is the thing I wrote down as a suggestion for what should replace the free energy.

KAUFFMAN: But that wasn't a flow. This is a different point. If the things have to eat the conserved quantity, then there's something about getting to a critical state which is optimizing how fast you're turning it into garbage.

BAK: That's something you can only do for systems where you have some kind of front. I have no idea how to use this flow argument for the game of life, for example. (Comment: or for economics).

KAUFFMAN: But if there is a mapping from bacteria who are mutualists and exchange red and blue stuff at an optimal rate to optimize their mutual growth (and I want to say again that this is the analogue of price), then we have an almost physical notion—because we can talk about it in connection with Peter's replicating RNA molecules that you can do in a pot. So, if we get an analogue of price for molecules in a pot, we can maybe drive these ideas all the way up, although at some point it becomes a little weird—what do I mean by wealth because we don't know what money is?

COWAN: This is the kind of discussion that can best go on in the smaller groups. It's threatening to go somewhere important and should be continued. Now it's John's turn.

HOLLAND: I have two questions I'd like to suggest to anyone who will discuss them with me. The first is "How does one model the process of adaptation?" I distinguish between this process from adaptation as an endpoint which biologists have done for a long time and very carefully. The other question I'd like to discuss in a small group is "Can usefully predictive social models be built?" Some people suggested yesterday that perhaps you are better off without a model than with a model but to me that seems very strange. I'd like to discuss that.

COWAN: There is an important school of economics that asserts exactly that. The "invisible hand of God" and all that.

HOLLAND: But then they allow you to introduce some kind of rationality.

WALDROP: Of course, once you make the model, you change the process you're trying to model.

EPSTEIN: The other thing is that everybody has models. The difference is only between those who write them down and those who don't.

HOLLAND: Exactly. That's the old story. If you don't have a model, you are really using a model that's out of date by a century.

GOODWIN: There is one topic that is extremely conspicuous by its absence In the context of economic modeling, I know that everyone is concerned about environmental costs, sustainability, limits to growth, and so forth, but we've heard absolutely nothing about their relationship to economic models and I'm a bit curious to know why it has been so totally absent. I know that it's an enormous issue and I just want to know whether it's included at all in the agenda for our discussions or is it hermetically sealed and going on somewhere else?

COWAN: Would you repeat the topics that have been left out?

GOODWIN: Sustainability, environmental costs, limits to growth (Comment: getting the price right). An economic model that is not based on these concepts now seems to me to be pretty irrelevant.

GELL-MANN: Maybe I should say a word or two about what we have done on that. The economics program has not addressed these issues although most of us here are very much aware of their importance and some of us spend huge fractions of our time on them. But Brian Arthur and I did organize a brief meeting a few years ago in which precisely those issues were discussed by people from a variety of disciplines: economists, including agricultural economists, some biologists interested in conservation, some ecologists, anthropologists, and so on. The result of that meeting was that it appeared clear that there were certain important matters like the rate of discount and national accounting systems that could immediately be tackled. These were not particularly deep issues, only places where one could identify simple parameters in which important sets of values directly emerged. But then we, together with the World Resources Institute and the Brookings Institution, have organized a study called Project 2050 with an initial proposal for eight million dollars in funding for the first three or four years, of which we have two million in hand right now, with the idea of adding to the project people from all over the world from developing countries, from ex-communist countries, and so on. We will try to study the whole world problematique, very crudely, including all these issues but also including military, diplomatic, political, and ideological considerations and not just environmental, demographic, and economic factors, because they all interact with one anothe. We'd like not to have to issue pronouncements such as have led to the discrediting of this kind of thing before. We would attempt to formulate some of the work in terms of game so that people could change the assumptions, change the mathematics, fool around with all the levers. The initial task consists of assembling real data, projections into the middle of the next century, problems of food requirements for the population at that time, the energy requirements, what might be the technological changes, and so on. Those questions would be addressed initially by the two other organizations, whereas the questions of imaginative gaming and so on would be addressed under the leadership of this organization. But we would try to coopt people from all over the world.

COWAN: Can you describe the disciplinary emphasis or the elements of various disciplines that would be combined in this study?

GELL-MANN: I think it would involve nearly every discipline.

COWAN: It embraces everything?

GELL-MANN: I think it would embrace all respectable elements.

COWAN: OK. I wanted to say also that George Gumerman will present a talk at SFI, perhaps in August, in which he will be prepared to describe how cultural anthropologists would divide their discipline into highly interactive subsets, including economics as one of the subsets, and how the subsets interact with one another. Part of the question you raise is "What does this generation owe future generations?" Is that an economics question, an ideological question, religious, or what? I suspect that this kind of question might very well enter into a cultural economics framework. If Gumerman's talk suggests that economics is a subset of culture, as George has already suggested, we might be able to find a name for that kind of activity which sets economics somewhat more in a cultural context. You might want to discuss this, Brian, with George, and see whether it's likely to suggest a useful departure in the Santa Fe economics program.

GELL-MANN: Interestingly enough we did have a meeting, one of three meetings on Southwestern prehistory, that was devoted to resource limitations and their relation to evolution of Southwest society. The speakers were extraordinarily reluctant to discuss questions of anthropogenic resource depletion. Most of them didn't want to discuss it at all. I found that very strange. I thought it was going to be the main emphasis of the meeting. It was otherwise a very nice meeting but this central issue was somehow avoided.

GUMERMAN: One of the main reasons why it was avoided was that, while we could see that there was going to be resource stress, the distinction between environmental cycles and the use and abuse of resources was very difficult to distinguish. Was there a dry period which reduced the amount of firewood and wood for construction or was it overuse on the part of the inhabitants? Making those kinds of distinctions became very difficult.

GELL-MANN: Same for mammal populations in the vicinity of a large pueblo.

GUMERMAN: What was part of the regular environmental cycle and what was anthropogenic?

ARTHUR: A quick comment on Brian Goodwin's question. A number of us in this room are concerned about the questions you raised as to population, the environment, the economy, sustainability, limits to growth, and so on. Many of us have very good credentials in that area. I spent ten years of my research life working on population and the economy in Third World countries. But to put the question back to you, you ask why these topics have not been mentioned in this conference. I see no reason to unless you can give me one. This is a workshop about integrated themes having to do with complex adaptive systems. If you can give me a good reason why we should discuss questions of the discount rate or questions having to do with population and the environment as an integrative theme in complex adaptive systems, I'd be glad to hear about them and we'd all be glad to discuss them. But, at the moment, I see that as a sort of application of some of these ideas, not as an overall theme.

GOODWIN: Well, perhaps I can just respond briefly to that. As I mentioned to you after Kenneth Arrow's talk yesterday, I was surprised that he didn't mention these issues as part of the development of relevant economic models. In other words, wherever we enter into a discussion of economics, and we're constantly on that boundary, because we're using biological models, ecosystems, ecological perspectives as a kind of informant of what might be happening in the economy, the similarities and differences. Certainly it is not possible to discuss ecological systems now without considering the environmental problems and, therefore, any discussion of complex adaptive systems, I would have thought, would include precisely these issues of sustainability and values and would bring in some theory of human nature. This is John's point which he is constantly insisting upon. We are very narrow in our terms of reference with respect to ecology, anthropology, etc. in relation to economics. In other words, economics is not a separate discipline as you certainly well know. It is an integrated entity within a larger system which is cultural.

ARTHUR: I couldn't agree with you more but I don't see what place this has in a workshop on integrated themes in complex adaptive systems. I'm willing to concede it may have but under the heading of applications.

COWAN: It calls attention to the great interconnectedness of all of these topics and that is part of the conference theme.

FRAUENFELDER: When I look at Murray's outline, I see a complex adaptive system here, the workshop itself. The first point is information gathering. I think one of the things that would be useful in the book that is supposed to come out is to have a relatively brief collection of experimental data—that is, what do we know in the various fields, what really are the observed parallels and differences between the various systems, and what do we consider systems. Now each of us has a different understanding of what a complex adaptive system is. Each of us has a different knowledge of facts and almost everything I see here is essentially models and theories. Underlying all of it are the facts so a collection of the facts, without models which should be excluded so that we know what the underlying facts are, what we really want to describe, would be probably useful.

PINES: I think you're right. What you are asking for are John Holland's reality checks, namely, a set of facts that are relevant to candidate models, a set of facts that can check on the grand schema that Murray has put forth or that can check on pieces of what we've been talking about. We've been focusing on schemes and phenomenological descriptions, a little bit on possible theory, and hardly at all on the underlying data.

EPSTEIN: I want to respond to both Brian's remarks. It seems to me that one point you may be making is that by restricting our discussion of social science essentially to economics, we seem to be implicitly agreeing that there is one social behavior, that is the invention, pricing, and exchange of commodities is really focal. There are obviously other behaviors that are important—war, expansion, the subjugation of other populations—lots of other behaviors that might be worth studying. And from an organizational viewpoint, the corollary is that the organizations that are interesting are markets but not military alliances or religious organizations. But those are equally amenable to rigorous study and equally important. If economics is part of this discussion, I don't see any reasons why the other things shouldn't be and it may well be that our methods and techniques may be much more natural as applied to questions of religious formations and imperial collapse and so forth than the economic questions. There may be a very elegant fit between the rise and fall of empires and avalanches in sandpiles.

GELL-MANN: Those of us who have been fighting for broader inclusion of these topics have not yet really...

EPSTEIN: But this is from a scientific perspective. We're not talking about...

GELL-MANN: I'm talking about the same set of subjects that you're talking about.

EPSTEIN: I wasn't suggesting otherwise. I was saying that there's a humane appeal to those topics but I wasn't making that appeal although I recognize it. I was making a solely scientific argument for undertaking these studies.

COWAN: You call it Project 2050, so happily it doesn't have the word sustainability in it. It's really about possible futures and that is a very broad, all-embracing topic.

GELL-MANN: It's focused on sustainability but defined very broadly, including sustainability with respect to avoiding catastrophic wars...

COWAN: Desirable sustainable futures, which is already different. I'll now call on Stu to begin the discussion of item 3, the formation of hierarchies, etc., and try to stop chairing and let the discussion organize itself.

KAUFFMAN: Brian Goodwin has repeatedly asked, "What would a theory of the organism look like?" Are there natural orthologies, in some sense? What would a theory of the life cycle look like and this is much the same as what Le Buss talks about. Then there is the units of selection issue which is well known in the evolutionary literature which is what...

MAYNARD SMITH: ...I was alluding to. The problem fundamentally is that you have what can be called replicators if you wish—there are some worries about calling things replicators. But now replicators at one level, RNA molecules, which come to collaborate with one another, making hypercycles and other kinds of higher level organizations, essentially coevolving communities of the replicators. They can be within a unit or they can be multiple units in a colony, and the problem becomes Leo Buss' problem, why should I give up my right to life as a reproducing cell in order to become part of a multicellular organism, leading to the major transition issue that John has raised for us which is why do major transitions occur, which is tied in with the whole issue of why do things become complex and why do things become hierarchical. They seem to. Is it merely a matter of that the possibility is there and if you start simple, then there is only one way to go; or maybe it's often advantageous to be complex. For example, we have the intuition in an arms race that fighter pilots are always pushing the envelope of what their machines can do to try to get tactical advantage.

EPSTEIN: If you got the human out of there, the plane could do much more. The pilot blacks out at nine Gs, the plane can pull twenty Gs. The pilot is a bureaucratic aberration. The Air Force says you can't pin wings on a missile.

KAUFFMAN: Anyway, we have countervailing ideas about whether or not things get more complex, and, if they get more complex, why they get more complex and I don't think we have a coherent theory or way of framing that set of questions. Meanwhile, I think there is this global/local issue that Chris Langton was trying to describe that we all jumped on him for but there is a way that once you have hierarchically organized entities, the higher level entities have their own dynamics and the dynamics among the higher level entities, in fact, influence what you see at the lower levels. What I was trying to point out is that there are a lot of bovine proteins around these days and not a lot of trilobite proteins because the interacting entities or organisms got their proteins. . .

COWAN: We've gone beyond the very next topic. David, what's your suggestion for the very next topic?

PINES: I'd still like to know what are the models that have been worked out in some detail which we could hear about, just listing those that are relevant to this issue, if any, and/or a discussion of what is the experimental or observational information which people around the table have that bears directly on that issue. I would strongly encourage my fellow discussants to avoid further examination of this interesting concept which is important and wonderful. I think we've had a delightful week doing that but it is time to be specific about what works and what doesn't. . .

COWAN: You're suggesting the question "What is the list of models and is there a set of useful and relevant data in the real world that can provide reality checks or whatever?" Murray.

GELL-MANN: We've had some discussion about whether biological evolution works with what it has so that the steps are, in some sense, small. Jim Brown and I were discussing the question of wings and of flight, which has developed several times in biology. Each time evolution was working with an organ that was already there and with behavior that was already there. People argue about the evolution of birds, whether it was arboreal or cursorial activity that was leading to flight but it was something preexisting. It has been suggested, and it may or may not be true, that in thought one can make bigger elementary leaps than in biological evolution but, whether you believe that or not, the fact is that, in biological evolution, there seem to be things that are even more significant than, say, the evolution of wings and flight, namely, major changes in organization, like the evolution of eukaryotes or of metazoa, and so on. Now, in reference to that, theorists here mention some comparatively ambitious theoretical ideas about hierarchy of organization. Walter Fontana and Leo Buss and several other people mentioned Chomsky hierarchies. And then we can go back and look outside of biological evolution and say that in the other kinds of CAS that we are talking about, there too major changes in organization may occur. The first firm or company would be, I suppose, an analogue in economics. I thought that this would be a good place to stop. All of the things that Stuart mentioned are related.

HOLLAND: I'd like to expand a little on what Murray said. I disagree with Murray that there is a big difference between the kinds of things that go on in biological evolution and the kinds of things that go in the thought system. The reason that I do comes back to hierarchy. I think there are, almost automatically but not quite, different time scales associated with different levels of the hierarchy. If I look at the biological systems and at some of these big changes that Murray was talking about, I see a much broader span of time and I say, oh yes, it looks like this and then, some 105–107 years later, it looked like this. That's one kind of change. Then I look at wings and that's a bit shorter and so on. I think very much the same thing is going on in the central nervous system and I see things that I call big changes, but in fact they occurred over a rather long period of time relative to those day-by-day thingswhich go on. It may be subjective. To go from there. . .

SIMMONS: Can you state the nature of the disagreement one more time?

HOLLAND: Yes. The disagreement is that underneath both biological evolution and the thought processes that, in fact, in these systems that I know anything about, in all cases there are different time scales associated with different sizes of jumps and that, when we're talking about the central nervous system, times run from milliseconds to months to years and perhaps centuries if I'm referring to culture. In biological evolution, I suppose the shortest time span is something like a century or a bit less if we talk about the moth in Britain to very long time spans and so we quickly talk about this big change in going to the metazoans but, in fact, that's over a longer period than some of the other changes we've talked about. So all I'm saying is that the idea that the brain makes big jumps fast may be a subjective thing and that underneath the process is very much the same.

SIMMONS: You're saying that there's a sort of natural time scale for the set of processes you're talking about and that on that time scale...

HOLLAND: Let me go one step further and then I'll close this off. I want to talk briefly about one model that I know well in which hierarchy has emerged from no hierarchy and so we can actually watch a hierarchy emerge from no hierarchy. In that model, which is just the classifier system, where you start with a set of coarse-grained rules which have a high sampling rate, these rules make small requirements on information; they simply say, if an event in this huge equivalence class occurs, then do something. And by the very nature of the way that this goes in classifiers, then the something that is done has only a slight probability higher than chance of being effective. It is an edge but not much of an edge.

Realize that we're talking about competitive systems here. That is, getting a bit better than your competitors so that, if you compare the thing that has an effect slightly better than chance with something that's only chance, naturally the first will prevail.

Now, once I've got such a thing, that's a zero-level hierarchy, a rule with no hierarchy. The natural thing is to develop second-order rules which become exceptions to very coarse rules and correct that rule in specific situations. So if the first rule is "If there's a moving thing out there, flee," then the second-order rule may be "But if the the thing out there is small and of the right color, then approach." Those two rules are contradictory in the situation where there is a moving object but the one overrules the other. That's the first level of the hierarchy, which we've seen actually occur in classifier systems.

Notice that there's a point here which is critical to time scales, which is that the second rule is sampled much less frequently than the first. In other words, as things come in, it gets tried much less frequently than the other. If I use any of the normal notions of learning rules that depend on frequency of use, then clearly the first rule is confirmed much more quickly than the more specific rule and, in fact, from a Bayesian or almost any other statistical point of view, it's no use to create a highly specific rule before I've begun to build up enough information to provide reasonable confirmation.

To close that issue, I would look at this relation, a critical relation, between adaptation and the general notion of modeling adaptation and hierarchy. It's important to look at them first. To look at them after the fact is going to be misleading, I claim. First, it's important to look at how they emerge. Second, the notion of graceful occurrence where you learn crude things first and then more specific things after, with appropriate time scales, is a natural for studying hierarchies and adaptation, and there is a natural time scale there where one can make explicit leaps in terms of classifier systems. And third, and this goes back to CAS in general, note that hierarchy is important to the extent that we always talk about aggregation, whatever our crude elements are, whether they are elements of selection or whatever, they become aggregated and they become building blocks for the next level of the hierarchy, and that's where sampling comes in again. That is, the building blocks I build first are the one that have high sampling rates and recombine in various ways to the next level of the hierarchy and that is where these two things work hand in glove.

So, the essence of hierarchy is that I try to find building blocks at one level that constrains the set of possibilities and I recombine those and try them out at the next level, then those in turn become building blocks for the level above and so forth. This is very much the story that Peter and Manfred have talked about. There are various ways to look at this but I think that this is an essential feature of all the CAS that I know much about.

COWAN: John, the idea has come in more than once here where we talk about an exogenous world and God suddenly changes the rules, God being you in the ECHO system, and something needs to be done that's quite different, faster, or a lot more radical than has occurred in the natural evolution of the hierarchy that you're talking about—is that another set of circumstances that needs to be considered?

HOLLAND: I was delighted with Jim's stuff because what happened there is that you remove some elements from the system and you don't get a collapse; what you get is a change in the form of things. If I've learned how to play a certain kind of game, now I'm talking conceptually, central nervous system kind of stuff, and someone gives me a different game—so I've learned how to play checkers and I've never heard of reversing and someone comes along and says, "Today we're playing reversing and here is how the rules are changed: I don't go back to zero. I've got certain default building blocks that still apply to games. I recombine them and I form a new way. I use some of the information that I had." It's graceful, it's not catastrophic.

EPSTEIN: I think that John has given a very elegant answer to the question I was about to ask so let me see if that's true. We've talked a lot about jumps in human knowledge of something and I keep getting confused about whether these jumps are in what is known—jumps, for example, in physics—or jumps in actual neurophysiology of the human.

GELL-MANN: We're talking about biological evolution on the one hand and then we're talking about different things, which are also biological, on the other hand, and one of those is human thinking.

EPSTEIN: But what does that mean? Does that mean "what is known"? Does the wiring change?

GELL-MANN: Not wiring. Wiring goes by biological evolution.

EPSTEIN: Let me finish. I'm still confused as to what is the entity that's making the jump. Is it in the body of knowledge or is it the actual brain?

GELL-MANN: Ideas, not just knowledge, but understanding.

EPSTEIN: In Holland's examples, the objects are the rules, the rules are the entities, the creature is the rule, and there's no confusion. That's what I hear Walter saying, that the object is the function. The objects are the rules and here there is no confusion as to what's jumping. I keep getting confused about what it is that's doing the jumping. In this context there's no confusion.

SCHUSTER: I'd like to make a couple of points. The first point is that we have to be very careful what we mean by jumps and by big jumps and by small jumps. I want to illustrate how I see it from the biological point of view. In varying the same theme, in varying already-present solutions, we have big jumps and small jumps. The small jumps are frequent events and the big jumps are rare events. This you can already see in every system. Now, when you want to build up a hierarchy, that's a big jump of a different kind because you go from one form of organization to another and, as John has said already, Manfred Eigen and I described a model in 1978 which I think is still correct in its details.

I want to add one thing. You have to have present entities which would otherwise compete because, otherwise, you can't operate without that. And what we have suggested is they are just members of one mutant distribution. If you replicate and mutate, it is the most straightforward way to have competitors present at the same time. Another way would be spatial heterogeneity that suppresses competition. Then you have to find some new kind of interaction and that was our proposal. It is a kind of coevolutionary, symbiotic interaction which in the simplest form gives rise to some kind of hypercyclic coupling which just makes a cycle and everybody benefits from its precursor in the cycle. There are other ways to do that. They are more complicated. Now if you introduce this new functional entity, you immediately realize that you have to solve another problem because now there is no competition. There is just a big soup and anything is going on within that soup and you cannot optimize any further. So the third step is that you have to have the formation of spatial boundaries. . .

KAUFFMAN: Sorry, just make that last point clear.

SCHUSTER: The first step is that you have to have competitors present for some mechanism; the second step is to integrate them functionally. This is the hypothetical argument that if you have catalysis, for instance, of these molecules, you have to be sure that it creates a new form of organization. It's very much like what Maynard Smith said yesterday on his symbiotic model. The next step is that you still have one big soup of your functionally coordinated systems, but they cannot optimize in a mutation selection way because everybody's present in that soup and there is no improvement. Let me say that if one molecule finds a better solution, then all the other molecules would profit from that solution. So you need isolation of this functional entity and it is provided either by spatial boundaries or organizational boundaries or perhaps communication boundaries if you think about social systems. If you have these spatial boundaries, you end up at a higher level where you can again have improvement by mutation and selection.

HOLLAND: But in my context, this coarser rule and then the rule that built up at the next level, they still are in competition. In other words, in your terms, if the entity that makes up the hypercycle is still out there, it still has to be competed with. This hierarchy is not that isolated. You are still competing with these other simpler guys and, if they can beat you out, then there's no reason for you to exist.

SCHUSTER: Yes, but you can show, if you do it with chemical kinetics, that what is needed is to pass the threshold concentration. When you pass the threshold concentration, the catalytic terms dominate. So the competition is suppressed just because of mass action.

GELL-MANN: In the history of life on earth, this has happened several times in a row. The first one, if you accept the RNA idea, is the transition from RNA to DNA-based genetic systems. The second one is the prokaryote-eukaryote transition. Then there is the transition from unicellular to multicellular organisms. And, if you want to go further, solitary individuals to societies.

KAUFFMAN: Peter, can you go back and explain one thing? In the soup you said that if something good is done by one hypercycle, it would benefit all.

SCHUSTER: You make one mutation and this mutant has a better product. The mutant is a member of the hypercycle. I want to add something that I think is necessary for this system to work.

PINES: I have one question. John raised the very interesting issue of time scales. Do you see some relevance of his question about time scales to the four transitions that you described?

SCHUSTER: Yes. The replication-mutation systems occur on a time scale that is determined by the error rate of the system. In functional integration, the time scale is highly variable because it is a higher order reaction and depends on the concentration. You start out with a slow reaction and, as material accumulates, it becomes faster. Spatial isolation which leads to the entity which is then selected at a higher level is a slow process.

COWAN: Does isolation lead to higher concentrations in local regions?

SCHUSTER: Not really. You get locally higher concentrations but, in the system as a whole, the concentration remains the same.

KAUFFMAN: Two mutualists might become symbionts? So in a hypercyclic organization where the RNA molecules are red and blue bacteria—I'm helping you and you're helping me—what do you think determines whether we stay at the same level of organization and collaborate or whether we make a higher order unit to do the job in-house?

SCHUSTER: Whether or not we are able to build a boundary around it.

KAUFFMAN: But what drives it? Obviously you have to create such a unit, but why should you?

SCHUSTER: Let me say, you have the blue and the red. Then the red can change to violet and violet and blue is better than red and blue. But if red changed to violet and it's all in one soup, the product of violet will also be profitable for red. So there is no advantage for violet. If we have isolated the system, then we have competition between red and blue and violet. Then, the better one will, by selection mechanisms...

KAUFFMAN: Let's talk about hypercycles. I have two molecules, each one of which replicates by itself. Now we make each one help catalyze the other as we have done in the model of what I was so glad to participate in, so now molecule 1 catalyzes the formation of molecule 2 at some cost to itself and, vice versa, molecule 2 helps catalyze the formation of molecule 1. They can remain mutualists forever or they can switch over to one unit that is now spatially compartmented. What drives the decision, this is also a question for...

MAYNARD SMITH: Why should two mutualists who are perfectly happy about getting the advantages of trade, decide that it would be good to become symbionts and live together forever with its problems?

SCHUSTER: If you have the two in a compartment, and 1 changes to 1' by mutation, and 1'×2 is better than 1×2, there is selection. If you are in solution, there is no way to select. Now, if you don't find a mechanism for spatial isolation, you stay within the same soup and you have a disadvantage. If you find the mechanism, then you have an advantage.

MAYNARD SMITH: I do see a pretty sharp distinction between those major transitions which I believe Leo was talking about, where there is a change in the way in which genetic information is transmitted, and those changes, which can be very important as in the invention of feathers and wings which led to a whole class, did not depend in any way on a change in the method of genetic information transmission. The latter is very well understood and the former is very little understood. There aren't very many of the former kind of transitions, the big transitions, perhaps a half-dozen, not hundreds. The one most worked on by population geneticists is, oddly enough, the origin of societies—animal societies, not human. There is a big body of mathematical work going back to Hamilton (the whole theory of kin selection) and there is an enormous amount of empirical evidence on whether the actual measured values of fitnesses explain what happened. Although I wouldn't want to say everything fits, it is a pretty extensive body of theory plus population data. On the question of why there are living compartments at all, I think we understand that rather well. I don't see any very deep problems with that. I can think of almost no work, other than purely verbal descriptive models, of the next two transitions from prokaryotes to eukaryotes, and from single-celled to many-celled organisms. Roughly the same is true of the origin of sex. Not the maintenance of sex but the actual transition that led organisms to fuse and then disjoin again. There is a verbal model with diagrams but, again, no mathematical modeling of any serious kind.

I have been working on the question: why have chromosomes? The problem here is you've got the replicators in a cell and there are different kinds of replicators, A's and B's and C's; a cell that has an A and a B and a C is more likely to grow and divide and make two cells than a cell with several A's and several B's but no C's. The genes are synergistic in their effects. But should they want to tie themselves together? In that case, when the cell divides, each daughter cell is sure to get at least one A, one B, and one C. From the gene's point of view, if I'm an A, I'd like to be tied to a B because I know that in the next cell generation I will have a B with me, whereas if I'm not tied to a B, there may be no B around. But there is a disadvantage because it takes twice as long to make a new chromosome as it would to make an isolated gene. So we've made a little simulation, and it turns out to be surprisingly easy to get chromosomes provided you don't have too many molecules of each kind in a cell. If you've got less than ten replicators in a cell, chromosomes originate rather easily.

SCHUSTER: There is an example of that. There are influenza viruses, such as influenza A, which have eight separate genes, whereas the closely related influenza B has a single genome. Thus, the transition from split genes to a genome occurs already in viruses.

I think what is necessary to initiate a major transition is some kind of stimulus because the replicating molecules would be perfectly happy to continue. I think the stimulus may either be, as Murray says, a barrier that breaks down and gives you an abundance of material that was not available before or a shortage or some problem with the environment as with the prokaryote/eukaryote which liked the change to more oxygen in the air that jeopardized the existence of most organisms. So I think such a stimulus, such a change in the environment, is necessary to make this model start.

COWAN: In order to get the semantics straight, we got into a little confusion before as to whether this sort of thing is an endogenous change or an exogenous change.

GELL-MANN: The change in the amount of oxygen was almost certainly biogenic.

GOODWIN: John, did I understand you to say that we have a relatively good understanding of how the limbot arrived or how the brain gets more complex or how the prokaryotic cytoskeleton transforms to the eukaryotic cytoskeleton? That those changes within lineages are relatively well understood?

MAYNARD SMITH: They are relatively well understood as far as the selective population aspects are concerned. What is not understood is the developmental mechanics whereby a change in a DNA molecule causes a change in a virus—we know how a change in DNA causes a change in a protein, but we do not at all know how a change in the protein affects the limbot in changing from five fingers to three, so I don't at all pretend that we have a good understanding of that branch of biology. We don't.

GOODWIN: Or the transitions that occur from a fish limb to a tetrapod limbot?

MAYNARD SMITH: Yes, I'm taking development as a black box. That's not to say that we don't want to open the black box and understand what's inside it. We want very badly to understand that but, curiously enough, we don't have to understand it in order to understand the selection mechanisms.

GOODWIN: To know what selection acts upon, we do have to know it. It's essential to know what the outcome is.

MAYNARD SMITH: Sure.

FONTANA: As Leo said several times in his book, the language that we talk frames the way we think and Murray pointed out several times in this conference that sometimes the most dramatic progress is made by getting rid of vetoes in thinking. It seems to me that there is one big veto here; namely, you shall not think of anything else but genes. The work that Leo and I are doing started out with objects that are allowed to copy themselves. What you get at that level is hypercycles all over the place. It is very easy to get this type of organization. This is not a transition within levels of organization in Leo's and my model. We would claim that a hypercycle is at the same level in our model.

GELL-MANN: Is that what you call the zero level?

FONTANA: Yes. We would say that a single self-copying entity as well as hypercycles are at level zero. The levels are not sheets. They have a thickness, so to speak, within which you might, with another model, put in several sublevels, no doubt about that. What your model parses out of nature depends on what you declare is important in your model. So we are just trying to draw attention to the fact that there might be a different definition of level. There is a different type of coarse graining than the one that has been usually made. And that coarse graining you get at level zero. Those are all ecologies in which the only action between entities is a copy action. This is epitomized in the theoretical world by the replicator equation. Now it was a fact of observation in a mathematical model based on very precise mathematical machinery that once you inhibit the most essential interaction that makes up that level, namely, copying, you get something else, something that is fundamentally different than what you see at level zero.

GELL-MANN: You mean once you prohibit copying under certain conditions? You don't prohibit it altogether?

FONTANA: Oh, yes, none at all. This does not mean that the physics has changed. The physics remains the same. It's just that we logically ask what if nature, or some other means, makes copy actions that are no longer possible to carry out. We don't say how this occurs. We just ask what if that occurs. It's at that point that we get a transition to a totally different level that strikes us as being a different organization. The question then arises, "What do we mean by organization in the first place?" I think organization should not emphasize the nature of the object. It should put emphasis on the relationships among the objects. In this respect I believe that the group, for example, the natural numbers with respect to modification, is a beautiful example of an organization. A group structure is an organization. It emphasizes the relationships among objects rather than the nature of those objects.

GELL-MANN: What do you have here actually? A semigroup in your model?

FONTANA: What we have are structures that you can't describe in a compact way like a group but in which the interaction activity is lacking, it is not associative.

GELL-MANN: So what is the name of it?

FONTANA: We don't know. It's an algebraic structure but I don't know if the algebra that we find has been looked at. What we observe is two things. First of all, when you inhibit copy action you get collectives as objects that organize at two levels. The organ is at the object level; that is, if you look at how these objects are made, purely syntactically, you realize that they share a common architecture. In this sense, they are a form of language. Soon you begin to walk up the Chomsky hierarchy because we see regular language. The difference from the Chomsky hierarchy is that we have not just words that are passive but we have words that are active because they act upon other words and transform them into new words. That's what we'd call function. Therefore, you cannot only try to characterize the systems according to how the objects are structured—are they polymorphs or sugars or something like that—but you also try to find an organization at the level of the relationships which these objects use among each other—how they act upon each other—and there it turns out that these relationships can't be described in a very compact way as algebraic structures. For example, the most familiar example would be a group but they lack associative activity. So it's different and you can generate a whole wealth of such examples.

GELL-MANN: This is level one? (Right.) You say that in their syntactical relations, having a common architecture, your objects form what some people call a language, a formal language, and in that way you can work your way up this Chomsky hierarchy although you don't know the mathematical meaning. Do you, in fact, work your way up that hierarchy?

FONTANA: No, working up is, perhaps, not the correct word. Under certain conditions, you look at the syntactical relations among the objects and you discover that they are a regular language. You change the boundary conditions and you discover that the syntactical relations among the objects are a language where they weren't before. Initially we start always with a random type of initial conditions. So if the boundary conditions are changed, we can change the grammar that describes the syntactic relations.

GELL-MANN: When you do that, does the algebraic structure of the relationships change? (Yes.)

PINES: These are computer simulations that you are tuning on a set of bit strings?

FONTANA: Yes. The mathematical machinery is the lambda calculus. We are altering the formula and the interaction among the formulas is the functional application.

Notice, however, that our reactor is a finite system so that at any time you see, at most, a thousand objects. If I told you that the syntactical relations among these objects are a grammar; the grammar specifies an infinity of objects, all objects that are built in that particular way. In fact, if you look at our reactor at time t_1 and then one million steps later...(Question: Is this a mathematical chemostat? Yes.)...your object might have changed completely but not the relationships among them. So that's what prompted us to think that this is a very sensible notion for an organization. You are hitting in some abstract space an invariant subspace, invariant in the sense that as soon as you are in that subspace, you might move around in it but you move in such a way that you don't change the syntactical architecture of your object. If you are in a particular regular grammar, you stay in that grammar and you don't change the algebraic structure. It's as though, if a firm were to change its entire roster of employees, it would do so without changing the organization of the firm. So this prompted us to think that we were dealing—it's a purely observational fact, we don't know whether this has anything to do with nature—with definitely another level of organization, genuinely different from what we call level zero where you cannot talk of an organization in our definition of an organization. You can absolutely talk about an organization in the definition of

Peter's work which is, I believe, an attempt to understand in much more detail what is going on at level zero.

GELL-MANN: Can we find out in more detail what level two and level three mean?

FONTANA: I'd like not to talk about level three because this would be the subject of an hour-long discussion and I don't think it's ready to be discussed. Level three is what John would call multicellularity and we can't, at the moment, add anything very intelligent to that definition. Level two is what we would call eukaryote. It's where you stay in this invariant subspace which you call an organization. The next natural question is what happens if you take two such organizations and let them interact with each other. Clearly, once you are in such an invariant subspace, the game's over. You just stay there. The only way to get out is either to perturb it with random functions or bring it into contact with another organization of level one at which point, as you remember with eukaryotes, you get a new, self-maintaining entity that contains the two organizations that were brought in to make it. But they do not outcompete each other because cross interactions lead to a metabolism, to an additional set of objects whose interactions among themselves as well as with the subsystems stabilize the coexistence of these subsystems. So, if you want to bring chloroplasts together, you had better take care in a metabolic way to have them live together in a civilized fashion.

PINES: Are there time scales involved with the various emergent systems?

FONTANA: Very good question. The quick answer is no. We threw out everything here that did not come out naturally from the mathematical machinery. We went to great lengths not to overstrain the mathematical machinery with arbitrary modeling aspects. All the reactions that occur in these systems occur with unit rate constants so there is no notion of computational resources. Functional applications occur with unit time.

HOLLAND: But you use them when you go to normal. If you simply look at how long it takes to use your normal, that's a measure of resources.

FONTANA: That's a measure of resources but, as a matter of fact, we do not take this into account.

HOLLAND: You run the simulation with data that you can take.

FONTANA: That is true but, for the questions we focus on, whether you put in rate constants or not, it doesn't change the meaning of level one or level zero.

HOLLAND: I didn't say rate constants; rather, there are certain computational steps that are an indirect measure of the complexity of each agent, if you will, and the complexity is the time it takes to reduce it to normal form.

FONTANA: Yes. One could look at that. Let me make one brief additional comment. This relates to John Holland. The hierarchy here is not of the type you discussed. It would be a lousy model for an economy because molecules don't learn, because the objects that are the basic actors in our system are not objects that have an internal structure that endows them with learning capabilities. They are just molecules. You could, however, make a model in which the notion of an object is much more sophisticated, where it becomes, perhaps, an economic agent endowed with learning capabilities. You could model such an object with a John Holland classifier.

GELL-MANN: That would be level four and five: four would be the self-aware intelligent agent and level five would be the society of these agents.

FONTANA: Right. So, at this point, your entire theory kicks in because you are looking at internal structures in a much more sophisticated way than we do. So, as time evolves, we might examine how they look internally and then we might discover that they are structured hierarchically with default hierarchies, for instance. Then it might become a good model for an economy.

HOLLAND: Quick comment on the comment because I want to make the point again about recombination and give an example that is a little off to the side so that I don't talk about my own work. If I look at a neural net as it's usually done, at feed-forward neural nets which are like most neural nets until you get up to the Hebbian level, what you don't have here is the opportunity to form these building blocks and then try them in various ways. And that process, emphasized by Walter at his level two, is very important. The fact that you can form something and make potential building blocks one, two, three, and four and get all the combinatorics working for you to try to recombine them in different ways, that's just the thing you can't do in some of these other systems because the building block once formed is fixed in place.

FONTANA: The very fact that, at the syntactical level, our objects can be described by grammar, at level one, tells you that they are made of building blocks. Otherwise you could not specify a grammar. In your case, the building blocks are much more sophisticated.

BROWN: David called for empirical reality checks. Let me share an insight that comes from my reading of natural history and biological evolution as I see it and that is that I think Murray is right. I think that there are two really different types of fundamental steps that we're talking about here. One of them refers to situations where different units get together and the second to which similar units stay together. Let me illustrate that. Different units getting together are things like the origin of the eukaryote cell from previously free-living individuals: corals, lichens, etc. Here we have a situation where the division of labor preexisted and we have a coalition that is formed on the basis of trade. In the second case where similar units stay together, situations like the evolution of a metazoan body, plants originating from the coloniality itself, and the evolution of coloniality in bees and ants, in this case what happens is that there's some sort of emergent potential from these units sticking together. In both cases there is mutualism versus competition playing themselves out.

SIMMONS: You described the system at first as situations in which different units get together and the second you described as what?

BROWN: Similar units staying together so that the metazoan organism forms as a colonial association of initially very similar cells. An ant colony is formed by the babies staying with Mom. I think you can see analogues of both of these processes in human and economic systems.

KAUFFMAN: What is the advantage—why does this happen?

BROWN: This is what is most fuzzy to me. Similar units stay together, in some ways because the new unit has created some additional potentiality. It can do somethings that the individual units could not do on their own and, in John's sense, the advantage to those individuals and/or to their genes outweigh the advantages of each going on their own. I think that we are really talking a little bit about apples and oranges here and I think it is important to keep clear which of the two kinds of organizational changes we are discussing.

COWAN: For one thing they can process more information; like, in the ant colony, they can specialize and communicate with each other with pheromones. Erika?

JEN: I was impressed by Walter's emphasis on the learning facility because it helped clarify something for me—as I was thinking about Murray's question about leaps in biology versus leaps in thought—I then remembered Thomas Kuhn's model of transitions in scientific evolution. We can disagree with many aspects of this model but it certainly is a paradigm for transitions, a paradigm for transitions in thinking and, in particular, for substitution of one idea by another idea and, oftentimes, for a transition to complexity. I think that it actually represents a lot of what Jim Crutchfield's model does, and that what the Kuhn model is saying, as you all know, is that you have some particular set of rules that you go through, you find that you have to add more and more new rules to explain things and more and more exceptions to those rules, and eventually your set of rules falls apart. There are obviously things we don't like about it but there are some basic ideas there. Eventually you have to come up with a new set of rules that are oftentimes more complex and may be different. It's something like moving up the hierarchy. It should also apply a lot to what Stuart is thinking in terms of pole vaulting agents who form models of each other's behavior, and when those models have to get substituted by other models. Why doesn't this model also apply to some of the things that Peter has been talking about and that John has been talking about? It really does have something to do with what Walter has been saying that at the RNA level you're not actually trying to learn something about your world. Something happens that is not exactly at the individual/society point. I'm not sure when it happens that people try to start to learn things and to build models but something happens when they start to build models. You have this process where you find that you're getting too many rules, you're picking up regularities, but then there are too many exceptions, as in the classifier system which is an example of that. So there is something funny going on.

FONTANA: I have a comment. There is no noise in this system in the sense that Peter, when focusing on level zero, looks at RNA models that can't even learn because they can't (major transition and gap, 12:13 P.M.) they are replicated to mutases and so forth and so they can vary their performance so you can argue that they can adapt to some extent and that they can learn something about their environment. However, this type of learning does not permit you to leap to level one or to level two. Going from zero to level one is driven by switching off copies in our system. We cannot say in what particular ways nature switched off copying action. It might be, as John said, about changing the mechanics of replication. I don't know, but it had to do something about switching off copies effectively.

SCHUSTER: I like Walter's model very much. It's very good for chemistry without copying and is everywhere a very good model where you don't have a copying mechanism. But I have a strong gut feeling that nature never switched off copying. We always have competition between these entities which are integrated in units. You can see that, if you look at mitochondria and the nucleus, they still compete in a certain way because chains have been transferred from the mitochondria to the nucleus in order to have control despite the fact that they are copying all the time. You get malignant cells because they escape this control. So I think that the important feature of biology, and I don't say that it works for economy and I don't say that it works for social science, but for biology you have to build your organization of levels despite the fact that every piece of DNA and RNA is ready to copy.

FONTANA: Obviously I agree because these are the facts, but I am emphasizing a way of looking at the facts. We can go to great lengths to describe the extent to which you have tight control over the copy action of each individual cell in your body or your genes, to what extent this is different from free selfish copying. I could say that what you need to switch off is free selfish copying in order to get to level one. Since you have a much more detailed notion of how copying is done, you say that you really don't switch off copying but you put such a tight control on copying that I would say it translates in our model to effectively switching off uncontrolled selfish copying. I'd like to add that the model prompts you to think that copying is suddenly transferred to a different level, is transferred to the organizational level, and it emphasizes the importance of the vehicle that the genes construct, to the phenotype or the organism, as Brian has pointed out several times, rather than to the genes which are just a housekeeping device. Leo constantly points out, if you look at the history of life what stays constant over geological time scales is the organization, not the gene. Genes change so quickly that you sometimes have difficulty in understanding the group relations. But at the organizational level, the genetic code, the whole translation machinery is conserved at the level of the living, the eukaryotic architecture is conserved at the level of the kingdoms, etc. So our model tries to make you think that the copy action that counts is what is important, the copy action at the level of the organization. There might be a copy action in there in which the genes are copying in a controlled fashion, etc., that might still be true but the essential point is that you must switch your attention to the other copy action, the action at the level of the organization.

WALDROP: To return to Walter's points here, what I see is an absolutely beautiful model of a self-supporting web of relationships and functional actions, and that's clearly very important. But we also have a notion of organizations as entities that try to achieve some purpose. I don't see this in Walter's model, that is, the concept of how it is that a collection of entities that are interacting acquire the ability to embody some kind of purpose, to embody a schema or to substantiate an if/then rule or a whole bunch of them.

FONTANA: It's not a question that we are presently pursuing.

GOODWIN: I want to reinforce Walter's whole approach to the problem of organization. What we are dealing with here is the emergence of the different categories of organization but at the same time working on the equivalence question. My particular preoccupation is to classify the different kinds of morphology that can be generated and to identify the equivalence classes that generate the taxonomy of morphological space which is the domain within which evolution takes place. We don't know what that domain is. That's something that is essentially missing from evolution. As I remarked before, organisms have disappeared as conceptual entities from contemporary biology. They need to be brought in again, not simply because that's what we deal with experimentally but because conceptually they are absolutely fundamental. The theory of organisms was a generative theory where we are actually trying to produce models and to see what actually happens with a preoccupation with the logic and the transitions in this logical framework. This contrasts with the preoccupations we have, that are somewhat complementary in that we are trying to see what these different types of organisms are generating in the way of patterns and forms.

Mitch's question about agency is a fundamental one and is something that we haven't really worked on much here in terms of trying to find out where the notion of agency comes from and at what level is it appropriate to think of purposive agents. I think what we end up with is the problem of causation in organisms and the fact that once you have self-organizing, self-maintaining, self-replicating entities, you do have a different kind of causation than billiard ball causation, which is the dominant notion of efficient cause in science. You have what Johnson called immanent causation because you can't separate cause and effect . If you have a system that is the cause and the effect itself, it is an agent and you can then talk about purpose within the context of that self-organizing, self-replicating entity. I think that that is where agency begins to come in. But you always have to be very clear about the context. There is a kind of holistic context in which that notion of purpose is relevant and outside that context is not relevant. So you can ask "What is the purpose of the heart in relation to the organism?" "What is the purpose of the organism in relation to the ecosystem?" These are different levels of order and you can go on up to the planetary system. I think that's a question that is very fundamental but we haven't really engaged in that discussion of causality and what we mean by agency. I think it is fundamental to clarify the wealth of misconceptions about adaptation and where it's relevant to use the term.

KAUFFMAN: I have several comments in support and some criticisms of what Walter is saying. There is a strong relationship between the model that Walter and Leo are working on and the model of autocatalytic self-reproduction which I described to you and which a number of us here have worked on, including Walter, and which bears on an important point related to what Peter said. Peter pointed out to us that copying is very difficult to do. I think that experiments now are accessible concerning whether, when you put down a complex mixture of organic molecules of some appropriate type, you can get template replication or collective autocatalysis.

Julius Rebek just gave a beautiful talk at the A-Life conference for a wide variety of nuclotide/nucleotide like molecules, where A and B can be ligated together, A by B and B by A as in nucleotides, then in those cases, and there are several examples, a little system can take those parts and glue them together and can, in fact replicate. Thus, it is an experimental fact that we are making self-replicating pairs of molecules of a variety of types. When you do that, you find that not only can you make those self-replicating pairs but you tend to make long sloppy chains in which things overlap one another in funny ways. Its an open question now as to whether the only forms of replication that you can get will turn out to be such template replications. In these cases it's very important that chemically what's happening is that you have something like a tetranucleotide or hexanucleotide that ligates together to a trinucleotide. It's not actually acting as a general polymerase in replicating those polymers. It's acting as a ligator which ligates two chemicals together. So it will be a matter of experimental proof that, I think will come in the next five to ten years, as to whether or not systems of complex polymers form collectively self-reproducing organizations and when and where it may be the case that template replication takes over. It's worth pointing out that, experimentally, there is a difference in organization between template replication copiers and collective reproduction of cells or polymers.

COWAN: Could we stop here to ask Peter if he accepts your distinction between template and collective reproduction?

SCHUSTER: No. As long as Stu is talking about RNA molecules, it's all right. I have always objected to the idea that you can do the same thing with proteins. That's something I don't believe unless somebody has done an experiment that shows you can do it.

KAUFFMAN: It's now an experimental question as to whether you can take nucleotides and nucleotide-like things with amino acids hanging off the sides, for example, as Rebek is doing, and the only form or reproduction you get is something that looks like a template mechanism or whether it turns out that more general models of collective reproduction can make polymers of the kind that Walter is looking at in his level one. I think we're going to know in a few years. Both are conceptually possible.

The second point has to do with the stress that Walter places on syntactic boundaries. The general notion that boundaries are important is very deep. The question is whether or not syntactic boundaries (building blocks for John and a syntactic substring in a string for Walter) in Walter's organizations are characterized by the fact that there is a crispness to the algebra. Once you have a syntactic boundary and are in it, you can't ever get outside of it.

FONTANA: If you take two words in a grammar and you let them act upon each other, then you don't leave that grammar.

KAUFFMAN: Walter and Leo appropriately force us to look at boundaries but I don't think that these are the only form of boundaries that are out there or that the most important will turn out to be syntactical boundaries. Murray, referring to the fact that there is closure under the operation of the grammar, I don't think that that is the most fundamental thing that organisms do. I think that they create, for example, physical boundaries which are called cell walls that prevent things from interacting with things in space. Things form physical boundaries or are careful not to be present at the same time. Things make boundaries in a whole variety of ways and I'm also worried about whether these physical objects respect syntactic boundaries that are so critical in an algebra. An example of that is the idea of the shape space that Alan Perelson has introduced us to. If I take an antibody molecule and I look at its shape complement, namely, an anti-idiotype antibody, then I take the shape complement of the anti-idiotype and go back and forth between shape and shape complement; because of imprecision in shape matching, you will eventually diffuse all over shape space and cover the whole thing. I agree with Walter very much in focusing on the boundaries but I think that the physical mechanisms that organisms use may be different.

FONTANA: Shape recognition adds nothing to the story I have described. We have been testing that. These strings are not linear strings. They are actually tree structures. We can begin to make a mock model of the surface of these strings and then introduce shape recognition between strings as a precondition for their interaction and then look at the entire system in this way and it doesn't change any of the results.

HOLLAND: I should mention, Walter, that in a somewhat similar vein, when Kosa does this with his strings, his tree manipulation does make a difference. Whether you can recognize a particular kind of subtree or not does make a difference.

FONTANA: Kosa's system is fundamentally different in that Kosa takes a list expression and then makes genetic algorithms. Here we take a list expression and let it act directly on another list expression. We feed a program with a program and the output is another program.

HOLLAND: I was trying to point out that that may be an artifact of the difference between the two approaches.

PINES: Let me attempt to summarize what I heard happening about twenty minutes ago. Peter Schuster was saying that you don't have copying in your model. Therefore, it is difficult to see how it applies to real biological situations. I didn't hear everybody say "Aha." What I concluded, therefore, is that this is a fascinating branch of mathematics probably related to some very interesting geometric constraints which is at the heart of what Walter is doing with his computer simulations and his lambda calculus. It should be viewed as such but it is exceedingly premature to discuss it as a biological paradigm.

GELL-MANN: Would you accept only local restrictions on copying?

FONTANA: We didn't do this in our model. We prohibited copying completely. It is a very good question. Why don't you, instead of assuming a probability of zero for copying, decrease the probability of copying and then look to see if you still get a transition with some non-zero probability? This obviously will be tested. But let me point out once more that we tried to think about what is a useful definition or a model or an abstraction of an organization. If you present offspring with something, you don't present them with a bunch of chromosomes. You present an organization.

To put it molecularly, Peter and I did an interesting experiment that was published in '89. Peter told you about folding of RNA molecules. Here you have a string and it folds to a secondary structure. That structure then determines the rate of replication of that string due to the loops and the helices and so forth. We then came up with an heuristic function of how the structure translates into kinetic rate constants and we could derive how the structure must be made so that it maximizes replication. Then we kept a couple of computers busy for several months to compute all the possible configurations, to fold all the strings up to a certain length to see how often that structure is realized. The answer is that it was never realized. In other words, if you go to the phenotype from the genotype, you have one mapping but, if you have a mapping from the phenotype to fitness, when you evaluate what that phenotype does, then the second mapping is utterly different from the first mapping and operates on totally different rules. Therefore, collapsing the genotype to a fitness description in just one mapping is simply a logical misunderstanding. How the phenotype gets out of the genotype is very, very important.

SCHUSTER: A very brief comment. Walter said what the parents give to the children is not a grammar, it's much more. Now the majority of organisms, except the mammals, do nothing more than putting the chromosomes with a bunch of nutrients in a proper organization to get into an egg.

FONTANA: The point that I was making is that your organization is a phenotype and that identifies a real need, that our theoretical framework is in a total vacuum with respect to how a phenotype gets out of a genotype. So long as we don't begin to make mathematical proposals about what a phenotype is, meaning what an organization is, we won't make much progress in understanding what evolution does.

CRUTCHFIELD: I feel a sense of frustration in trying to shift from a comparative discussion of different examples of complex evolving systems to an integrative discussion. It is very difficult to find the new language that is necessary for this purpose. I see my comments as an attempt to be integrative although it sometimes seems merely to add new terminology, another class of model, to the mathematical experiments that we have heard about. My interest is in the mechanism of the emergence of structure and in trying to detect when that has occurred. I see that as the goal of the workshop.

To emphasize the notion of emergence, what I said the first day is a formalization of the topic I was just discussing. In moving from one level of representation to a more powerful one, the complexity changes. If you look at the relationships between different model classes, in moving from one to another there seems to be a phase transition. You can give in a number of cases an analysis of how, at the lower level of transition, a model becomes baroque, like Brian Arthur's description of technological encrustations at some point. Then, a specific proposal for a mechanism of emergence is that we look at the available set of models we have and look for commonalities between them. The question of emergence to new levels is the question of how do we realize what the equivalence relationship is that gives us new classes upon which we build the new language. The problem of dealing with the notion of integration is that if you move yourself back to more and more primitive mathematical concepts, you necessarily describe more and more.

QUESTION: Are you really saying less and less about more and more?

CRUTCHFIELD: That's the question as I see it. When we look at systems that we have been able to analyze that exhibit this transition, like the onset of chaos, by examining cellular automata models and structure, we see that the mechanism of the higher level organization is associated with a change in the computational capability. This observation has occurred in a sufficient number of different cases now that I'm sort of suggesting that it may be a general property. This is relevant to David's question: Can you identify change in organization and the emergence of a new level in the structure in terms of a change in the computational capability of the system? The difficulty I have is that there's a lot of homework to do. If I say go identify the equivalence class of the relation I'm talking about that distinguishes it before and after its emergence process, it's going to be a lot of hard work to do that for RNA replication or for Walter's system but I still think that, at this point, it is progress to say that that's the problem to work on.

PINES: Point of clarification. When you speak of before and after, would you tend to identify emergence as you are using it with the giant steps which were being discussed as part of the hierarchy model? Is there a general sense that this is, at least, a useful poetic way of talking about such things as emergence, innovation, giant leaps?

CRUTCHFIELD: The specific suggestions, I hope, have the benefit of possibly being wrong. I would love to see a counter example, a system tha goes through a transition where there is a very different structure and there is no change in the computational capability. Then I would repair to my lair and sulk for a bit. But I'd actually be happy about that. In the examples I have analyzed, transitions are always associated with a change in the computational capability. (Q: Do you mean an increase in the rate of computation?) Yes. Always an increase.

KAUFFMAN: I think this is a fascinating suggestion. Can you carry out such a mapping from replicating RNA molecules to hypercycles or collectively reproducing things?

SIMMONS: Stuart and Walter and Peter have all used the term hypercycle. What does hypercycle mean that cycle doesn't?

SCHUSTER: If you have any kind of replicating entity—we think about RNA molecules—they couple to each other by having catalytic activity so A replicates but also has a catalytic capability for synthesizing B. These things happen in nature. We know that.

SIMMONS: Not with any kind of self-replicating system but some kinds.

SCHUSTER: Yes. We are looking at those that have this property. The general feature is that self-replicating entities are competing and all disappear except one and its mutant. Now assume that you have coupled sets with these molecules. It turns out that these networks are highly unstable except for few classes. By unstable I mean they lose members. Stable ones are if you can form a cycle, A helps B, B helps C, and C helps A. If you happen to form such a cycle, you get a stable form of organization.

SIMMONS: That's a cycle.

SCHUSTER: Yes, but it's not a cycle of simple reactions but a cycle of already autocatalytic, replicating entities. A makes copies of B and B makes copies of C and they are coupled together in such a way that they don't compete.

PINES: That's a hypercycle; in other words, it sounds like what Jim was describing, a real shift in how the system is organized, going from competition to cooperation, and is capable, therefore, at some level, of more computation. Brian and I were accepting the fact that the hidden agenda for people like Jim and me, and those who tend to follow this line of thinking, is that we like to think that we represent a higher level of computational capability than do chimpanzees or mice or whatever.

FONTANA: I should like to make two comments. One is that to go from competition to cooperation is certainly a shift in organization but there are other shifts in organization. Leo and I are emphasizing different shifts in organization that have nothing to do with going from competition to cooperation. Structurally they are totally different. There is a shift in organization in what Peter describes but there are others as well.

PINES: To follow your suggestion, Walter, in the new forms of organization that Leo and you are looking at, do they meet Jim's notion that the new forms have greater computational capacity?

FONTANA: I tend to think of computation in different terms but that would mean a long discussion. I'd like to make another point. We had a very interesting discussion a few nights ago when we paid attention to the deterministic cellular automaton as something that recognizes regular languages. Suppose now that the thing has states that are nodes in the automaton and there are transitions between states. Now suppose you put into each of those states a deterministic cellular automaton. Then you have a context for language. You have a machine that recognizes the next level in the hierarchy. Isn't it funny that you didn't have to invent very much except putting in the transitional structure that you have at each node that made up the organizational structure before. It is very illuminating that you can set up such a hierarchy without invoking any magic. The nodes are just states. Now, instead of a primitive state, you have an entire automaton. You're into the next level. The next kind of language is called context-free. There are many other ways of defining context-free languages that do not clarify this architecture, where you are completely lost because you wonder, "How the heck am I going to switch languages, how do I invent a new language that is utterly different from the one I had before?" In our representation, this is very straightforward. You just plug in that which was your entire organization at the first level and it now becomes a new atomic limit for the next level.

GELL-MANN: Does the whole Chomsky hierarchy, so-called, get built up that way?

CRUTCHFIELD: Yes, that is the basic idea, in addition to the complicated picture that I put up which has a lot to do with compiler design. What I take from it is just the hopeful interpretation that, given that they can develop these classifications and understand the relations between the model classes, then we, looking at nature, might also do the same.

GELL-MANN: But this would be an illuminating way to describe the hierarchy.

CRUTCHFIELD: To say the least, if it's really true, if what Walter says is literally true.

LLOYD: What kind of machine corresponds to a hypercycle?

SCHUSTER: If you wish, you can construct a parallel computer that works in that way. It would contain three different units constructed in such a way that they interact by message passing.

KAUFFMAN: It is fascinating that it is possible to construct a molecular machine, that something more powerful can be made from molecular machines, and that a parallel can be drawn between such machines and nature's steps in evolution.

CRUTCHFIELD: The approach that I'm focusing on is to look at nature. We have this very impoverished list of what computation is; just some aspects of what nature does which seem to be useful to measure the change in organization. I'd like to look at hypercycles instead but these MPs don't have that level of organization. They are all finite automata equivalent. After they have interacted for a while, there is a higher level organization and I can write down some context-free grammar that is, by appealing to the theorems that exist, a more descriptive and more powerful representation. Therefore, the computational capability has changed.

GELL-MANN: I gather that all of these things become rigorous, as is usual in this field, when n goes to infinity, where n is the number of sites, or whatever. I wonder whether, in the discrete case that you're always talking about, this doesn't have some relation to the taxonomy of transfinite numbers. Because you have aleph-null, the number of countable things, and now Walter said that you take aleph-null sites and at each one you put in something with aleph-null states and you get, perhaps, the next transfinite number depending on the famous axiom, and so on. So is there a correspondence at these levels to the different transfinite numbers?

CRUTCHFIELD: Sure. But that gets a little beyond even Chomsky. . .

GELL-MANN: But in the Chomsky characterization, as I understand it, it is very hard to see why the next one is such a big deal as compared to the previous one. But if you look at it in terms of counting the transfinite numbers, aleph-null, and then aleph-null to the aleph-null, and so on, then it becomes very clear-cut.

FONTANA: I don't know that the small area with transfinite iteration is correct. We have to turn to the mathematicians for that. But the basic message you are giving is correct, yes. I was proposing, too, that we not write in terms of Chomsky hierarchy people who are into complex systems but in terms of people who are interested in compiler design. Maybe it would look much more transparent.

CRUTCHFIELD: We have to present this to the theoretical computer scientists. They appreciate the historical structure of that hierarchy and they say, "You can go out and look at nonlinear systems but what better computation then results? Give us the different hierarchies."

MOORE: This is the kind of conversation I've been wishing we would have around the different generalisms that people are proposing. I side with Jim that his generalism, the computational paradigm, is a very good one. We have referred to several other generalisms like the edge of chaos and self-organized criticality. I'd like to suggest that people who advocate a particular generalism would do well to actually tell us some context in which they think that generalism does not apply, of several contexts in which they think that their grand idea fails. I can think of several contexts in which my grand idea fails. There are several ways in which I see generalisms failing. One is when the quantities they refer to cannot be clearly defined. For instance, I can imagine a lot of contexts in which it is not clear to me in what sense the system is either doing computation or how I can derive a computation from it. Another way these generalisms can fail is when the quantities are well defined and your statement is wrong, the sort of counter example Jim was asking for. A third way is when you're right but in a way that strikes the people who are working on the problem as trivial or counter-logical, which I think may be the case in some contexts with the edge of chaos thing, whereas in other contexts it may be very useful. If I'm going to propose a general idea, it's part of intellectual honesty and also part of what would contributes to constructive discussion for me to say where I think the limits are and then constructively to try to push those limits. For instance, computation is very hard to define in a continuous context so, as someone who is very interested in that idea, I am trying to define it in a continuous context. But I want people to know that I don't think it extends well beyond discrete symbolic situations. So I'd like to request that, as we review these other generalisms, I'd like to hear from Stu a few examples of contexts where he doesn't think that the edge of chaos is that useful. I'd like to hear from Per Bak a few examples where he thinks self-organized criticality is useful or isn't true.

KAUFFMAN: Chris, is it easy to say why computation doesn't extend to continuous systems? And I'm wondering if that's related to my worry and, with respect to Walter's ideas, if whether real molecules will be as crisp as grammatical and algebraic structure.

MOORE: Well, my personal view is that they're not, partly because of Peter Schuster's point that molecules have reaction rates.

KAUFFMAN: I'm asking about computation in continuous systems. Why doesn't that work?

MOORE: There are concepts of what a computable real function is. Computation theory was born around the integers, around what kind of functions there are for integers, and with set theoretical notions where we are dealing with sets and sequences of symbols. There are a number of concepts for computation in a continuous state but, in my view, they all have visible drawbacks or arbitrary limits and none of them have achieved the kind of primacy that, say, the Turing machine has in a discrete case. There are many concepts for machines, all with various limitations on applicability, whereas the Turing machine has really been enthroned for a lot of good reasons as a model of discrete computation.

LLOYD: I want to pursue Chris' point. The Chomsky hierarchy is a hierarchy of languages. A language is simply a collection of strings, of symbols. Now I accept fully that the Chomsky hierarchy is a very good way of analyzing collections of one-dimensional strings of symbols. However, it appears to me that it's not at all a good way of analyzing, say, different sets of chemical reactions. It's completely unclear to me how to make the transition to one-dimensional strings of symbols and also that, if one were to do that, one would get a reasonable sort of answer. So my question to people who are attuned to the linguistic hierarchy is "Do you think that there exist different contexts where the complexity that you are trying to analyze is not a string of symbols but, rather, clusters of reactions or expressions of RNA molecules, etc.? Do you believe in a hierarchy of this sort? Do you believe it might be useful?"

FONTANA: I think it would have to be invented.

MOORE: There is a theory of computation on graphs and on trees. There are tree grammars. That doesn't seem to be the problem. They deal with inventing a notion of structure, what I would call states. You have to lay out a language that's appropriate to the problem at hand. What computation theory does, and the most general lesson to take from it, is that it keeps track of how things are structured. Computation theory has made a lot of progress because of the discrete mathematics but there are certainly a lot of things we have to invent anew and just develop by analogy with computation theory for analyzing structures.

PINES: Following up on this group's fascination with using molecules for computing, if you combine that with Jim's remark about examples in nature, and borrow from Murray one of his favorite people to quote, Pogo: "We have met a molecular computer and it is us."

FONTANA: The hierarchies that we are looking at are not simply a Chomsky hierarchy, they are hierarchies that try to point out what are relationships among atoms, algebraic structure, that, once you have words in the language that are active, they do something, that bite, not passive words but active words. You not only buy a Chomsky hierarchy in terms of grammatical structures but you also buy algebraic structures as well because now these things are put into new relationships among each other. Why do we claim level one, in our case, is a hierarchy different from level zero? It is because an entire description of level one can be done without ever referring to a level zero description. You have laws by which you can completely state the symmetries, the invariances of your system at that level, without referring to the symmetries and invariances at the level zero which will be different. This takes us back to a remark that Phil Anderson made in his paper some time ago: emergence is when suddenly you have a system in which you're forced in the description to change equations in order to capture totally different symmetries from where they were before. In other words, if we have a pot of water, and you might use a Navier-Stokes equation to describe the pot of water, then as soon as you lower the temperature, it goes to the solid state, and then you are no longer well off with the Navier-Stokes equation because totally different symmetries became relevant and you have to switch your level of description and you are forced to switch your equations. So that is what happens at level one.

QUESTION: How does that come about? You can say it can happen but what forces it? That, it seems to me, is the question that emerges.

FONTANA: In this case, we have a complete map from level zero to level one. Everything that happens is controllable. We can look into the computer and retrace the history of what is happening. I mean mechanically it is very simple. It means that such structures have small generators and all you have to do is to produce two or three elements of the generator and bang, there you are.

GELL-MANN: Can we go on now to discuss these transitions between order and disorder? One application is to variation of schemata, as in a simple case where the mutation rate may adjust itself to a point between order and disorder. Regarding the use of the schema in the phenotypic arena, I think the adjustment occurs there also. Then in the selection pressures back on the schemata from the phenotypic arena—I'm sure that's a place where people claim to have encountered it. And, finally, the fourth case is a collection of complex adaptive systems coevolving and making models of one another.

KAUFFMAN: Could you explain the two middle ones?

GELL-MANN: Yes. The schema is like a genotype. It's what Walter calls the object as information. It stimulates some activity in the phenotypic arena. In the case of Tom Ray's model or of the RNA theory of the origin of life, the phenotype is physically the same. In all the other cases it's physically distinct as far as we know. The activity consists of description, prediction, and/or behavior. There is then some feedback. It shouldn't be thought of in general as a deterministic procedure or a fitness landscape because those are too special. But there is some sort of tendency for the description/prediction/behavior to affect the success of the schema in competition with others. Failure or success needn't be thought of as extinction versus survival. It could be just demotion in a ladder versus promotion lower population versus higher population, or whatever. The third case, then, is that of selection pressures back on the competition of schemata from the phenotypic arena.

ARTHUR: Do cellular automata exhibit this sort of thing?

GELL-MANN: The cellular automata studied by Norman Packard and others—I never understood exactly where they fitted into this series of stages. What they did there, as I understand it, was something very simple; they just took cellular automata and tried to get them to perform some absurdly simple task, like converting a string of zeroes and ones into a string of all zeroes or all ones, depending on which was in the majority. And they found that when the initial cellular automaton was in a particular Wolfram class, in between the ordered and the "chaotic regimes," that was a good place to start the learning process. They then saw a tendency for the automaton to move toward that class.

ARTHUR: But I'm not clear on what you would mean by schemata in the cellular automaton case.

GELL-MANN: But I would rather apply the idea to some other field like biology or economics.

KAUFFMAN: Let me say something about the parallel processing network. The first step is when Steve Wolfram came up with rule classes, type 1 behavior, type 2 behavior, and 3 and 4, where type 1 is a point attractor, 2 is a periodic attractor, 3 is a chaotic attractor, and 4 is these long, funny filigree transients. The question was "If you think of a rule space, how is it partitioned?" And it turned out that 4 is in a region that separates 1 and 2 from 3.

GELL-MANN: And then he noticed that 4 has a certain correspondence to universal computation.

KAUFFMAN: That's right. So let me draw this. This is one of the spots where I think we almost have something coherent, almost believable. Of all of the cellular automata types that Erika was telling us about, each of which she is exploring in great detail, it turns out there is very rich behavior in the class 4 region. Chris Langton embedded his lambda parameter, which is the same as Bernard Derrida's p parameter. Let me explain what it is. You have a Boolean function like the "or " function.

QUESTION: It's the fraction of ones in the rule?

KAUFFMAN: It is the deviation from 50–50 for a binary variable which is the fraction of ones in the rule. What Chris looked at is the behavior of systems where you tune p and it turns out that if it's mostly ones in the rule table, the system freezes in. This corresponds to the frozen components in Boolean networks that I told you about, the little green island. Then there is a chaotic region where there is a green twinkling sea which is the phase transition region containing Chris' measure of mutual information. So, there is now the argument that parallel processing networks in this regime can carry out the most complex coordinated computation. In the fully chaotic regime you can't coordinate things because of noise. There are two different points here. One is that the most complex computations can be carried out here and the second is the claim, first made by Norm Packard, that natural selection to carry out complex computations will actually get you there.

There are two pieces of evidence: Norm's experiments and the experiments that I've tried to do having Boolean networks play games with one another. There's quite good evidence for one class of games that Boolean networks, under selection to change their internal structure and logic, do actually move to that phase transition.

GELL-MANN: What are those cases where it has been shown?

KAUFFMAN: The existence of a phase transition has been shown analytically by Bernard Derrida using an annealed model. The trick that Bernard used was to imagine that you have a parallel processing network and that at every moment in time the connections among the light bulbs and the Boolean functions governing the light bulbs are randomized. (Q: The annealed version of the NK model?) No, it's in the old version of the Kauffman model. I gave you the technical name that has been applied to it but not by me. Here's what Bernard showed. You picture the following thing: think of an infinite succession of networks with n light bulbs and at every moment the light bulbs maintain their identity but the connections among them are switched and the logic of each light bulb is switched. Bernard took the following approach: Take two initial states, submit them to the network, and look at their two successor states. If you look at the Hamming distance between the two initial states and the Hamming distance between the two successor states divided by the number of light bulbs, you get a mapping of the current distance between the two states into their distance one moment from now. In the ordered regime, the return map is always below the mean diagonal. In the chaotic regime the return map goes above the diagonal for small distances, meaning that the nearby states diverge. Bernard showed that if you have two input networks, you're always in the ordered regime and if you have K greater than two, you're in the chaotic regime. That is the first proof of a phase transition in this class of models, technically chaotic, a discrete dynamical system containing an orbit that is closed and one-dimensional and, therefore, a limit cycle. But it's chaotic in the sense that nearby states diverge. (Q: So it partakes half of one cycle and half of the chaotic behavior?) Yes. These are finite-state automata so eventually they have to cycle.

After Bernard showed that, he went on to show that if you look at networks with a large number of inputs per light bulb and vary what he called p, you, in fact, shift through the phase transition from the chaotic regime to the ordered regime. It's a phase transition in the annealed model. If you sample the expected behavior by making a specific Boolean network in the quenched case, the numerical experiment that I've done consists of taking a population of Boolean networks in the chaotic regime playing games with one another. The game we are playing is that each has some input and output light bulbs; I sample your outputs, and you sample my outputs and I want mine to be as different from yours as possible. It's a mismatch game. What you find is that if you start with networks in a chaotic regime and let them evolve, they change their internal structure and approach that boundary. If you start in the ordered regime, they do the same. So we are beginning, therefore, to have decent evidence that parallel processing networks actually, by performing one complex game—the mismatch game—approach this transition. It cannot be universally true. If I use Boolean networks to play a game where their fitness did not depend on what they did, obviously there would be no selection.

So for what class of games will it be true that parallel processing networks will -be attracted to the transition? It certainly is not universal because it is trivially true that, if my score does not depend on what I do, there will be no selection principle. So, for what class of games is it true that there is an attraction to this boundary?

GELL-MANN: Is there evidence for attraction to the transition region in cases other than collectivities or complex adaptive systems which are trying to model one another? Are there simpler cases where you see attraction to the transition region?

ARTHUR: There is the one you mentioned, the Packard computation. Imagine that suddenly you say that your light bulb systems are cellular automata and use them as coevolving systems so that each of the elements is trying to adapt to the others, when are such systems likely to coevolve or coadapt and when will they always be mutually upsetting each other and is there a state of transition where they may or may not coevolve?

KAUFFMAN: I think that the answer to that is that there is such a regime, that there is a transition regime in a coevolutionary context, and that one can identify the selection forces that will take a coevolving system of independent agents to that point.

ARTHUR: I'm not asking about selection forces. I'm thinking about the kind of treatment we find in Langton's model of approaching chaos.

MAYNARD SMITH: What little I know about this comes in very relevantly here. It is clear that we do not have a satisfactory theory of coevolution of species. The best serious attempt to model this was done by myself and others. The basic idea is that the physical environment is constant. For species in the ecosystem, their environment is constituted by all the other members. The assumption is that for species x, its fitness surface is determined by what all the other surfaces are but it's a smooth surface with a peak. If everything else stays still, it would gradually evolve to that point. And that's true of every species in the system. Now Nils and I, foolishly perhaps, attempted to analyze this analytically rather than by simulation. We had a matrix, Δij, where we ask "If I move, how much difference does that make to your peak?" And we found conditions for ij which distinguished between two possible solutions, namely, the solution where every species gradually moves to its own peak and everybody thrives and stays there in equilibrium or the alternative situation where everybody moves further and further away from their peaks and, as long as you kept everything linear, everything got worse. That's called the Red Queen model where you have to run as fast as possible to stay in the same place. The condition in our case is simply the condition on the Δij's: how much effect does it have on you if I change?

GELL-MANN: There is a transition regime and there could be some situation under which the system might be attracted to that regime.

MAYNARD SMITH: We don't really have any theory about that. If we insist on linearity, then you get a system which doesn't blow up but stays within a very complex limit cycle.

MOORE: I want to introduce a mathematical reality check, a result from nonlinear dynamical theory. Imagine a family of systems; anytime the attractor of a system changes its structure qualitatively, you have to pass through a state of neutral stability and, therefore, in that state you will have arbitrarily long transients. You have to look at the structure of the process at the boundary between these two states.

GELL-MANN: You're talking about the Lyapunov exponent, negative or positive, so that when you go from one state to the other you pass through zero?

MOORE: You can go from a negative to another negative, say, from period three to period seven. You've changed the qualitative structure. This is a basic element of bifurcation theory. What is distinctive here is the structure of the memory and the computational net. You have to ask what the computational structure is. So to round this out, when Chris talks about the structure of the CA space in terms of a diagram that says there is universal computation here at this particular lambda value, there is absolutely no mathematical necessity in that statement. That's not the case. A large amount of memory does not mean that you can use universal computation. Universality require a very particular structure. The memory has to be organized in a particular way to do a particular kind of computation.

KAUFFMAN: There are two regimes that come up commonly in these models, Nash equilibria or evolutionarily stable strategies that John invented and Red Queen behavior which, in nonlinear systems, means that things will keep oscillating in chaotic ways. I will now talk about the NK landscape models. Here is the general idea. People have looked at coevolutionary models that are in the ordered regime. And they have looked at others that continue an ongoing dynamics that never settles down. I made models of fitness landscapes that can be tuned from very smooth and single-peaked to very rugged, so they're not John's linear case; they are multi-peaked surfaces. The two basic things that go into this entire class of models are: what's the structure of my landscape if everybody else didn't move, and how badly is my landscape deformed when the other guys move? What emerges from this entire class of models is that if you tune those parameters, you pass from a chaotic regime to an ordered regime. The ordered regime is John's ESS. The chaotic regime is the Red Queen. This is high-dimensional chaos where things are wandering all around in a big space. If you look at this class of models, there is a third level of the game where the players can change the games they play. That means they can change the structure of their landscapes and how those landscapes are coupled to one another. Murray, you use the same description when you say there are selective forces with no landscapes.

GELL-MANN: I say that that can happen even when you don't have mutual effects. Your assumption is always that, if you turn off the mutual effects, you will have a landscape. It's a false assumption.

KAUFFMAN: Of course, for general dynamical systems that's not true. We're describing adaptive agents that are getting better and better if nobody else moves. Evolution is supposed to make me get better and better, absent recombination and Marc Feldman, in either order. It then turns out for this wide class of models in the chaotic regime, the Red Queen regime, everybody has low fitness because nobody's near a peak. Deep in the ordered regime, everybody has low fitness because the peaks are crummy in this class of models and, in the phase transition region, people have the highest peaks. You do coevolve to a phase transition between order and chaos because it's to my advantage to change the structure of my landscape and how it's deformed; the evidence in favor of it is that you get a power law distribution in avalanches that looks like the power law distribution of extinction events.

GELL-MANN: So it has attraction, it has a critical point with a power law?

KAUFFMAN: Yes. The claim is the following: John has worked for years on coevolving species in which the game is set, they are playing whatever game they are playing and we watch them evolve their genotypes but there's nothing here that fixes the game I am playing with other species. If I change the kind of thing I am, I interact with other things in other ways. So we really have to consider this third level of dynamics in which the players change what they are and how they interact with one another. There's tentative evidence that if I change the kind of thing I am and, therefore, the kind of landscape I'm on, and, therefore, how sensitive I am to you, then that that kind of system can evolve, for at least a wide class of models, to this phase transition region. It shows up because there are power law distributions of avalanche regions and that's the answer to your question, Brian.

MAYNARD SMITH: Intuitively, I think it's fair to say we do not know in biology whether species in a uniform environment...supposing God said let there be no more Ice Ages, no more meteorites, I'll keep everything constant...we do not know whether evolution would gradually fade out and we'd get stasis or whether it would stay in some kind of equilibrium. Most of my paleontologist friends are persuaded that there have to be continuous interruptions in the physical environment in order to keep evolution happy.

SIMMONS: What you say is that if you hold the physical environment constant, evolution would go to a fixed point, to some stable condition.

MAYNARD SMITH: That is the view held by most paleontologists, I suppose.

SIMMONS: Meaning that there would be absolute stasis.

MAYNARD SMITH: That's right.

GELL-MANN: Then they don't know about Tom Ray's model.

MAYNARD SMITH: I'm not convinced that there's any hard theoretical reason why they have to be wrong. Maybe there is.

COWAN: The oceans provide a relatively unchanging environment. Does biological stability of species in the oceans help support the stasis idea?

MAYNARD SMITH: True. But when there was a major collision at the K-T boundary, there was a major reorganization of species within the oceans.

GELL-MANN: But the terrestrial paleontologists went berserk at first over this suggestion. It was only the marine paleontologists who reacted favorably.

MAYNARD SMITH: Yes. They really did know that there had been a massive extinction at the K-T boundary. The land people. . .look, there's always a lot of irrationality. Many paleontologists did not want to believe in this extinction.

KAUFFMAN: John, I recently found out that during the Cambrian explosion there is an extraordinarily high rate of speciation but there is also an extraordinarily high rate of extinction. So that during a period of 100 million years or so, the total species abundance is increasing. There is a very high species turnover which then gets quieter and quieter through the next 100 million years as if the thing were in a kind of chaotic regime and are learning how to live with one another and settling down.

MAYNARD SMITH: Things were put into motion by some sort of disaster and then gradually recovered.

KAUFFMAN: But it's over 100 million years.

MAYNARD SMITH: Well, it takes a long time.

KAUFFMAN: But they're also speciating and learning to live with one another.

GELL-MANN: Well, this attraction can't be very powerful if it took hundreds of millions of years to return to normal.

MAYNARD SMITH: After the K-T boundary, after massive extinctions, the numbers of species starts increasing; then it drops away and stays still. That's the total number of species, not which species are present.

GELL-MANN: We can imagine that maybe, if you looked at it in terms of a Pareto distribution of some kind, the power law would have been disturbed 65 million years ago and then would have returned after, say, 20 million years to what it was before the extinction, or conceivably to a different power-law.

KAUFFMAN: About the power law of avalanches, Murray, there's both the power law distribution of the size of avalanches and then there's the interesting observation that Jim reminded us of, which is that some species command a lot of the available resources and some don't command many, and that looks like a power law distribution, too.

GELL-MANN: I'd like to ask the following question. Let's abolish for the moment the competitive modeling carried out by many complex adaptive systems in an ecology and go back to a single one in a constant time series environment, but very far from equilibrium. In that case, how many of these features would still be there? Would there still be a technical transition? Would that transition still be associated with efficient adaptation? Would there still be attraction to it? Would it still be associated with the possibility of high complexity? And would it still be associated with the possibility of universal computation?

KAUFFMAN: There is one extraordinarily tentative piece of evidence, an inconclusive body of work that Mark Beaumont just told me about and that I don't really believe yet. What he has is a population of Boolean networks which are sitting on landscapes where they can sense the direction to walk, as though they were sitting on hillsides with slopes and each little network has to decide where to move. The actual physical hillside is rather rugged. There is very tentative evidence that if you let Boolean networks evolve so that they control their motion on the piles, going from low parts to high parts, that is, if you start deep in the ordered regime, they become less frozen and, conceivably, if you start deep in the chaotic regime, they may be approaching this boundary and they are also, in a nice happy way, building up correlations between the input variables and the output sites. They are increasing the mutual information between inputs sensing the environment and output sites that affect the environment. But otherwise there is no answer to your question.

GELL-MANN: We've been looking at a collection of complex adaptive systems, coevolving and modeling one another, and at the existence of self-organized criticality with a transition region, a critical point with a power-law distributions of avalanches and maybe power law distribution of other things as well, attraction to that regime from both sides, and, at that regime, the possibility of high complexity and the possibility of universal computation, the latter because of long transients in space and time associated with a self-organized criticality where all scales are present. Then I asked, suppose we abolish some of these conditions and go back to a single complex adaptive system with a constant time series environment (not a solved problem, like tic-tac-toe, etc., but some unsolved problem). How many of the previous features would you still have, if any?

KAUFFMAN: Here's, I think, the way that one could pose that question before I try to answer it. Suppose that I've got a system that's trying to match or predict an environment. Then the way we might ask the question is that as we vary the complexity of the time series into our evolving automaton, for what classes of inputs like, for example, a steady hum or a Bach sonata that it's trying to classify, like looking at the process of games that networks play and how they have to evolve, the way to ask that question is, for what types of inputs to which our system has to respond in some way, if you ask the system to evolve by changing its internal structure and its logic by mutation and selection, could it be, for a stationary time series of some complexity, that such a system would evolve so that it would go to Langton's phase transition? I bet the answer is that, for a sufficiently complicated time series that has some mutual information and correlations in it to be compressed, the system has a good chance of getting to the phase transition. For a totally chaotic time series, then it couldn't. If it's totally stable, it wouldn't bother. So it's an answerable question but there are no data on it and it suggests a fine set of numerical experiments to be carried out.

GELL-MANN: It's funny that the experiments on the competing systems should already have been done without the experiments on this much simpler question.

ARTHUR: Stuart, I simply don't get your dynamics. You're talking about a metadynamics in which you say, well, on the one hand, there is a deeply ordered state where the entities that are coevolving couldn't be doing too well; on the other hand, they would be disrupted all the time and their average fitness would be pretty low. So there is some sort of metadynamics that imaginably brings you to a state where things are better off. Now I think that that is highly conjectural and remains to be proved, not just proved on the output side by the existence of power laws and avalanches but as a conjecture for any particular system. Why should the dynamics care one bit about how well off these entities are? Suppose they are locked into a mutual Nash attractor deep in the ordered regime and it's relatively stable and they can't jump out of it. I can't see why any dynamics would care to get them out. It's quasi-teleological.

KAUFFMAN: No, no. But I didn't want to bore you once again with the NK model. Let me recall it to you again.

ARTHUR: I know the NK model and you don't need to recall it.

GELL-MANN: Let us say the Kauffman model, a different model.

ARTHUR: All right. Stuart, suppose I have these systems, one is deep in the ordered regime, one is in the middle, and one is at the far chaotic end. Why is the one in the middle to be preferred?

KAUFFMAN: I think I know the answer although, of course, it may be wrong. I have a class of landscape models, I've got N sites, and each site is affected epistatically by K other sites. As you tune the amount of epistatic action, these are just Murray's spin glass models, these are just k-spin models. When k is zero, you have a ferromagnet, where each site is independent of every other site. When k is very large, you have a highly coupled system and a very complex potential. The key thing to realize is that we can play with these landscapes, they are just toys. In this class or family of landscapes when k increases, the number of peaks goes up but the height of the peaks goes down because the system is more frustrated. So if we are evolving on landscapes and I'm on a very rugged landscape and so are you, when I move on my landscape, I deform your landscape and vice versa, I rapidly get to a peak and so do you, peaks that are mutually consistent, and we stop moving. So we are in the analogue of a Nash equilibrium or John's evolutionary stable strategy. The entire ecosystem goes into an ordered regime but fitness is low because the peaks are low. In the chaotic regime where k is low, the peaks are very high but, when you move, you move my peak away from me faster than I can catch it.

GELL-MANN: But you're doing coevolution.

KAUFFMAN: Yes, it's coevolution.

GELL-MANN: The question was about what happens when you have a complex adaptive system.

KAUFFMAN: I am answering Brian's question, not yours, Murray. Now, in the chaotic regime, my fitness is low and so is yours. It turns out that the average fitness is highest if you're at the phase transition. Now we need the next step which gets us into the following: If we're in the chaotic regime and I change the structure of my landscape so that my peaks are a little bit lower and the sides are steeper, does my fitness go up? The answer is yes. If we're deep in the ordered regime and I change the structure of my landscape so that it's smoother with higher peaks, does my fitness go up? Yes. So that there's an individual selection operating on me, one of the players who can change the structure of my landscape for my benefit and it looks like the attractor of that dynamics happens to be the phase transition.

ARTHUR: If I were in a Nash game and it were my turn to move and we're at some low-level equilibrium and I perturb the game slightly, I might move to some neighboring low-level equilibrium but it doesn't mean that I necessarily select that.

KAUFFMAN: Brian, it's true in this class of models, in the sense that there's a selection force that looks like it's taking you there, to this phase transition. I haven't actually shown that the real dynamics will get you there...the right question to phrase is "How broad is this family of models that you're talking about?" I believe you for the NK models...

ARTHUR: It isn't the right question to phrase. All you've said so far is that individually things don't do too well in the ordered regime and that individually they don't do too well in the chaotic regime but that in the middle they do better; so what spurs the dynamics to make the individuals want to change the game?

KAUFFMAN: Well, I've just told you. In this model, if I change the k couplings among my sites so that my landscape is different, so that if we're all deep in the ordered regime, I'm at a local equilibrium but my fitness is low. But if I change the structure of my landscape, the fitness of me and my progeny goes up.

ARTHUR: But if I'm in a Nash equilibrium, presumably...

KAUFFMAN: You're flying in the face of evidence. I've done the simulations in this class of models and that's what it does.

ARTHUR: I take your word for it but, in a Nash equilibrium, I'm doing as well as I can relative to everybody else. What you're telling me is that if I change my connections, I'm likely to do better.

KAUFFMAN: No, I have not said that. If I change the structure of my landscape, I'm changing the game I am playing. I'm not moving within a game. I'm moving between games.

ARTHUR: Yes, but everybody else gets to move between games as well.

KAUFFMAN: That's right.

COWAN: We will return now to the discussion centering on models.

HOLLAND: I'll lead off by repeating the point that John Maynard Smith made. Then I'll go to the social aspect of this work. The first point: If I'm going to talk about adaptation, I have to have both the system and the environment. I am free to draw the boundary between the two. If I'm looking at a species, say, humans, I can draw it at what seems to be a natural place, say, the skin, and I can say that everything inside the skin is the human and everything outside is the environment. But I can change that. I can draw it at the liver or I can draw it at the nervous system and then say the environment of the central nervous system is all the rest of the human plus the outside. Or I can draw the boundary much further out and have a whole group of humans and say that's what I'm going to study, that population. In each case, I ask a different question. And so, to a degree, where I draw the boundary and what I call the environment is going to determine what question is reasonable for me to ask. So when I talk about the process of adaptation, there is this rather arbitrary aspect of it which is really a matter of what I want to learn. My boundary is greatly dependent on that. Now, the other view that's being put forth here is that I don't want an environment. I want to consider a total system. Then, again as John [Maynard Smith] was pointing out, I have to talk about something like development and, since there's no environment, adaptation doesn't make much sense. I would claim that that is the point of view that Per Bak would like to take. He wants to look at the whole thing with nothing outside. When it comes to social systems, to carry this point a bit further, and discuss to what degree we can model these higher level interactions that have phenomena that might be called social phenomena, then I think these same kinds of considerations apply. If you ask what part of the system is adapting, then you have to separate out the part whose adaptation you are asking about and what acts as the environment to that. I do those things and I see, from my own point of view in this particular area, to me, in the social systems I know, the attractors have a lot less to do with what's going on than the properties of the trajectories. An awful lot of the analysis that we have done doesn't bear on the properties of the trajectories. And yet trajectories have properties just as attractors do and one can do analysis on properties of trajectories. There are various ways to do this. People haven't done it much but it's possible. When we start building models of this kind, we have to keep in mind what the relevant questions are and what the relevant analysis of the trajectory might be. I want the model-building process and the analytics to be informing each other so that I can keep my options open as long as possible.

MAYNARD SMITH: The point you're making is that the social systems are far from equilibrium and to look at a stationary space is rather boring.

ARTHUR: There's something even worse and that is that all economics has been able to do for the last fifty or one hundred years is to look at systems with very strong attractors, not even talk about how an equilibrium point is reached but simply point out that there is an equilibrium and that if we were there, there would be a tendency to stay there. How we get there is not well defined in economics. A point on doing science comes up. If we were to take seriously a lot of the work that's done in Santa Fe, most of the issues that people have been talking about is that the economy is indeed a complex, evolving, adaptive system in which no trajectory may ever repeat; new things are coming along all the time. There's a problem of style in doing science. A lot of economists would ask how do you talk about a system like that which is ever changing, ever emerging, what I call non-repeating. This goes back to an argument we had two or three years ago. I was stymied then. I said I don't believe in equilibrium and I don't want to work on that any more but I don't know how to talk about something that's always changing . And then John came along and said that you don't have to worry about that too much because even if something is changing, like the polity and the international scene which are always changing, nevertheless historians and political scientists feel they can talk in reasonably powerful terms about many of these features. Although these systems never repeat perfectly, and you may never be able to predict perfectly, nevertheless there are common features, there are patterns that you can talk about, not just restrospectively. You mentioned the weather. No cold front is the same as any other cold front, yet we know that there is an overall pattern that you can isolate and talk about and use for some kinds of limited predictions. I find that useful. What we are missing in economics is a leap of intellectual courage, because this is what it takes to move away from talking about static equilibrium solutions to talking about things that may never repeat, where there is no overall answer, and you're just talking about something that changes like the natural ecology.

COWAN: Don't we all strive mightily to create strong attractors? They don't necessarily exist spontaneously. They can be contrived by cultures, by laws, by custom, by ideologies. The whole point here is to stabilize an otherwise unstable system rather than put up with a high-dimensional system that wants to fly apart. The only reason for persistence, perhaps, is the creation of these apparent artifacts.

ARTHUR: I think I can give you one good example which Ken Arrow mentioned the other day, that the standard theory of the stock market shows an attractor in the market called fundamental values. Any stock is worth the discounted value from this extreme. That's probably correct to the first order but, if you adopt the line that John and Richard Palmer have taken, we're in a world where people form hypotheses about what the worth of that stock will be and those hypotheses depend on other people's hypotheses so we're in Murray's world of schemata or in an adaptive complex world where we have satisfied ourselves by simulation experiment that there is no time series pattern of stock prices that ever seems to settle down. The time series is always nonstationary. If I said that to Ken Arrow or any one else in the economics profession, I can guarantee that this would be viewed as an unpublishable result at this stage. I'm hoping that in ten or fifteen years time that will have changed, long after I've left the field.

GELL-MANN: What is unpublishable?

ARTHUR: The notion that the market could settle down to generating a times series of prices that is nonstationary. That's unpublishable because...I could say that it settles down to a stochastic equilibrium, one that sums a stationary time series, I could describe a generator for that stochastic time series... that would be publishable because that's something that you can grab and that stays still. But if it's always moving and always changing and never settles down, then that's not considered to be science in economics.

GOODWIN: But you can take heart from evolution. That's always changing, never the same, yet there is a possibility of discovering a taxonomy of forms that is, to some extent, invariant. So you can get generic properties in a transient system.

ARTHUR: I've thought that myself. The problem is not so much that the economists have totally closed minds...

GELL-MANN: It's just that they suffer from physics envy rather than biology envy.

ARTHUR: Precisely. And before I came to the Institute, I wouldn't have known what to do differently.

GOODWIN: I want to remark on John's comments. This resolution of the problem of adaptation between the internal model and absorbing that into a more extended dynamics point of view, that is a really important resolution to have reached. I think that the other question which it raises is the one that Ben was talking about the other day in relation to internal models. I think that once again the internal model approach is something that can disappear, if you like, when looked at from a different perspective. It can be used to considerable value if you're taking a particular point of view with respect to what's going on. But if you want to dissolve that away into a total dynamic, you can do that. I wonder if you would agree that this is another duality that can be either preserved or dissolved depending on the point of view.

SIMMONS: There is a question I wanted to ask Brian Arthur which is, in a way, slightly off the point. Is the little stock market model, which you and John and Richard made an example of, a system with such an attractor, which the system more or less found? You defined an attractor, a fundamental value, which the system found?

ARTHUR: No, we didn't define it. We let it emerge. To first order, that attractor emerged.

SIMMONS: But you put in the return rate.

ARTHUR: We put in the return rate. The adaptive agents had then to discover that they could put their money in the bank or put it in a stock and drive the price to its value.

HOLLAND: But we could have put in other agents and gotten rid of that.

SIMMONS: The question I wanted to ask is somewhat technical; namely, if you had defined that fundamental value, either with another agent or just exogenously, it's something that changed with time. Presumably, the agents would have changed their learning behavior over time and you could have explored the rate at which they learn. Have you done that? If you change the fundamental value, will they successfully follow the change?

ARTHUR: We allowed the dividend stream to change in a trend way and in a stochastic way and in a nonstationary stochastic model. In all cases we found that the price tended to track the dividend stream because that was a fairly strong attractor. But, we subtracted that from the price and you get variations that never settle down. You get many bubbles and crashes. Overlaid on the fundamental price was the risk psychology.

PINES: Let me try to put the discussion in a little broader context, trying to carry out my homework assignment. As George has said, when a group of us self-organize, we should report back to the larger group. Some of us had an interesting half-hour discussion. Beginning with terminology, and just to remind you, we decided to focus on adaptation at the level where individual agents—either people, institutions, societies—are constructing models or schemata. Ben Martin pointed out that what we call learning or adaptation is called tuning by cognitive scientists and that restructuring is, in the cognitive science trade, equivalent to innovation and is isomorphic with large jumps. That's just by way of background and language. You may remember that Ben talked to us about cognitive science within this framework. George Gumerman raised a very interesting question: what is known in cognitive science about the ability of individuals to give up a short-term advantage for the sake of the larger good of the society? (For the society or for the individual?) Either or both. Put most simply, as Bill Bradley said last night at the Democratic convention, we all have to give up something if we are going to get on with the health of the whole of society. Ben was hard put to come up with an experiment involving rats in which this could be demonstrated. It's not that psychologists haven't tried to carry out the experiment in all manner of ingenious ways. But it's still regarded as an open question.

SIMMONS: Do you remember the wonderful experiment of the monkeys and money that was described at the workshop last spring in which they did learn to forego immediate reward in order to amass money?

GELL-MANN: Yes. That was done in the '30s and has not been repeated. There wasn't enough money left over to repeat it. The monkeys had it all.

MARTIN: To the extent that psychology has something to say here, it tells us that it depends on the representation that the agent is able to construct. To the extent that you can induce a representation that takes account of long-term payoffs and makes them somehow commensurate with short-term costs, you can induce people and animals to behave in ways that seem prudent or as if they involve planning for the long-term good. Likewise, I think Murray's question is critical, whether it is in the interest of the individual or in the interest of the collective. If you can induce a representation of a problem that makes it somehow commensurate to weigh the interest of the individual against the interest of the collective, then you can induce that kind of behavior, and unless you can induce that kind of representation in agents, I would say that psychologists would argue that can't induce that kind of behavior. It's true only to the extent that you can induce that kind of representation.

PINES: Sort of the General Motors principle.

GOODWIN: How do you test the principle of the interest of the community against the long-term interest of the individual?

MARTIN: I would say again that's a matter of definition. I would suppose that economists believe they have some kind of representation when they make a distinction between people who...

GELL-MANN: Well, people get themselves killed sometimes for causes.

GOODWIN: I was thinking of psychological tests where the phenomenon of learning exceeds the lifetime of the individual.

MARTIN: Psychology only makes this sort of philosophical question something that you can ask about individuals empirically. You don't really get any innovation here. What it says is: if you choose to look at the problem in one way, you'll get one kind of behavior. If you look at it in another, you get another. The problem is "How do you look at it?" Do people look at a problem in altruism as a problem where they are gaining something? In some sense, it's to their greater glory or it dignifies their existence to act in a certain way or to sacrifice themselves. To construct those representations as an individual is to engage exactly in this game of asking whether you are maximizing your own utility or maximizing the larger good, and psychology doesn't have an insight as to when you're doing one and when you're doing the other. It just points out that what the individual considers himself to be doing is critical to understanding how they're going to behave in given situations.

COWAN: I'm a little concerned about equating learning with adaptation because there are two kinds of learning. One kind is related to what might be called maintenance of the system, to make sure that things don't change, to stabilize the system. In fact, perhaps most of learning has to do with that aspect and relatively little with the capacity for innovation or creativity.

GELL-MANN: It's still learning.

COWAN: But not adaptation.

GELL-MANN: Yes, it's still adaptation, quietly adaptive.

HOLLAND: You want to separate two things. The question of whether learning is adaptive is separate from the question of whether I can use a common formal framework to discuss both adaptation and learning. Those are two very different things.

PINES: John expressed his mild sense of optimism that it should be possible to construct models of appropriate systems, of institutions, that would have some predictive capacity, or which would, at the very least, play the role that Josh suggested, play the role of being humility-injectors. And then we talked in that context about two things, one that is the equivalent of the experienced pilot for your humility-injector, probably a good politician, someone who has been around for a while, because they have become tremendously skilled in the art of the possible, in knowing how to deal with tradeoffs. So much depends, in the utility of these simulations, on understanding what is possible and so much depends on defining adaptation and having some sense of how to deal with it. I think that we all agree that probably the key element for Project 2050 is going to be to have a clear understanding of what is possible, given existing limits of time scales for individuals to adapt and change institutions within the society, states within the society, and so on. John went on to say that he was sort of encouraged, for example, by Axelrod's model of how alliances might form and how, with a comparatively simple set of rules, he was able to generate a set of alliances that have actually formed as one of the solutions to his comparatively simple mean-field equations.

GELL-MANN: May I inject something here? Someone said this earlier, that we have to be terribly careful to avoid overstating the capabilities of our models These are caricatures that might be prostheses for the imagination. They aren't really predictive models. The uncertainties are enormous, there are a number of things not taken into account, not even imagined. They're so great that we will get into terrible trouble if we pretend that we're predicting or have something like a flight simulator. So I would like to maintain my agreement with what John Holland says without adopting some of the language because I think the language can be dangerous. So we have to be very careful with our disclaimers, or say that what we can do is just to present some ways to play with trajectories to the future and some probabilistic information about them but certainly not to follow trajectories in a mechanical way.

HOLLAND: You want to separate the notion of flight simulator as predictor from the notion of flight simulator as a neat interface. Those are two different things. We need a flight-simulator-like interface so that people who are not experts in programming can play with the model. There's no disagreement there.

PINES: Before I get to Josh's homework problem, let me add one point: one of the homework assignments on the way to developing the model for Project 2050 might be seeing if one could develop a model that explains the growth of the underclass in this country. Again, presumably a comparatively simple model might have a go at that. Then George Gumerman said to me later that he thinks there's a real possibility that an underclass, representing a real disparity between rich and poor, may be a characteristic of any developed civilization. He makes a distinction between a tribe and a civilization with the point of transition being the Chaco Canyon tribe which never quite made it to a civilization but was on the verge of doing so. He argues that if you're trying to take this broader view of society from the point of view of a complex adaptive system that's a well-defined feature of any civilization and if it's something you want to try to explain, why is this (the development of an underclass) something that goes with the development of a civilization?

GELL-MANN: It's very interesting that he mentioned the Chaco system because, so far as one can tell from the archaeological evidence, the Chaco system had its center in the Chaco Canyon which was perhaps partly trade, partly cere-monial, but with service by a gigantic number of people who carried in wood, food, pottery, and so forth from twenty-five, fifty, a hundred miles away in order to fuel whatever kind of mystification was going on there. You can say they didn't have an underclass but they managed without one.

PINES: There is another interesting thing (when you talk about the underclass in the country) that one might want to put into the studies going into Project 2050. That relates to the fact that this country is a kind of open system to the extent that we have permitted large-scale immigration. We are always bringing in new members of the underclass and, until recently, we seem to have been pretty successful in getting them up through the system. We still are successful with some groups in that underclass but with other groups it's gone quite the other way. My own sense is that it would be a lot simpler to deal qualitatively with this problem, in the sense that Murray keeps saying is what we're about, a qualitative understanding based on simple models and have a go at dealing with some of the major features of our own country before we have a go at dealing with the system as a whole.

GELL-MANN: I think we have to include something about our borders if we are going to study the development of an underclass.

WALDROP: One of the most complicated things here is John Holland's question, borders with respect to what question? There are multinational corporations that are ostensibly functional parts of this system but whose borders do not match neatly with national borders.

PINES: Are they, in fact, a part of the environment in which the rest of us who are not multinational are operating?

HOLLAND: It's the other way around, David; we're part of their environment.

GELL-MANN: The U.S. Southwest in prehistoric times is often regarded as a distant, distant relative of Mesoamerica.

EPSTEIN: Then there's the inverse fitness problem, which is simply: if you have a system that's evolved to some steady state, it would be interesting to try to extract what was the fitness function that was maximized in the course of the optimization procedure. If you think of a string of rules, it would be interesting to know that you are observing a system that is well characterized by this string of rules, and then ask under what fitness function did evolution lead you to that string. John, do you think that that's a well-defined problem?

GELL-MANN: It's the many-to-one problem.

EPSTEIN: It may not be many-to-one. Why is it many-to-one?

GELL-MANN: There may be a lot of possible functions...

EPSTEIN:　　　　　It's not clear that there would ever be a unique solution. I'm just throwing it out as a thought.

ARTHUR:　　　　　There was some very nice work done about twenty years ago by Dan MacFadden at Berkeley, a review of preferences in a government bureaucracy. He looked at California data and tried to find the objective function of the state bureaucracy that would have an inordinate fondness for highways. So you can easily get yourself into studies that say the objective function of this evolving system seems to be highways but, actually, it was a coevolutionary system where the political setup was very much an interactive system that emphasized highways, water, agriculture, and so forth. Coevolving systems are games with many players. The problem is that there is no objective function, there are multiple objective functions; and I believe that it would be incredibly difficult, if not impossible, to extract the payoff function for the single groups.

EPSTEIN:　　　　　But then there was no convergence to some fixed point in the process.

ARTHUR:　　　　　There may be convergence to a Nash equilibrium between the growers, the highway construction people, etc.

EPSTEIN:　　　　　I'm simply asking "Are there inverse problems that are interesting?"

GELL-MANN:　　　　　Our educational system, for example. You could add this kind of question. Suppose you were a Martian investigating this system; what would it seem to be for? I've always said that the number one conclusion would be that it was for baby-sitting, to keep the children away from home under a different kind of care for a certain number of hours so that, if someone had invented a way of conveying all the information in five minutes per day, school would not shrink nor would the amount of information increase. They would simply keep the kids there because that was their primary function. The secondary function is to break their spirits. The third function is keeping them out of the job market until a certain age so they don't compete. And so on. Finally there is education.

EPSTEIN:　　　　　But that might not be uninteresting, demonstrating that it isn't so obvious what it's all about.

ARTHUR:　　　　　There was some work done recently at Stanford by Steve Derlauf. In looking at a stochastic system in which there is a continuum of people, and all these people have enough income for education but, if you put some of them in a situation where they have more access to education or their children have, they are more likely to value education and spend money on it. It turns out that you can start everybody equally with the same education and roughly the same income level but, with a few perturbations, you find that the ensemble starts to go in different directions and, so to speak, starts to bifurcate or, at least, spreads across a wide continuum and an underclass develops. And that underclass tends to get into a vicious cycle with low income, low value on education, low education, then lower income. Funny enough, during the thirties and forties Myrdal came to this country and wrote a classic book on sociology talking about the black underclass (*The American Dilemma*) dealing with precisely the same thing. What Derlauf has done is to describe a formal system that develops this dynamic. So I did the obvious thing. I said, "Give us your paper and we'll print it." It will shortly appear.

GELL-MANN: But I bet this is profoundly altered and exacerbated by ethnic traditions. You have ethnic groups that love education and propagate it.

HOLLAND: It shows up within ethnic groups like, say, Jamaicans. When I was in New York recently, a black panhandler approached my driver. As we drove off, he said, "They sure make it easy for us Island boys."

MAYNARD SMITH: There is another aspect that's more obvious in Europe than it is here. People seem to pick up sides and these things go on forever. The boundary between the Croats and the Serbs goes back to the Roman Empire. It's incredibly stable and I don't understand how people decide which side to pick up, what criteria they use, why they would have done so, and why it's incredibly difficult to change.

(In Germany, where the old Roman wall used to cross, you find primarily the Lutherans to the north and the Catholics to the south.)

ARTHUR: In watching BBC one night, they showed a riot and stuck a ten-year-old kid in front of the camera, a Catholic. They asked the ten-year-old, "Why are you taking part in this riot?" And he looked into the camera and said, "These people have been oppressing me for fifty years."

BROWN: I have a comment that you might think changes the subject but I'm leaving soon. If I look at the evolution of biological systems (this may apply to other things), there's sort of a problem that, on the one hand, the systems obey certain conservation laws or, at least, they almost obey conservation laws; to some extent, we can think productively about the problems of trade offs and the fixed pie, how that pie gets divided up. But, on the other hand, they don't quite do that because a lot of these innovations we've talked about expand the pie and a lot of innovations in biological evolution do the same. The things in the soup that are feeding on other organisms, some of them figure out how to feed on sunlight and some how to feed on land and so forth, and I think these analogues go through. Both are going on and influencing in different ways the structure and dynamics of these systems that we're talking about. And I think this is really relevant to issues like getting to sustainability because we have to obey these conservation laws but we also go through a series of transitions that change the way we use fundamentally limited things.

PINES: I think you're quite right. A closely connected question is that, when you're doing your model of adaptation and evolution in a cultural and political sense, you really need to distinguish between a model in which the landscape, which may be chaotic, still is not being changed by the actors, and the models in which the actors are, in fact, changing the landscape. A simple example: Murray was saying yesterday that Hubler and I should not put so much emphasis on the fact that you could, as an agent, control the environment, but there is the obvious counterexample of how some firms create a niche for themselves by influencing Congress to pass laws that influence the environment in which they are operating. This is always something you must pay attention to when you start to model real-life systems.

GELL-MANN: Even there, the lobbyist might get sick on the day they vote. It's probabilistic and not absolutely deterministic.

HOLLAND: As Jim and I said in a conversation about conservation laws, these laws simply point out that conservation laws are a classic example of something that's a property of a trajectory, not the endpoint. The endpoint is incidental so that you can derive the trajectory, or parts of it, by knowing that there is a conservation law. So that's a classic way of attaching properties or constraints to a trajectory.

GELL-MANN: Then there are adiabatic invariants which are kinds of things you are much more likely to run across than real conservation laws. There are questions of time scale. If you average over certain time scales, then on the remaining time scales you may have an approximate conservation that is adiabatically invariant. That's much more likely in the kinds of things we're talking about.

ARTHUR: I want to ask the two biologists a question. (Comments: You'll get two answers. You're modest. There will be three.) When you talk about adaptation, you tend to think of the entity as under some pressure to adapt to the changing environment. I've been wondering if the pressures against adaptation that might work something like this.... Suppose I'm a predator and I have a stable prey and that prey tastes good. Were that prey to find somehow that I have the ability to change what I do with respect to the prey, it's in my interest to see to it that the prey doesn't adapt too far. (Domestication?) That case is a very good example: we don't want cows to stop giving milk. Are there examples in biology of such pressures against adaptation exerted by one species on another?

GELL-MANN: What happens in colonies that keep domestic creatures? Aren't there aphids and things that are kept in ant colonies?

MAYNARD SMITH: Yes, ant colonies farm fungi, for example.

HOLLAND: What Brian is asking is "Because they do this, can we compare fungi that are farmed by ants with closely related fungi and find that the ants are retarding any further adaptation by their fungi?"

MAYNARD SMITH: Actually, the fungi have adapted to the ants. The ants disperse them. The fungi are quite happy to be found by the ants.

HOLLAND: Oh, so the ants are the underclass here?

BROWN: Let me deal with it first at the level of individuals because I think that, if you look at individual foraging behavior, you do find that they seem to not specialize too much. They continue to devote a certain fraction of their time to sampling, which enables them not to get locked into some narrow feature of the environment that might change and strand them on a peak when the environment changes.

PINES: That's very interesting in the light of Hübler's remark that it's always a good idea to have a somewhat noisy environment because you can sample it more easily.

BROWN: For the second answer, I want John's input on this, I think this is much more speculative, but there has been a suggestion that there is a sort of inertia built into the genome so that certain kinds of organisms, and some of this might come from species-level selection, so that if organisms respond too rapidly to environmental change, they have the same problem. I don't know how good that evidence is but I've heard that comment made by evolutionary biologists, that certain kinds of genomes seem to have some sort of regulator that keeps them from adapting too quickly, which is more or less in the same vein.

MAYNARD SMITH: I'd want an awful lot of persuading to accept that.

ARTHUR: You notice that trees in traditional Chinese paintings are always quite crooked and quite beautiful. An interesting story I heard about that from a tree geneticist was, if you actually go to China, you will notice that the trees that grow there are indeed crooked just like the ones in the paintings. The reason is that all the straight trees have been logged and are not leaving their progeny around. So the trees have adapted and tried to become totally useless.

MAYNARD SMITH: I must tell you a story that probably all of you know and that is that the theory of evolution by natural selection was first put forward in the appendix of a book on naval architecture. Now this may seem a puzzling place to publish it. It was published by a guy named Matthew (I've been making desperate attempts to demonstrate that Matthew was a relative of my wife who was named Matthew before we were married). He was worried—this was in the days of wooden ships—that there were no longer big, straight trees because people were cutting them down and leaving all these little scrubby trees and this was leading to the evolution of scrubby little trees and, by the way, he said that this will explain the evolution of species. Now obviously he hadn't mastered the first art of being a successful scientist, to publish in the right place at the right time. He was upset that Darwin didn't acknowledge him and he wrote a furious letter saying why didn't Charles Darwin acknowledge my contribution and he published the letter in *The Gardener's Chronicle*.

HOLLAND: John, you're not pulling our leg; this is actually true?

MAYNARD SMITH: Oh, yes! He was actually a very distinguished guy. He wrote something about the law of diminishing returns. He not only preceded Darwin, he preceded Marshall.

ARTHUR: In the tree case, adaptation to becoming straight again is being ruled out by that.

MAYNARD SMITH: We would just interpret that by saying that if predators have certain preferences, then indeed you evolve to avoid those preferences. Often you do it in ways that are difficult to see. But I'd be very doubtful that you do it by slowing up your rate of evolution. I should have mentioned earlier, in connection with models of coevolution, a simple model that has some facts in it and some predictive power. This is the May and Anderson model of coevolution. As you know. this virus was deliberately introduced into Australia to kill rabbits, which it did. However, it has become steadily less virulent. It now takes a year to kill a rabbit instead of two weeks. What they say is that if you are a virus and get into a rabbit, you want to maximize the number of rabbits that will be infected from the rabbit that you are in, and so you don't want to kill it. But, of course, it's critical that it not be infected with the more lethal form. It's no good leaving the rabbit alive if the other form kills it within two weeks. This is a nice model and it's well known in the literature and is concerned with something like symbiosis.

BROWN: There was a plague of rabbits. Has that leveled off at a lower level?

MAYNARD SMITH: There was a high degree of infection and they got pretty sick but now they're coming back up.

MARTIN: I'd like to go back to John's earlier question about why ideologies are so stable. It gets back to where you choose to draw the line around the system you're studying because it's only mysterious if we're the agents operating in the complex adaptive system. If you think about the evolution of ideologies or ethnic groups in which we are just the substrate in which the complex adaptive system is developing, it makes quite a lot of sense that the ideologies that hang around the longest are ones that somehow encode their own virulence into the units that are vacant so, given the fact that the ten-year-old is so upset and yelling about oppression for fifty years is an example of the ideology expressing itself in that individual in a way that leads to the ideology's survival so that the unit of adaptation which is evolving is the ideology.

MAYNARD SMITH: This is what Dawkins would call a meme. The idea is that ideas are replicated in our heads. Genes are also replicated. It's a cultural gene. The ones that are good at getting replicated are the ones you will find, but I haven't applied it in this context.

MARTIN: The general point here is it matters a lot how we choose to bound the system we are studying. The point I made in my talk with regard to language is that it's critically important to know the difference between language as a complex adaptive system, and brains as complex adaptive systems that do something called "use of language." Here is another example of the fact that two ways of looking at something as a complex adaptive system will lead to very different conclusions. In one case, it seems perfectly obvious why they should be stable. In the other case, it's quite mysterious.

MOORE: They're like genes that have evolved in one of these arms races that John Holland is fond of describing. You have two different sets of memes that are engaged in an arms race and move themselves up to being Serbs and Croats.

GOODWIN: I want to reiterate a point that John made earlier about alligators and crocodiles that really haven't changed much over the years. In response to Brian's question about whether there are instances where organisms don't change, you were asking about selection pressures. You don't really need mechanisms to do that because there are so many examples of organisms that don't change, like a panda that is eating vegetarian food but should have a carnivore's diet. Just remember that there are plenty of instances where, despite selection pressures, there has been no change.

GELL-MANN: I listed three candidate reasons for maladaptive behavior at the phenotypic level and I imagine that there are many more that people here can suggest, probably better ones. One of them was time mismatch; that is, the conditions change more rapidly than the evolutionary procedure is able to catch up. Second was the selection pressures being elsewhere, mostly, as with language, where it's not so much the success of the language as the military success of the tribe that determines whether a language survives or not.

PINES: Brian, I have a question for you. Would you take the point of view that adaptation really matters when you are trying to develop a dynamic picture of the economy? Thus, does the ability of agents to adapt to the environment make a real difference in the nature of the evolution? The relative positions of the agents in the economy depend on their ability to adapt. If so, the kind of economy which evolves is influenced by adaptation. Put another way, the kind of economy that would evolve in a country with a set of rigid rules is quite different from the kind of economy which evolves in a country where the only rigid rule is "thou shalt not eat a cow" and that is, in turn, different from the kind of economy that has evolved here. That has to do with the environment influencing the adaptation, adaptation being the key ingredient in determining eventually the kind of economy that emerges.

ARTHUR: Malcolm Marshall keeps getting mentioned here. He wrote a very influential book on the principles of economics in 1891. He was very much influenced by the Darwinian point of view and so there seems to be something on this that dates back at least a hundred years in economics and picked up by others to mathematize, to reduce to algebra and calculus, Adams' point of view on supply and demand equilibrium. There's another side of Marshall that looked upon firms as adapting. Some firms were very fast at adapting and others were not. You have to remember that in his day these were mainly family firms and Marshall did see the economy as always adapting, essentially biologically, as he put it, and after two or three generations, these family firms were not much good at adapting. They tended to solidify, the people running them weren't very smart, so they went from open to closed in three generations. So there's been about a hundred years of intermittent modeling. In the last major effort along those lines, it was found that you can get firms that adapt very rapidly and firms that are not able to adapt. In an evolutonary system, the firms that are not very good at adapting go out of business. DEC, for example, might be such a firm. It did extremely well with minicomputers. Now they are no longer a big thing and Digital doesn't seem to know what to do next or how to get into the work station market. If you're asking whether it make a great difference if in one country there are firms that are more highly adaptive than in another country, I don't know. It's very hard to say that one would be better off than the other. One thing that I've learned from economics in coevolutionary terms, looking at game theory which is eventually coevolutionary, very quickly what you learn from game theory is that people can coevolve to a state where nobody is better off. You can set up two games where the agents learn very fast and a neighboring game where the agents are learning relatively slowly. It's not clear to me that, on the average, one set of agents is better off than the other set of agents because, if I'm a faster learner, I may be doing you down at a faster rate whereas, if we're all pretty stupid, we may all be funky but I'm not doing you down as fast and you're not doing me down as fast. Consider an academic department in which everyone is very smart; the department may be top rank, but life might be miserable for most of us. It's hard to say whether the economy is better off under rapid adaptation. So, yes, there are several answers and economists have repeatedly thought about it. What people are thinking about now is learning games and we're starting to be able to choose the degree of learning.

PINES: Which might well become computer simulations as soon as we learn how.

ARTHUR: Oh, yes, with artificial agents. I was going to say that if I had to design an economy, I would design one where the agents didn't learn that fast, for a very specific reason. If everyone is mutually adaptive and goes to a Nash equilibrium, generally speaking, there's not much profit to be had. I'm doing as well as I can and everybody else is doing as well as they can and usually that means that there's not much left over so I pay my workers out of a lean profit. Albert Hirschmann from the Institute for Advanced Study at Princeton did nice work in the '50s and '60s. He pointed out that a lot of inefficiency is really very good because if I can get ahead and nobody can catch up to me, then I get a big pot of cash. I can send that cash to a Swiss bank and do nothing useful with it or I can start another company hoping, thereby, to acquire another monopoly, thereby growing the country. But Hirschmann purported to show that you don't want developing countries to be too efficient too fast, modulo dictators and Swiss bank accounts. As long as you have a lot of cash sloshing around, there's a lot of extra profit available for putting into new companies.

GELL-MANN: By the way, in connection with cow taboo, there is a very interesting book by an anthropologist, Marvin Harris, chairman of the department at Columbia, on all these things. He shows that cow love in India, pig hatred in the Islamic lands, pig love in New Guinea, all make a certain amount of sense given the local ecological system.

George A. Cowan, Chairperson
Santa Fe Institute, 1660 Old Pecos Trail, Suite A, Santa Fe, NM 87501

Summary Remarks

COWAN: It's obviously impossible to summarize in any detail the massive amount of material that was discussed here. On the other hand, I think that each of us can say a few things about what they feel the high points were, and make at least an initial stab at their own summary statements.

I think we might do it in the style which I suggested to John Maynard Smith; I said, "Why don't you start off by saying (with the right intonation) "'What the hell are you doing here anyway?'" We should try to explain to our friendly critic how we have spent the last eight days but try also to summarize what we think are the conclusions, or the important elements of progress, that have been made here this past week. Murray, will you kick it off?

GELL-MANN: I could say a couple of words.

I continue to maintain that not only within the framework of Project 2050, but in general in our more relaxed scientific work, we need to pay much more attention to society, to social science, to try to find those few social scientists who don't suffer from crippling math phobia but are, nevertheless, not the kinds who trivialize social problems by mathematizing them. We have to keep searching for such people, and I think the economics effort needs to be informed by being embedded more in the study of society in general, and culture in general. A number of people have suggested that.

As to the study of complex adaptive systems, I listed there on the board the three things that I found most fascinating and about which I wanted to learn more First of all, what is the generalization of neural nets (based on an analogy with the human brain) and genetic algorithms (based on the experience of biological evolution)? These belong to some large class of interesting adaptive computing systems, with their software and hardware. And what is that class, and what are some other fascinating members of that class?

COWAN: Can you state what the minimum requirements of the general class have to be? Clearly, they have to contain interconnected neurons, or whatever you want to call them...

GELL-MANN: They have to be learning systems. They're complex evolving systems, or, if you prefer, complex adaptive systems. And I think there's some fascinating work being done on that and that over the next few years we will learn a great deal more.

COWAN: Learning and response; they do have to respond.

GELL-MANN: Well, they do these things that we listed. They perceive regularities in a data stream, and they compress them into a schema, and then they test the schema.

COWAN: They exhibit what they have learned and what they are doing by responding in some, presumably, appropriate way.

GELL-MANN: Right. Now, genetic algorithms are famous for being used on optimization problems and problems with an exogenous fitness like winning at chess, or something like that. They don't have to be used that way; it just happens that in many uses they were. As Melanie Mitchell emphasized at the Science Board symposium, you can use them in other ways. Neural nets, too, can be used in a great many different ways. It's just that certain habits have grown up in connection with each method.

COWAN: One of the things, Murray, that's puzzled me a little bit about these systems is that I'm never clear to what extent they are able to filter out noise. We do it. I mean, if we listened to all the information out there, we'd be totally poisoned all the time.

GELL-MANN: Well, filtering out some noise is part of the coarse graining at the beginning, and then regularities are identified by filtering out what is perceived to be random. Now, in fact, we always make mistakes. We mistake randomness for structure, since we're pattern recognition devices and that gives rise to superstition. Because evolution of complex adaptive systems favors a middle course, there's also the opposite error of denying regularities that are there. And I think that we can predict that on every planet where there are complex adaptive systems, they will make both kinds of error.

There's also the opposite error,

COWAN: There's also the error of leaving out sampling at all in regions that should be of interest. That is, there is so much information that you can't use it all, you can't process it all, and you may be frozen in to sampling something that is no longer relevant.

GELL-MANN: Yes. The coarse graining, in other words, of the universe may be inappropriate, and evolution operates on that, too.

KAUFFMAN: Murray, one puzzle I think we were going to get back to, and perhaps you can solve it: Is a protein sitting in a heat bath, or in a fluctuating thermal environment or in an ionic environment that's fluctuating, that's being driven through microstates of the kind that Hans told us about. Does it have a description or a schema of some kind? The point that I think is worth focusing on here is the following: I think that we're importing some ideas when we use these phrases and we should understand where that happens. For example, I think it's quite clear that a protein like that sitting in a membrane of *E. coli*, which helps *E. coli* make its way in its environment, is something that I'd be comfortable saying is an adaptation, in precisely John's sense—the standard biological sense—I'd be perfectly comfortable in saying it's a schema of some kind...

GELL-MANN: Well, maybe; I don't know that much about it.

KAUFFMAN: ...but it's not clear that if a protein is just floating around in a pot, it has that. And so when you're saying schema—it's a conceptual point, as well as a factual point—when you're saying schema, for whom or what to use in what way, or is it simply the case that...

GELL-MANN: That's not the way to state it. The way to state it is that under certain conditions there are certain selection pressures on the varying schemata. There's no such thing as purpose in the universe...

KAUFFMAN: What do you mean by selection pressures?

GELL-MANN: The feedback on the various schemata...

KAUFFMAN: Do they have to be better or worse ones?

GELL-MANN: The pressures that favor the promotion of certain schemata and the demotion of certain others determine the gradual evolution of the schemata, and that's it. There's no such thing as a purpose in the universe.

KAUFFMAN: Of course not. But it's a question of where we'll describe it. So under that description, Peter Schuster's RNA molecules evolving in a pot with Qb would satisfy...

GELL-MANN: They satisfy the criteria. The RNA molecules in the RNA theory of the origin of life (assuming that's right) satisfy it. And they satisfy it in the degenerate limit in which the phenotype and the genotype are physically identified. But, as Walter keeps telling us, they are not conceptually identified, because the genotype is the information in that thing—RNA, or a bit string, or whatever—and the phenotype is its activity in acting as an instruction.

The second question I should like to emphasize is the one we discussed a lot yesterday, and maybe today we can still be enlightened a little more about it. It's this whole question of the order-disorder transition and how it works in all the different kinds of complex adaptive systems, and at which stages it operates, and which ways I listed some features that have been proposed for such a transition region:

- as an attractor;
- as a source of efficient adaptation;
- as a critical point with a power law distribution (or several power law distributions, because you could have one law for avalanches, and another one for a Pareto distribution of resources, and so on);
- all scales of space-time at such transitions leading to a possibility of universal computation;
- the possibility in that region of having high complexity; and so on.

That's a second field in which a lot of interesting things are being done. One of the most interesting is the little paper that was reported by Arthur, Holland, and Palmer on complexive adaptive systems trying to predict one another's predictions of stock prices and that sort of thing.

The third one that I listed is the matter of innovation in organizational structure and how that is to be mathematized and how it is to be described in all the different fields, how it's to be distinguished from somewhat more routine developments in learning or evolution. We heard a lot that is instructive and there's obviously going to be much more to do on it.

In addition, we discussed a little bit about Project 2050, in which we, and other institutions around the world (if we can get further funding), are going to try to learn something about the future of the human race and the biosphere by modest gaming of the future in a search for paths toward a desirable sustainable future. But we have to realize that we're not dealing with Newtonian dynamics in this kind of thing, and we just have to be very modest about what we can do. What we can accomplish is to explore possibilities, including qualitative structures, in connection with the quantitative data and extrapolations that are obtained in the usual ways in looking toward the future. We can contribute the former, and the other organizations will contribute mainly the latter.

I should say one more thing to be fair: I've pushed very hard for including nonadaptive complex systems here, because they are part of the subject matter of the Institute and there's an enormous amount of interesting material on that subject as well.

One thing that we haven't mentioned at all here, but which is one of the most exciting subjects that the Institute has worked on, is the connection between, on the one hand, the fundamental quantum-mechanical field theory of all the elementary particles of the universe and the boundary condition on the universe, giving the fundamental laws of physics, and the quasiclassical domain that we deal with here, on the other.

FRAUENFELDER: Clearly, what I learned is the surprising range of problems that are in one form or another on the blackboard. When I looked at that, my main impression is that what the Institute needs is more contact between experimentalists and theorists, and that it is all too easy to go into a theory and make it bigger and bigger and greater and greater. It may be correct, but it may also lose contact with reality. That leads me to another question, namely...

GELL-MANN: Do you think, Hans, that we do that, that we have theorists here who are not interested in comparison with observations? Everyone I know here, as far as I can tell, is driven by the desire to understand certain observations.

FRAUENFELDER: Yes, but what I miss is the contact with the experimentalists who can have a feedback...

GELL-MANN: Oh, them. [Laughter]

FRAUENFELDER: Yes, who could have a feedback, because it leads me to a language difficulty. We had one meeting at a more restricted level here in which we discussed glasses, spin glasses, and evolution. And after about two days of good exchange, we discovered that we were totally talking past each other. I think it was Bernard Derrida who talked about Hamiltonians and such and I stopped him for a second and asked the evolutionists, "Do you know what a Hamiltonian is?" They said, "No." So there was no exchange; while apparently there was a very good interaction, the languages went totally past each other. One has to try to have feedback to find when information is really exchanged and when it just appears to be exchanged between the different fields. I think once you've worked with somebody for a few years, you begin to understand that you talk about either the same or very different things, but at the beginning it can be quite misleading.

You know, there are two famous statements, ascribed to different people. One is "God is in the details," and the other is "The Devil is in the details," and unfortunately, both are true. And the difficulty is that only when you talk about the real details do you find out whether you talk about the same thing, and that's very difficult in a group as big as this. By big I don't mean in numbers, but in the range of fields. What I'm suggesting is that, from time to time, one should have an attempt at a feedback mechanism in which one takes one little field and tries to see whether one really talks about the same problem.

GELL-MANN: That's a good idea.

COWAN: Do you think conferences like this one, a several-day conference, helps construct the language?

FRAUENFELDER: Yes, but then one has to pick up one part and get people together. You know, one thing which I missed hearing—which Murray will tell me is, of course, trivial: experiments do have surprises which may tell us that some part of what we are looking at may be very different from what we think now. What I'm really suggesting is, you know, that I'm here as a token experimentalist, that we make a conscious effort to find people who could add knowledge that may exist but hasn't made it into the literature where it's available to a wider range of people.

COWAN: Hans, could you identify the appropriate fields or bodies of experimental data? For example, should we look in the social sciences for bodies of experimental data?

FRAUENFELDER: I can only do it in the field in which I'm working, because what one has to do is deliberately go out and find in each field whether there is a body of knowledge that we haven't tapped yet.

GELL-MANN: Would you accept an emendation, that instead of using experimentalists, use people who collect observations? Because in most fields, there aren't experiments.

FRAUENFELDER: Oh, I agree with that. No, experiment in the broader sense, whether they're historical or observations.

COWAN: The historical record...

FRAUENFELDER: ...represents an experiment. Yes.

COWAN: But we, as you know, get into numerical simulation, which is not theory, and which is not real-world experiment either. And so, where do we class numerical simulation, or do you just disregard it?

FRAUENFELDER: Oh, I'm quite afraid of it, in one sense. Because, you know, I've heard more than once at a conference somebody say, "I did the experiment; it shows that..."

MAYNARD SMITH: It really struck me that a fact at this meeting has often been a simulation.

FRAUENFELDER: Yes. What did Peter say? Peter's simulation. And the simulation may be perfectly great within what the assumptions were. It may have very little to do with the actual part that they want to simulate. I'm not saying that numerical simulation is not important, but it should be distinguished as a third leg.

COWAN: I'll bet John has a good word. How do you distinguish between the controlled physical experiment and the numerical simulation, John?

HOLLAND: I don't know whether I've got a word, but actually I would adopt Brian's notion: If you talk about calibration of simulations, where you run the simulation and then compare to some set of data, not produced by some other simulation, then that becomes the critical step, and until you've done that. . .That's why I was so taken by Jim Brown's offer to supply me with data for ECHO, because until I've run ECHO against that data, it's a lot of fun for me, and it may be a metaphor, but it's not more than a metaphor until you do that.

GELL-MANN: What about factoid, John?

JEN: It's actually sometimes referred to as experimental mathematics, which is an appropriate term.

ARTHUR: It's a bit like "truthoid."

PINES: There is a well-defined term which is used in many fields of science. It's called "Computational X," "Y," or "Z." One can talk of experimental mathematics, but one also talks of, for example, Computational Physics. . .

COWAN: And does that invariably mean numerical simulations?

PINES: What it means is: taking a system for which you've carefully specified the laws of interaction among the agents or particles which you run on a computer, it is considered much closer to a branch of experimental physics than it is to theory, in the sense that in the design of your numerical experiment you have to do as much as you can to take advantage of the existing technology. You have to have a very clear sense of what it is you want to get out, because you're always making choices of what you're going for. And when you get a result, you then face, as does any experimentalist, the question of, in fact, "Have I written the program correctly"—so have I carried out the experiment correctly, or "Have I made a mistake somewhere along the line and the results really are not to be believed."

COWAN: So calibration is a key issue.

PINES: You're doing all of this. But the next step, which is the most important one, is understanding what your experiment means. You have to approach it, as we do any experiment, from two perspectives. One, just a phenomenological description of the results of your computer simulation, or your computational physics experiment, and then second, try to relate all of that to some sort of fundamental theory. But this is an extremely well-defined subfield of physics, and chemistry, at present. We're a very long way in moving to other parts. You not only talk now about computational physics but about computational statistical mechanics; you take a subfield of physics. And, people give courses of lectures; it really is a major frontier field of science, "Computational X," "Y", "Z." I don't see anyone else here apart from me who's really had a hands-on contact with the power of such an approach to doing science and also with the computational techniques. I think people are far more aware of the problems than they are of the power and success of good simulations.

COWAN: We've seen horrible examples of misuse, and that's the problem. How do you avoid that?

PINES: Used properly—just like any kind of an experiment—used properly, it's an extraordinary adjunct and may turn out to be one of the significant tools as we go forward.

GELL-MANN: But, there may be a problem with the use of phrases such as the NSF-like "Computational Biology," "Computational Social Science," "Computational Economics," because it implies that you're actually working on biology in some way, or that you're studying social science in some way, whereas what you're doing is very interesting modeling, speculation, and simulation. I'm all in favor of such activity, and I'm in favor of SFI doing it, too. But I think the use of that language outside of physics and chemistry is dangerous.

MAYNARD SMITH: Aren't you doing biology or physics? I mean—but surely you are doing—if you're doing computational work, you are doing physics.

PINES: When you're doing computational physics, you're doing physics.

GELL-MANN: I agree with that. But when it comes to biology, economics, and so on, where the laws that you should put in are a little bit vaguer and less well understood, there may be a problem with calling the results of these simulations, necessarily, science. They may or may not come out to look like a contribution to science.

COWAN: There must be a good word in the English language for "exploring a range of possibilities," which is generally what these computations do. They don't point at the fact, but sort of a spectrum of what is possible.

HOLLAND: I think there's a very important thing here about the notion of "model," and there's more of a distinction made here than I would want to make, in fact. Because I do a set of PDE's, as a physicist or a biologist, or, you know, come up with Lotka-Volterra equations and so on—that's a model as well. And whether I do it computationally or in terms of standard mathematics I'm doing essentially the same process.

COWAN: And so you don't want the word "computational" being a necessary qualifier?

HOLLAND: Yes, all I'm saying is that because I'm using a different tool and a different way in using algorithms instead of PDE's, doesn't mean that. . .

JEN: I think that what George is pointing to is the fact that much of what we do at SFI actually has not yet reached the plateau that I think you and David are talking about, and which I think is actually the direction we should be going in—computational physics, biology, whatever—and I actually don't think that the distinction is in the field itself. Even within physics at this point, if you look at the methods that people are using, oftentimes they are adopting something quite analogous to Walter's approach. A fundamental characteristic of Walter's approach is that it is an axiomatic system and is actually a mathematical system. And I think it's appropriate in such cases to term them "experimental mathematics," which is the commonly accepted term at this point in this field for models that have not yet gotten to the point where you have a particular problem with numerical or experimental data against which to benchmark your model.

ARTHUR: We've had absolutely dreadful experiences in the social sciences with simulation. Just think of "Limits to Growth." Sloppy thinking with sloppy models and then hyped, to the degree that social scientists started to talk about Newton's Fourth Law of Simulation. Namely, every simulation is accompanied by an equal and opposite amount of dissimulation. So this is the way it stands, unfortunately, in economics, yet I find myself absolutely in sympathy with what David has to say here. We need more of this.

I believe Santa Fe is a pioneer here—I very much favor this approach. So what I would ask for is that this approach be held to the same standards of rigor in statement as, say, mathematical theorems, namely, that under conditions A, B, C, D, E, F and under this well-specified model, which would be reproducible in some other computational platform, we found outcomes E, F, G. And, rather than just saying that we know that E, F, G happens, this is stated as a computational theorem, if you want.

GELL-MANN: And then you get the result that John was speaking of, namely, that this is then on the same footing as other kinds of theory.

ARTHUR: Yes.

KAUFFMAN: There are two points that I think that we should remember. The reasons we're doing simulations is that many of the problems we want to look at are ones for which closed-form mathematics is not available, and won't be. And precisely because we know that there can be simulations for which there is no shorter description, if we want to ask what these kinds of systems do, we have to do this. Which leads to the deeper thing that we're really about.

We're playing with model systems, and what we're actually trying to do, our points 1, 2, and 3, are very serious intuitive attempts to formulate out of this welter of games that we're playing, what looks like they might be general principles or general laws. In fact, this is a very important point for us: you don't know what experimentalists to talk to yet (although, of course, we do in many areas). What we're trying to do is find our way to some very general principles which we glimpse. We haven't nailed any of it, but I think that one begins to see a framework here and, if so, this is a very large achievement.

If we can now, by our numerical simulations, gather numerical evidence, in a variety of contexts, for robust results and so have the framework to go out and touch data, touch real observations, touch real experiments, in a whole variety of arenas, we have done a quite wonderful thing. But let's stay aware of the fact that what we're actually looking for are principles.

JEN: But, Stuart, what you're saying is still the first step. There's such an enormous gap between what you're saying and what David's saying. You're identifying the principles but you're saying that you're doing it through numerical simulations, in part because the sort of problems that people here are interested in are in the realm of, I guess, what people call descriptive science, historical science, whatever; evolution-type things. There are no control parameters, there usually are not numbers that you can compare your simulations with against their simulations, or ways that you can vary those parameters. Also our theories are usually—at least in the cases that I'm aware of—such that they do not actually define numbers in the theory that are in close correspondence with numbers that you can measure from any physical phenomenon. And so it's very, very likely that we will identify, as in the case alluded to by Murray during Jim Brown's talk. It's quite possible that we will find ourselves in the position that the Chaos people found themselves in, where they could identify all these wonderful principles and yet the parameter regimes in which these wonderful phenomena emerged were not the physically realistic parameter regimes that you actually see in real phenomena.

And so then you ask: whose responsibility is it to actually go quantify these principles? And I think it really is the theorists. Your work is an incredibly wonderful example of collecting observations and formulating the principles, but I don't think it's reasonable to expect Peter Schuster to translate your stuff into something that he can then measure. I think it's actually your responsibility.

KAUFFMAN: Of course.

FRAUENFELDER: There's actually a very clear case where one can show the problems in a beautifully simple situation, coming again to the protein. The protein is a relatively simple system compared to what you're talking about. One has now calculations which start essentially with Newton's Law, and with the potential function between all the atoms. And you can then start with an x-ray structure and you let it run and you see what happens. The problem is twofold: One problem is that the elementary step is 10–15 seconds in the calculation. So even with the best present-day computers, you can get to something like a nanosecond, which is 9–10 seconds; that's all the way at the extreme limit. Now, that's fine. But now there are two problems. One is that we can assume that we get much better computers, which permits extension of the calculations instead of 9–10 seconds to seconds—where essentially live things happen—or milliseconds. We then have so much information that Murray's problem of coarse graining becomes important. Because you're totally swamped with data, there's no way to see what really happened. The only way I believe that one can really see what happened is to compare with experiment, because the data will be so incredibly large.

The second problem is that by that time we have taken the potential function, which isn't too well known, and stretched it over fifteen orders of magnitude in time. So even a very small mistake in the beginning gets you incredibly far off. What many of you do is to work only on the first problem. You have modern computers, and what I would claim is that you are working in a range that doesn't get you to any place where you can compare to reality. So the games you are playing are in a part of the universe that is really not yet relevant for reality, and doesn't permit good comparisons and benchmarking.

PINES: I think, Hans, it's very field specific...

FRAUENFELDER: It's very field specific.

PINES: I'll put it another way: Intuition plays an enormous role in science, and if your intuition is poor and you're computing on something on which there is no hope to compare with experiment, or with observation, then you're just another misguided soul. And you're saying that there are undoubtedly a lot of misguided souls out there. On the other hand, one can work in all of these fields, one can carry out excellent work, if your intuition is good and you have a sense of what is a relevant computer experiment. Relevant in terms of one that is going to turn up information which can be compared with real data of one sort or another, either acquired observationally or acquired in a standard laboratory experiment.

MAYNARD SMITH: You have to ask yourselves—I mean this is an institute, after all—why be interdisciplinary? I mean, why want to talk about social sciences, biology, physics, and so on, in the same institute? If all you're going to do is to do computational physics, you should be doing it in a department of physics, where you're constantly exposed to physicists. So unless Stuart is right—and I don't know whether he is or not—but unless it is true that there are general principles about complex systems which you can discover, which apply to all complex systems, then you simply shouldn't be here.

GOODWIN: I want to amplify what John has just said. It seems to me that the study of complexity really is an opportunity for approaching the problem of nondisciplinary science. We all grow up in particular disciplines, and we're bringing particular skills into this forum of dialogue and discussion. But really, what complexity is moving towards is a nondisciplinary, integrated science which actually goes beyond science. It goes right into the so-called social sciences, and now I think that it goes into the arts as well. But that means that we are really pushing at a totally new type of educational process, and we're bringing with us these particular skills that we're used to. We don't know how to transfer those skills into this new arena, because we don't want to lose any of the rigor, the skills, the testing; that's what we're insisting on maintaining as standards. But we're not all sure how to use those in the context of what is an extremely important educational development which is setting the scene for a new pattern of education. Murray has pointed these contexts out very well, that they are very broad, but somehow we have to maintain standards and rigors.

JEN: When you talk about education, I want to bring up something that there may be general opposition to at this table, which is that I think this conference has had an enormous impact on the younger SFI researchers who have been attending this conference. And they are certainly an important component of this educational process that you've been talking about. Certainly they've been fascinated by people talking about specific problems. In particular, the sort of work that you do and Hans Frauenfelder does, I've heard many comments about how interesting it is because they see actually very little of that if they are in permanent residence at SFI. They've also been very impressed by how Stuart's system and John's system actually reflect the range and the diversity of these phenomena. I think this has been a true education for many of them, in some ways positive and in some ways negative.

One of the most interesting comments I heard was that some of them had never, ever heard expressed the idea by anyone that measures of complexity are not necessarily quantitatively and theoretically interesting. Phil Anderson, at some point, said we find the idea of measures of complexity qualitatively interesting, not necessarily quantitatively interesting, or theoretically interesting, and that opinion had never been heard by this particular individual. To me, it is shocking that he had never heard this opinion.

The people who are actually in the trenches of the Santa Fe Institute are in danger of becoming just interdisciplinary, without the discipline. There is the potential of becoming a self-referential, autocatalytic community of people who are working on general principles without the grounding that you're talking about, not even being aware of the fact that you need the rigor to look at specific problems. Every senior person here has come to complex systems after having already worked on specific systems for a very long time. But for young people, in terms of this educational process, it's a huge problem.

COWAN: Good point.

GELL-MANN: Yes. Very interesting point.

BAK: As an outsider, that's precisely the feeling that many people at SFI have, that it's at the center of science and that they're out in front, whereas in fact it may be provincial and esoteric.

JEN: Yesterday, some people were standing outside saying, "We never knew that there was an old school." And I said, "Well, what's the old school?" and they said, "Well, these are the people who say they're doing complex systems but actually they're just doing the old-fashioned stuff." And I said, "Well, what's the new school?" and they said, "Us." There's a real problem!

COWAN: David, can we get to your statement?

PINES: I think this meeting will go down as a defining moment in our attempts to do what John Maynard Smith said we should try to do, which is to see if there are common features, if there are integrative themes—a set of what one might call "grand ideas"—out there, which may turn out to be relevant to almost all of the specific fields of science that we've been discussing. My own sense is that it has been great to do this, but I hope we don't try to do it again for a long time.

That is, I feel saturated with grand ideas and concepts, and integrative themes, and hierarchies, and so forth. I'm really excited about the possibility now of getting on with science. Grand ideas, etc., do play a role, but science is a lot of little incremental steps in which you work out not a grand theory—you work out a little theory. You take a toy model, which you specify carefully as a theorist. You try to make a minimalist theory, one that incorporates some subset of the essential conditions for complex adaptive systems. But we're trying here to define how we go as a community, and how we go as a scientific institution. One very fruitful approach is to develop, explore, understand comparatively simple, rough toy models which, however, incorporate enough of the aspects of a complex adaptive system that the results of the investigation might be relevant.

At the same time, it is probably more important for us to be aware of the relevant facts in terms of experiments and observations that may relate to, on the one hand, the grand ideas and, on the other, toy models that you can carry out and look at and say, "This is a result. I now have something that I can compare it to."

I am not unoptimistic that we can go forward in this mode. I think we've heard at this meeting a number of preliminary attempts both at theories that might be relevant, and a range—from very simple things that Hubler and I have been trying to do, to the more sophisticated things that Per Bak has been doing, to the whole apparatus of spin glasses which is in one way an interesting complex physical system which is reasonably well understood, and the results of which have been applied by a number of us in a number of different ways in trying to understand complex adaptive systems. There is a great range out there.

I think we are at a position to start it, and I think there are a whole range of problems, however, that aren't going to be handled analytically. And this has everything to do with the social sciences, and probably a great deal to do with biology, ecology, etc. And there I think our responsibility is to really define what we mean, if we write "computational social science." Not try to avoid the word, but take advantage of the opportunity to define it and to make it as respectable a branch of science as computational physics is. And I'm just delighted to watch Brian's [Arthur] body reaction—he's nodding his head—I think we're capable of doing that.

I think a lot of the people in this room are capable of contributing to developing that kind of approach. And, were I to try to pick out one sort of niche for the Santa Fe Institute in which it could play a really substantive role, given the relatively small resources that we have available on the one hand, and on the other hand our extraordinary convening power, I would say it's in trying to do a proper job with computational ecology, computational social science, computational economics, and so on. I think we're capable of doing it, and I hope very much we'll take advantage of this niche.

COWAN: But you've borrowed Brian's [Goodwin] suggestion. We establish standards of rigor, and make it very clear: if A, B, C, D, then we observe E, F, G.

PINES: Any paper written under the auspices of the Santa Fe Institute which involves a numerical experiment should say at the beginning exactly what has gone in to the computation and what has come out, and be rather open about the fact that it's not going to be easy to understand what has come out. Thus we might say that there are some interesting qualitative features, etc.; and that this is how we are going about it.

Because this is, I believe, the future of social science: To do carefully designed numerical experiments that can be reproduced by anyone else. I think it enables one then to handle the problems that are before us. It's going to take a very long time, and only the very tiniest first steps have been taken. I see the Arthur-Holland-Palmer set of computer experiments as an extremely promising first tiny step along that line.

COWAN: But you also, I assume, accept Stu's grand vision, that we are pointing toward general principles, we're not there entirely to...

GELL-MANN: It's not his grand vision, it's a grand vision...

PINES: Not Stuart's grand idea, not Murray's grand idea; Murray and Stuart, each in his way, have played a seminal role in developing the list of grand ideas on the board, and if one worked very hard, one might possibly establish a tiny subset of those ideas which could uniquely be said to be Stuart's or Murray's. I would hate to be the historian of science who took on that job...

GELL-MANN: The main thing is that we not be forbidden to search for general ideas and for links among the different fields, because that was the original point of the Institute.

COWAN: We're never forbidden to, it's just that we start to make our grand vision the development of numerical simulation, and that's not our grand vision.

KNAPP: What we've actually been doing is in the direction of getting some tools which will allow, perhaps, some of these things to get started. Now would everybody agree with that?

PINES: I would agree, but we have the tools at hand. The adaptive computation program is trying to develop better tools. There are plenty of tools out there. In terms of computer simulations you can do an extraordinary amount with a Silicon Graphics machine. You can do even better—if you really know exactly what you want to do—you can do better with a Cray, but for most things you don't need anything that powerful. We don't even begin to know—and John made this point very well—we don't even begin to know what we're capable of doing with thinking machines, with parallel processors, because it is going to be a major research project to sort out the interface in order to make effective use of the power of machines like that to solve the problems we're interested in.

GELL-MANN: But there's a problem, David, with this idea, I think. At least if it's exclusive, or the principal drive...

PINES: No idea put around this table is exclusive.

GELL-MANN: No, but even if it's the principal one.
The problem is that scientists are driven by curiosity, and that one of the interesting questions to ask is where that curiosity is directed. And there is certainly room for people whose curiosity is directed toward trying to see what happens in a certain numerical run, but some of the effort of the Institute has to be devoted, I think, to work that's motivated by curiosity about what's actually happening in human society, and what's actually happening in biological evolution; what's actually happening in thinking, what's actually happening in all of the other subjects that are relevant to this. And contact with observation, as Hans mentioned, is part of that, but also real curiosity about the description of what is actually going on in those places. And a huge amount of the computational whatever-it-is, as it is today

and I think as it will be in the future, is not so motivated. It's motivated more by curiosity about what will happen if you put in certain things into a computer, what will come out. Now the two are perfectly capable of supplementing each other and helping each other, and that's the ideal way in which we can work. But if there isn't a great deal of emphasis on curiosity about all these systems, then I think we're not really doing science.

COWAN: Murray, my experience has been that the people who are interested in what is actually happening should be looking at responsible simulations, because these are bookkeeping systems. It focuses discussion. Once you have a model, there's no better way to focus a random discussion than by having that model to argue about, to throw rocks at it. In fact, the models are almost useless unless they are being used in that way.

GELL-MANN: Absolutely. These two approaches have to be synergistic with each other.

ARTHUR: I want to start first with a comment on what David had to say about computational approaches. I'm all for them, and I want to give a reason—we haven't really given a reason why we ought to carry them out here. Well, several people have, but I want to give one that hasn't been mentioned. Namely, that our standard approach in economics and in physics before these computational approaches, and in other subjects, has been to look at problems that are analytically tractable. So this gives us a bias against looking at problems that might be more realistic but are not analytically tractable. So for me, what a computational approach would bring is, it would get rid of that bias and it would allow us to look at problems that may be beyond the bounds of standard analytics. Yet if we do it rigorously, and we are careful about the statements we make, we should be able to come out with the same degree of rigor as an analytical approach. In economics, you know the old cliché, that we're looking for the keys where the light is. For my money, these computational approaches will allow us to look at a much, much wider range, and I hope Santa Fe pioneers this.

Let me talk about three or four things I've learned from this meeting.

One is that I came here as interested as anybody else in this region of criticality that's been referred to under many names, but what interests me in particular is the notion of the edge of chaos. I listened to Per, and to Stuart, and a number of others. I'm very much persuaded that this edge of chaos is where a lot of the action does lie. I had a skepticism that Stuart and I have sorted out since, toward the details of evolution to the edge of chaos. So I want to make a proposition and I think this is something Stuart would concur with. What we need at this stage is a detailed set of metadynamics that are well specified for many of these models that tell us how you reach the edge of chaos. Not sweeping statements, or not pious hopes, but actually well-specified metadynamics. I'm persuaded in some instances listening to Stuart that there may be [metadynamics] for systems on this metalevel to evolve to the edge of chaos, but I'd like to see that spelled out; I'd like to see the experiments...

BAK: We have five specific models that do that.

ARTHUR: You have self-ordered criticality, I'm sure, and that's one set. But Stuart has alluded to a much, much wider range, [which] personally I find fascinating, but we are missing the metadynamics that are well specified to see that we get there...

BAK: Well, we have specific models and at least we had some rather more complicated models we're doing, and that's yet to be done, and we're doing precisely what you suggest. But we have specific models.

ARTHUR: But I don't see that happening in some of these other...

BAK: Right. But that may happen later. These are just bigger problems. That's why I'm here.

ARTHUR: Let me say it again, that in many cases that Stuart alluded [to] that I find absolutely fascinating, I would like to see the metadynamics well specified and then carried out, either analytically or computationally. I think that's the next step and, as Per said, it's been taken for several models so far.

Number two thing that I've been thinking about a lot is bounded rationality. When I first came here, I thought that that's the next step for economics. My thinking before I came to this conference, and it hasn't changed, is that the major obstacle economics faces is: how do you do economic theory with agents who can think about as well as human beings, not as well as superhuman beings.

I came here also thinking I had the beginnings of an answer to that. In fact, I've been working on this the last two years. Stuart, I think, is homing in the same region. John Holland knew it all along anyway...

HOLLAND: These gratuitous insults must cease!

ARTHUR: Now I'm convinced it's a wider problem. That is, it's a problem that runs through many of these sciences. The problem is this: How do you reason in non-well-defined problems? There's two ways to look at it. You could take a very complicated problem like chess and say that problem is well defined, but our reasoning about chess is not well defined. Or you could say that we know how to think with certain algorithms: means of thinking, habits of thinking. I'd be glad to lay some of them out, but actually Ben Martin did more than anybody else here. And for us, the moves of chess might be well defined but what our opponent's about to do is certainly not well defined, and how the game will unfold is not a well-defined thing.

So, what we do is exactly what Murray said on Day One: we form internal models, or schemata, or we form hypotheses. They may be multiple hypotheses, they may be mutually supportive, some may be mutually contradictory—and we act upon one or more of those hypotheses. And this is very much what Murray was saying on Day One.

The funny thing is that it carries over to this problem of bounded rationality. We strengthen those hypotheses—that's called accretion by Ben Martin—or we refute them and throw them out if they're not performing very well. I see this all the time in the actual stock market. Traders say, "Moving averages aren't working this month—out they go." Then you take up something else, maybe the stochastic, or something, is another trading method. You redo your hypotheses—some are strengthened and acted upon—and in John Holland's term, you wind up with an ecology of hypotheses. And that ecology has sort of a succession to it—it may never settle—and, where you have coevolution, you get a sort of ecology of ecology of hypotheses which gives you a truly complicated setup. Now as I say, I'm fascinated that several people around this table are homing in on this. Stuart says—he met me on Day One—"I've solved bounded rationality." I said: "Oh, so have I Stuart." And funny enough, our solutions are much the same. I was very much inspired or guided by John's classifier system that I don't think John has used for quantitative purposes, but certainly he shows the way qualitatively.

I think this is something the Santa Fe Institute needs to do. If you want a phrase for it, it's a shift from deductive reasoning to inductive reasoning, and we are starting to get very good at theorizing in this dimension. Moreover, what I'm convinced about is that you can build analytical models of this sort of thing, or computational models, or both, and rigorously investigate the implications of this whole approach. I've learned far more at this meeting about this just by sitting and thinking than I did even in the last six months when I was writing up my paper on this.

Number three is the question "Does complexity in the form of complication tend to grow as evolution or coevolution take place over time?" I came here convinced that it does; I'm still convinced. I gave some reasons in my talk, and John Maynard Smith yesterday gave some reasons why he thought the question had not yet been answered.

GELL-MANN: It's the envelope of complexity that's growing. Isn't that it?

ARTHUR: In my talk I said there might be two different mechanisms. One is that the degree of complicatedness at the interconnections—what might be called the biodiversity of a system—tends to grow, simply because the new entrants to that system that are brought in create fresh niches for further new entrants. And I then said that there are temporary collapses at unspecified times. I was fascinated by Jim Brown showing something like that in a working ecosystem, or referring to that in his experiments. And yet, there may be diminishing returns so that these things might settle down.

Number two mechanism I referred to in my talk was the endency of individual organisms to accrete subfunctions, subparts, submodules, subsystems. And there I referred to the jet engine. And I believe that you can see this happening over time in biological organisms as they accrete more and more complicated functions to perform better. And yet again I'm perplexed and fascinated by John's counterexample of crocodiles that have stood still, or any number of other organisms. Maybe social insects of various types, and so on.

GELL-MANN: That's because the envelope is growing, but you can still have simple things with simple niches.

ARTHUR: Certainly. I think this is a whole area that's unsolved here.

COWAN: I'm confused. Is it growth or nongrowth of complexity with time, or with adaptation, or with accretion of parts?

ARTHUR: Over time, yes. I'm talking about over time. And again, this is a theme that's run through Stuart's research for years. Stuart's been talking about things that switch on, things that switch on other things, in Stuart's very simple language. But I do think Stuart's talking the same language. I'm not sure we've nailed it down; we have for certain examples that Stuart has shown. I was very much taken by Kristian Lindgren's simulation that I think shows these two mechanisms: coevolutionary increases in sort of web complication and, secondly, individual strategies, in which the entities are getting more and more sophisticated. And then a sort of to and fro between the two. And I don't think we've settled the issue by any means.

KAUFFMAN: What we need here is the theory of product differentiation. We need to know why a set of entities create niches that drive the formation of still more entities, which drives the creation of still more entities—because the niches are exploding faster than the things filling them. And then yesterday I learned for the first time something that made me understand at least the inklings of what drives hierarchy. It's what John Maynard Smith and Peter Schuster was telling us: if you have advantages of trade, it's awfully good to get married to your trading partner and form a higher-order entity, and that drives integration to higher units. So this is maybe well known to John, and to certain aspects of the biological community, and I ought to have known but I really didn't. So this ties together very very nicely.

ARTHUR: It seems to me when we're talking about emergence, whether in biology or in economics, that there's a number of mechanisms that we haven't even begun to catalog.

We are becoming more aware of the difference between coevolution and evolution. In economics it's a difference between simple optimization of what one's doing, one's own performance, versus being in a game. There's no reason to believe, when players in a game reach a Nash equilibrium where they can't mutually improve, that anybody is at any performance peak whatsoever. This is something that most of us are well aware of. Coevolution and evolution have different rules or, if you like, different ways. Coevolution is not simply evolution multiplied by n.

On a different subject, I've been asking myself what was I most fascinated by at the conference. Certainly I was taken by John Holland's ecosystem, Tom Ray's Tierra, Kauffman's examples, and Kristian Lindgren's evolving strategies.

What fascinates me about these models is that besides creating theory, besides corroborating our theories with real-world data, besides coming up with new principles, we are also beginning to develop metaphors. And I have a very strong belief that science and thinking progresses not so much by theorems but by metaphors. And what fascinates me about what John Holland is doing is that if I really get into his systems, and really look at them and watch them evolve, I will have a deeper understanding or a deeper metaphor for the creation of life and what life is all about, in the sense of interactive life. That's also what interests me in Tom Ray's evolutionary system. These examples are metaphors that we absorb, that go in deep, that we digest, perhaps also consciously forget. But two years later you start to write about evolution in the economy and you're deeply informed about how evolution actually takes place. In these simple examples we are not abstracting evolution or adaptation in the form of Lotka-Volterra equations or population genetics, but actually watching evolution *in situ*, or *in vitro*, or *in silico*, actually unfold. You actually see evolution and more than that, you feel it. Now, Murray might deny that this is science but I think that it's what keeps us going...

GELL-MANN: No, I don't deny that at all. I think it's wonderful.

ARTHUR: So I would say this. I think that this Institute is certainly in the business of formulating theory, principles, and doing experiments. But at a deeper level we're in the business of formulating the metaphors for this new science, metaphors that, with luck, will guide the way these sciences are done over the next fifty years or so.

GELL-MANN: Can I say a word about this?

I loved Brian's remarks, and the last one was particularly interesting, about being fascinated by these models. I'm fascinated by them, too, and I think they're among the best things we do. But I think one reason why we're fascinated by them is that they show what you can get along without. That is, you really learn by Tom Ray's Tierra that phenomena can take place in evolution without A, B, C, D, E, which you might have thought were essential. For example, some people have argued that the change in the environment is very important. Well, there's no change in the environment. Some people may argue that you require some sort of fitness. Well, there's no fitness.

MAYNARD SMITH: There is fitness.

GELL-MANN: Any fitness there is emergent.

ARTHUR: That's coevolutionary.

GELL-MANN: Coevolutionary. Exactly.

All of the things that he was going to put in later are examples. This was a debugging exercise, and he was going to put in later five or six complicating features that would make it a real system that would be interesting. Every single one of those is something that you might have expected to need, and that you don't need. So I think that's one reason why it's so exciting, and ECHO has similar properties.

WALDROP: Murray, it's the converse of that, too, because there are certain things that won't happen in Tom's Tierra system. I've talked to him about this. But this also tells him what he needs to put in to, say, get sex or multicellularity, or things like that.

COWAN: Can we go to John Holland?

HOLLAND: First of all, I would just underline what you said, George, and which, I think, in fact is something that both Mitch and Murray, and Brian were saying: that, models and questions serve to focus discussion. And one of the real difficulties in interdisciplinary work is the discussion becomes very diffuse and very highfalutin' and it doesn't meet almost any of the criteria that you would want. So I feel that this notion of the model—and especially the well-formulated model that tells you what's been left out, and what you can do with that, and then suggests what you might want to add to go the next step—as a kind of focal thing is very important and not done very effectively in a lot of places, but is being, I would claim, done pretty effectively at the Santa Fe Institute.

That leads me to this business of the formulation of questions, and I think that, to me, is much more important than the formulation (at the stage we're at) of this or that set of important facts or important models. The formulation of questions seems to me very important and we know in many fields a good list of questions. I think of Hilbert's questions in mathematics, [which] directed a great part of the field for the next fifty years. So one thing I would like to hear at a second round here is what two or three questions that people—and with some care about formulating the questions—that is, what two or three questions this Institute should be directing a major part of its activity toward over the next five or six or whatever years.

COWAN: In the near future.

HOLLAND: I think somewhere along the line we're going to have to focus on what questions are central, what kinds of activities are going to receive core funding here. I think that's very important. While it sounds bureaucratic, I think in terms of our scientific mission, and especially in this interdisciplinary area, it's just critical.

COWAN: When you say central, do you mean central to our scientific endeavor or central to the concerns of the people who have money?

HOLLAND: Central to our scientific endeavor. But related to the other as well.

COWAN: That's a real problem. You mentioned funding in the same breath. You've introduced the problem. What is central to our scientific concerns, what is central to the concerns of the people who have the money?

HOLLAND: I agree, George, and being on the scientific side, I would say that our activities have to collapse to the point that we can get them funded rather than making them match what our funders will give.

GELL-MANN: What is it that serves both ends?

HOLLAND: I think well-formulated questions will help us attract funders but I think to the extent that the funders aren't attracted—but that we think the questions are important—then we follow the questions, not the funders.

COWAN: I'm glad to have that on the record. I mean, it may not help Ed. . .

HOLLAND: No, I'm sure it makes for difficult decisions.
Now, from that notion of question and the notion of model let me point out that some certain types of standard, simple models can be very critical to the development of areas. I think of something like the two-armed bandit problem, to pick something that most people know about. It's generated a lot of mathematics, a lot of thought on economics, a lot of thought in study of certain kinds of ecosystems. You see what can and can't be done but, moreover, the results are very clear. And so I think it's important that we work towards finding such models, and in lots of fields. You know, we've all heard things like Wicksell's triangle, the two-armed bandit, and overlapping-generation model. They serve this metaphor purpose in a very important way. We have perhaps rather less of those so far that we've formulated than we might be able to produce if we thought about it. And I think about how models like that lead to some very clean experimental results, like the Kahneman-Tversky stuff, which is closely related to what's rational and irrational in situations like that. Broader concepts, like bounded rationality, inform those concepts and at the same time lead to certain kinds of experiments.
If I were to talk about themes that I find both interesting and would advocate—and certainly not exclusively, but just the ones I find interesting and would advocate—certainly the notion of internal models, and all that centers on that. The ability of systems to anticipate seems to me to be one of the things that actually is pretty pervasive through the complex adaptive systems I know. And this leads us back to this discussion we've had between the whole-system approach and

the sort of system-environment approach. You can do both things in this context of internal models.

I can approach chess from the whole-system viewpoint and I get things like Von Neumann's minimax theorems, which ignores all of the interplay between opponents and so on, and tries to talk about whole systems. That's a very useful result. It has guided us over a long period of time. On the other hand, if I try to look at this, as Brian was suggesting, in terms of an opponent who's trying to model what's going on, I get a different outlook on this thing. It tells me different things; I try to do a different kind of research.

I think those are equally valid, but one should be aware that neither one displaces the other, and you're going to get different kinds of answers depending on which one you take. And, you know, there's all of game theory—that was essentially the whole-system approach to games—and then there's the stuff that's coming along now in cognitive science which tries to take the system-environment approach. So that's one thing: internal models. In other words, I'm saying some substantial part of our effort should be on the system-environment side, not the whole-system approach, so that I can study internal models and what they do.

I think another thing that's somewhat related to that is the notion of organization and hierarchy: how do these things come about? And this is of great interest. We had a workshop here somewhat earlier which did involve, for a change, quite a few social scientists. They considered this a critical problem. And Brian has certainly emphasized it. Where do firms come from? Why? This is a matter of organization. To me that's an important topic that offers a lot, and about which we know very little, I think.

And finally, much more vague, is this whole notion of innovation. I think trying to understand how innovations come about, and how they affect systems, is just critical to so much that we want to know about complex adaptive systems. There my only sort of suggestion, that you've heard over and over and over again, is that the notion of building blocks and recombination is central to the things that, say, Per would talk about. The fact that sometimes we move a short distance, and sometimes longer, and sometimes still longer; that recombination can supply that kind of dynamics; and that this notion of building blocks is not at one level. It's when I finally get building blocks at one level, then the recombinations yield building blocks at still a higher level. That kind of thing, in my mind, actually tells me a lot about everything that I want to call adaptation, or learning. That's the reason I tend to treat them the same. Because I think that one is this process of recombination on a millisecond time scale, and at the other extreme where I'm talking about evolution I'm talking about this same kind of process but now on a one-century to one-millenium time scale. But I think formally there are real similarities, and that's more an article of faith right now than anything else.

COWAN: If I'm learning the solution to a problem that's already been solved, is that the same kind of learning? I mean, if I'm just being told what the solution is to a problem?

HOLLAND: People call that learning. Psychologists would. But it's not the kind of learning I want to study.

COWAN: That's what I thought. Because that's most of what we call learning.

HOLLAND: Clear back in the '50s, when Art Samuel was studying the checker player, he called one rote learning—and that's where I accumulate game boards that I've seen in the past, where I say, "This looks like that game board that I've seen in the past"; that was one kind of learning that he studied. The other kind that he studied was this notion of changing this linear form which directed their strategy.

COWAN: If you're talking to a teacher, they talk about rote learning, usually, without creativity or anticipation.

GELL-MANN: Yes, but what happens to those pupils, George? What they're doing is formulating schemata for dealing with the teacher. Jonathan Holt, in *Why Children Fail*, describes a phenomenon that I saw repeatedly when I was a child, among other pupils in the classes. In arithmetic, they would be given a variety of these word problems, of which everyone is so terrified, because supposedly you have to think in a word problem, before you get to the mechanical operation of adding, or whatever it is. Holt heard the children saying exactly what I heard them say. They would exchange schemata. One of them would ask, "Now let's see, in a 'John-does-a-job-of-work' problem, you add the first number to the second and divide by the third, is that right?" "No, that's in a bathtub problem!" [Laughter.]

HOLLAND: It's sad to report, Murray, that some odd years later, my youngest daughter is going to a very good school and she's going through the same damn thing. It still is a major part of education in mathematics.

COWAN: We have to turn our attention to that, by the way, sooner or later, I think. Because if we allow ourselves to get confused between learning by rote—or what some people call "maintenance learning"—and real learning, I think that we tend to go off in the wrong direction.

WALDROP: I just had one thing to add to what John said about the formulation of questions. I was very struck a while back when I asked Melanie Mitchell how she had enjoyed her summer here last year, and she said the thing she thought was most important about being at Santa Fe is that you could be talking to someone, trying to formulate a problem, and someone else from a completely different field would come in, and ask a question you never would have thought to ask yourself. And that to me goes right to the heart of why it is that this is supposed to be an interdisciplinary institute. In part it's not just finding a common language, finding a common culture, but also the bringing of questions you never would have thought to have.

KAUFFMAN: The first thing I want to say is that the reason that we wanted to have this meeting was to ask ourselves whether or not there was a common core about which the Institute could focus. Our worry was that we were creating departments, all of which were studying the same thing, but with different names. And I think that the meeting that we've had buttresses the view that there really is a core; there really are common themes. The Institute should assure itself, and its sponsors as it proceeds, that it should keep looking back at this core, to pursue it in as cogent a way as it can. I think that involves precisely what both David Pines and Brian Arthur were saying. We need to look for overarching sets of principles, we need to find the baby models that embrace them, and we need to nail the issue of which principles are applicable.

Second, what we have on the board really is the start of a framework to think about complexity and complex adaptive systems. I'm charmed that Brian and I are trying to find the same way towards bounded rationality roughly as a coevolution of the schemata that we have of one another, and that may go through the same phase transition; it may emerge that criticality is the general solution to a lot of things. And in this sense, I think that what I find myself doing is unpacking Per Bak's intuitions, over and over again. I think that he's close to right.

I want to remind you of what Phil Anderson said. What we want to understand as our primum question is: how do you get from physics to the Boston Symphony? Namely, physics to reality. And the reason this is so important is the following: There are an infinite number of dynamical systems, or algorithms, or whatever, that we could play with, all of which are complicated and do intriguing things. Some subset of those point towards whatever are going to be the really critical, important processes that have to do with how one got from quarks to jaguars to the Boston Symphony, and that's what we have to find our way to. And I think we must have taste.

The third issue I want to raise has to do with historical contingency. When we're looking for general principles—which I, for one, am spending my life doing—we have to bear in mind that sciences such as biology, or geology, or a variety of other arenas, are absolutely rich in historical contingency. All the frozen accidents...Jacob wrote that evolution is tinkering, that organisms are tinkered-together contraptions. And there's an enormous amount of truth in that. History itself is filled with contingencies. I think that one of the things that we have to try to understand is how will we relate the emergence of law in arenas where there are alternative ways that the system under question could have gone. Lots of basins of attraction, lots of minima, lots of ways that Walter Fontana's grammar models can explode in different directions, or in which polymer sets will go—that if you rewound the tape over and over again, it would come out differently every time, yet there are patterns in the emergence. What will be the relationship between law, chance, contingency, and design? In biology, we really have contingency and design driven by natural selection but without too much faith that there really are some general laws beyond natural selection and genetic transmission. I hope that there are such laws.

GELL-MANN: That may be a big issue in fundamental physics. It's possible that the fundamental properties of the elementary particles in the universe, may be contingent. That is, there may be a probabilistic distribution of universes with different symmetry breakings, different parameters for the system of particles. It's not impossible.

VOICE: It's perfectly plausible that there really are some general laws that emerge in the patterns. For example, patterns of phylogeny, branching, and radiation and stasis, and the beauty of the fact that the Cambrian explosion occurs with the species that found the phyla giving rise to the species that found the classes, giving rise to the species that found the orders, and so on down. And filling in the taxa from the top down is a lot like technological evolution of major innovations that are the kinds of things that we want to understand.

But, because there's are so many frozen accidents, there are so many broken symmetries in the way that systems actually unfold, there's always room for the naturalist to ask the phylogenetic question "How did this particular, quirky thing come to be?" I mean, why do we have radiators, why do screw caps turn to the left, not to the right? Why does the aardvark have a nose that's shaped in a particular way that depends upon the peculiarities of its particular history, and whatever design principles were around? We have to figure out how to marry all of those.

And then the final point that I want to make echoes the sense that we want to be able to reach out to social systems. There's something very appealing to me about the relationship between polycentric political systems, or economic systems, which stumble their way forward and coevolve with bounded rationality, with lots of alternative hypotheses about what a good thing to do is, and the way that our small model systems are beginning to tell us what you have to do if you want to coevolve successfully. If you have a dictator, you get stuck on some really stupid solution. So as we've been saying for years, totalitarian regimes are wonderful in the short run but in the long run they fail. And I think that the kind of work that we're doing has a real chance to illuminate some of these issues. For example, why political parties are useful: because they have to compromise a certain number of constraints internally and then across their boundaries they can fight. But if you had a thousand political parties, there'd be so many single-action units that they could, in fact, never reach sensible compromises, and so on.

COWAN: John? And leave your modesty at the front door.

MAYNARD SMITH: Oh, I certainly don't intend to be modest. But I do think there are different styles of doing science. I do find that my style of doing science is mainly very different from most of the things I've heard people talking about. Don't say it's better or worse, it's just different.

I would argue that at least in biology, most of the significant, important general ideas have emerged, not really because somebody was looking for a general idea, but because somebody—usually a naturalist, and I glad you admit there's room for naturalists, because that's really what I see myself as—is puzzled by some specific thing. For example, the origin of the most important idea of the nineteenth century, Darwin's theory of evolution by natural selection, emerged because he was fussed by the animals that he saw on the Galapagos Islands. And he was a passionate naturalist; that's why he was there.

It's not universally true that the important breaks have been made by people who were naturalists. It's not true of Haldane, Fisher, and Wright. That particular breakthrough—it's a very important one—was made by three men who became naturalists later, so to speak. But I don't think natural history was the source of their drive. If you ask, for example, why did Bill Hamilton develop the kin-selection theory of social behavior, and Haldane didn't (because Haldane had the idea, in a sense), the reason is very simple. Bill Hamilton wanted to understand ants. And Haldane didn't. It's as simple as that, I believe.

Of course, you don't choose any old detail. If I can be personal again, perhaps my main contribution to evolutionary biology was the development of evolutionary game theory. I did that because I was puzzled about certain things that spiders and other animals were doing. I couldn't understand why they were fighting in the way they did. But, of course, I didn't choose that at random. I chose it because what they were doing appeared to contradict the accepted theories that people were working by. So there was a sort of a taste for choosing that particular problem rather than any other.

What I think is, first of all, that at least in biology, going from the particular to the general has been a procedure that usually has worked. Not always. So what's the point of knowing anything other than natural history? Well, obviously there's a point. But I think the main point, oddly enough, is the one that Brian Arthur mentioned. We need what he called metaphors, I tend to call them analogies. When we look at something before we can get an idea about it, we have to say it's rather like so-and-so. You explain something you don't understand by seeing the analogy with something you do understand.

To give a trivial example, once you've seen one case of simple harmonic motion—you see the weight bobbing about on the spring sinusoidally—next time you see something behaving like that, you have at least an idea for the mechanism. To give a rather more significant one, perhaps, I think that the whole development of genetics, during the last fifty years anyway, and perhaps right since 1900, has gone the way it has because we live in a world in which we are surrounded by information-transducing machinery. And basically genetics is a problem of the transmission and translation of information. And if we didn't have that analogy very firmly in our minds, genetics simply wouldn't have been the same, and I think it wouldn't have happened.

So the main thing I could hope to get out of listening to you people talk would be other fruitful metaphors. And I have to say the main thing I've got out of this meeting was to hear Per Bak talking about his sandpiles. I think his notion of how to apply them to biology is not right, but it's a lovely, metaphor, analogy—even if I can't do the algebra in any particular case, and I'm not sure one can anyway. But what's special about it, why it's appealing, is that it has—rather like sinusoidal motion has—mathematical regularities that a mechanism of that kind might be expected to give you. Namely, power laws, and so on, of avalanche size. And so I know that from now on I will be on the lookout. If I see something like this in biology, a bell will ring, whereas before I'd heard his talk the bell wouldn't have rung, because I wouldn't have had the analogy ready to use.

Let me say one thing which I'm sure you're all going to think is totally crazy, but let me talk for one moment about what influence you might have applying your ideas in society. And perhaps at this point I should reveal a fact that some of you know, but for most of you, there's no reason why you should know. In common with many young Englishmen of my generation—now old Englishmen—I had a period of Communist Party membership, and Marxism, and all that. And if that happens to you between 18 and 22, it has a big influence on the way you see the world afterwards.

One of the things that puzzled me then, and puzzles me now—but less now than then, I think—why did Marx and Engels decide to be dialectical materialists, for God's sake? I mean, one could understand why there was a socialist movement in the nineteenth century, but why did it get linked up with dialectical materialism? I think it was a disaster that it did, but I don't think that Marx and Engels were either stupid people or malevolent people. I think they were both extremely intelligent, and, on the whole, well-wishing people, even if the results of their activity were disastrous. You could say, "Well, they were both Germans so they were exposed to Hegel at school, it was an historical accident, one of these contingencies that Stuart mentioned." And maybe that's right, but I don't think it is. I think the reason why they were dialectical materialists was that they were trying to understand complex systems. It was a genuine attempt to understand highly complex systems in a world in which there was no mathematical language of any kind that they could use to describe them.

They did not want to accept the kind of linear extrapolation notions, whereby you foretell the future by putting a line on what's happened and just continue it to infinity. Because they had this sort of gut feeling that there was going to be revolutionary change. I think it's fair to say that if there had been a mathematics of bifurcation theory then, they would not have needed to saddle themselves with the transformation of quantity into quality, because they would have had a more precise and useful mathematical language to use. Even if they could not have written down the equations that led to the bifurcations—because that they certainly could not do—they would have had a mathematical metaphor that they could use. And they didn't.

The question now is: are you people going to provide mathematical metaphors which people thinking about society and its likely future can use as Marx and Engels tried to use changes of quantity into quality, and interpenetration of the opposites?

Maybe they would have been better off if they could have used bifurcation theory. Maybe that's what you should be trying to do, to provide metaphors we can use when thinking about complex systems...

GELL-MANN: Or trying to avoid.

MAYNARD SMITH: Or try to avoid. That's right. That's the question, isn't it? Because the Marxism business was clearly disastrous.

Let me describe something that is terrifying me at the moment in the political scene. Socialism did represent genuine idealistic feelings. Those of us who were in the Communist Party weren't doing it for any other reason than that we genuinely did want to benefit our fellow man. We really did, believe it or not. That enormous idealism was distorted and led to the most ghastly results through the adoption of a false philosophy. I think it could be said that, at the moment, there is a terrible danger that another idealistic movement, for which I have great sympathy—namely, the Green Movement, and the environmental movement—may, at least in Europe (I don't know whether it's happening here), be distorted by an equally dangerous false philosophy, which is Gaia, which is just simply fallacious. It's not wicked—I mean, Lovelock is neither a stupid man, certainly not a malevolent man, any more than Marx and Engels were. It's not all that relevant to what you're doing here—but perhaps it is.

It's really quite important that idealism be informed by the correct and modest—and I'm glad to hear some of the modesty that I've heard from you—philosophies, and not incorrect and immodest philosophies.

HOLLAND: Some of us would genuinely like to see our notions of complex adaptive systems used out there, but with the same dangers.

PINES: We have to use the utmost caution in tossing out our metaphors. I really get frightened by the economists and the social scientists who were embracing spin glasses as a model, because it's very appealing, and you can calculate it; it's neat mathematically. But I think it's as wrong as the general equilibrium model and is unlikely to be productive in the long term.

GELL-MANN: Could I comment, briefly, on that?

I was very interested in your remarks, John, all of them. I just want to comment on the last one, though.

There are terrible examples of philosophers preaching what sound like reasonable doctrines that lead to disaster. It's true that the recent ones have mostly been Communist in one form or another, but they're not all mainstream Communists. In Cambodia, Pol Pot and his henchmen were driven by philosophical ideas circulating at the university in the capital about the importance of rural life, the corruption resulting from urban life, and so on and so forth—things with echoes of some reasonable criticisms of society. And, of course, it led to forcible deurbanization with millions of deaths. Guzmán, in Peru, was a philosophy professor at the University in Ayacucho, and we're seeing some similar things there.

So I think you're absolutely right, both on the positive and on the negative side, to say that our metaphors may end up being passed through journalists and philosophers, totally deformed. And that's the way things go these days, because science writing and most communication with the public is done by people outside the scientific community who make heroes and villains and distort the scientific results. People often generate philosophical ideas that are loosely driven by scientific discoveries. And when it's a vague field like ours, with lots of metaphors—as David was saying—it's even easier to do that. So there's hope and also fear to be attached to the dissemination of our metaphors. I think that's absolutely right.

COWAN: Thank you. We come next to Mitch.

WALDROP: Many of you, I suspect, have read David Marr's wonderful book about the visual system, which he published just before he died in 1980. One thing he said in there about how one goes about making theories has always stuck with me, and I think it might be relevant to some of the things here. Marr divided kinds of explanations into sort of three levels, one of which he called, I believe, the functional level, the functional definition of the system. He was, of course, talking about the visual system, filtering, and so forth.

I think here it would apply in the list that Murray calls stages of complex adaptive systems. That is something approaching a functional definition of what a complex adaptive system is, the kinds of things it has to do, such as the coarse graining, interpreting the environment, and so forth. We could argue about the details of the list, but that is essentially a functional definition of it. Now Marr also pointed out—the distinction, I think, is important—that there are other levels, too. There is the algorithm by which you accomplish this and, as John Holland pointed out when he first got here (I believe it was on Saturday), you can have a functional definition of something (he was talking about functional definitions of computational machines), yet the algorithm by which you actually realize it can make some things very easy and other things very hard, and it makes a great deal of difference. And I'll come back to that.

But the lowest level that Marr defined was what he called the hardware level, or the wetware level, that is, the things out of which you make this stuff. And here, now that we're talking about making cells out of molecules (or cells out of organelles, perhaps), making human organizations out of people. The whole issue of how you take a large collection of things and get them to organize or self-organize themselves into some kind of higher level system is looking at things on the hardware level. I've noticed on occasion some confusion about the levels here, that I just thought this distinction might help a little bit. So I just want to throw that out.

I want to say one thing very briefly in regard to what Erica said about particularly some of the younger people thought about what they were doing at the Santa Fe Institute.

After my talk Saturday night when I basically just threw out a comment, almost a throwaway, that I didn't think that this effort in complexity was a paradigm shift in the Kuhnian sense. And I meant that on purely technical grounds, that it wasn't replacing anything old, it was just adding to it. I mean, we aren't saying that quarks are wrong, or don't exist, we're just saying that you should look at bottom-up emergence, and how things self-organize, and so forth. And I had a couple of people come up to me afterwards who seemed to be greatly disturbed by that, as if I was attacking a vision they had of themselves. And I think it may relate to what you were saying, Erica, that people may be seeing themselves as transcending the disciplines, and I don't know if it's healthy or not.

RASMUSSEN: I want to comment, because I think you totally misinterpreted me.

I don't have any great thoughts about what I'm doing myself, but I really feel that everything within this broad field of complexity has to be interpreted, in technical terms as Kuhn defines them, as a paradigm shift, and it does not have anything to do with what I'm doing.

MARTIN: It's certainly true in psychology that connectionism has represented a shift in the paradigms that we use to look at memory. I would argue that across various fields you'll find this, that the paradigm shift is in some sense distributed. These ideas are general enough that they're causing shifts in the ways we look at things in many fields.

COWAN: I'll ask you, Mitch, to speak for a couple of minutes, and then shift to George Johnson—who's going to be surprised by my pointing him out.

One of the problems that's constantly presented to us by lay people is, "What are you guys up to? How do you explain to the public, how do you capture and embrace, what's going on here?" And, of course, we all launch into pages of a seminar on the subject, and the usual response is—I guess the most extreme response I got is—"Can I put that on a bumper sticker?" You understand the problem very well because you're standing at that interface, and so is George. Can you put it on a bumper sticker?

WALDROP: As I said Saturday night I had despaired of ever being able to do that. I had tried several formulations and the best I could ever come up with was, "The Unified Theory of Holism," which I knew was a joke; but some people I talked to didn't.

It may well be that, as this flurry of books comes out that are trying to paint broad-brush portraits of what this kind of research is about, and as people become more used to it, the very word "complexity" will take on itself a meaning that it doesn't now have. It will, in much the same way that chaos now does, pull certain mental levers, when people hear it, as Murray has complained himself. And, in fact, I ran into the same thing. I would try to describe what I was doing and people would say, "Oh, chaos," and I never used the word. And I'd have to mumble, "Yeah, well it's a tool people use, you know, (mumble, mumble)." So I think it will happen to a certain extent spontaneously, and perhaps after enough popularization the word "complexity" could be put on a bumper sticker and people will know what it means, sort of. They'll think they know what it means.

GELL-MANN: Well, they will not.

WALDROP: They will by definition not, but there will at least be some kind of vague mental model that will evoke. So there's that.

People ask me, "Well, what good is all this?" And I'm not quite sure they know what they're looking for, and I'm not quite sure they are either. But perhaps the good that comes out of it is what Brian calls "metaphors," and an example is this whole issue of global sustainability, which you've now started calling 2050.

I know that for me, and I suspect a great many other people who have thought about these issues, for a long time I had a tacit assumption that there exists some state out there—we're in state "A" now, there's some state "B" that's sustainable out there, we just need to get to it and then stay there and we'd all be happy. And somehow held rigidly in place. I know there are certain environmentalists who feel that. (This may relate to what John Maynard Smith was saying about Gaia.) After listening to this, and I hope after the public starts hearing more about this, I have a new metaphor, which is...call it "the edge of chaos," or "self-organized criticality," or just in general the sense that a state that's static is dead, that the natural state of nature is not stasis, but continual evolution. That changes the kind of framework people have for asking the questions and seeking the solutions for these large-scale...

GELL-MANN: That's what we've been telling our partner organizations.

WALDROP: Exactly. I think that kind of metaphor shift can be profoundly important. So if that's addressing your question...I hope it is.

COWAN: I call it "moving from the snapshot to the movie." George, do you have some comment to make about this interface between the very little that we understand of what we're doing and the even less that the public understands. Do you see this as important, while you've been sitting here and taking it all in? You plan to write something on the subject so presumably you feel that it merits attention.

JOHNSON: First, I should say that I don't feel particularly comfortable commenting at a conference that I'm probably going to be writing about, because I feel fairly strongly about the importance of maintaining a divide between journalists and the people that the journalists are writing about. Which is, you know, very difficult in a field like this which is very interesting, and I'm very caught up in it, and becoming friends with a lot of the people here.

But it is extremely hard to write about a subject that is in such a nascent state, and sometimes I wonder if it's really ready to be written about. I probably will anyway. [Laughter.] I'm really at the beginning of what's going to be a fairly long stay out here for the rest of the year, and I feel like I'm just getting my feet wet and starting to wade into the field, so I don't think I or anyone, but least of all me, is in any position to reduce it to a bumper sticker. We're already accused so much of oversimplification that that would be a big mistake.

I would like to say that I've been impressed by how most of the scientists at the Institute are extremely helpful and patient at working with journalists, and explaining things, and seem to truly appreciate the role that science journalists play. At lunch the other day Murray Gell-Mann was talking about some of my colleagues and what idiots they were and that he thought that such science coverage was completely worthless.

GELL-MANN: Not completely. I didn't say completely.

COWAN: That should put you at your ease; he was just trying to be charming.

JOHNSON: I asked Dr. Gell-Mann if there was a single science writer whose work he respected, and he said "No." I then asked him if there was a single physicist he knew who he thought could write, and he said "Yes, but I've yet to hear the name." We're all eagerly waiting to hear, to read Murray's book I'm sure, which is bound to be clear.

I didn't think I had anything to say but I'd like to end by saying that most of us are just as serious and proud of what we do as you are about what you do. And we consider ourselves very good at it, and other people seem to as well, including scientists, many of whom I think have approached both Mitch and me, and Betsy Corcoran when she was here, to offer very helpful comments and compliments on stories—not necessarily ones we did, but just about the field in general. That's all I have to say.

GELL-MANN: There was a time when scientists themselves did this kind of work and, as very eloquently described in that book *How Science Lost, and Superstition Won*, the results were much better. It was, of course, a time when there were a number of literate scientists.

MARTIN: Just as a point of interest, I recall a study that I read about in *Science*, six months ago or something, that looked at the citations to medical research, and found that medical research that was described in *The New York Times* was two or three times more likely to be cited in medical journals by medical doctors doing medical research than articles not cited. And to try to control somewhat whether they were really looking at *The New York Times* effect, they looked at the period of *The New York Times* strike. There was a list of articles that *The New York Times* kept of articles that they would have cited, if they had been publishing, [and those] were not cited with the same increased frequency that other articles were. So I think even though you may bemoan the state of scientific journalism, there are purposes to which it is put. I mean, it serves a function, even for scientists, to have these kinds of things around as a kind of a rough cut of what might be interesting to look at, in some cases. And I think it definitely does influence science, maybe for good, maybe for bad—maybe both.

EPSTEIN: I'd like to reinforce something David Pines brought up. He expressed two views. The first I do not associate myself with; I'm theoretical.

But, on the second point, that social questions are where this group may make, I think, very impressive important inroads—that I'd like to simply restate. And as I've said many times, I think there are huge questions that are central questions in social science—that really are not economics—and that may be approached at a whole variety of levels and that may yield work that has something to do with actual data. I mean, can we build simple simulations of the sort that John, Chris Langton, and others have built, that will produce aggregation into groups, the emergence of group conflicts, evolving patterns of alliances, patterns of wars and other interstate conflicts—that actually do mimic statistical distributions of wars, interstate conflicts, in their sizes, and their time distributions? There may be power laws here; that's a very interesting thought. Can we study, obviously, the coevolution of actual arsenals—not biological arms races, but actual arms races as instances of coevolution of security regimes, and so forth? What is the effect of injecting various levels and types of cooperation into those dynamical systems?

I've been studying these questions for around ten years with toy models; toy models of war, toy models of arms races, and so forth. And I think that can be quite illuminating concerning the qualitative behavior of systems of this sort, but it would also be nice to try to grow these things from the ground up. And it seems to me that threshold events in society are the ones that attract the attention of social scientists other than economics, or including economists: the question of imperial formation, or imperial collapse. It may well be that there's some type of self-organized criticality going on here as a metaphor. When the Soviet Union collapses, there are little avalanches of all scales—this is very fuzzy talk now, but in fact the KGB and internal security apparatus fails at the national level, at the regional level, at the local level, in the town, and so forth. It's the same for SAVAK in Iran, or other regimes when they collapse. It would be interesting to try to apply these ideas to areas of that sort.

As I say, the toy approach has led me in the direction of studying whether the mathematical theory of epidemics and other explosive biological phenomena might mimic in interesting ways phenomena like revolutions and the like. Some of the things that Brian Goodwin has been talking about, the sort of closed form PDE analytical approaches to morphogenesis—that strikes me as also very interesting; the whole use of reaction-diffusion equations, and the like, to study the formation of social patterns, ideological groupings, and so forth. Marcus Feldman's work clearly could be tied in to this.

So these are sort of sketchy remarks, but I think Murray's point is also very apposite; [it] is that it may be possible to "grow" these social patterns, these social forms, dynamical relationships, alliances, empires, their collapse, their replacement, and so forth, with local rules that are shockingly simple. That has not been tried, and it seems to be a very attractive and interesting line of work.

COWAN: I will express my personal prejudice here. I think that this is a very useful exercise but only if all of your assumptions are explicit, exposed to people's examination, and you get into a reiterative circle with such examination and so forth.

EPSTEIN: Right. I think that's crucial, and Brian's point is well taken. But I must say that Brian is saying we should be honest and scientific, and I've done three books using war models to look at things, and all the time half the books are: "Here's exactly the data that was used, here's exactly the model that was used, here's where to call if you want the software." You make everything completely replicable, and then you're clear. You're in the clear. You say, "Here are my assumptions, here's what follows from those assumptions; I don't know that these assumptions are right and I'm not prepared to act on the deductions from those assumptions"—and then you're completely in the clear.

COWAN: But I wanted to go beyond that, namely, make it an iterative look in which we cycle and recycle, and then it becomes really useful. Brian?

GOODWIN: I come at this through a fascination with developing organisms, as you know. And so my take on evolution is to say, well, you know, we really do need a theory of organisms because that's undergoing transformation during evolutionary processes. Now what I find as the problems that are associated with that is understanding, essentially, process. Brian Arthur has mentioned this; we're making a transition from trying to understand things at equilibrium to understanding things in process. And this requires that we step away from all these theories of thermodynamic equilibrium and stasis and so on, into the far-from-equilibrium domain. And the most illuminating metaphor that I know in relation to that is Per Bak's work on self-organized criticality. So once again I resonate with John on this. The interest of this particular model, as a way of understanding organization patterns on all space and time scales, and applying that specifically to understanding the way developing organisms are organized, and how this hierarchical system expresses itself on all these different space and time scales simultaneously, and how the energy levels are populated—it's not Boltzmann distributions we're dealing with; we're dealing with all energy levels equally populated, effectively—and what the consequences are for that.

The other area that I find particularly interesting is the work that Walter Fontana and Leo Buss are doing on, essentially, theory of organisms—a logical theory of organisms—which is complementary, in a sense, to the work I'm interested in, which is the actual dynamics.

But this resonates with an awful lot of themes that have come up. Just to pursue briefly the notion of metaphors, a lot of the ideas in biology are informed by metaphors that form a very coherent whole. There are metaphors of functionalism, optimization, opportunism, games, winners, losers, consumption, and so on. In other words, we see things as adaptive and optimizing certain patterns, and certain qualities, and certain quantities. As we understand organisms more completely, as agents of a particular kind—the notion of internal model is also informed by a particular view—embodying particular kinds of agents in action, and a particular form of immanent causation, this will lead to a holistic theory of the organism. Now holism is a term that often arouses intense reactions...

GELL-MANN: Especially if spelled with a "w"...

GOODWIN: It's interesting that John has introduced Gaia, and one of the things that really gets John's back up—apart from what he sees as inaccurate elements in the model—is the whole notion of a kind of holistic approach. And yet I think that this is precisely the challenge: to do it properly, to do it rigorously, to understand organisms as wholes, to understand ecosystems as wholes (if, indeed, they are in some sense wholes), to understand the planet as a whole—for me these are urgent questions. They're not just interesting, fascinating, important scientific problems; they are actually urgent questions, and they do require changes of metaphor. And the metaphors that, of course, I want to pursue are ones that are somewhat at variance with some of the dominant metaphors expressed here, which accounts for the conflict.

It's at the metaphorical level that we actually have these most intense conflicts, because when you get down to a specific example, we always agree, no problem. But when we move away to the generalities, that's when the perspectives, and the orientation, and the emphasis diverge.

Now, just one final comment. I think Erica's observations were extremely interesting. She has gathered a certain amount of interesting reaction from various people in relation to younger people experiencing the conference as an opening of ideas in certain ways. This whole question of how we develop a new educational method in relation to an integrated form of science, and beyond education. I'd like to suggest that we consider, in relation to any publication that comes out of this meeting, writing in a manner that is not addressed to our own research colleagues but is addressed to students, the young members of the research community. In other words, that we try to make it something of a pedagogical volume, in which the ideas are presented as clearly as we can possibly make them...

WALDROP: Brian, aren't you coming dangerously close to science journalism there?

GOODWIN: Maybe. I wouldn't want to get into that one, but it may well be the case. Maybe we should train ourselves in the style of science writing that Murray feels has been lost, that we should do that. We should all do it. We should all become science journalists, and transmit our ideas to the younger generation, through these publications. And so perhaps that's something we could consider when the instructions for the publication are distributed.

COWAN: The market seems to say it's becoming an increasingly profitable thing to do, if you look at some of the recent successes. Steen?

RASMUSSEN: Let me say a few things about some central aspects of our activities, try to make some connections between what some of us call core science, and the naturalists, and also a little bit about funding.

I think what's different, at least from what I've experienced before I came here, is that we take concepts or observations—big questions from the real world—and then focus on them. Let me give some examples.

For instance, this notion of open-ended evolution, a notion of how can we formalize processes that will invent new processes all the time. There is the problem "How does compartmentalization occur in the real world?" and the problem about the edge of chaos—is it really true that out at the edge of chaos systems are more adaptive, and so forth. The unique thing is that we are studying these questions *in abstracto* in some sense. We take them from the real world and then we embed them in these computational models, and then we get some spinoffs. And I really love Brian's notion of "we create metaphors." That's really one of the very important spinoffs.

But also, sometimes I think that we are able to solve bits and pieces of some of the questions. We can maybe pose some of the questions in mathematically tractable ways so we can solve them. I think there's some examples. For instance, I think that we have an arsenal of open-ended evolutionary histories, for instance, Lindgren's example and Tom Ray's example, and a lot of other people's examples. There are a really large class of systems that can produce open-ended evolution. We don't still know how we can get successive evolution of new hierarchical organizations. This is a question we still have to struggle with for a while, I think.

So there is, on one hand, these problems, and then we try to attack these problems interdisciplinarily, and also invent, I think, new concepts and new methods— these self-constructive or self-programming systems—and so forth. I should mention a few more examples of these mathematically tractable problems. I think what Jim has done about actually being able to making this constructive approach to using this machine construction approach to say something about how complicated dynamics is, is an example of actually looking at a particular kind of dynamics and then actually being able to say something sensible in mathematical terms, something which will stay forever. It's not just a metaphor, it's turned into being a solid theory.

There's a third thing, which has to do with glueing all this stuff back to the real world, and I think that this is maybe one of our weakest points. We are really good in producing metaphors, and I think we have quite a few success stories about turning some of these big question, being able to solve bits and parts of them, and make mathematical statements about it. But it's in particular sad that we don't take the third part in here—the attacking of these problems—because it's really when we go back to the real world that we can get money. It seems as if we are all excited about these grand theoretical questions, and we don't pay too much attention to actually taking these grand questions and applying them.

Let me take one person, who's also associated with the Institute, but it's a very clear example of what the Santa Fe Institute maybe is doing. Brosl Hasslacher was one of the inventors of the lattice gases. And right now, after the lattice gases were established as a real tool, then Brosl went off to other things again. And now we have this industry up at Los Alamos where there's this huge group working on lattice gases, and able to attract funding and so forth. We don't have this place to put our good ideas that might be able to attract funding. And I think we should maybe think about, I don't know, creating a device. . . .

GELL-MANN: Send it to a university and tax it, that's what we really need— take a cut.

RASMUSSEN: There are these three spinoffs, and the last one we're not focusing very much on. And another weakness, I would say, is that it's true that all these big questions about open-ended evolution—how do successive evolution of new hierarchical organizations occur—are coming out of the real world. But still, working at the Institute, and to some extent also up at Los Alamos, it's really wonderful when a real naturalist comes in and says, "Hey, is that really what you mean, and are you really sure that can interpret this in the correct way?" So these interactions with the naturalists could maybe also be a little stronger than they are right now.

COWAN: Jim, do you have something?

CRUTCHFIELD: I guess the last few days I've been listening to all the different theories, and certain models and metaphors, and in a sense considering them as data in my own thinking of what possible integration there might be. And one thing that keeps coming back—I think at some point we all have to decide a notion of mechanism for these processes, particularly as a basis for explanation of the phenomena that we're going to see. The notion of mechanism that is used is always informed by the particular science, and in trying to integrate these very different disciplines—and trying to pull out common themes—I think we're going to need to agree on what a good explanation is. I particularly mean coming up with a definition, an agreeable notion of what mechanism is. I see that from just having a background in physics: as soon as you can demonstrate the mechanism, everyone sort of nods their head, "Yes, I understand what you have discovered, what the process is, what causality is," and so on. So I've labeled "Notion of Mechanism" as one thing we might want to think about.

Along those lines, the other comments I wanted to make have to do with what you might call mathematical methods for complex systems. In particular, listing these mathematical methods is a way that I see as formulating a notion of mechanism. There are three areas which I mentioned in my talk, just general areas that essentially use mathematical methods. Namely, dynamics (and I mean dynamical systems theory), as a source of diversity and different kinds of behavior; geometric methods of analysis, notions of qualitative dynamics separate from being very quantitative about things; understanding solutions of equations without closed forms; and so on. And then another element of these mathematical methods is computation theory, as a way of articulating what these mechanisms are and analyzing their structure. This is different from dynamics. And then, connecting things back to the real world, the third category under mathematical methods would be broadly labeled statistics, which often has a very bad name. In particular, I mean by this the mathematical methods that one finds in statistical inference and, say, also statistical mechanics. Those two areas overlap quite a bit.

Those are the three main areas—dynamics, computation, and mathematical methods of statistics—that I think would form the basis of this. It would be—say in some future time—a course, just like in most sciences. Mathematical methods of biology, chemistry, and physics—everyone takes a course like that in their discipline. At some point there should be a course "Mathematical Methods of Complex Adaptive Systems."

COWAN: Are you optimistic about developing a robust statistics for high-dimensional systems, for complex systems? Multimodal, and so forth?

CRUTCHFIELD: I'm very optimistic about that.

These three different categories are pursued in great depth by what you might call technical specialists—and I don't mean that pejoratively, I just mean relative to the level we're discussing things now—and these specialists don't appreciate that there are broader questions in which the cross-fertilization between these different areas could be quite fruitful, and bring in new problems. As a methodology for the Institute, I would certainly want to encourage enlisting other disciplines, trying to advertise these general problems and indicate their commonality. Of course, we all know the specialization forces in academia, but this is something we have to struggle against.

COWAN: You're going to be invited, as will people who are not here, to contribute summary statements. And I hope you take this seriously. I think this exercise has been extremely fruitful.

Let's go to Seth, who hasn't had a say, and to Per, who even when he was out of the room was quoted so frequently that I felt he was here.

LLOYD: I found this conference both very exciting, and very frustrating. The frustration is a continuation of frustration I felt as a physicist. I began studying high-energy experimental physics. And when I realized that I would not get a Ph.D. until age 40, and would not be in charge of my own experiment until I was dead, I went to high-energy theoretical physics. But then I realized that because everybody else my age had the same feeling about high-energy experimental physics, I wouldn't get a theory that would explain anything—at least not for another forty years. So I realized that this was also a mistake, and when I started to do physics of information and complex systems, I felt, "Great. There are many complex systems, there is lots of interesting physics of information. Here's something where I'll find out something about the real world—make models which I can see if they work, or not." And the parts of this conference which I found most exciting were exactly those parts where I saw results and models.

I found Peter Schuster's result that, exploring the space of self-replicating proteins, RNA, one could find many proteins within a particular area that were functionally effective...I think that that's something that's always bothered me about evolution on high-dimensional spaces, and random walks on high-dimensional spaces, and it's obviously the key to understanding how life could have evolved in the first place. Maybe it shouldn't have been a revelation to me but it was, and I found that extremely interesting. I've heard several models of mechanisms at the genetic and organismal level which could be falsified. They may not yet have been falsified but they could be, and that was great.

As all of you know, I'm not averse to theory. The purpose of science is not to explain data. There's vast amounts of data out there, millions of books and computer disks full of data that nobody ever wants to explain, and you'd be a fool to try to do so. And if you were to do so, nobody would be interested. And one of my paradigms for a very fine theory—general relativity—has, so far as I know, only two pieces of confirming data. That is, for most of its life as a theory, it only had two. And that was fine by everybody. So it's great to theorize. And we've heard of theorizing about how adaptation works here, we've heard lots of talking about methodological questions, about how we should investigate complex adaptive systems, and I think that's good. But I found it frustrating that much of this talk did not seem to be directed in a fashion that would allow me to make models that I could test in the real world.

I do not require a well-written, beautifully phrased, true article on why I should work on complex systems. Look, I mean, the question why are things complex, what makes things complex—anybody who has any scientific interest at all is interested in those questions. Nobody needs to be convinced to go into this field by people saying, "These are great problems, problems that need to be solved, problems that need to be worked on." I would like to reiterate Steen's point that if people are going to go into this field, they need to be paid...[Laughter.]

SIMMONS: That's a nontrivial point.

LLOYD: No, it is a nontrivial point. And, thank God, the Santa Fe Institute is here for people like Steen, for people like me, for people like Murray when he wants to say things that at Caltech they'd shoot him for. I thank God for that. But the Santa Fe Institute is only a drop in the ocean of research.

COWAN: Erica, you spoke before but you haven't had your five minutes. Would you care to?

JEN: I will make one very specific recommendation which is that one of the things that I've not been part of but I have liked about the SFI have been the competitions. In particular, the double-auction and the time-series ones, both of which I think had serious problems in their organization and in particular in their subsequent analysis. But I liked the idea of having the competitions, and I thought that it was an excellent sort of activity for the Institute in terms of getting outside people in, really emphasizing the point of a need for rigor and connection with data. I'm bringing this up because the question was also raised as to what should go on at the Institute.

I differ from a lot of you guys here in that I don't visit the Institute for a block of time; I go maybe once every two weeks, or something. And for me, if it's this question of what do I find going on at the Institute when I go there. Part of it, of course, is nice workshops, but I like things like the competitions where the people at the Institute are really actively involved. It's very difficult to plan these competitions, and then to do the analysis is really hard, so I was wondering if you—in particular, Stuart, Brian, John. and some other people—could think of one for coevolving agents. You were certainly doing that with the double-auction thing, but it is a pretty well-defined problem: to guess each other's program, or something like that. And I actually really like those competitions, and would like to encourage you guys to do another type. . .

ARTHUR: It was Tom Sargent who thought up the double auction years ago.

JEN: I didn't like the particular implementation of that competition, but now that some of these ideas have crystallized, more ought to be done. Because it's very much different from Prisoner's Dilemma type of thing.

ARTHUR: It's a very good idea, and I think a lot of that was inspired by Bob Axelrod's original competition.

COWAN: And experimental economics which flourishes at Carnegie-Mellon and so forth.

ARTHUR: Nobody's mentioned that and I'm glad you brought it up. I think it's excellent. And actually it's still going on at the Institute, the double-auction thing, in terms of workshops and write-ups, and so on.

COWAN: Yes, the effort to do a controlled experiment in social science continues. George, and then Ben?

GUMERMAN: In some ways I feel a little bit like the illegitimate child at the family reunion here, but it has been a marvelous week. And what I have to say is not cosmic, and some of you may consider quite trivial, but, first of all, I have no fear that what has been said here this week will be distorted and change the world for the worse. After all, I may be wrong. Lincoln did say, "The world will little know nor long remember what I said here." [Laughter.] And he was wrong.

But rather than say what I learned, I'd like to emphasize perhaps what I didn't learn—and yet I feel was here. Last night I tried to go over my notes, and I agree—as many of you participants have said—that this workshop really is a complex adaptive system, and like all the complex adaptive systems I've heard about, and explored myself, I realize I'm missing a lot of the connections here. There's a lot here, a lot went on this week that I feel needs to marinate for a while, and I need to learn a little more of the language—and some I never will learn. After all, I want you people to know that you are listening for the first time for George Gumerman ever to put these two words together: "spin" and "glass." So there's a lot that we don't command.

And so what I'd like to suggest—and I don't know if this has already been discussed by the steering committee or this group, but I feel there's so much here that's been said that hasn't been explored in terms of the connectivity. So I'd like to propose a very short term goal for the Santa Fe Institute, which I would hope might lead to long-term goal. If I've heard one criticism that's been consistent of the publications that come out of the SFI, in terms of the Proceedings, is that they're a number of papers put together between two covers, without much integration. And I would suggest that while it may take a little longer, it might be very profitable to have a very small group of people representing, say, the biological sciences, the physical sciences, and behavioral sciences, with perhaps somebody more attuned to mathematics or computation, to go over the results of the proceedings—the tapes, the transcriptions—to see what kinds of connections can be made, so that those things that aren't readily apparent throughout the week, but the common themes, the common integrating themes, really can be explored in a synthesizing chapter for this volume. So I would suggest that that be a very short term goal for the Santa Fe Institute, and perhaps out of that short-term goal might come better directions for long-term goals for the Santa Fe Institute. Thank all of you.

COWAN: Ben?

MARTIN: I addressed my comments on Saturday towards something like this question of how complex systems applied to the work I did and how I thought psychology might apply to the study of complex systems, so I think it would be redundant to say most of the things that I'd be interested to say.

But I'd like to comment on this idea of metaphor that's been coming up over and over again. There was some discussion earlier of rote learning, and here I want to address metaphor both as we use it as scientists, and also metaphor literally, as a psychologist would study it.

Murray's example of word problems—you know, the bathtub problem where you take the ratio of the first thing and the second, and then add the third thing, or something like that—brings up the idea of schemata in rote learning in a sense. The question is "When do you apply the schemata and how accurately should they be constrained by the facts that you have at hand?" In a sense, there's not that much difference between the bathtub schemata (or other schemata like that), and what we do when we look at Per Bak's work on particles of sand interacting and apply that metaphor...

COWAN: That's a sandpile problem.

MARTIN: Right [Laughter]...and trying to understand your memory or something like that. The difference seems to be the control of some kind of a parameter, how loose we're willing to be in making our analogy, and then in how strong we want to be in making claims about what follows from the analogy. What's exciting about this field is that it seems to be at this sensitive point, this almost mystical kind of point where these loose analogies end up having consequences that are really interesting, not just sort of trivial. Or wrong. So it seems in this case, using the bathtub problem or the sandpile problem metaphor actually works in a lot of cases. And that, I think, is for interesting reasons. It's not obvious why it should, and yet in many cases it seems to.

But there's a danger in this, which is: if you're adjusting this parameter that's telling you when it's a metaphor and when it's silly, on one end you're looking for almost completely surface detail matches between the thing you want to understand and the problem schema that you're going to apply to it. And that's dangerous, because often then you'll be led astray. You'll pay too much attention to surface features and miss out some important underlying regularity. On the other hand, if you go too far in the other direction, you run the risk of seeming to be a crank, where you're seeing analogies everywhere, where there really aren't any. (Murray and I talked about this a little bit last night.) There's this idea of "connections" that are somehow mystical, that have value that you can't really explain but you know they're there, but you feel them very strongly. And I would say when we're operating in that mode, we've got our tightness parameter for metaphor set too far in the other direction.

Moving around in this parameter space, from paying too much attention to the surface details to making odd, mystical leaps to things that don't really match— that's the business of science overall. When we build scientific theories we're in that parameter space. We're moving around in there and we'd like to find what seems an apt place to be, and the agreement or disagreement of our peers, I think, is what pushes us in those directions. But I think we should be very sensitive to that parameter, because people are apt to call you a crank if you're too far in one direction, and people are uninterested in your work if you're too far in the other direction. And part of the business, I think, of the Santa Fe Institute is to try to understand where that parameter should lie, in regard to applying complex systems to real problems. I don't think we should ignore that risk. It's a very real risk, that we could end up being dismissed because our parameter setting was too low and we saw analogies everywhere, and it was this sort of mystical fusion that didn't really tell us very much. So a note of caution, and also a note of optimism, I hope.

COWAN: I want now again to repeat my thanks to everybody who has come here and participated in what is, in my own academic, intellectual, scientific experience, a unique occasion—to have this many really good people from different disciplines talking to one another was just absolutely the most wonderful thing that I've had the occasion to be a part of. I might say that in putting this meeting together, I've had a great deal of help. I'm not sure whether I should tell a story that comes to mind. Well, I will.

There was a boy scout who came late to the boy scout meeting—he was fifteen minutes late or so—and his clothes were torn, hair rumpled, and scratched, and a little bloody. And the scoutmaster said, "My gosh, where have you been?" And he said, "I've been helping a little old lady across the street." The scoutmaster said, "But—this result?" The scout said, "She didn't want to go!" [Laughter.] I'm reminded of this story not because I'm the boy scout. I'm the little old lady. [Laughter.]

I have to say that much of my enthusiasm for this kind of thing is to see how to reunite the intellectual agenda with a language that we all share, and if it's metaphorical, fine; if it's going to be partly metaphorical, partly mathematical, and so forth, even better. And I think that what we have been doing this week is a necessary part of that exercise, but obviously it's going to take a long time. Let us hope that it continues.

HOLLAND: George, along such lines and perhaps as a suggestion, I would suggest that a list of well-formulated questions would be enormously useful, more than an overview, I suspect.

SIMMONS: The key is really the phrase "well-posed." If it's "What is adaptation?," it's not a very useful question.

COWAN: You are all invited to submit such a list. It would be an invaluable exercise. Once again, let's look forward to the next time; but taking note of David Pines' statement, let's not do it again real soon. We do have a lot to think about. This was the first real effort to examine what we've been doing for the last five or six or seven years. I hope that it will be sooner than five years from now, but perhaps three or five years from now we'll get together again...

GELL-MANN: It's eight years since the founding workshop, nearly eight years.

COWAN: Eight years! My God. You haven't changed a bit! [Laughter.]

GELL-MANN: I've learned a lot though.

COWAN: One thing that's come out of this is that we've all gotten to know each other a little bit better, and I hope that that promotes communication. That's always been an article of faith here. Let's plan to come back to the Santa Fe Institute frequently. All of you.

GOODWIN: George, can we thank you very much for your efforts? [Applause.]

COWAN: I try to do things that I enjoy so much that I would pay for the privilege of doing them. This has been at the top of my list of such occasions.

Afterwords

George A. Cowan† **and David Pines**‡

†Santa Fe Institute, 1660 Old Pecos Trail, Suite A, Santa Fe, NM 87501
‡Department of Physics, UIUC, 1110 W Green Street, Urbana, IL 61801

From Metaphors to Reality?

GENERAL CONCEPTS

The research described in this book focuses principally on descriptions and dynamic simulations of various aspects of the macroscopic "real" world, the topic of greatest interest at the Santa Fe Institute. This world is comprised of living, self-organizing systems which occur at all of the many levels that constitute a hierarchy of complexity extending from macromolecules to the largest organizations.

Metaphors are commonly used to express descriptive views of these aspects of reality. Most of them refer to the mind and brain, vision, and genetics. Their vocabularies tend to come from biology, biochemistry, and psychology. More quantitatively, researchers use a repertoire of mathematical formulations which represent the trajectories of objects in motion and the statistics of ensembles of objects.

Formal research does not effectively embrace data from all of our senses, each with its own library and range of memory and emotion. The human brain integrates all sensory information, providing additional dimensions to our perceptions of the real world and enlarging our ability to characterize our environments. Unfortunately, not all of this information can be expressed in dynamical and statistical terms and essential elements of difficult-to-quantify data are commonly omitted from analytical and numerical models of the complex systems of interest.

It should be noted that our formulations of reality, however adequate for particular purposes, may also be dangerously incomplete. Reality ranges far beyond what is immediately accessible to human physical senses. It includes a near-infinity of loosely to strongly coupled events on every scale of time and size. In contrast, the word "reality," as applied to our understanding of the nature of any system more complex than, say, the simplest cell, usually refers to what Gell-Mann calls a "coarse grained" view based on an eclectic sample of some set of features, appropriate to a particular time scale, which are used to describe the system. Even the most sophisticated descriptions tend to be metaphorical and, except for unrealistically constrained systems, are inadequate to permit long-term predictions of a complex system's behavior.

The most evident challenge in research on complexity is to find models simulating the behavior of complex systems that can be used to predict their dynamics. Anderson described a number of mathematical formalisms for describing the behavior of real systems including Turing/von Neumann/modern complexity theory, information theory, and ergodic theory. Explicit analytical solutions can be forced by constraining the systems so formulated to low-dimensionality and simple central attractors. Such solutions have only a very limited relevance to real systems. Achieving comprehensive "stereoscopic" views of most complex systems usually demands that they be described in more than three dimensions which, as in the case of "spin glasses," invariably produces a potential for a very large variety of metastable outcomes requiring more time and effort to fully describe than is realistically available. Numerical techniques, largely dependent on computers, are necessary to apply this multidimensional approach. Anderson included descriptions of the major numerical-simulation methods: spin glasses, genetic algorithms, neural networks, and even more empirical descriptions with local rules and scaling.

The research presentations called particular attention to the shared properties of living complex adaptive systems. Some of their features are summarized in the remainder of these remarks.

In order to function, biological systems must be endowed with a library of heritable information which is incorporated in the genotype. The organism which houses the genotype, provides energy to extract, process, and generate information, and transmits its heritage to future generations, is the phenotype. Although in principle the genotype and phenotype might be incorporated within the same cells, they are observed to be separate with the single exception of viruses where the phenotype and the genotype are essentially the same.

Biological, complex, self-organizing, adaptive systems dissipate energy and must metabolize, i.e., they must be coupled to an external, more or less stochastic environment in order to extract the energy and information necessary to operate and to excrete their waste products to an essentially infinite-capacity dump. Survival of any given system rests on four necessary but not sufficient conditions. These are an ability of the individual phenotypic members of a species (1) to extract an adequate supply of energy and information from their environment; (2) to distribute energy and information internally; (3) to avoid being exterminated by predator systems in the external environment; and (4) to sexually or asexually make sufficient copies of the species genotype so that the replicating members continue to exist in the quantity necessary for the species to survive.

Gell-Mann described major features of a complex, adaptive system, beginning with its ability to "coarse grain" and to "chunk" information about itself and its environment. It perceives regularities on some scale of size and time, neglects random effects, and constructs a schema, a model of its environment. The schema may be highly compressed in which case it is highly complex. Through a dynamic, never-ending process of evolutionary adaptation, including mutation, the genotype of the species develops a utility function which, to use Anderson's term, represents a "frustrated" solution to the problem of satisfying conflicting conditions for survival. Persistence of the phenotype rather than achievement of arbitrarily defined optimality in a hypothetically stable environment is the most realistic measure of success.

Given sufficient memory, a flexible condition most nearly satisfied in human beings and less fully in other living organisms, an individual phenotype may modify, expand, or compress its schemata and subsequent behavior during its lifetime. A fraction of these modifications may be transmitted to its descendants through a social memory or cultural organization, a property which has given rise to the term "culture gene." However, the species continues to depend on variability in the genetic blueprint for some part, presumably a major part, of its adaptive properties.

Particularly when humans are involved, the construction of models involves psychological factors which depart from the precise analytical formulations controlling the behavior of perfectly rational and perfectly informed economic agents. Martin examined the notion of the schema in research on human behavior, emphasizing the ability to affect schemata by learning within a given lifetime, thereby incorporating information within the culture rather than through the generations-long process of incorporation within the genome. In this respect, it was noted in the discussion that a science of economics which focused on what people actually do as economic agents would presumably be embraced within the broader field of cultural anthropology.

Each of the conditions for survival of any given system requires that the system possess one or more sensors to monitor the external environment for information related to its metabolic needs, to avoid predators, and, in the case of sexual replication, to mate successfully. The phenotype strives to optimize its utility function, particularly with respect to successful replication which is usually defined as the major necessary condition for a complex, adaptive system or for the emergence of a life form.

There are special cases. Viruses represent the closest approximation to a life form in which the genotype and the phenotype are identical, replicating but dispensing with a metabolic system by deriving their energy needs from a host, and subject to predation only by agents generated within hosts capable of sustaining an immune system. In theory, nonreplicating systems which metabolize and can maintain themselves for a long time might also be defined as complex adaptive systems. "Artificial life" programs fall into a special category in which energy is provided by a machine. They self-organize, maintain themselves, may or may not replicate, and adapt to local rules. In a future world, at a higher level in the hierarchy of complexity than has yet been seen on our planet, an intelligent race of beings that learned how to avoid senescence and death would be complex and adaptive, would metabolize, presumably would have to depend on replication only for occasional replacement of losses, and would depend chiefly on self-maintenance and learning to survive and adapt.

At every level of adaptive complexity beyond the prokaryote, nature has invented an almost countless number of solutions to meeting the conditions of survival. None of them can be called the best possible solution in environments which, on one or another timescale, are inevitably subject to change. Relative survival rates are the only readily available external measures of success in systems which may otherwise be far from optimal.

The conference agenda might well have devoted much more time to consideration of the observed chemistry and overall metabolic system requirements in complex systems. "Housekeeping" energy requirements are much more demanding than the energy required for the information processing part of complex systems. Metabolic disequilibrium of particular species was an essential precursor to life before replication appeared. Thus, Kauffman suggested that the initial chemistry in the primeval soup, which led to the appearance of complex organic molecules with the ability to operate out of thermodynamic equilibrium with their environments, required an increase in the rates at which certain endothermic reactions occurred compared to the rates at which other reactions competed for a limited energy supply. Acceleration of these rates can be regarded as improvements in metabolism which, in a limited source of nutrients, leads to selection pressures and survival of the fastest "metabolizers" and, eventually, to their replication. Heterogeneities, provided by membranes, vesicles, and cells, and autocatalytic species spontaneously arise in the environment and can give rise to favored rates for the reactions of particular molecules.

Once replication appears, self-organizing factors may continue to operate. These factors are defined by local, endogenous rules and increase the importance of exogenous or Darwinian selection which is generally considered to be the most important single factor in evolution after the onset of replication. The continuing influence of self-organizing processes is particularly emphasized by Kauffman. The organizing factors are hypothesized to reside in the chemistry of complex organic molecules which can serve as templates for their own replication and for incorporating changes in the proto-genotype.

Notions of how order might arise from local rules are extended by Bak who introduces the idea of "self-organized criticality" (SOC) into concepts of how large dynamical systems may further organize their behavior. Energy is pumped into the system from an exogenous source and is dissipated somewhere in the system. The system is driven by this energy into critical states with a wide range of length and time scales which seem to have a scale-free structure. The various critical states may be thought of as attractors for the dynamics. The phenomenon seems to be universal and has been described in sandpiles, turbulence, earthquakes, volcanic activity, solar flares, and $1/f$ noise and is looked for in biological evolution, economics, and elsewhere. In the critical state in biology, individual species would interact to form a single, highly coherent biology. In economics a single demand at a critical state might produce an avalanche and an aggregate demand with a non-Gaussian, Pareto-Levy distribution.

A second candidate metaphor, which was frequently discussed during the meeting, was the idea that for a given system the region which lies between order and disorder provides an optimal environment for learning and adaptation. Often (but incorrectly) described as adaptation at the edge of chaos, the metaphor is based on work by Norman Packard and Chris Langton which suggest that a phase transition existed between ordered behavior in cellular automaton systems and chaos. It is an appealing concept: an ordered system is, almost by definition, one that stays put and cannot adapt, while a highly disordered system is incapable of learning; between these extremes may lie the possibility of sorting through a wide variety of different ordered states in a comparatively short period of time and selecting that response which is most appropriate. Whether it is of widespread applicability is, however, an open question. As discussed below, computer simulations by Mitchell, Crutchfield, and Hraber do not fully support the Langton/Packard hypothesis while Hübler and Pines, in their computer simulations of agents responding to a chaotic environment, find that for two competing agents, optimal adaptation does occur when the environment is entrained in a pattern near the edge of chaos.

MODELS AND APPLICATIONS

A small number of computational models have been proposed which simulate various aspects of real systems. This area of research has been given considerable support at SFI. Predictions were made that extensive development and proliferation of increasingly useful models should occur within a relatively brief period of time. The models discussed here include:

1. Computations for exploring the possibility that relationships between theory of computation, particularly dealing with computation in cellular automata, and dynamical systems theory might identify a dynamical systems regime in which the possibilities for computation are maximized:

Two very general theoretical frameworks which can demonstrate how coherent global behavior might emerge from the behaviors of relatively simple, locally interacting elements are dynamical systems theory and the theory of computation. Mitchell, Crutchfield, and Hraber have examined relationships between computation and dynamical systems which bear on the usefulness of computation in cellular automata to simulate some properties of complex systems. They paid particular attention to work by Packard and Langton which suggested that a phase transition existed between ordered behavior in cellular automaton dynamics and chaos. Their results did not support the hypothesis but also did not disprove the notion that computational capability can be correlated with phase transitions in cellular automata rule space. More particularly, they did not preclude the frequently cited possibility that complex adaptive systems tend to move to a dynamical regime popularly called "the edge of chaos" where computational capability might be maximized.

2. The further development of the genetic algorithm with classifier systems as a computational model for a complex adaptive system (CAS):

   One of the more widely used classes of models for simulating the behavior of a complex adaptive system is based on the genetic algorithm with a classifier system. Holland includes in his general description of a CAS the need for large numbers of diverse agents which incessantly interact and, as specified by Gell-Mann, construct internal models or schemata which direct their behavior. It is the aggregate behavior of the system that is of greatest interest and this is not predictable from the sum of behaviors of individual agents. All such systems develop highly complicated hierarchies in which a diverse spectrum of agents exploit all opportunities for existence. Classifier systems simulate the development of hierarchies. A simulation system must also create equivalence classes (coarse graining) for dealing with individual scripts so that regularities are identified and appropriate rules for behavior are formulated, Internal models foster the creation of rules that correspond to building blocks that can be used in a variety of ways. A current set of models designed to facilitate exploration for mechanisms that generate phenomena such as diversity and internal models is called ECHO which Holland describes in detail.

3. Hübler and Pines describe a model of a complex adaptive system containing a set of K_m fixed functions each with a variable weight which relate the actual state of a map to a future state. The values of the weights evolve in a random fashion driven by the dynamics of the map within a nonstationary network. The network provides the environmental dynamics. The weights are extracted through a maximum likelihood estimation from the most recent history of the map.

   This system adapts to an evolving chaotic environment. When it is tuned for optimal performance, it demonstrates properties of optimal rationality and optimal complexity which are small in a rapidly changing environment. A particularly interesting feature is that when two adaptive predictors are made to compete, a comparatively stable configuration is achieved when one of the adaptive predictors imposes a weakly chaotic dynamics on the environment and the other predicts this controlled environment; i.e., a leader-follower relationship emerges. Additional properties of optimal adaptive predictors of individual maps are described for a randomly evolving network of weakly coupled logistic maps.

4. The neural network model has proved to be a useful and widely used machine learning or adaptive computation algorithm. Its application in biology by Lapedes is mentioned later in terms of its application to an understanding of protein folding.

5. Ray describes recent progress on his Tierra model of the living world which he designed to test Darwinian evolution. He defines the fundamental elements of evolution as self-replication with heritable variation. His model consists of a self-replicating machine language in which the machine makes occasional mistakes in replication. The mistakes may be bit-flip mutations or small errors in calculations or in the transfer of information. A single "digital organism" will give rise to entire ecological communities which illustrate well established principles of evolutionary and ecological theory. The model provides a computer-based experimental facility for the study of evolution.

6. A λ-calculus model has been proposed which simulates the chemical behavior of reacting molecules in a context-free system:
Fontana and Buss have developed a mathematical formulation based on a formal λ-calculus dealing with objects (syntactical structures) and functions that can represent the reactions or chemistry of the objects. They distinguish a level zero system dominated by self-copying systems, a level one system in which self-copying or replication is forbidden which is then dominated by self-maintenance, and a level two system in which a variety of organizations formed in level one systems can collide, merge and form meta-organizations. Replication can emerge from level two. As noted by Schuster, a difficulty in this line of argument is finding a feasible way to forbid replication in the earliest processes. If it is always present, the "selfish gene," operating in a limited resource environment, will forbid the rise of diversity unless some other mechanism, not yet proposed, can explain how such diversity arose in the presence of replication.

BIOLOGICAL AND SOCIAL APPLICATIONS

Participants in the discussions constantly returned to the necessity to calibrate models and their parameters against observation of the real-world systems they purport to simulate. Questions were raised and left largely unresolved about the potential usefulness and hidden dangers of models as "flight simulators" which could provide a feeling for complex behavior to policy makers who might insert their own assumptions in numerical simulation models and observe the range of effects.

The agenda included a number of examples of applications of models and of the behavior of real systems. Here is where the greatest divergences in views of complexity and the need for "reality checks" emerged most visibly. The discussion involving these contributions can best be summarized in terms of its emphasis on increasing, wherever and however possible, the amount of "hard" data that can be used to test the validity of models.

Complex systems approaches to describing the behavior of biological systems spanned the hierarchy of increasing complexity from descriptions of the early role of replicating DNA in evolution (Schuster) to an outline of major transitions and mechanisms in evolution that led ultimately to human society and language (Maynard Smith). Considerable attention is paid to the possibility that the space of possible biological forms constrained the development of morphology and behavior of organisms to a considerably smaller volume than the space of all possibly viable genetic programs which might have been explored by mutation and evolutionary selection processes (Goodwin). The use of neural network algorithms to predict secondary structure with consequent inference of tertiary structure in folded proteins is described by Lapedes. Complex systems approaches to theoretical immunology and to estimating the complexity of the brain (*"the* most complex of the complex systems") are outlined by Perelson and Stevens.

At the boundary between biology and social science, Brown discusses ecological organizations and the features that they share with other complex systems. These include the diversity of parts, the openness of the systems and the fact that they are far from thermodynamic equilibrium, their adaptation to the environment by Darwinian processes and by behavioral adjustments of individuals, their irreversibility in time, and their rich variety of complex, nonlinear dynamics. Examples were presented of deductive, reductionist, and experimental approaches to study the structure and dynamics of these systems.

Feldman, Cavali-Sforza, and Zhivotovsky describe a complex model for transmission and evolution of culture based on analogies to biological evolution, particularly the well-developed theory of population genetics. They attempt to classify and quantify modes of cultural transmission in order to find the relationship between these modes and evolutionary processes of culture. Their model is illustrated by two examples in which predictions are made on the basis of the model: the interaction of cultural transmission of sign language and genetically transmitted deafness; and the interaction of a cultural sex bias and the primary sex ratio (PSR) which is measured at birth.

Much of the resources of the Institute are devoted to studies in economics which attempt to include the effects of limited information and a more realistic approach to rationality in economic agents. The agenda included a brief review of the strengths and weaknesses of general equilibrium theory and some of the current variations in approaches to this science by Arrow. Lichstein led an informal discussion of banking in a rapidly changing global economy. Arthur discusses the role of nonequilibrium systems with many interactive elements in economic processes and commented on the gradual growth of complexity in some of these systems as they continually evolve and discover new behavior. He also reports on computer experiments with Holland and Palmer in which the stock market is simulated with artificially intelligent computer programs who buy from and sell to each other.

In summary, it is evident that the SFI community does not lack for candidate metaphors for complex adaptive systems and that a considerable degree of consensus has emerged on the major elements that must be incorporated in efforts to capture their behavior in toy models. Whether, over time, a synthesis or, perhaps, a distinct scientific subfield emerges from the attempt to identify integrative themes and common features of disparate complex adaptive systems, and to establish their universality through model-building, observation, experiment, and computer simulation, remains to be seen. For now, we suggest that *Complexity: Metaphors, Models, and Reality* provides an appropriate title for these proceedings and for the existing scientific programs at the Santa Fe Institute. For the future, a challenging title for the scientific goals of the Institute might be: "Complexity: From Metaphor to Reality."

Index